AF355999

Recycling and Resource Recovery from Polymers II

Recycling and Resource Recovery from Polymers II

Editors

Sheila Devasahayam
Laurence Dyer

Basel • Beijing • Wuhan • Barcelona • Belgrade • Novi Sad • Cluj • Manchester

Editors

Sheila Devasahayam
WASM: Minerals, Energy and
Chemical Engineering
Curtin University
Kalgoorlie
Australia

Laurence Dyer
WASM: Minerals, Energy and
Chemical Engineering
Curtin University
Kalgoorlie
Australia

Editorial Office
MDPI
St. Alban-Anlage 66
4052 Basel, Switzerland

This is a reprint of articles from the Special Issue published online in the open access journal *Polymers* (ISSN 2073-4360) (available at: www.mdpi.com/journal/polymers/special_issues/recycling_resource_recover_polym_II).

For citation purposes, cite each article independently as indicated on the article page online and as indicated below:

Lastname, A.A.; Lastname, B.B. Article Title. *Journal Name* **Year**, *Volume Number*, Page Range.

ISBN 978-3-7258-0426-9 (Hbk)
ISBN 978-3-7258-0425-2 (PDF)
doi.org/10.3390/books978-3-7258-0425-2

Contents

About the Editors

Sheila Devasahayam

Dr. Sheila Devasahayam is a senior lecturer at the WA School of Mines: Minerals, Energy, and Chemical Engineering, Curtin University, Australia. Sheila's current research focus is on carbon dioxide reduction through advanced energy conversion and utilization technologies and energy and emission reductions towards sustainable processing of materials. She has co-edited *Sustainability in Mineral and Energy Sectors*, CRC Press 2017, *Nano Tools and Devices for Enhanced Renewable Energy (Micro and Nano Technologies)*, for Elsevier, 2021, and *Towards Hydrogen Infrastructure: Advances and Challenges in Preparing for the Hydrogen Economy*, for Elsevier, 2023. She is the co-guest editor of *Special Issue: Recycling and Resource Recovery from Polymer 1*, which has been published as a book, and *Recycling and Resource Recovery from Polymer 11*. Sheila's research interests span many aspects of the mineral sector, including acid mine drainage, hydrometallurgy, pyrometallurgy, mineral processing, and coal processing, with a focus on green chemistry and engineering as well as soil, water science, and polymer science.

Laurence Dyer

Prof. Laurence Dyer leads the Metallurgical Engineering program from Curtin University's historic Kalgoorlie campus in the heart of Western Australia's Goldfields. Originally trained in chemistry and physics, he has spent more than 15 years in education and industry-based research in extractive metallurgy and has a passion for developing solutions that both improve business and sustainability outcomes. Laurence's primary drive is to continue to produce high-quality, job-ready graduates to service the industry, and his research is currently focused on the development of new metal extraction processes, water treatment, and the treatment of low-grade materials and industrial waste streams. He sees great potential in the adaptation of primary production technologies in recycling applications and a greater overlap between extraction, processing, materials science, and manufacturing design for more robust and efficient value chains and minimizing waste.

Preface

As the world grapples with the challenges of environmental sustainability, the imperative to transition towards a circular economy becomes increasingly evident. This compendium encapsulates a diverse array of research endeavors that collectively contribute to the ethos of circularity, focusing on critical themes such as the plastics economy, energy recovery, and emissions reduction.

The Journey Towards Circular Economy: Central to this volume is the exploration of circular economy principles, where the traditional linear model is supplanted by regenerative practices. The chapters delve into innovative approaches in waste management, emphasizing the transformation of materials, particularly plastics, through processes like gasification, carbon capture, and catalytic reactions.

Plastics and Sustainable Materials Processing: A significant emphasis is placed on the plastics economy, investigating methods for monomer recovery, hydrogen energy systems, and feedstock recycling. Contributions extend to sustainable materials processing, offering insights into the use of polymer wastes in geopolymer concrete, a pivotal consideration in the construction and built materials sector.

Technological Advancements and Economic Perspectives: Cutting-edge research on thermal and mechanical degradation of recycled materials, upcycling of pine branches, and thermocatalytic conversion of plastics into liquid fuels is presented. The integration of novel catalysts, process modeling, and economic analyses further enriches our understanding of the technological landscape.

Multidisciplinary Perspectives: Spanning across diverse disciplines, the compendium encompasses topics ranging from fabricating recycled polycarbonate fibers for thermal signature reduction to the valorization of waste subproducts in composite materials. The chapters collectively underscore the multifaceted nature of circular economy endeavors.

Table of Contents: The volume opens with a deep dive into chloride permeability coefficient prediction in rubber concrete, setting the stage for subsequent explorations. From investigating recycled PLA filaments for additive manufacturing to the pulverization of waste PVC film for alternative fuel, each chapter offers a unique perspective on the journey towards sustainability.

Conclusion: The journey towards a circular economy is a collaborative effort, and this compendium serves as a testament to the diverse research initiatives contributing to this global endeavor. The pursuit of sustainable practices, economic viability, and technological advancements converges in these pages, offering a valuable resource for researchers, practitioners, and policymakers alike.

Sheila Devasahayam and Laurence Dyer

Editors

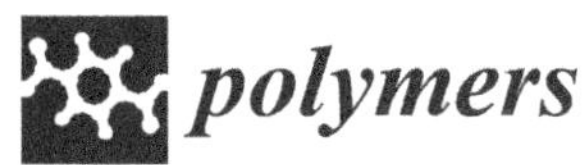 

Article

Chloride Permeability Coefficient Prediction of Rubber Concrete Based on the Improved Machine Learning Technical: Modelling and Performance Evaluation

Xiaoyu Huang [1], Shuai Wang [2], Tong Lu [2], Houmin Li [1,*], Keyang Wu [2] and Weichao Deng [1]

1   School of Civil Engineering, Architecture and The Environment, Hubei University of Technology, Wuhan 430068, China
2   Wuhan Construction Engineering Company Limited, Wuhan 430056, China
*   Correspondence: lihoumin2000@163.com

Abstract: The addition of rubber to concrete improves resistance to chloride ion attacks. Therefore, rapidly determining the chloride permeability coefficient (D_{CI}) of rubber concrete (RC) can contribute to promotion in coastal areas. Most current methods for determining D_{CI} of RC are traditional, which cannot account for multi-factorial effects and suffer from low prediction accuracy. Machine learning (ML) techniques have good non-linear learning capabilities and can consider the effects of multiple factors compared with traditional methods. However, ML models easily fall into the local optimum due to their parameters' influence. Therefore, a mixed whale optimization algorithm (MWOA) was developed in this paper to optimize ML models. The main strategies are to introduce Tent mapping to expand the search range of the algorithm, to use an adaptive *t*-distribution dimension-by-dimensional variation strategy to perturb the optimal fitness individual to thereby improve the algorithm's ability to jump out of the local optimum, and to introduce adaptive weights and adaptive probability threshold values to enhance the adaptive capacity of the algorithm. For this purpose, data were collected from the published literature. Three machine learning models, Extreme Learning Machine (ELM), Random Forest (RF), and Elman Neural Network (ELMAN), were built to predict the D_{CI} of RC, and the three models were optimized using MWOA. The calculations show that the MWOA is effective with the optimized ELM, RF, and ELMAN models improving the prediction accuracy by 54.4%, 62.9%, and 36.4% compared with the initial model. The MWOA-ELM model was found to be the optimal model after a comparative analysis. The accuracy of the multiple linear regression model (MRL) and the traditional mathematical model is calculated to be 87.15% and 85.03%, which is lower than that of the MWOA-ELM model. This indicates that the ML model that is optimized using the improved whale optimization algorithm has better predictive ability than traditional models, providing a new option for predicting the D_{CI} of RC.

Keywords: machine learning; rubber concrete; prediction; algorithm; chloride permeability coefficient

Citation: Huang, X.; Wang, S.; Lu, T.; Li, H.; Wu, K.; Deng, W. Chloride Permeability Coefficient Prediction of Rubber Concrete Based on the Improved Machine Learning Technical: Modelling and Performance Evaluation. *Polymers* **2023**, *15*, 308. https://doi.org/10.3390/polym15020308

Academic Editor: Antonio Pizzi

Received: 22 November 2022
Revised: 3 January 2023
Accepted: 4 January 2023
Published: 7 January 2023

1. Introduction

Concrete is one of the most widely used building materials today [1,2]. With the economic boom, the disposal of used tire rubber is becoming a significant issue for urban development [3]. Developing concrete from tire rubber is considered to be a viable technical solution contributing to resource and environmental protection [4–10]. With the rapid development of marine technology, concrete structures have been widely used in coastal projects which has led to widespread interest in the structural durability of concrete [11–13]. The structural durability of concrete plays a vital role in sustainable development. Chloride ion attack is one of the main factors affecting the durability of concrete structures [14–17]. Chloride ion attack can cause structural failure of concrete, which can lead to many problems. Rubber is a hydrophobic material and has a high resistance to permeation. Chloride ions are transported in concrete using water as a medium, so adding rubber can improve the concrete's resistance to chloride ions [18–22]. The chloride permeation coefficient (D_{CI}) is one of the three main indicators of concrete durability design, and its permeation process is also an effective way to understand chloride ion attack on concrete [23,24]. It is currently difficult to measure each project's D_{CI} and permeation rate of concrete [25]. To this end, several empirical models and equations have been developed to determine the D_{CI} of concrete [26–28]. However, chloride permeation is a complex and time-consuming process, and these traditional methods cannot consider the influence of multiple factors and have low predictive accuracy [29,30]. Therefore, it is particularly important to find a quick and accurate method for determining the D_{CI} of rubber concrete (RC).

Machine learning (ML) techniques have good non-linear learning capabilities, which can learn from given data and make accurate predictions through complex systems [31]. ML technology has also been applied in the study of RC. For example, Nyarko et al. [10] used 457 sets of RC data to build a 9-3-2 deep neural network (DNN) model to predict the strength of RC and demonstrated that the DNN accuracy was high at 0.9779. Gupta et al. [32] successfully investigated the mechanical properties of rubber concrete at high temperatures using a multi-input and multi-output artificial neural network (ANN) model. Ly et al. [33] used DNN to predict the strength of RC successfully and found the best structure to be 12-16-14-3-1. However, existing research has mainly focused on the mechanical properties of RC. Research on the chloride penetration of RC is limited.

Compared to traditional feed-forward neural networks, Extreme Learning Machine (ELM) can set the input weights randomly and obtain the output weights by least squares. This network has a better generalization capability and faster iteration speed, as the entire iteration process is not required, while the probability of the local extremum and overfitting is lower [34,35]. Since the input weights are random, there is a problem with blind iteration and low accuracy [36]. The random forest (RF) model was proposed in 2018 [37]. The advantages of the RF model are controllable generalization error, fewer parameters to be adjusted, suitability for high-dimensional feature vector spaces, and protection against overfitting to a certain degree. However, its parameters still have randomness and limitations [2,38]. Elman neural network (ELMAN) is similar to artificial neural networks in structure. The difference is that the ELMAN can store information. The output of the previous neuron is stored to guide the prediction of the next neuron. Therefore, the ELMAN has better dynamic prediction capability. However, due to the influence of weights and bias, it is prone to gradient explosion, resulting in lower prediction accuracy [39,40]. The three models have shown extraordinary capabilities in solving prediction problems and have been used in various studies [2,39–42], but the research applied to RC is still limited. Therefore, these three models were chosen for this study.

Determining the ML model parameters is crucial, as they directly affect the model's predictive performance [43]. The parameters can be optimized by intelligent algorithms [44]. Intelligent algorithms can improve the model's predictive performance by searching for optimal parameters [45]. The main idea of the whale optimization algorithm (WOA) is to simulate the humpback whale foraging process, proposed in 2016. The WOA has fewer parameters and is also quite competitive compared to other optimization algorithms [46], so it is widely used in various fields [47–49]. However, the standard WOA suffers from slow convergence and local optimal solutions, so improvements are needed. Tent chaotic mappings have a more uniform distribution and are often used to optimize the initial populations of WOA [50,51]. Therefore, in this study, tent chaotic mapping was chosen to optimize the population of WOA. Since the linear weight adjustment strategy of WOA is not adapted to the constantly changing population, an adaptive weight adjustment strategy was introduced. The Probability threshold value was introduced to enhance the algorithm's global search ability [52]. At the end of the iteration, to avoid the WOA falling into the local optimum, an adaptive t-distribution dimension-by-dimensional variation strategy was used to perturb the optimal individual to increase the WOA'S ability to jump out of the local optimum [53]. There is limited research on applying improved algorithms to optimize ML models to predict D_{CI} of RC, so this approach is used in this study.

In summary, this study chose the ELM, RF, and ELMAN models to predict the D_{CI} of RC, while the WOA was improved to form a new mixed whale optimization algorithm (MWOA). The MWOA algorithm can improve the accuracy of machine learning models. The optimized model also has advantages over the traditional model. Thus, data were first collected from the published literature and created a database for analysis. Secondly, using the three models to predict the D_{CI} of RC, the three models were optimized using MWOA. Thirdly, we evaluated the model performance and found the optimal model. Fourth, we conducted sensitivity analysis for models. Fifth, the representative prediction result of the optimal model was compared with the actual value. Finally, using the optimal model compared with the traditional model.

2. Database Description and Analysis of Variables

Since the experimental data were collected from the published literature, processing is required to allow the model to learn better. This study collected 88 sets of RC mixed ratio data [22,54–59]. Three ML algorithms using nine input variables were used: (1) measurement method, (2) cement content, (3) water reducing agent content, (4) water content, (5) water to ash ratio, (6) fine aggregate content, (7) coarse aggregate content, (8) rubber size, (9) rubber content. The D_{CI} is the only output variable. Due to the different methods used to measure the D_{CI} of RC, this study distinguishes between the measurement methods. Two measurement methods are included in the collected literature, the rapid chloride permeability test (RCPT) method and the rapid chloride migration Test (RCM) method. Therefore, the RCPT method is defined as 1, and the RCM method is defined as 2. Considering the different types and sizes of rubber, this study distinguishes between rubber by size under different measurement methods. The RCPT literature method includes two rubber sizes, 0–1 mm and 1–3 mm. Therefore, the two rubbers were recorded as 1 and 2 by size. The RCM method literature includes five types of rubber with dimensions of 0.063–0.6 mm, 0.25 mm, 0.6–0.7 mm, 1–2 mm, and 4–10 mm. Therefore, the five types of rubber are noted as 1, 2, 3, 4, and 5 by size. The RC samples selected for this study met the 28-day curing period. Cement substitution materials were not used as an input variable in this study, as they are rarely added in the published literature. Figure 1 represents the hotspot plot of the correlation coefficient between the variables. The correlation coefficients between the variables are all less than 0.8, as seen from the graph. Some studies suggest that the correlation between variables should be less than 0.8 to reduce the effect of multiple collinearities [60,61]. Figure 2 represents the input and output parameters' frequency distribution histogram. The statistical analysis of each variable is shown in Table 1. Stdd denotes overall sample bias, and Stde indicates sample bias.

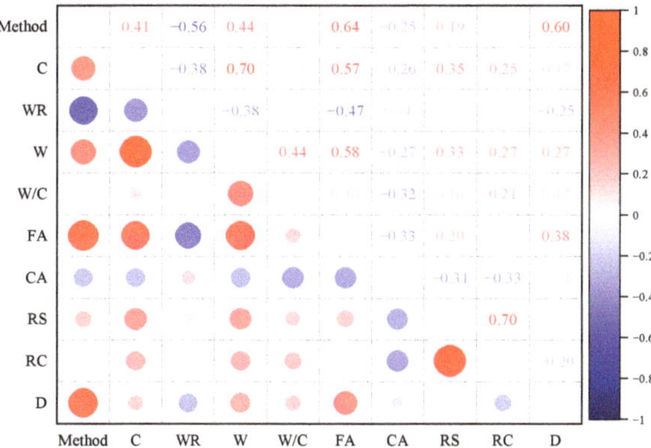

Figure 1. Heat map of correlation coefficients for input and output variables.

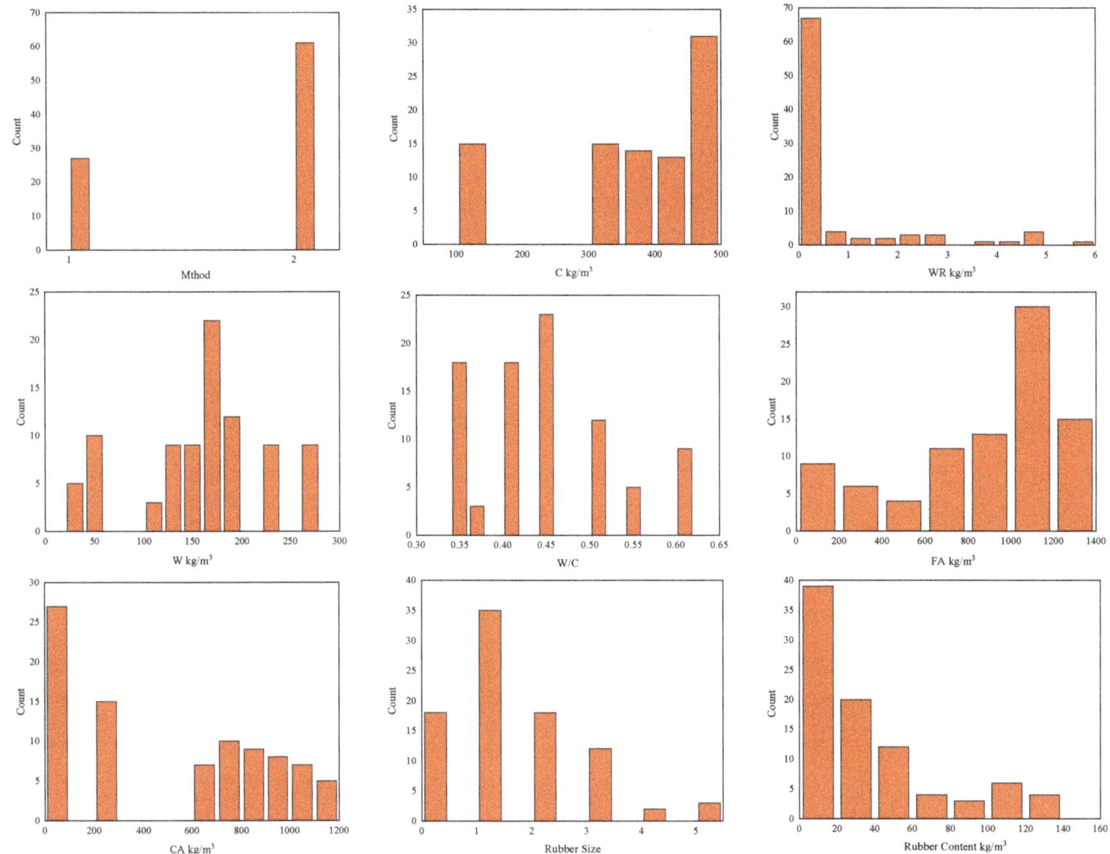

Figure 2. Histogram of frequency distribution for input and output variables.

Table 1. Statistical analysis of input and output variables.

	Max	Min	Average	Median	Stdd	Stde
Method	2.00	1.00	1.6932	2.00	0.4612	0.4638
$C \cdot kg/m^3$	457.00	100.00	350.69	387.50	123.16	123.87
$WR \cdot kg/m^3$	5.82	0.00	0.62	0.00	1.34	1.35
$W \cdot kgm^3/$	272.00	35.00	157.32	161.90	65.19	65.57
W/C	0.60	0.35	0.44	0.45	0.076	0.078
$FA \cdot kg/m^3$	1360.00	174.00	862.76	1005.00	365.72	367.82
$CA \cdot kg/m^3$	1124.00	0.00	504.95	607.00	410.24	412.59
Rubber Size	5.00	0.00	1.4773	1.00	1.215	1.222
Rubber Content kg/m^3	138.40	0.00	35.89	22.00	37.55	37.77
$D_{CI} \cdot 10^{-12} m^2$	18.55	1.07	8.39	9.74	3.88	3.90

3. Method

3.1. Whale Optimization Algorithm

Humpback whales are herd animals because they can only hunt small fish and prawns. They have developed a unique way of hunting known as bubble net hunting, where the term WOA comes from [46]. The algorithm is divided into three main parts: encirclement predation, prey predation, and prey search. The specific process of the WOA is as follows:

3.1.1. Encirclement Predation

In this process, humpback whales randomly search for prey based on the positions of each other in the population. Since the location of the best target has yet to be discovered in the search space, the WOA assumes that the location of the best target is within the search range. Once the position of the best target is determined, other populations will approach the best target and update their positions. The mathematical expression for this process is as follows [62]:

$$X(v+1) = X^*(v) - A{\cdot}D \tag{1}$$

$$D = |CX^*(v) - X(v)| \tag{2}$$

where v indicates the number of iterations; A and C are vector coefficients; $X^*(v)$ is the location of the best target; $X(v)$ is the current location; D is the process quantity; the expressions for the calculation of A, C are as follows [62]:

$$A = 2a \cdot r_1 - a \tag{3}$$

$$C = 2r_2 \tag{4}$$

$$a = 2 - \frac{2v}{V_{max}} \tag{5}$$

where v indicates the number of iterations; V_{max} indicates the maximum number of iterations; r_1 and r_2 are both random numbers belonging to the range [0, 1]; a decrease gradually from 2 to 0.

3.1.2. Prey Predation

The bubble net foraging method is a unique hunting method for humpback whales. The WOA is a simulation of the spiral bubble net foraging strategy for optimization. A total of two methods were designed to simulate this behavior:

(1) Shrinkage envelope mechanism: This is achieved by reducing the value of a in Equation (3). Other targets will move closer to the best target when the best target is identified. The current position (X, Y) is gradually contracted to the optimal target position $(X*, Y*)$.

(2) Spiral update mechanism: The distance between any whale (X, Y) and the optimal target position $(X*, Y*)$ is first calculated, then spiral update equations are created to simulate the whale's hunting motion. The main expressions are as follows [62]:

$$D' = |X^*(v) - X(v)| \tag{6}$$

$$X(v+1) = D' \cdot e^{bl} \cdot cos(2\pi l) + X^*(v) \tag{7}$$

where b is a constant and indicates the parameter for the shape of the spiral; l denotes a random number between [–1, 1]; D' suggests the distance between the best target and any whale.

These two mechanisms co-occur, with a probability of 50% each. The expression of the equation is as follows [62]:

$$X(v+1) = \begin{cases} X^*(v) - A \cdot D & P < 0.5 \\ X^*(v) + D' \cdot cos(2\pi l) \cdot e^{bl} & P \geq 0.5 \end{cases} \tag{8}$$

3.1.3. Prey Search

Whale populations randomly search for prey, and the mathematical expression for this process is as follows [62]:

$$D_{rand} = |CX_{rand}(v) - X(v)| \tag{9}$$

$$X(v+1) = X_{rand}(v) - A \cdot D_{rand} \tag{10}$$

where $X_{rand}(v)$ indicates a randomly selected location in the current population of whales.

The WOA, like other intelligent algorithms, suffers from the problem of falling into local extremum. Therefore, improvements to the WOA are needed.

3.2. Improved Whale Optimization Algorithm

3.2.1. Tent Chaotic Mapping Initializes Populations

Chaos is a complex, non-linear state that exhibits irregularity and randomness [63]. Therefore, chaotic mapping can be used to improve the algorithm's performance. The two commonly used chaotic mapping sequence models are Logistic and Tent. Compared to logistic mappings, Tent mappings have a more uniform distribution, allowing the algorithm to have a wider search range [50]. Therefore, Tent chaotic mapping is used to initialize the population in this study. The expressions are as follows [51]:

$$x_{n+1} = \begin{cases} 2x_n, & 0 \le x_n \le 0.5 \\ 2(1 - x_n), & 0.5 \le x_n \le 1 \end{cases} \tag{11}$$

The expression after the Bernoulli displacement transformation is as follows [51]:

$$x_n = 2(x_n)mod1 \tag{12}$$

3.2.2. Adaptive Adjustment of Weight

The inertia weight is a crucial parameter in the WOA. Appropriate weight values can improve the algorithm's performance, since the original WOA did not consider that the prey would guide the whale for position updates during the iterative process. Therefore, an adaptive weight formula is established in this paper. The specific expression is as follows [52]:

$$w = d_1 * (P_{worst} - P_{best}) + d_2 * \left(x_i^{upper} - x_i^{lower}\right)/t \tag{13}$$

where t indicates the current number of iterations; x_i^{upper} and x_i^{lower} denote the upper and lower bounds of x_i respectively; d_1 and d_2 represent constants; P_{worst} and P_{best} denote the worst and best positions of the current population respectively. Thus Equations (1) and (7) can be improved as:

$$X(v + 1) = w * X^*(v) - A \cdot D \tag{14}$$

$$X(v + 1) = D' \cdot e^{bl} \cdot cos(2\pi l) + w * X^*(v) \tag{15}$$

With the introduction of an adaptive adjustment weights strategy, the algorithm can adaptively change the weights' size according to the whale population's current distribution. At the beginning of the algorithm iteration, if the whale population falls into the local optimum and the difference between the optimum and the worst solution is not significant, the value of $d_2 * \left(x_i^{upper} - x_i^{lower}\right)/t$ is not affected by the population distribution. At this point, obtaining a large value of weights is still possible and avoids the algorithm falling into a smaller search range at the beginning of the iteration. As the whale population iterations increase, the value of $d_2 * \left(x_i^{upper} - x_i^{lower}\right)/t$ decreases, and the effect on the weights decreases. If the algorithm does not obtain an optimal solution, $d_1 * (P_{iworst} - P_{ibest})$ can play a dominant role in the weight and can make the algorithm find the optimal solution in larger steps. The adjustment of these two components makes the inertia weights highly adaptive and strengthens the algorithm's optimization search capability.

3.2.3. Adaptive Adjustment of the Search Strategy

To prevent the algorithm from falling into the local optimum, a Probability threshold value Q is introduced to update the expression of the random search. The expression for Q is as follows [52]:

$$Q = \frac{\left|\overline{f} - f_{min}\right|}{\left|f_{max} - f_{min}\right|} \tag{16}$$

where $\overline{f}$ indicates the average fitness of the current population; f_{min} indicates the current best fit value; f_{max} indicates the current worst fit value; for each whale, a $q \in [0,1]$ is compared with Q value. If $q < Q$, the randomly selected individual whale updates its position according to Equation (17), and the other individual whales remain unchanged [52]. Otherwise, other individuals update their position according to Equation (10). This allows the algorithm to generate a set of random solutions globally with a greater probability in the early iterations, reducing the likelihood of population diversity decline and enhancing the global search capability of the algorithm.

$$X_{rand}(v) = X_{min} + r * (X_{max} - X_{min}) \tag{17}$$

where r is a random number between [0, 1]; X_{min} and X_{rand} are the maximum and minimum values of X_{rand} respectively.

3.2.4. Adaptive *t*-Distribution Dimension-by-Dimensional Variation Strategy

Population diversity declines in later iterations of the WOA. This leads to the algorithm being prone to fall into the local optimum. Therefore, this study introduces an adaptive *t*-distribution dimension-by-dimensional variation strategy to perturb the individuals with optimal fitness and improve the ability of the algorithm to jump out of the local optimum. Depending on the size of the degree of freedom n, the *t*-distribution curves show different patterns. When $t \to (n \to \infty) \to N(0, 1)$, $t(n = 1) = C(0, 1)$, where $N(0, 1)$ is a Gaussian distribution and $C(0, 1)$ is a Cauchy distribution. This shows that the two boundary special cases of the *t*-distribution are the Gaussian and the Cauchy distributions [64]. The dimension-by-dimension variation is calculated as follows [53]:

$$X_{new}^{d} = X_{Pest}^{d} + X_{Pest}^{d} \times t(iter) \tag{18}$$

where *iter* indicates the current number of iterations; $t(iter)$ denotes *t*-distribution with the degree of freedom parameter t. The flow chart for the MWOA is shown in Figure 3.

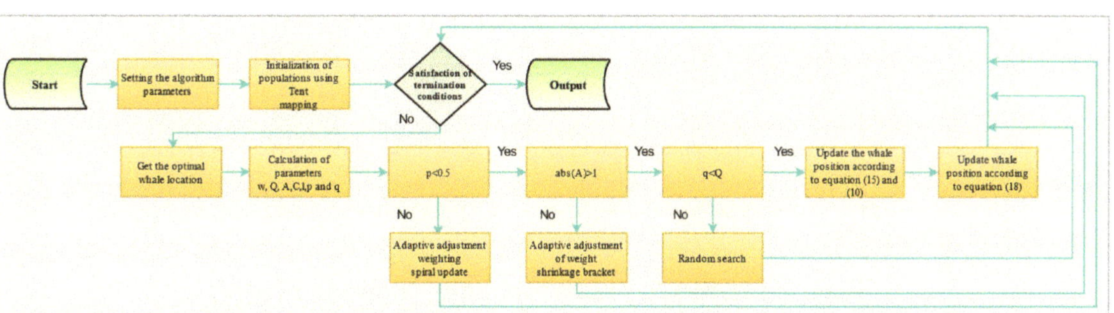

Figure 3. The flow chart of MWOA.

3.3. Machine Learning Models

3.3.1. Extreme Learning Machine

The Extreme Learning Machine is a new neural network learning algorithm proposed by Professor Guangbin Huang in 2004 [65]. The Extreme learning machine also evolved from the feedforward neural network, which can randomize the input weights, bias, and the number of hidden layer neurons and then obtain the output weights by least squares without the need for the entire iteration of the network [34,35]. ELM is widely used in various fields such as pattern recognition, image processing, signal processing, combinatorial optimization, and prediction [66–69]. The structure of the ELM is shown in Figure 4. The primary calculation process for ELM is as follows:

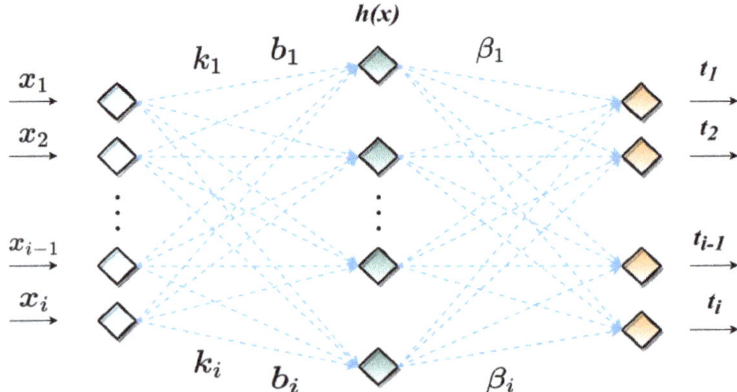

Figure 4. Schematic diagram of the ELM structure.

For N arbitrary samples (x_i, t_i), where $x_i = [x_{i1}, x_{i2}, x_{i3}, \ldots, x_{in}]^T \in R^n$, $t_i = [t_{i1}, t_{i2}, t_{i3}, \ldots, t_{im}]^T \in R^m$. Assume that the number of hidden layer neurons is $\widetilde{N}$, and the activation function is $h(x)$, the standard single hidden layer feedforward neural network (SLFN) expression is as follows [70]:

$$\sum_{i=1}^{\widetilde{N}} \beta_i h(k_i \cdot x_j + b_i) = y_j, \quad j = 1, \ldots N \tag{19}$$

where $k_i = [k_{i1}, k_{i2}, k_{i3}, \ldots, k_{in}]^T$ denote the vector of weights connecting the ith hidden neuron to the input neuron; b_i denotes the bias of the ith neuron; $\beta_i = [\beta_{i1}, \beta_{i2}, \beta_{i3}, \ldots, \beta_{in}]^T$ denotes the weights connecting the ith hidden layer neuron to the output neuron. $k_i \cdot x_j$ denote the inner product of k_i and x_j; The activation function is usually Sigmoid, RBF or Sine, and in this study the activation function is Sigmoid.

A standard SLFN with $\widetilde{N}$ hidden layer neurons and activation function $h(x)$ can approach this N samples with zero error, where $\sum_{j=1}^{\widetilde{N}} \left\| y_j - t_j \right\| = 0$. Thus, the following expression exists [70]:

$$\sum_{i=1}^{\widetilde{N}} \beta_i h(k_i \cdot x_j + b_i) = t_j, \quad j = 1, \ldots N \tag{20}$$

The above N equations can be written as [70]:

$$H\beta = T \tag{21}$$

Where $H(k_1, k_2, k_3, \ldots k_{\widetilde{N}}, b_1, b_2, b_3, \ldots, b_{\widetilde{N}}, x_1, x_2, x_3, \ldots, x_n)$

$$= \begin{bmatrix} h(k_1 \cdot x_1 + b_1) & \cdots & h(k_{\widetilde{N}} \cdot x_1 + b_{\widetilde{N}}) \\ \vdots & \ddots & \vdots \\ h(k_1 \cdot x_N + b_1) & \cdots & h(k_{\widetilde{N}} \cdot x_N + b_{\widetilde{N}}) \end{bmatrix}_{N \times \widetilde{N}}, \quad \beta = \begin{bmatrix} \beta_1^T \\ \vdots \\ \beta_3^T \end{bmatrix}_{\widetilde{N} \times m}, \quad T = \begin{bmatrix} t_1^T \\ \vdots \\ t_N^T \end{bmatrix}_{N \times m}$$

H is called the hidden layer output matrix of the neural network [71,72]. The ith column of H is the output vector of the ith hidden neuron concerning the input $x_1, x_2, x_3, \ldots, x_n$.

When the input layer weights and the hidden layer bias are determined, the hidden layer output matrix H can be obtained by following the input samples. So, the final conversion is to find the least squares solution for $H\beta = T$ [70]:

$$\left\|H\left(K_1,\ldots,K_{\widetilde{N}},b_1,\ldots,b_{\widetilde{N}}\right)\hat{\beta} - T\right\| = \min_{\beta}\left\|H\left(K_1,\ldots,K_{\widetilde{N}},b_1,\ldots,b_{\widetilde{N}}\right)\beta - T\right\| \tag{22}$$

the least squares solution of Equation (14) is as follows [70]:

$$\hat{\beta} = H^{\dagger}T \tag{23}$$

where: $H^{\dagger}$ is the Moore-Penrose generalized inverse matrix of the matrix H [73].

Random input weights and hidden layer bias can lead to problems, such as blind iterations and accuracy degradation [36]. Therefore, this study introduces the MWOA into the ELM model to optimize the input weights and hidden layer bias to improve the model's accuracy.

3.3.2. Random Forest Model

The RF model is one of the most commonly used regressions and classification models proposed by Leo Breiman in 2001 [74]. The main idea is to train decision trees by taking n samples from the original data set N to form a new training set, and m random forests are created by these n decision trees at random [75,76]. Meanwhile, the predicted value is decided by the voting of these m random forests [77]. A mathematical model can explain the RF regression model, the leading theory being that X is the independent variable (input data) and Y is the dependent variable (output data). Assuming that the distributions of (X, Y) are independent, the randomly generated training set is Q, and the predicted outcome is $G(X)$, the mean squared generalization error is [78]:

$$E_{X,Y}[Y - G(X)]^2 \tag{24}$$

Assuming that there are h decision trees, the average of the predicted values $\{G(Q, X_h)\}$ of the h decision trees is the prediction of the RF regression. If $h \to \infty$, then the following equation holds [78]:

$$E_{X,Y}\left[Y - \overline{G_h}(X, Q_h)\right]^2 \to E_{X,Y}\left[Y - E_Q(X, Q_h)\right]^2 \tag{25}$$

where $E_{X,Y}\left[Y - E_Q(X, Q_h)\right]^2$ denotes the generalization error, noted as M. When h is infinite, the average generalization error of a single tree is noted as M^*. The expression for M^* is as follows [78]:

$$M^* = E_Q E_{X,Y}[Y - G(X, Q)]^2 \tag{26}$$

where Q satisfies the following expression [78]:

$$M \leq \overline{\rho} M^* \tag{27}$$

where $\overline{\rho}$ denotes the residual weighted correlation coefficient. The final RF regression function is as follows [78]:

$$Y = E_Q G(X, Q) \tag{28}$$

Since the number of forests and leaves in the RF model significantly impacts the model's performance, at the same time, they have randomness and limitations. Therefore, this study introduces MWOA to optimize these two parameters. The structure of the RF model is shown in Figure 5.

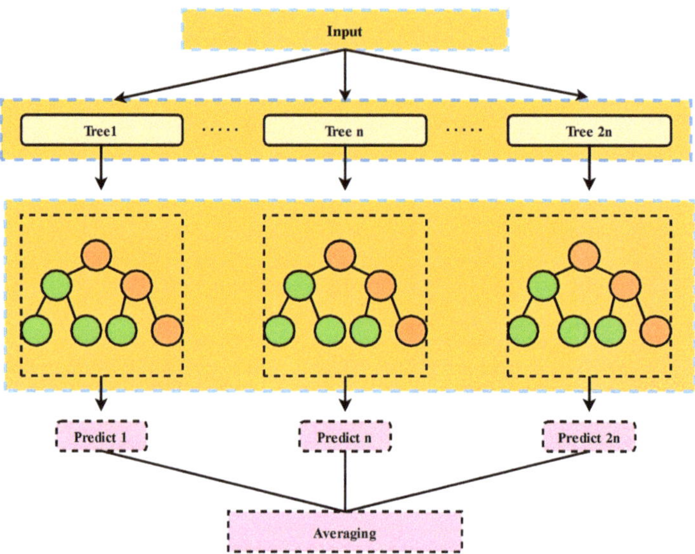

Figure 5. The structure of RF model.

3.3.3. ELMAN Neural Network

Elman neural network was proposed by ELMAN in 1990 [79]. ELMAN is a multi-layer dynamic recurrent neural network that can approximate nonlinear functions well and is therefore used in many industries [39,80,81]. Like artificial neural networks, ELMAN has an input, hidden, and output layer. The difference is that ELMAN has a unique storage layer. This particular storage layer, which acts as a delay operator, can store the output values of the neurons in the previously hidden layer, giving the network a memory function and improving the net work's ability to process dynamic information. The structure of ELMAN is shown in Figure 6. The expression for ELMAN at the moment t is as follows [82]:

$$x_j(k) = f\left(\sum_{i=1}^{n} \omega 1_{i,j} u_i(k) + \sum_{i=1}^{m} \omega 2_{i,j} c_i(k)\right) \tag{29}$$

$$c_i(k) = x_i(k-1) \tag{30}$$

$$y_j(k) = g\left(\sum_{i=1}^{r} \omega 3_{i,j} x_i(k)\right) \tag{31}$$

where $\omega 1_{i,j}$ denotes the weight of node i in the connected input layer and node j in the hidden layer; $\omega 2_{i,j}$ denotes the weight of node i and node j in the connection storage layer; $\omega 3_{i,j}$ denotes the weight connecting node i in the hidden layer and node j in the output layer; $x_j(k)$, $c_i(k)$ and $y_j(k)$ denote the output vectors of the hidden layer, the storage layer and the output layer respectively. f and g denote the transfer functions of the hidden layer and the output layer, respectively. The transfer function for this study is tanh.

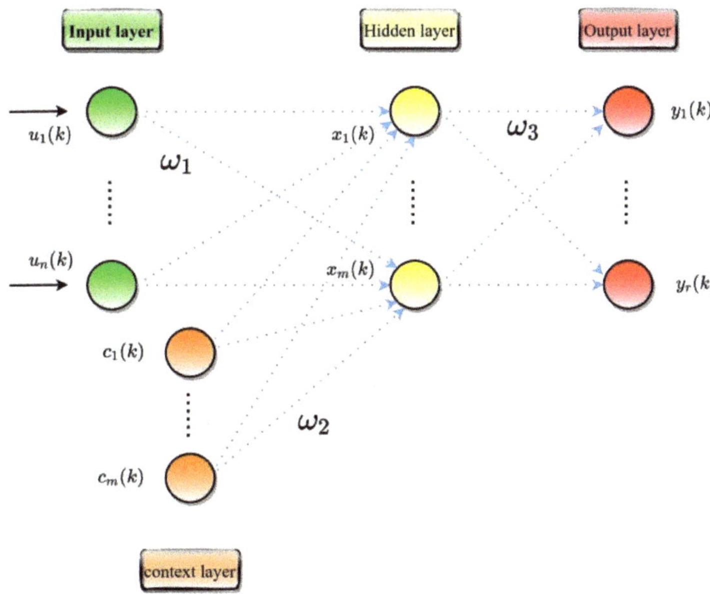

Figure 6. The structure of ELMAN.

ELMAN calculates the number of hidden layer neurons in the same way as ANN. The main expressions are as follows [31]:

$$h = \sqrt{m+n} + a \tag{32}$$

where m is the number of nodes in the input layer; n is the number of nodes in the output layer; $a \in (1, 10)$.

ELMAN's predictive performance is influenced by weights and biases. Therefore, the optimal weights and biases are found by optimizing the ELMAN neural network using MWOA. The flow chart of the research process is shown in Figure 7.

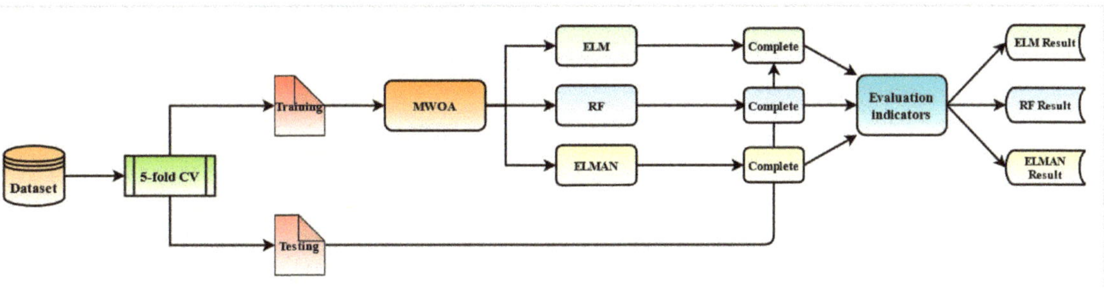

Figure 7. Flow chart of the research process.

4. Evaluation Indicators for the Three Models

This study uses root mean square error (RMSE), mean absolute error (MAE), mean absolute percentage error (MAPE), and coefficient of determination (R^2) to assess the performance of the model. R^2 is the metric used to evaluate the accuracy of the model's predictions [83,84]. The closer the R^2 value is to 1, the closer the MAE is to 0, and the more accurate the model will be. These four evaluation indicators are expressed as follows [85,86]:

$$R^2 = \frac{\sum_{k=1}^{N}(q_{0,k} - \overline{q_0})(q_{t,k} - \overline{q_t})}{\sqrt{\sum_{k=1}^{N}(q_{0,k} - \overline{q_0})^2 \sum_{k=1}^{N}(q_{t,k} \sum \overline{q_t})^2}} \tag{33}$$

$$MAE = \frac{1}{N}\left(\sum_{K=1}^{N}\left|\frac{q_{0,k} - q_{t,k}}{q_{0,k}}\right|\right) \tag{34}$$

$$RMSE = \sqrt{\frac{1}{N}\sum_{k=1}^{N}(q_{0,k} - q_{t,k})^2} \tag{35}$$

$$MAPE = \frac{100\%}{N}\sum_{k=1}^{N}\left|\frac{q_t - q_0}{q_0}\right| \tag{36}$$

where N indicates the number of samples; q_0 indicates the actual value; $\overline{q_0}$ indicates the actual average value; q_t indicates the output value; $\overline{q_t}$ indicates the output average value, $k = 1{:}N$.

5. Results of the Three Models

The objective of the computational analysis was to predict the D_{CI} of the RC using three ML models (MWOA-ELM, MWOA-RF, and MWOA-ELMAN). Therefore, the model optimized by WOA and the conventional model was also built for comparison. A cross-validation operation is also used in the calculation process, and the result is the average of a 5-fold cross-validation. This was done to make the results more realistic and to avoid chance. Therefore, the data set is divided into five groups by a 5-fold cross-validation operation. For each training session, one set was used as the testing set and the remaining four sets were used as the training set. This resulted in three 70 training sets, 18 test sets, two 71 training sets, and 17 test sets of data sets. Consistent data for each model during training and testing by programming. The models constructed and the computational results are described in detail in the following subsections.

5.1. MWOA-ELM Model Result

Like neural network models, ELM models need to determine the number of hidden layer neurons. In this study, the number of neurons in the hidden layer of the ELM model was calculated by the corresponding program and determined by the trial-and-error method to be 28. The parameter settings for the MWOA-ELM model in this study are shown in Table 2. The parameter settings for the WOA-ELM model are the same as the MWOA-ELM model.

Table 2. Parameter settings for the MWOA-ELM model.

Parameter	Setting
Popsize	30
Maxgen	100
d1, d2	1×10^{-4}
b	1
Hiddennum layer	28
Activation function	Sigmoid

The average results of the three ELM models under 5-fold cross-validation are presented in Table 3. It was clear that the MWOA-ELM model performs the best. Its test set R^2 improved from 0.6458 to 0.9971, while the other error metrics RMSE, MAE, and MAPE were all the lowest among the three ELM models. Figure 8a,b represent the Taylor diagrams for the training and testing sets of the three ELM models [87]. As can be seen from the graph, the MWOA was effective, as reflected by the fact that the MWOA-ELM model was closest to the optimal reference point for each indicator.

Table 3. Average of evaluation indicators for three ELM models.

		R^2	**RMSE**	**MAE**	**MAPE (%)**
ELM	Train	0.9602	0.7691	0.6233	0.1132
	Test	0.6458	2.6232	1.4539	0.3509
WOA-ELM	Train	0.9848	0.4518	0.3475	0.0619
	Test	0.9390	0.8810	0.6584	0.1155
MWOA-ELM	Train	0.9927	0.3287	0.2181	0.0353
	Test	0.9971	0.1911	0.1356	0.0212

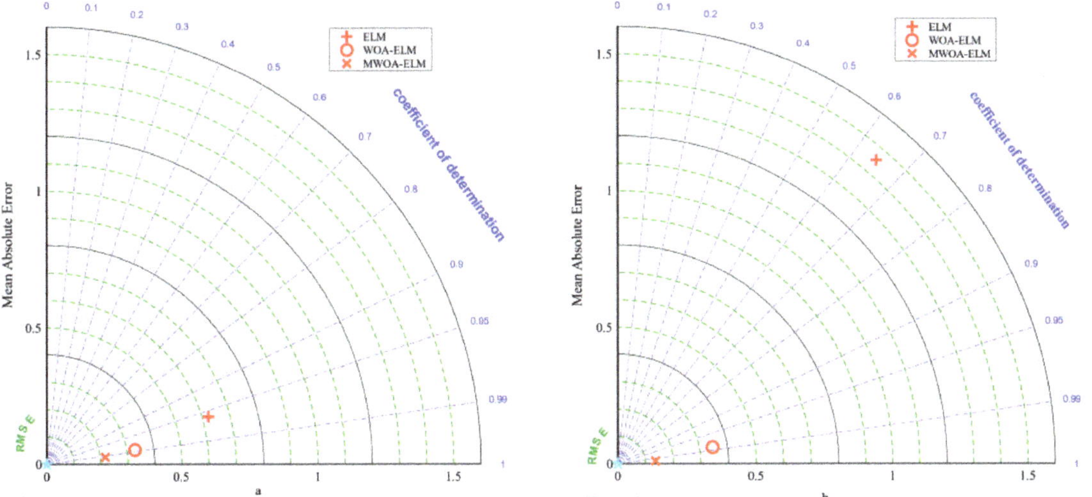

Figure 8. Taylor diagrams for the Training (**a**) and Testing (**b**) sets of the three ELM models.

5.2. MWOA-RF Model Result

Introducing the MWOA is to find the optimal number of forests and leaves for the RF model. The parameter settings for the MWOA-RF model for this study are shown in Table 4. The parameter settings for the WOA-RF model are the same as the MWOA-RF model.

Table 4. Parameter settings for the MWOA-RF model.

Parameter	Setting
Popsize	30
Maxgen	100
Forest size	24
Number of leaves	8
Number of cross-validation	5
d1, d2	1×10^{-4}
b	1

The results of the 5-fold cross-validation of the three random forest models are presented in Table 5. On the test set, the MWOA-RF model had the highest R^2 of 0.9341 and the lowest values of 1.0164, 0.6533, and 0.0962 for RMSE, MAE, and MAPE, respectively. Figure 9a,b show the Taylor diagrams for the three RF models on the training and testing sets. It was clear that the MSSA-RF model was closest to the optimal reference point among the three evaluation indicators of the Taylor diagram. Therefore, MWOA is effectively increased the probability of the RF model finding the optimal number of forests and leaves.

Table 5. Average of evaluation indicators for three RF models.

		R^2	**RMSE**	**MAE**	**MAPE (%)**
RF	Train	0.653	2.2463	1.2972	0.2766
	Test	0.5768	2.5709	1.7408	0.4015
WOA-RF	Train	0.9661	0.7709	0.5698	0.1009
	Test	0.8776	1.4409	1.0027	0.1941
MWOA-RF	Train	0.9870	0.4520	0.3152	0.0495
	Test	0.9341	1.0164	0.6553	0.0962

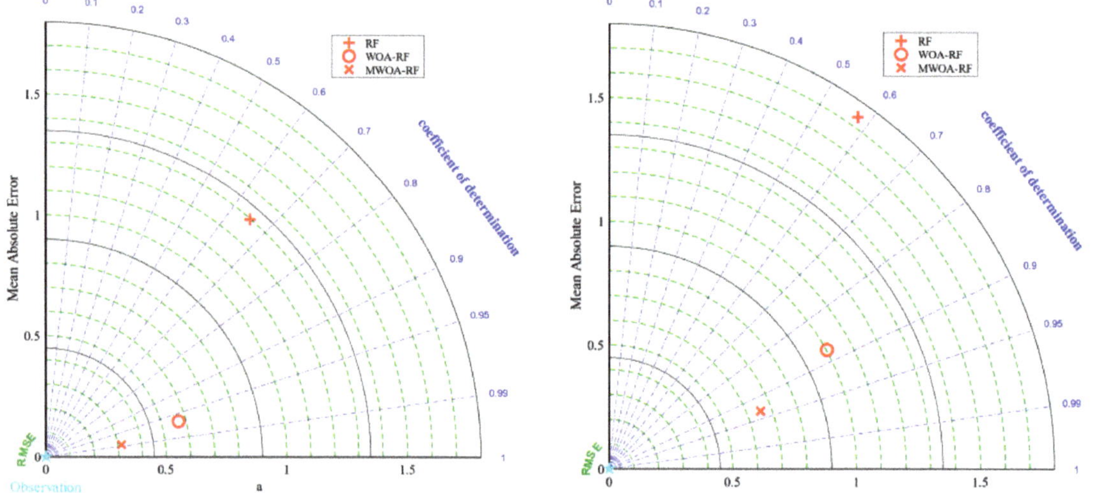

Figure 9. Taylor diagrams for the Training (**a**) and Testing (**b**) sets of the three RF models.

5.3. MWOA-ELMAN Model Result

A three-layer feed-forward MWOA-ELMAN model was established, and the optimal number of neurons was obtained by Equation (32) and trial-and-error method as 13. The MWOA-ELMAN model parameter settings for this study are shown in Table 6. The parameter settings for the WOA-ELMAN model are the same as the MWOA-ELMAN model.

Table 6. Parameter settings for the MWOA-ELMAN model.

Parameter	Setting
Popsize	30
Maxgen	100
Hiddennum_best	13
Number of cross-validation	5
Activation function	tansig, purelin
Training function	trainlm
d1, d2	1×10^{-4}
b	1

The 5-fold cross-validation results for the three ELMAN models are presented in Table 7. Similar to the pattern of the first two models, the MWOA-ELMAN model has the best results. Figure 10 represents the Taylor diagrams for the training and testing sets. From Figure 10, the results of the MWOA-ELMAN model were closest to the optimal reference point, indicating that the algorithm is effective in optimizing the weights and improving the model's prediction accuracy.

Table 7. Average of evaluation indicators for three ELMAN models.

		R^2	**RMSE**	**MAE**	**MAPE (%)**
ELMAN	Train	0.8275	1.6528	1.0820	0.1938
	Test	0.7108	2.1198	1.3609	0.2590
MWOA-ELMAN	Train	0.9783	0.5704	0.3207	0.0523
	Test	0.9390	0.8810	0.6584	0.1155
MWOA-ELMAN	Train	0.9883	0.4140	0.2261	0.0373

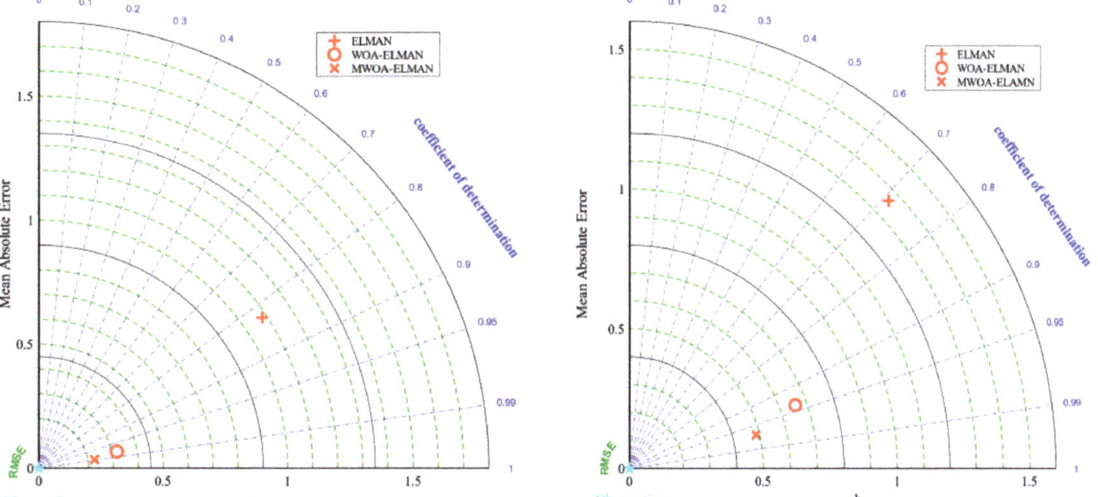

Figure 10. Taylor diagrams for the Training (**a**) and Testing (**b**) sets of the three ELMAN models.

6. Discussion

6.1. Comparative Analysis of the Three Models

In Section 5, MWOA is proven to improve the generalization of the three models ELM, RF, and ELMAN. It is also possible to prove that the three optimization models are the most exact (it has been shown in the literature that a model is highly accurate when its R^2 is more significant than 0.9 [88]). This indicates that ML techniques can meet the prediction accuracy. However, it is necessary to perform a comparative analysis to obtain the optimal model. Figure 11 represents the metric radar plots for the training and testing sets of the MWOA-ELM, MWOA-RF, and MWOA-ELMAN models. The figure shows that the MWOA-ELM model outperforms the other two models on the training and testing sets, with the highest R^2 and lowest RMSE, MAE and MAPE. Figure 12 represents the Taylor diagrams for the training and testing sets of the three models. The MWOA-ELM model performs the best, with the lowest error metric on the Taylor diagram, while being closest to the best reference point. Table 8 shows the average of the 5-fold cross-validation results for each of the three models. MWOA-ELM model outperformed the training process during testing. In contrast, the prediction accuracy of both the MWOA-RF and MWOA-ELMAN models decreased, with the MWOAA-RF model decreasing the most (by about 5.6%), indicating that MWOA-ELM is very stable. Figure 13 represents box plots of the training and testing sets for the three models. The bars indicate the mean value of each model. The training and prediction results of the models can be seen more visually in the box plots, where the MWOA-ELM model is not necessarily the best on the training set but has the lowest mean value. In the testing set, it is the best performer and relatively stable, with all R^2 around 0.99. This suggests that the MWOA-ELM is the best. It can also demonstrate that with the introduction of cross-validation, the model is very realistic in its calculations and avoids chance.

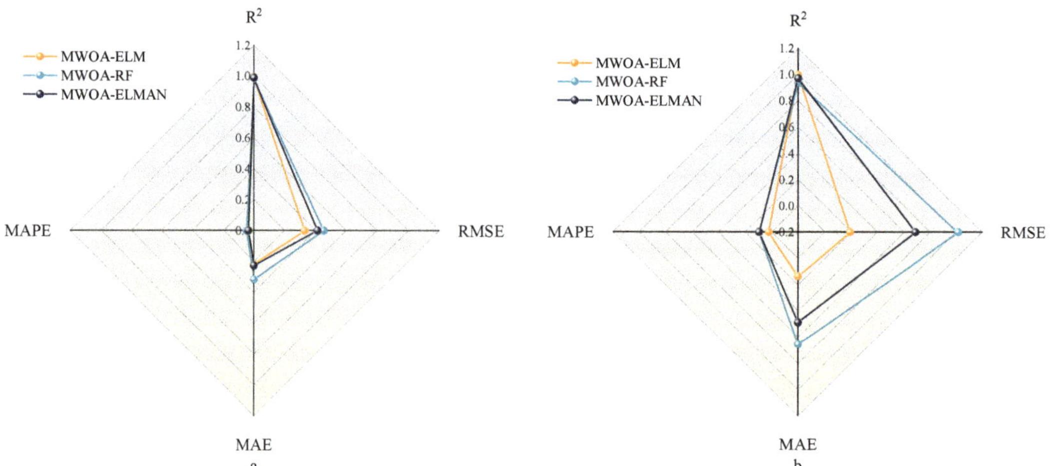

Figure 11. Radar plots of metrics for the Training (**a**) and Testing (**b**) sets for MWOA-ELM, MWOA-RF, and MWOA-ELMAN models.

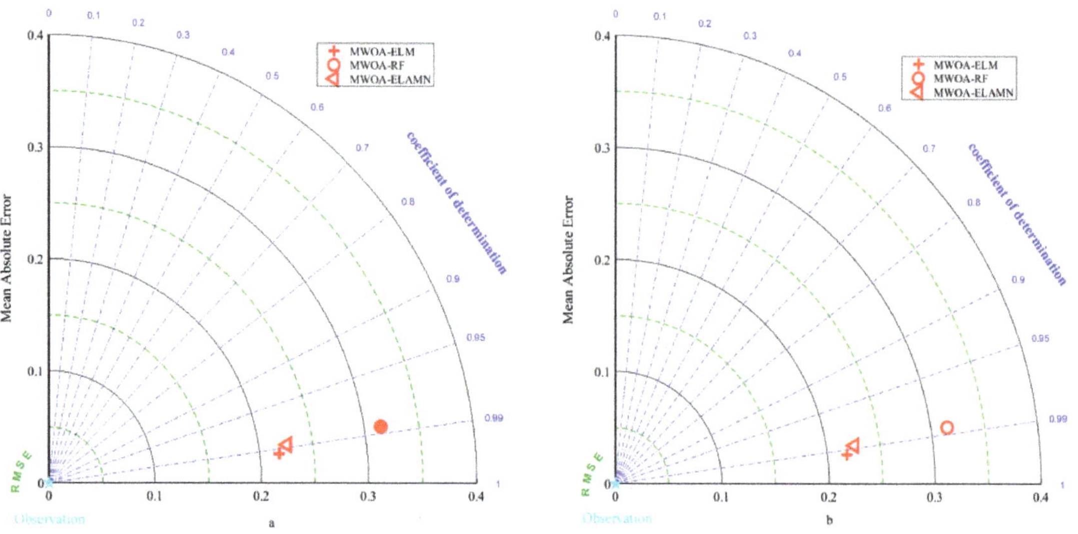

Figure 12. Taylor diagrams for the Training (**a**) and Testing (**b**) sets of the MWOA-ELM, MWOA-RF and MWOA-ELMAN models.

Table 8. Average of evaluation indicators for MWOA-ELM, MWOA-RF, and MWOA-ELMAN.

		R^2	RMSE	MAE	MAPE (%)
MWOA-ELM	Train	0.9927	0.3287	0.2281	0.0353
	Test	0.9971	0.1911	0.1356	0.0212
MWOA-RF	Train	0.9870	0.4520	0.3152	0.0495
	Test	0.9341	1.0163	0.6553	0.0962
MWOA-ELMAN	Train	0.9883	0.4141	0.2261	0.0373
	Test	0.9698	0.6870	0.4867	0.0947

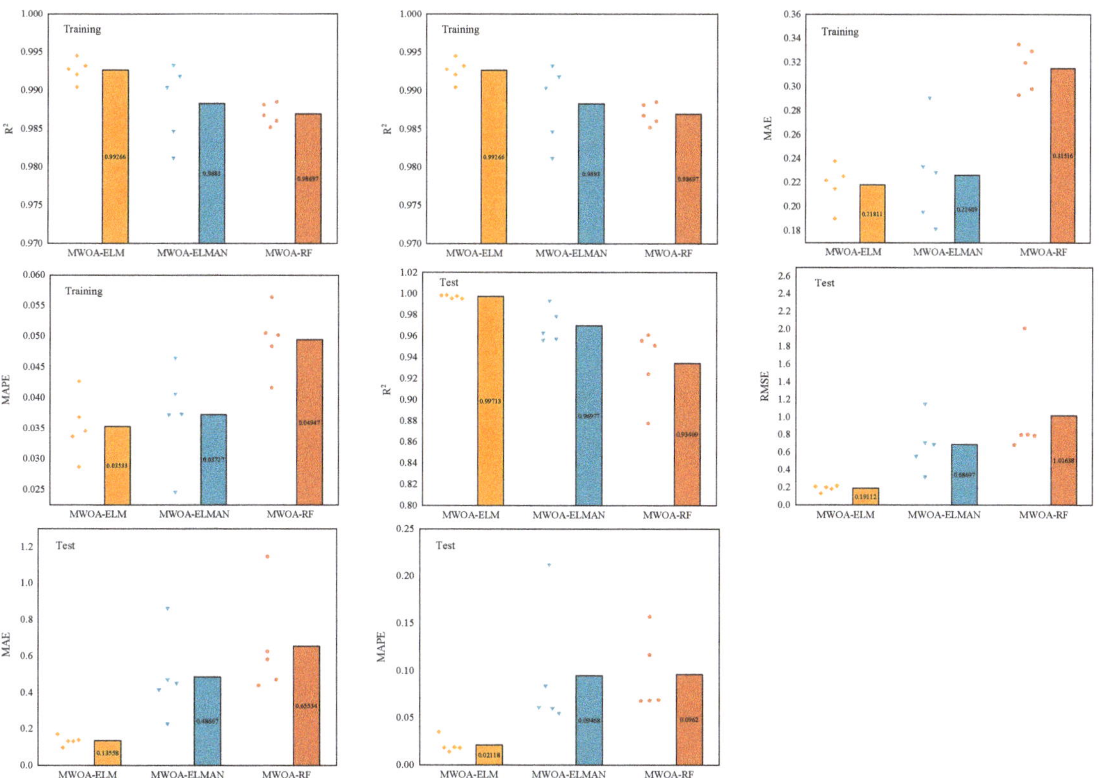

Figure 13. Box line plots of the Training and Testing sets for MWOA-ELM, MWOA-RF and MWOA-ELMAN models.

6.2. Sensitivity Analysis

Sensitivity factor analysis (SA) is an effective method for measuring the influence of model input parameters on output parameters. Sensitivity factor analysis provides feedback on the importance of the model input parameters. Therefore, this study uses the cosine amplitude method (CAM) [89] to perform sensitivity factor analysis on three models and an experimental model. The expressions are as follows:

$$R_{ij} = \frac{\sum_{k=1}^{n} x_{ik} x_{jk}}{\sqrt{\sum_{k=1}^{n} x_{ik}^2 \sum_{k=1}^{n} x_{jk}^2}} \tag{37}$$

where x_i denotes input parameters; x_j denotes output parameters; n denotes the number of data; R_{ij} denotes the strength of the relationship.

Figure 14 represents the strength factor of the relationship between each variable and D_{CI}. It can be seen that the three models show similar sensitivity to the experimental model, justifying the developed model. As seen from the graph, the measurement method has the most significant effect on the D_{CI} of RC, followed by FA, while the impact of WR is the least.

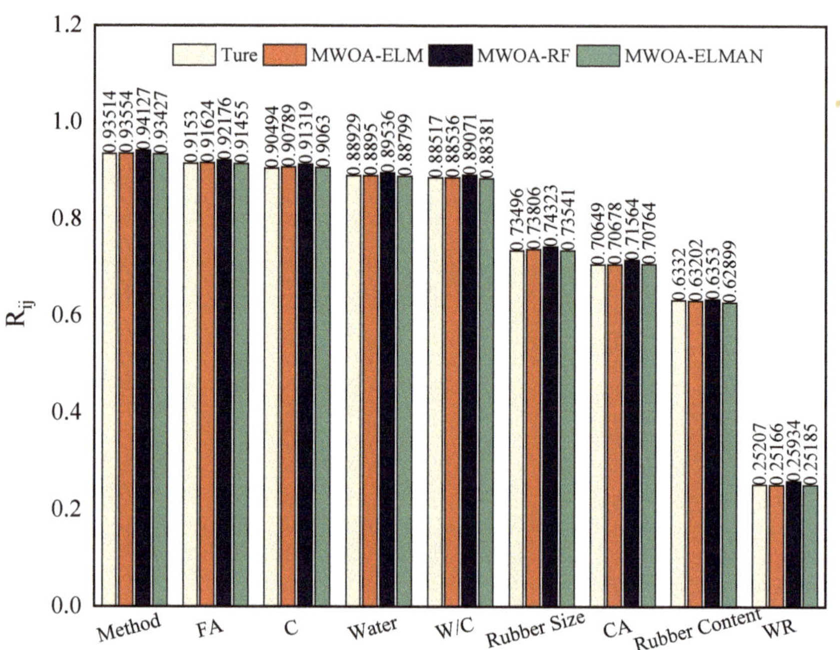

Figure 14. Sensitivity analysis of three models and experimental models.

6.3. Prediction of Typical Machine Learning Model

In Section 6.1, the MWOA-ELM model was proven to be the best model. Therefore, the typical predictions from the MWOA-ELM model are shown in this section. Figure 15 depicts the regression results for the training and testing set. It is important to emphasize that the MWOA-ELM model has strong predictive power. Its indicator values for the training and testing set were $R^2 = 0.9928$, RMSE = 0.3243, MAE = 0.2219, MAPE = 0.0287 and $R^2 = 0.9987$, RMSE = 0.1336, MAE = 0.0979, MAPE = 0.0187. Figure 16 shows the predicted values of the MWOA-ELM model compared to the actual values, with the error values included. The comparison results show that the predicted values of the D_{CI} of RC are consistent with the experimental values. It is worth noting that the error between the training set and the testing set is small, which indicates that the MWOA-ELM model can predict the D_{CI} of RC well. The above results suggest that predicting the D_{CI} of RC using the MWOA-ELM model is feasible. This may contribute to developing a numerical tool for determining durability indicators for RC. The application of intelligent algorithms is equally effective. In the future, consider increasing the amount of data and input variables, which would improve the ability of the MWOA-ELM model to predict the D_{CI} of RC. The weight matrix for the MWOA-ELM model's typical prediction result is shown in Appendix A.

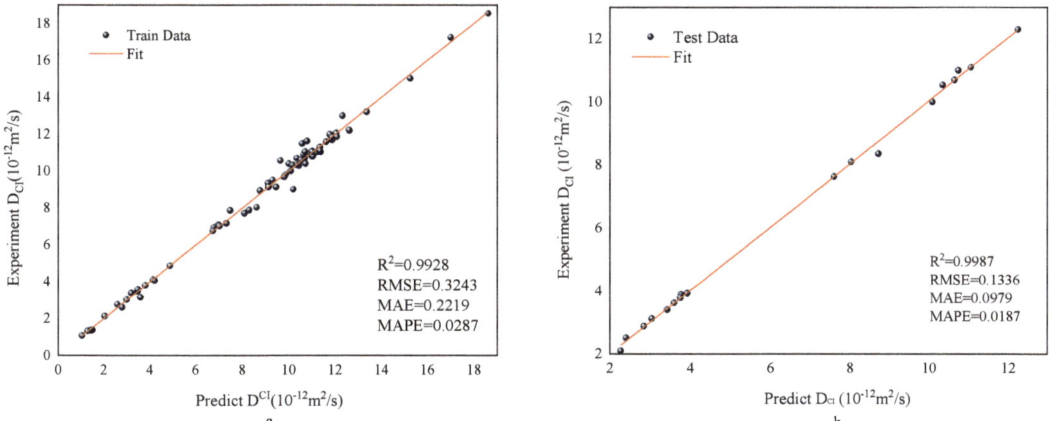

Figure 15. Regression results for the Training (**a**) and Testing (**b**) sets of the MWOA-ELM model.

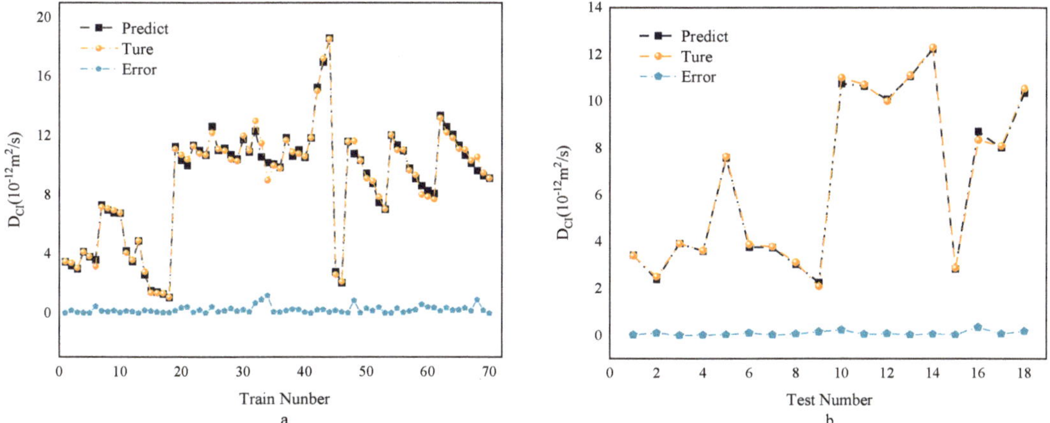

Figure 16. Comparison and error results between the Training (**a**) and Testing (**b**) sets of the MWOA-ELM model.

6.4. Forecast Comparison

The overall results of the representative predictions of the MWOA-ELM model are shown in Section 6.3. However, it is necessary to show the forecasting results within the model. This provides a more intuitive view of the model's predictions. Therefore, this section shows the prediction results of the MWOA-ELM model under different methods separately. Figure 17 represents the prediction result of the MWOA-ELM model for the literature [54,55]. The model's predictions can be judged from Figure 17, in general agreement with the experimental results. Figure 18 represents the prediction results of the model for the literature [57–59]. It can be seen from Figure 18 that the model predicts better results for the literature [58,59]. The predicted curves for the literature [57], showed some deviations, but the overall trend was consistent. However, the errors are acceptable in terms of the overall results of Section 6.3. Figure 19 represents the prediction results of the model for the literature [56]. From Figure 19, it can be observed that the prediction curves both show some deviations. This may be due to the algorithm falling into the local optimum when optimizing this part of the data, making the model learn insufficiently. However, in general, the errors are still acceptable. Figure 20 reflects the predictions of the literature [22]. From Figure 20, the prediction trends are consistent and the errors are relatively small. The above conclusions indicate that using the MWOA-ELM model predicting the D_{CI} of RC is feasible. The model can still be further optimized, which includes developing more powerful algorithms and increasing the amount of data.

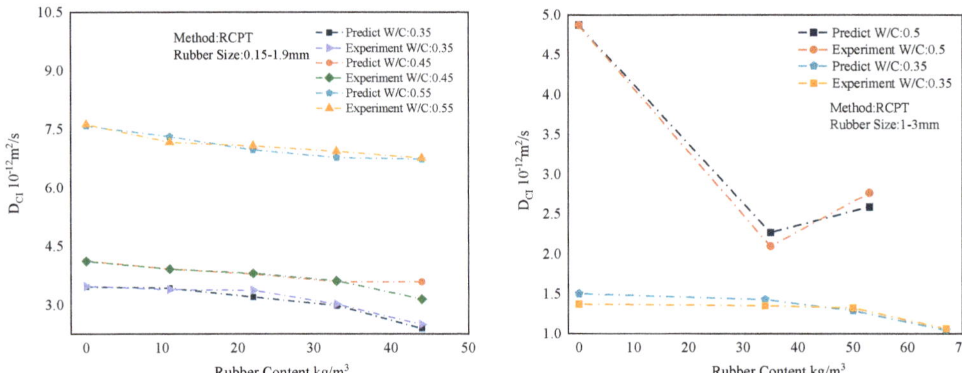

Figure 17. Prediction results of the MWOA-ELM model for the literature [54] (**Left**) and [55] (**Right**).

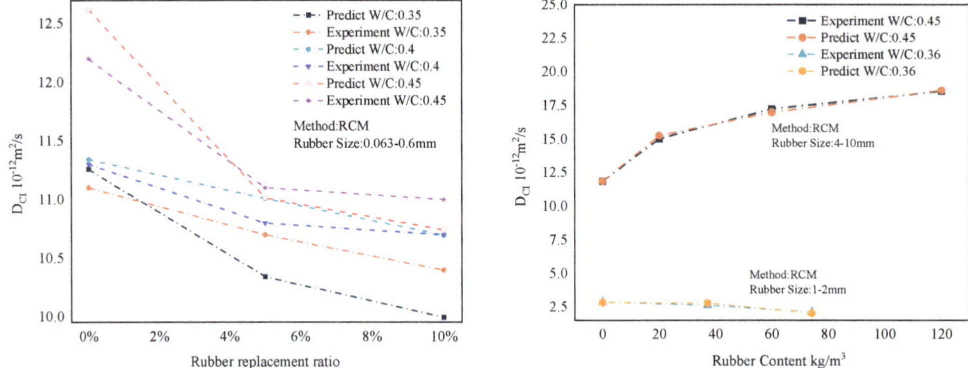

Figure 18. Prediction results of the MWOA-ELM model for the literature [57] (**Left**), [59] (**Right, Up**), and [58] (**Right, down**).

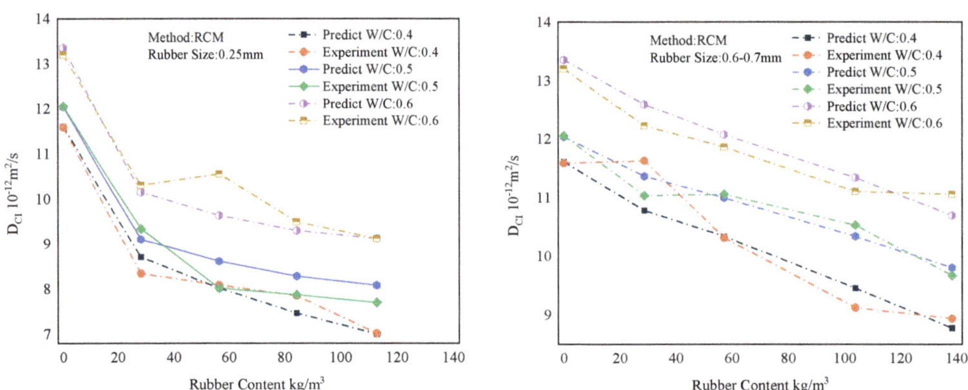

Figure 19. Prediction results of the MWOA-ELM model for the literature [56].

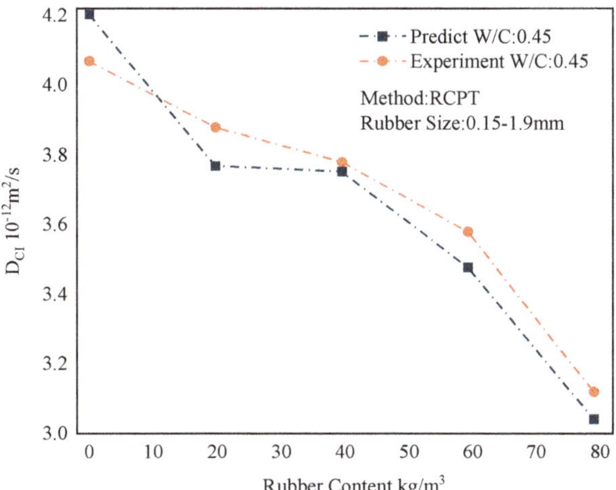

Figure 20. Prediction results of the MWOA-ELM model for the literature [22].

6.5. Comparative Analysis with Other Models

To further verify the MWOA-ELM model's reliability, this study introduces a multiple linear regression model for comparison [90]. Since the MRL model is similar to the ML model in that it also studies the effects of multi-factor interactions, it is compared with the MWOA-ELM model. Figure 21 represents the regression analysis results of the MRL model. The results of the evaluation indicators for the MWOA-ELM (Mean of overall model results under five-fold cross-validation) model and the MRL model are shown in Table 9. Obviously, MWOA-ELM is superior to the MRL model.

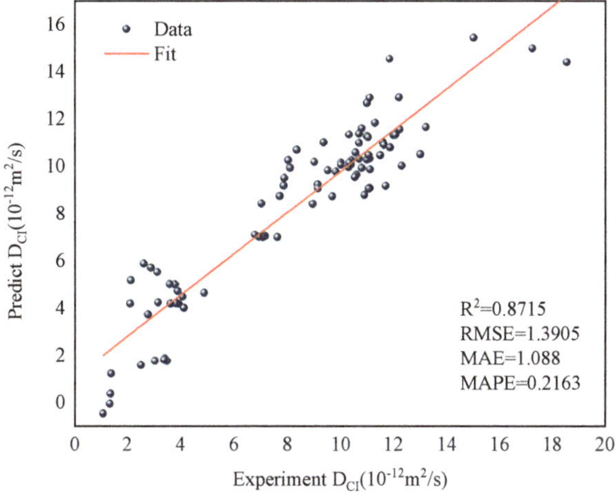

Figure 21. Regression analysis result of the MRL model.

Table 9. Evaluation indicators for the MWO-ELM and MRL model.

	R^2	RMSE	MAE	MAPE (%)
MWOA-ELM	0.9937	0.3064	0.2096	0.0325
MLR	0.8715	1.3905	1.088	0.2163

Comparison with other models is equally necessary. Ye [91] developed a mathematical model to predict the D_{CI} of RC, because the input variables for the mathematical model used are the water-cement ratio, the rubber admixture, and the rubber size. Therefore, the input variables of the MWOA-ELM model are replaced in the same way. Keeping the same input variables is better for comparison. The data were obtained from three randomly selected papers to avoid complex calculation [22,54,55]. Figure 22 represents the results of the regression analysis for the two models. The results of the evaluation indicators for the two models are shown in Table 10. From Figure 22, the regression analysis result of the MWOA-ELM model is better, with the R^2 of 0.991 higher than the mathematical model of 0.8053. Similar results are seen in the other error evaluation indicators in Table 9. This indicates that the MWOA-ELM model has better prediction and generalization ability.

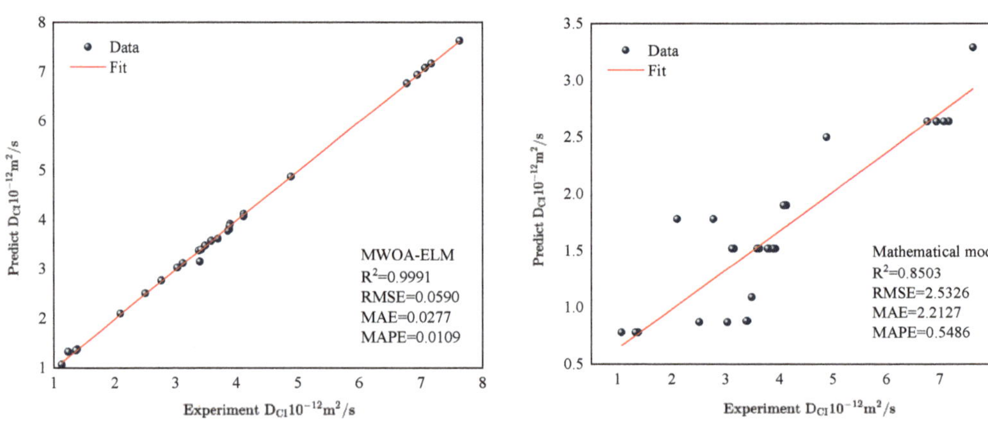

Figure 22. Regression analysis result of the MWOA-ELM and Mathematical model.

Table 10. Evaluation indicators for the MWO-ELM and Mathematical model.

	R^2	**RMSE**	**MAE**	**MAPE (%)**
MWOA-ELM	0.9991	0.0590	0.0277	0.0109
Ye [91]	0.8503	2.5326	2.2127	0.5486

7. Conclusions and Future Prospect

This study used ML techniques to predict the D_{CI} of RC. Three models, ELM, RF, and ELMAN were developed to investigate. The established MWOA was also used to optimize the three models. Four metrics, RMSE, MAE, and MAPE, were used to evaluate the performance of the models. According to the prediction results, the MWOA-optimized ELM, RF, and ELMAN models successfully predicted the D_{CI} of RC. At the same time, they had the highest R^2 and lowest errors compared to the unoptimized models. This indicates that the algorithm established is valid. Comparing the three optimal models, the MWOA-ELM model performs the best. The three models were shown to have similar sensitivity to the experimental model by the CAM method. This justifies the developed models. Comparing the typical prediction results of the MWOA-ELM model with the actual values shows that the prediction results generally agree with the experiment, while the error is within a reasonable range. Comparison with the MRL and published mathematical model show that the MWOA-ELM model performs the best. This suggests that the MWOA-ELM model can accurately predict the D_{CI} of RC.

In summary, this study successfully used ML techniques to predict the D_{CI} of RC while demonstrating the proposed MWOA is valid. This provides a new option for determining the D_{CI} of RC. However, observation of the results revealed that the proposed method could be further optimized to better understand RC's chloride permeation process. This includes developing more robust algorithms, increasing the data set, adding input variables, and enhancing the interpretable analysis of the model.

Author Contributions: Conceptualization, X.H. and K.W.; methodology, X.H.; software, X.H.; validation, X.H. and K.W.; formal analysis, X.H.; investigation, X.H. and W.D.; resources, X.H. and W.D. data curation, X.H. and W.D.; writing—original draft preparation, X.H.; writing—review and editing, H.L.; supervision, H.L., S.W. and T.L.; project administration, H.L.; funding acquisition, T.L. and S.W. All authors have read and agreed to the published version of the manuscript.

Funding: This research was funded by Wuhan Construction Engineering Group Co., Ltd. 2022 Annual Research Project, Grant number 3; Wuhan Urban Construction Group Co., Ltd. 2022 Annual Research Project, Grant number 5; 2022 Hubei Construction Science and Technology Program Project, Grant number 90, and 2021 Hubei Construction Science and Technology Program Project, Grant number 43.

Institutional Review Board Statement: Not applicable.

Informed Consent Statement: Not applicable.

Data Availability Statement: The data used in the article can be obtained from the author here.

Conflicts of Interest: The authors declare no conflict of interest.

Appendix A

MWOA-ELM Model:

$IW1=$

$$
\begin{bmatrix}
-5.89E-08 & -3.83E-04 & -1.34E-04 & -1.70E-04 & -7.72E-09 & -1.76E-04 & -1.13E-05 & -2.99E-05 & -1.12E-04 \\
-1.23E-03 & -2.42E-04 & -9.56E-09 & -3.21E-05 & -1.1E-04 & -8.49E-10 & -1.14E-04 & -6.21E-05 & -1.17E-04 \\
-4.32E-04 & -1.16E-04 & -2.07E-04 & -3.83E-05 & -3.04E-05 & -2.23E-09 & -6.17E-04 & -4.25E-04 & -1.4E-04 \\
-4.62E-04 & -1.29E-04 & -5.61E-04 & -2.23E-04 & -4.96E-05 & -6.35E-07 & -1.02E-04 & -2.60E-07 & -1.87E-04 \\
-5.3E-04 & -1.59E-04 & -4.83E-05 & -3.22E-05 & -1.26E-05 & -7.97E-05 & -8.61E-05 & -1.69E-04 & -5.17E-05 \\
-9.35E-05 & -2.04E-04 & -1.96E-04 & -6.98E-05 & -3.92E-04 & -3.25E-05 & -5.29E-05 & -2.92E-05 & -1.50E-04 \\
-2.03E-04 & -1.99E-05 & -8.04E-07 & -8.62E-05 & -1.93E-07 & -1.96E-04 & -1.54E-05 & -5.52E-04 & -1.09E-04 \\
-1.5E-04 & -3.19E-04 & -4.49E-05 & -1.87E-05 & -1.62E-05 & -9.87E-05 & -1.15E-04 & -7.21E-05 & -1.32E-04 \\
-8.34E-04 & -3.69E-05 & -4.42E-05 & -3.46E-04 & -8.46E-05 & -1.66E-04 & -1.58E-04 & -1.22E-04 & -1.46E-05 \\
-2.36E-05 & -1.67E-05 & -2.21E-04 & -3.93E-04 & -1.01E-04 & -6.5E-04 & -1.35E-04 & -3.88E-05 & -7E-05 \\
-4.57E-05 & -1.75E-04 & -6.48E-05 & -9.73E-05 & -1.53E-04 & -4.93E-05 & -1.49E-04 & -1.34E-04 & -6.58E-05 \\
-6.98E-10 & -1.09E-03 & -1.3E-07 & -2.22E-06 & -6.58E-05 & -2.63E-04 & -7.77E-05 & -1.2E-04 & -1.2E-04 \\
-3.74E-04 & -1.2E-04 & -1.43E-04 & -3.47E-05 & -2.52E-05 & -4.37E-05 & -1.49E-04 & -5.12E-07 & -3.2E-04 \\
-5.86E-04 & -1.03E-04 & -1.15E-04 & -5.44E-05 & -2.32E-05 & -2.57E-07 & -1.47E-04 & -2.62E-08 & -4.46E-04 \\
-1.22E-08 & -2.33E-04 & -1.08E-04 & -1.37E-08 & -8.57E-05 & -5.1E-05 & -7.4E-05 & -3.16E-04 & -3.93E-05 \\
-1.02E-05 & -9.2E-09 & -1.16E-04 & -7.36E-05 & -2.74E-05 & -1.97E-04 & -8.83E-05 & -2.55E-05 & -1.2E-04 \\
-4.87E-07 & -2.74E-04 & -9.85E-05 & -2.43E-04 & -2.36E-04 & -2.86E-04 & -9.91E-07 & -2.67E-05 & -1.77E-04 \\
-4.09E-04 & -5.15E-05 & -6.74E-05 & -2.71E-04 & -1.01E-04 & -1.3E-04 & -1.11E-04 & -5.24E-04 & -2.51E-04 \\
-4.42E-04 & -1.87E-04 & -2.43E-04 & -1.53E-08 & -4.16E-05 & -2.66E-04 & -3.43E-05 & -4.49E-05 & -1.93E-05 \\
-3.16E-04 & -5.35E-08 & -1.3E-04 & -1.02E-06 & -8.47E-05 & -2.33E-06 & -4.22E-04 & -1.02E-04 & -1.24E-04 \\
-2.93E-04 & -5.51E-05 & -6.66E-05 & -1.27E-04 & -6.37E-05 & -3.62E-04 & -1.09E-04 & -6.29E-05 & -4.33E-05 \\
-2.19E-04 & -2.28E-04 & -6.23E-06 & -1.05E-04 & -1.14E-04 & -1.51E-09 & -6E-05 & -1.02E-06 & -1.95E-04 \\
-1.21E-04 & -2.93E-06 & -1.29E-07 & -5.49E-05 & -6.5E-05 & -9.06E-05 & -1.24E-04 & -9.52E-05 & -1.39E-04 \\
-2.47E-04 & -2.33E-04 & -3.81E-05 & -2.98E-05 & -1.38E-04 & -9.32E-05 & -2.31E-04 & -1.33E-05 & -1.33E-04 \\
-1.64E-04 & -6.01E-04 & -1.16E-04 & -2.45E-04 & -9.14E-05 & -1.59E-09 & -4.81E-05 & -1.63E-04 & -2.26E-05 \\
-3.79E-05 & -2.37E-08 & -5.6E-05 & -1.75E-04 & -3.41E-04 & -3E-07 & -2.32E-05 & -2.73E-04 & -8E-05 \\
-4.62E-05 & -1.72E-05 & -7.78E-05 & -2.79E-05 & -1.7E-08 & -7.59E-05 & -7.74E-05 & -2.8E-04 & -3.48E-04 \\
-5.06E-04 & -2.63E-05 & -6.66E-05 & -7.55E-05 & -1.62E-04 & -1.37E-04 & -5.47E-05 & -1.97E-04 & -8.98E-05
\end{bmatrix}
$$

$B1=$

$$
[\,-3.11E-05\;-7.13E-06\;-1.01E-04\;-8.49E-05\;-2.3E-04\;-9.25E-05\;-1.48E-04
$$
$$
-1.28E-04\;-1.44E-04\;-1.19E-04\;-2.24E-04
$$
$$
-1.95E-04\;-3.91E-11\;-1.62E-04\;-2.03E-04\;-2.39E-04\;-6.9E-06\;-7.85E-05
$$
$$
-2.04E-04\;-2.79E-07\;-5.29E-09\;-1.3E-04
$$
$$
-1.04E-04\;-1.97E-04\;-3.65E-05\;-2.34E-05\;-9.78E-05\;-1.09E-04\,]^{\mathrm{T}}
$$

$LW1=$

$$
[\,-1.67E+11\;-7.39E+11\;-3.52E+11\;0.37E+11\;1.95E+11\;-5.82E+11\;1.71E+11\;-2.19E+11
$$
$$
-1.96E+11\;6.51E+11\;-4.07E+11\;-1.30E+11
$$
$$
1.13E+11\;-2.14E+11\;1.68E+11\;1.61E+11\;-2.45E+11\;3.63E+11\;-6.37E+11\;-3.95E+11
$$
$$
-4.37E+11\;8.69E+11\;-1.1E+11
$$
$$
3.79E+11\;1.42E+11\;8.13E+11\;-6.08E+11\;-9.04E+11\,]^{\mathrm{T}}
$$

References

1. Asutkar, P.; Shinde, S.; Patel, R. Study on the behaviour of rubber aggregates concrete beams using analytical approach. *Eng. Sci. Technol. Int. J.* **2017**, *20*, 151–159. [CrossRef]
2. Sun, Y.; Li, G.; Zhang, J.; Qian, D. Prediction of the strength of rubberized concrete by an evolved random forest model. *Adv. Civ. Eng.* **2019**, *2019*, 5198583. [CrossRef]
3. Sofi, A. Effect of waste tyre rubber on mechanical and durability properties of concrete—A review. *Ain Shams Eng. J.* **2018**, *9*, 2691–2700. [CrossRef]
4. Toutanji, H.A. The use of rubber tire particles in concrete to replace mineral aggregates. *Cem. Concr. Compos.* **1996**, *18*, 135–139. [CrossRef]
5. Skripkiūnas, G.; Grinys, A.; Černius, B. Deformation properties of concrete with rubber waste additives. *Mater. Sci.* **2007**, *13*, 219–223.
6. Batayneh, M.K.; Marie, I.; Asi, I. Promoting the use of crumb rubber concrete in developing countries. *Waste Manag.* **2008**, *28*, 2171–2176. [CrossRef]
7. Ganjian, E.; Khorami, M.; Maghsoudi, A.A. Scrap-tyre-rubber replacement for aggregate and filler in concrete. *Constr. Build. Mater.* **2009**, *23*, 1828–1836. [CrossRef]
8. Mohammed, B.S.; Azmi, N. Strength reduction factors for structural rubbercrete. *Front. Struct. Civ. Eng.* **2014**, *8*, 270–281. [CrossRef]
9. El-Khoja, A.; Ashour, A.; Abdalhmid, J.; Dai, X.; Khan, A. Prediction of rubberised concrete strength by using artificial neural networks. *Int. J. Struct. Constr. Eng.* **2018**, *12*, 1068–1073.
10. Hadzima-Nyarko, M.; Nyarko, E.K.; Ademović, N.; Miličević, I.; Kalman Šipoš, T. Modelling the influence of waste rubber on compressive strength of concrete by artificial neural networks. *Materials* **2019**, *12*, 561. [CrossRef]
11. Pradhan, B. Corrosion behavior of steel reinforcement in concrete exposed to composite chloride–sulfate environment. *Constr. Build. Mater.* **2014**, *72*, 398–410. [CrossRef]
12. Yu, Z.; Chen, Y.; Liu, P.; Wang, W. Accelerated simulation of chloride ingress into concrete under drying–wetting alternation condition chloride environment. *Constr. Build. Mater.* **2015**, *93*, 205–213. [CrossRef]
13. Andisheh, K.; Scott, A.; Palermo, A.; Clucas, D. Influence of chloride corrosion on the effective mechanical properties of steel reinforcement. *Struct. Infrastruct. Eng.* **2019**, *15*, 1036–1048. [CrossRef]
14. Lee, H.-S.; Cho, Y.-S. Evaluation of the mechanical properties of steel reinforcement embedded in concrete specimen as a function of the degree of reinforcement corrosion. *Int. J. Fract.* **2009**, *157*, 81–88. [CrossRef]
15. Song, H.-W.; Shim, H.-B.; Petcherdchoo, A.; Park, S.-K. Service life prediction of repaired concrete structures under chloride environment using finite difference method. *Cem. Concr. Compos.* **2009**, *31*, 120–127. [CrossRef]
16. Lu, C.; Yuan, S.; Cheng, P.; Liu, R. Mechanical properties of corroded steel bars in pre-cracked concrete suffering from chloride attack. *Constr. Build. Mater.* **2016**, *123*, 649–660. [CrossRef]
17. Jung, J.-S.; Lee, B.Y.; Lee, K.-S. Experimental study on the structural performance degradation of corrosion-damaged reinforced concrete beams. *Adv. Civ. Eng.* **2019**, *2019*, 9562574. [CrossRef]
18. Thomas, B.S.; Gupta, R.C.; Kalla, P.; Cseteneyi, L. Strength, abrasion and permeation characteristics of cement concrete containing discarded rubber fine aggregates. *Constr. Build. Mater.* **2014**, *59*, 204–212. [CrossRef]
19. Su, H.; Yang, J.; Ling, T.-C.; Ghataora, G.S.; Dirar, S. Properties of concrete prepared with waste tyre rubber particles of uniform and varying sizes. *J. Clean. Prod.* **2015**, *91*, 288–296. [CrossRef]
20. Thomas, B.S.; Gupta, R.C. A comprehensive review on the applications of waste tire rubber in cement concrete. *Renew. Sustain. Energy Rev.* **2016**, *54*, 1323–1333. [CrossRef]
21. Mao, L.-x.; Hu, Z.; Xia, J.; Feng, G.-l.; Azim, I.; Yang, J.; Liu, Q.-f. Multi-phase modelling of electrochemical rehabilitation for ASR and chloride affected concrete composites. *Compos. Struct.* **2019**, *207*, 176–189. [CrossRef]
22. Gupta, T.; Siddique, S.; Sharma, R.K.; Chaudhary, S. Behaviour of waste rubber powder and hybrid rubber concrete in aggressive environment. *Constr. Build. Mater.* **2019**, *217*, 283–291. [CrossRef]
23. Beushausen, H.; Torrent, R.; Alexander, M.G. Performance-based approaches for concrete durability: State of the art and future research needs. *Cem. Concr. Res.* **2019**, *119*, 11–20. [CrossRef]
24. Dierkens, M.; Godart, B.; Mai-Nhu, J.; Rougeau, P.; Linger, L.; Cussigh, F. In French national project 'PERFDUB' on performance-based approach: Interest of old structures analysis for the definition of durability indicators criteria. In Proceedings of the 16th fib Symposium, Concrete Innovations in Materials, Design and Structures, Krakow, Poland, 27–29 May 2019; Fédération de l'Industrie du Béton-FIB: Montrouge, France, 2019; p. 8.
25. Tran, V.Q. Machine learning approach for investigating chloride diffusion coefficient of concrete containing supplementary cementitious materials. *Constr. Build. Mater.* **2022**, *328*, 127103. [CrossRef]
26. Saeki, T.; Sasaki, K.; Shinada, K. Estimation of chloride diffusion coeffcent of concrete using mineral admixtures. *J. Adv. Concr. Technol.* **2006**, *4*, 385–394. [CrossRef]
27. Jasielec, J.J.; Stec, J.; Szyszkiewicz-Warzecha, K.; Łagosz, A.; Deja, J.; Lewenstam, A.; Filipek, R. Effective and apparent diffusion coefficients of chloride ions and chloride binding kinetics parameters in mortars: Non-stationary diffusion–reaction model and the inverse problem. *Materials* **2020**, *13*, 5522. [CrossRef]
28. Liu, Q.-f.; Iqbal, M.F.; Yang, J.; Lu, X.-y.; Zhang, P.; Rauf, M. Prediction of chloride diffusivity in concrete using artificial neural network: Modelling and performance evaluation. *Constr. Build. Mater.* **2021**, *268*, 121082. [CrossRef]

29. Van Noort, R.; Hunger, M.; Spiesz, P. Long-term chloride migration coefficient in slag cement-based concrete and resistivity as an alternative test method. *Constr. Build. Mater.* **2016**, *115*, 746–759. [CrossRef]

30. Wang, H.-L.; Dai, J.-G.; Sun, X.-Y.; Zhang, X.-L. Time-dependent and stress-dependent chloride diffusivity of concrete subjected to sustained compressive loading. *J. Mater. Civ. Eng.* **2016**, *28*, 04016059. [CrossRef]

31. Huang, X.-Y.; Wu, K.-Y.; Wang, S.; Lu, T.; Lu, Y.-F.; Deng, W.-C.; Li, H.-M. Compressive Strength Prediction of Rubber Concrete Based on Artificial Neural Network Model with Hybrid Particle Swarm Optimization Algorithm. *Materials* **2022**, *15*, 3934. [CrossRef]

32. Gupta, T.; Patel, K.; Siddique, S.; Sharma, R.K.; Chaudhary, S. Prediction of mechanical properties of rubberised concrete exposed to elevated temperature using ANN. *Measurement* **2019**, *147*, 106870. [CrossRef]

33. Zhang, J.; Zhang, M.; Dong, B.; Ma, H. Quantitative evaluation of steel corrosion induced deterioration in rubber concrete by integrating ultrasonic testing, machine learning and mesoscale simulation. *Cem. Concr. Compos.* **2022**, *128*, 104426. [CrossRef]

34. Huang, G.-B.; Zhou, H.; Ding, X.; Zhang, R. Extreme learning machine for regression and multiclass classification. *IEEE Trans. Syst. Man Cybern. Part B* **2011**, *42*, 513–529. [CrossRef]

35. Ding, S.; Zhao, H.; Zhang, Y.; Xu, X.; Nie, R. Extreme learning machine: Algorithm, theory and applications. *Artif. Intell. Rev.* **2015**, *44*, 103–115. [CrossRef]

36. Han, F.; Yao, H.-F.; Ling, Q.-H. An improved evolutionary extreme learning machine based on particle swarm optimization. *Neurocomputing* **2013**, *116*, 87–93. [CrossRef]

37. Li, C.; Tao, Y.; Ao, W.; Yang, S.; Bai, Y. Improving forecasting accuracy of daily enterprise electricity consumption using a random forest based on ensemble empirical mode decomposition. *Energy* **2018**, *165*, 1220–1227. [CrossRef]

38. Fan, G.-F.; Yu, M.; Dong, S.-Q.; Yeh, Y.-H.; Hong, W.-C. Forecasting short-term electricity load using hybrid support vector regression with grey catastrophe and random forest modeling. *Util. Policy* **2021**, *73*, 101294. [CrossRef]

39. Cai, C.; Qian, Q.; Fu, Y. Application of BAS-Elman neural network in prediction of blasting vibration velocity. *Procedia Comput. Sci.* **2020**, *166*, 491–495. [CrossRef]

40. Liu, B.; Zhao, Y.; Wang, W.; Liu, B. Compaction density evaluation model of sand-gravel dam based on Elman neural network with modified particle swarm optimization. *Front. Phys.* **2022**, *9*, 806231. [CrossRef]

41. Kang, F.; Liu, J.; Li, J.; Li, S. Concrete dam deformation prediction model for health monitoring based on extreme learning machine. *Struct. Control Health Monit.* **2017**, *24*, e1997. [CrossRef]

42. Falah, M.W.; Hussein, S.H.; Saad, M.A.; Ali, Z.H.; Tran, T.H.; Ghoniem, R.M.; Ewees, A.A. Compressive Strength Prediction Using Coupled Deep Learning Model with Extreme Gradient Boosting Algorithm: Environmentally Friendly Concrete Incorporating Recycled Aggregate. *Complexity* **2022**, *2022*, 5433474. [CrossRef]

43. Dong, L.; Shu, W.; Sun, D.; Li, X.; Zhang, L. Pre-alarm system based on real-time monitoring and numerical simulation using internet of things and cloud computing for tailings dam in mines. *IEEE Access* **2017**, *5*, 21080–21089. [CrossRef]

44. Hai-Bang, L.; Thuy-Anh, N.; Hai-Van Thi, M.; Van Quan, T. Development of deep neural network model to predict the compressive strength of rubber concrete. *Constr. Build. Mater.* **2021**, *301*, 124081. [CrossRef]

45. Liu, H.; Mi, X.-w.; Li, Y.-f. Wind speed forecasting method based on deep learning strategy using empirical wavelet transform, long short term memory neural network and Elman neural network. *Energy Convers. Manag.* **2018**, *156*, 498–514. [CrossRef]

46. Mirjalili, S.; Lewis, A. The whale optimization algorithm. *Adv. Eng. Softw.* **2016**, *95*, 51–67. [CrossRef]

47. Rathore, N.S.; Singh, V. Whale optimisation algorithm-based controller design for reverse osmosis desalination plants. *Int. J. Intell. Eng. Inform.* **2019**, *7*, 77–88. [CrossRef]

48. Mirjalili, S.; Mirjalili, S.M.; Saremi, S.; Mirjalili, S. Whale optimization algorithm: Theory, literature review, and application in designing photonic crystal filters. In *Nature-Inspired Optimizers*; Springer: Cham, Switzerland, 2020; pp. 219–238.

49. Qais, M.H.; Hasanien, H.M.; Alghuwainem, S. Enhanced whale optimization algorithm for maximum power point tracking of variable-speed wind generators. *Appl. Soft Comput.* **2020**, *86*, 105937. [CrossRef]

50. Teng, Z.; Lv, J.; Guo, L.; Yuanyuan, X. An improved hybrid grey wolf optimization algorithm based on Tent mapping. *J. Harbin Inst. Technol.* **2018**, *50*, 40–49.

51. Hang, X.; Zhang, D.; Wang, Y.; Song, T.; Fan, Y. Hybrid strategy to improve whale optimization algorithm. *Comput. Eng. Des.* **2020**, *41*, 3397–3404. [CrossRef]

52. Kong, Z.; Yang, Q.-f.; Zhao, J.; Xiong, J.-j. Adaptive adjustment of weights and search strategies-based whale optimization algorithm. *J. Northeast. Univ.* **2020**, *41*, 35.

53. Zhang, W.; Liu, S.; Ren, C. Mixed Strategy Improved Sparrow Search Algorithm. *Comput. Eng. Appl.* **2021**, *57*, 74–82.

54. Gupta, T.; Chaudhary, S.; Sharma, R.K. Assessment of mechanical and durability properties of concrete containing waste rubber tire as fine aggregate. *Constr. Build. Mater.* **2014**, *73*, 562–574. [CrossRef]

55. Noor, N.M.; Yamamoto, D.; Hamada, H.; Sagawa, Y. Study on Chloride Ion Penetration Resistance of Rubberized Concrete Under Steady State Condition. *MATEC Web Conf.* **2016**, *47*, 01004. [CrossRef]

56. Ding, X.C. *Study on Durability of Waste Rubber Cement Mortar*; Henan Polytechnic University: Jiaozuo, China, 2018.

57. Amiri, M.; Hatami, F.; Golafshani, E.M. Evaluating the synergic effect of waste rubber powder and recycled concrete aggregate on mechanical properties and durability of concrete. *Case Stud. Constr. Mater.* **2021**, *15*, e00639. [CrossRef]

58. Han, Q.; Wang, N.; Zhang, J.; Yu, J.; Hou, D.; Dong, B. Experimental and computational study on chloride ion transport and corrosion inhibition mechanism of rubber concrete. *Constr. Build. Mater.* **2021**, *268*, 121105. [CrossRef]

59. Nadi, S.; Beheshti Nezhad, H.; Sadeghi, A. Experimental study on the durability and mechanical properties of concrete with crumb rubber. *J. Build. Pathol. Rehabil.* **2022**, *7*, 17. [CrossRef]

60. Smith, G.N. *Probability and Statistics in Civil Engineering*; Collins Professional and Technical Books: London, UK, 1986; 244p.

61. Dunlop, P.; Smith, S. Estimating key characteristics of the concrete delivery and placement process using linear regression analysis. *Civ. Eng. Environ. Syst.* **2003**, *20*, 273–290. [CrossRef]

62. Al-Janabi, T.A.; Al-Raweshidy, H.S. Efficient whale optimisation algorithm-based SDN clustering for IoT focused on node density. In Proceedings of the 2017 16th Annual Mediterranean Ad Hoc Networking Workshop (Med-Hoc-Net), Budva, Montenegro, 28–30 June 2017; IEEE: New York, NY, USA, 2017; pp. 1–6.

63. Cai, D.; Ji, X.; Shi, H.; Pan, J. Method for improving piecewise Logistic chaotic map and its performance analysis. *J. Nanjing Univ.* **2016**, *52*, 809–815.

64. Zhou, F.-j.; Wang, X.-j.; Zhang, M. Evolutionary programming using mutations based on the t probability distribution. *Acta Electonica Sin.* **2008**, *36*, 667.

65. Huang, G.-B.; Zhu, Q.-Y.; Siew, C.-K. Extreme learning machine: A new learning scheme of feedforward neural networks. In Proceedings of the 2004 IEEE international joint conference on neural networks (IEEE Cat. No. 04CH37541), Budapest, Hungary, 25–29 July 2004; IEEE: New York, NY, USA, 2004; pp. 985–990. [CrossRef]

66. Kahramanli, H.; Allahverdi, N. Rule extraction from trained adaptive neural networks using artificial immune systems. *Expert Syst. Appl.* **2009**, *36*, 1513–1522. [CrossRef]

67. Zhang, D.; Wang, Y. Rough neural network based on bottom-up fuzzy rough data analysis. *Neural Process. Lett.* **2009**, *30*, 187–211. [CrossRef]

68. Ding, S.; Jia, W.; Su, C.; Zhang, L.; Liu, L. Research of neural network algorithm based on factor analysis and cluster analysis. *Neural Comput. Appl.* **2011**, *20*, 297–302. [CrossRef]

69. Ding, S.; Xu, L.; Su, C.; Jin, F. An optimizing method of RBF neural network based on genetic algorithm. *Neural Comput. Appl.* **2012**, *21*, 333–336. [CrossRef]

70. Huang, G.-B.; Zhu, Q.-Y.; Siew, C.-K. Extreme learning machine: Theory and applications. *Neurocomputing* **2006**, *70*, 489–501. [CrossRef]

71. Huang, G.-B.; Babri, H.A. Upper bounds on the number of hidden neurons in feedforward networks with arbitrary bounded nonlinear activation functions. *IEEE Trans. Neural Netw.* **1998**, *9*, 224–229. [CrossRef]

72. Huang, G.-B. Learning capability and storage capacity of two-hidden-layer feedforward networks. *IEEE Trans. Neural Netw.* **2003**, *14*, 274–281. [CrossRef]

73. Liang, N.-Y.; Huang, G.-B.; Saratchandran, P.; Sundararajan, N. A fast and accurate online sequential learning algorithm for feedforward networks. *IEEE Trans. Neural Netw.* **2006**, *17*, 1411–1423. [CrossRef]

74. Breiman, L. Random forests. *Mach. Learn.* **2001**, *45*, 5–32. [CrossRef]

75. Amato, L.; Minozzi, S.; Mitrova, Z.; Parmelli, E.; Saulle, R.; Cruciani, F.; Vecchi, S.; Davoli, M. Systematic review of safeness and therapeutic efficacy of cannabis in patients with multiple sclerosis, neuropathic pain, and in oncological patients treated with chemotherapy. *Epidemiol. Prev.* **2017**, *41*, 279–293. [CrossRef]

76. Hou, K.; Yang, H.; Ye, Z.; Wang, Y.; Liu, L.; Cui, X. Effectiveness of pharmacist-led anticoagulation management on clinical outcomes: A systematic review and meta-analysis. *J. Pharm. Pharm. Sci.* **2017**, *20*, 378–396. [CrossRef]

77. Tang, Q.Y.; Zhang, C.X. Data Processing System (DPS) software with experimental design, statistical analysis and data mining developed for use in entomological research. *Insect Sci.* **2013**, *20*, 254–260. [CrossRef] [PubMed]

78. Gao, A. *Research on Prediction Model of Tillage DepthBased on an Improved Random Forest*; Changchun University of Technology: Changchun, China, 2022.

79. Elman, J.L. Finding structure in time. *Cogn. Sci.* **1990**, *14*, 179–211. [CrossRef]

80. Mehr, A.D.; Vaheddoost, B.; Mohammadi, B. ENN-SA: A novel neuro-annealing model for multi-station drought prediction. *Comput. Geosci.* **2020**, *145*, 104622. [CrossRef]

81. Yolcu, O.C.; Temel, F.A.; Kuleyin, A. New hybrid predictive modeling principles for ammonium adsorption: The combination of Response Surface Methodology with feed-forward and Elman-Recurrent Neural Networks. *J. Clean. Prod.* **2021**, *311*, 127688. [CrossRef]

82. Cheng, Y.-c.; Qi, W.-M.; Cai, W.-Y. Dynamic properties of Elman and modified Elman neural network. In Proceedings of the International Conference on Machine Learning and Cybernetics, Beijing, China, 4–5 November 2002; IEEE: New York, NY, USA, 2002; pp. 637–640.

83. Menard, S. Coefficients of determination for multiple logistic regression analysis. *Am. Stat.* **2000**, *54*, 17–24.

84. Le, L.M.; Ly, H.-B.; Pham, B.T.; Le, V.M.; Pham, T.A.; Nguyen, D.-H.; Tran, X.-T.; Le, T.-T. Hybrid artificial intelligence approaches for predicting buckling damage of steel columns under axial compression. *Materials* **2019**, *12*, 1670. [CrossRef]

85. Ly, H.-B.; Le, L.M.; Phi, L.V.; Phan, V.-H.; Tran, V.Q.; Pham, B.T.; Le, T.-T.; Derrible, S. Development of an AI model to measure traffic air pollution from multisensor and weather data. *Sensors* **2019**, *19*, 4941. [CrossRef]

86. Pham, B.T.; Jaafari, A.; Prakash, I.; Bui, D.T. A novel hybrid intelligent model of support vector machines and the MultiBoost ensemble for landslide susceptibility modeling. *Bull. Eng. Geol. Environ.* **2019**, *78*, 2865–2886. [CrossRef]

87. Taylor, K.E. Summarizing multiple aspects of model performance in a single diagram. *J. Geophys. Res. Atmos.* **2001**, *106*, 7183–7192. [CrossRef]

88. Iqbal, M.F.; Liu, Q.-f.; Azim, I.; Zhu, X.; Yang, J.; Javed, M.F.; Rauf, M. Prediction of mechanical properties of green concrete incorporating waste foundry sand based on gene expression programming. *J. Hazard. Mater.* **2020**, *384*, 121322. [CrossRef]

89. Jahed Armaghani, D.; Hajihassani, M.; Sohaei, H.; Tonnizam Mohamad, E.; Marto, A.; Motaghedi, H.; Moghaddam, M.R. Neuro-fuzzy technique to predict air-overpressure induced by blasting. *Arab. J. Geosci.* **2015**, *8*, 10937–10950. [CrossRef]
90. Hong, F.; Qiao, H.; Wang, P. Predicting the life of BNC-coated reinforced concrete using the Weibull distribution. *Emerg. Mater. Res.* **2020**, *9*, 424–434. [CrossRef]
91. Ye, W.C. *Experimental Study on the Durability of Rubber Concrete*; Shenyang University: Shenyang, China, 2013.

Article

Thermal and Mechanical Degradation of Recycled Polylactic Acid Filaments for Three-Dimensional Printing Applications

Dongoh Lee [1], Younghun Lee [1], Inwhan Kim [1], Kyungjun Hwang [2] and Namsu Kim [1,*]

[1] Department of Mechanical Engineering, Konkuk University, Seoul 05029, Republic of Korea
[2] Gangneung Science and Industry Promotion Agency, Gangneung 25440, Republic of Korea
* Correspondence: nkim7@konkuk.ac.kr; Tel.: +82-2-450-3434

Abstract: The recycling of filaments used in three-dimensional (3D) printing systems not only mitigates the environmental issues associated with conventional 3D printing approaches but also simultaneously reduces manufacturing costs. This study investigates the effects of successive recycling of polylactic acid (PLA) filaments, which were used in the printing process, on the mechanical properties of recycled filaments and printed objects. The mechanical strengths of the printed PLA and the adhesion strengths between 3D-printed beads were evaluated via the tensile testing of the horizontally and vertically fabricated specimens. Gel permeation chromatography analysis revealed a reduction in the molecular weight of the polymer as a result of recycling, leading to a decrease in the mechanical strength of the 3D-printed product. Additionally, scanning electron microscopy images of the cutting plane showed that the fabricated beads were broken in the case of the horizontally fabricated specimen, whereas in the case of the vertically fabricated samples, the adhesion between the beads was weak. These findings indicate that the mechanical strength in the in-plane and out-of-plane directions must be improved by increasing the mechanical strength of the bead itself as well as the adhesion strength of the beads.

Keywords: poly-lactic acid; 3D printing; thermal degradation; mechanical degradation; recycling; molecular weight; additive manufacturing

check for updates

Citation: Lee, D.; Lee, Y.; Kim, I.; Hwang, K.; Kim, N. Thermal and Mechanical Degradation of Recycled Polylactic Acid Filaments for Three-Dimensional Printing Applications. *Polymers* **2022**, *14*, 5385. https://doi.org/10.3390/polym14245385

Academic Editors: Sheila Devasahayam, Laurence Dyer and Beom Soo Kim

Received: 3 September 2022
Accepted: 25 November 2022
Published: 9 December 2022

Publisher's Note: MDPI stays neutral with regard to jurisdictional claims in published maps and institutional affiliations.

1. Introduction

Additive manufacturing (AM) using three-dimensional (3D) printing is an emerging technology that offers considerable design freedom, particularly in the design of rapid prototypes, personalized medical components, and parts that cannot be mass-produced [1–4]. Similar to other AM techniques, components or systems used in 3D printing are fabricated via the deposition of successive layers, rather than the removal of material from a larger piece, as in subtractive manufacturing (e.g., a lathe) [5]. Among the currently available 3D printing technologies, fused deposition modeling (FDM), in particular, has been widely utilized in various applications owing to its low cost. In 3D printers that employ the FDM method, a thermoplastic filament is heated to temperatures close to its melting point, followed by layer-by-layer extrusion to create 3D objects. This technology has been employed in a wide range of applications; however, its increased use has raised several environmental concerns. Therefore, the production of filaments from eco-friendly recycled materials has been investigated to mitigate environmental issues. Recently, the recycling of polymer waste to alleviate the environmental issues caused by polymeric waste accumulation has received significant attention. Moreover, most polymeric materials are produced from oil and gas; therefore, the increased production of polymer-based materials ultimately increases the exploitation of natural resources [6].

Currently, various raw materials are used in FDM-type 3D printers, such as acrylonitrile butadiene styrene (ABS), nylon, polycarbonate, high-density polyethylene, high-impact polystyrene, and polylactic acid (PLA). Among these materials, PLA has received considerable attention owing to its relative abundance and cost-effectiveness as well as its relatively low melting point (150–160 °C), which reduces the energy required for printing. Additionally, PLA is an important biodegradable polyester that can be used in biomedical and pharmaceutical applications, such as implant devices, tissue scaffolds, and internal structures. Therefore, numerous studies have been conducted for characterizing 3D-printed PLA materials, such as fatigue, anisotropy, crystallinity, heat conduction, composites, dimensional accuracy, and dielectric property analyses [7–14]. Additionally, PLA is generally derived from renewable resources, such as starch and sugar, and is therefore a safer alternative to potentially toxic ABS plastics [6,15–17].

The increase in the use of filaments in 3D printing applications has led to an increased interest in recycling strategies, which can reduce the cost of feedstock, manufacturing cost of fabricated products, and generation of waste [18,19]. In addition, recycling reduces greenhouse gas emissions and lowers the environmental impact of products fabricated using 3D printers [18–22]. Several previous studies have examined the recycling process for 3D printing applications. Baechler et al. developed an extruder to produce filaments from polymer wastes, such as high-density polyethylene (HDPE) from waste bottles and laundry detergent containers [23]. McNaney reported on Filabot, which is a plastic filament producing company that utilizes post-consumer plastic waste, thereby reducing the manufacturing cost of 3D-printed models. Kim et al. designed a recycling system that includes a shredder that crushes the output of 3D printing and any generated waste as well as a spooler that enables the recycled filament to be used directly in the 3D printer [24,25]. Considering other thermoplastics, mechanical (or physical) recycling is more commonly used for PLA than chemical recycling and reuse. This process involves mechanically grinding the plastic into small pieces and subsequently reprocessing and compounding them at elevated temperatures to produce a new component or filament. This process is typically conducted using a recycling system; the steps involved in the recycling process are illustrated in Figure 1 [25]. However, thermal and mechanical degradation, similar to that observed during injection molding and extrusion, might occur as a result of this process [6,15,26–28]. Thus, understanding the degradation mechanism of recycled PLA filaments, as well as any object produced using them via FDM type 3D printing, is critical.

Figure 1. Steps involved in the recycling process of initial waste materials to form filaments (outside view) and an inside schematic of the filament recycling system, consisting of a shredder, extruder, sensor, and spooling systems.

Several studies have been conducted to investigate the mechanical properties and degradation mechanisms of 3D-printed specimens and to eliminate the degradation caused by recycling. Lanzotti et al. reported the impact of process parameters, such as layer thickness, infill orientation, and the number of shell parameters, on the mechanical properties of 3D-printed PLA specimens [29]. These studies clarify the impact of process factors on the mechanical properties of 3D-printed specimens. To improve the thermal stability and mechanical properties of PLA, Yang et al. employed crosslinking via the chemical treatment of the melt by adding small amounts of triallyl isocyanurate and dicumyl peroxide as crosslinking agents [17]. The results showed a decrease in crystallinity and a significant improvement in the thermal stability, tensile modulus, and strength of PLA. In another study, the dopamine coating of the PLA pellets used in the 3D printing process, to improve the adhesion between layers fabricated by the 3D printer, was investigated [30]. The mechanical properties of the recycled specimen improved after polydopamine coating. Thus, to improve the thermal and mechanical properties of 3D-printed parts, understanding the degradation mechanism of materials and the main cause of failures is critical.

This study investigates the feasibility of recycling commercialized PLA filaments by employing a distributed recycling system which was designed and assembled in the laboratory. Furthermore, the mechanism of degradation resulting from the recycling process is presented. Compared to centralized recycling systems, distributed recycling is advantageous when the amount of waste materials for recycling is relatively small. Tensile tests were performed to compare the mechanical properties of the 3D-printed specimens fabricated using pristine and recycled PLA. Further, gel permeation chromatography (GPC) was performed at each recycling stage to better understand the degradation mechanism and measure any change in the molecular weight of the PLA, which is directly related to the mechanical strength of the polymer. In addition, the mechanical strengths of the horizontally and vertically fabricated specimens were also tested and compared. The fracture surfaces of the specimens were observed using scanning electron microscopy (SEM) to investigate the fracture mode.

2. Materials and Methods

The recycling of the PLA filament products was conducted using a custom mechanical recycling system consisting of a shredder, extruder, spooler, sensor, and controller, as shown in Figure 1 [24,25]. Using this approach, failed or broken parts fabricated using a 3D-printer were broken into small pieces using a shredder comprising an auger driven by an electric motor, followed by heating to temperatures close to the glass transition temperature for softening. The optimized recycler setting for extruding filament was summarized in the reference which was published by same authors [25]. The resulting product was forced through a die to extrude the PLA filament. The temperature of the heating zone and speed of the extruder were determined by recycled materials and controlled via a closed-loop controller to regulate the diameter of the extruded filament. The details of this process can be found another reference [26]. To investigate the mechanical performance of the recycled PLA filament, an open-source FDM type 3D printer (Cubicon 3DP-110F, Hyvision System Corp., Seongnam, Republic of Korea) with a 0.4 mm-diameter nozzle was used to fabricate Type-5 tensile test specimens, as shown in Figure 2, using a PLA filament with 1.75 mm-diameter, as per ASTM standard D638 for the tensile properties of plastics. To investigate the impact of recycling on the mechanical properties of PLA, the pristine material was compared to samples recycled once and three times.

Tensile testing was performed using a universal testing machine (Instron 5569, Instron Corp., Norwood, MA, USA) with a grip speed of 5 mm/min and a grip distance of 25.4 mm. The data of at least four samples for each test were used to calculate the average value to ensure consistency. All tensile tests were performed at 25.2 °C and 45.5% relative humidity using the two types of test specimens, as shown in Figure 2. The specimens fabricated horizontally (Figure 2a) were used to investigate the mechanical properties of the material itself, whereas those vertically produced (Figure 2b) were used to investigate the mechanical properties and adhesion between the fabricated beads. In both instances, the temperature of the nozzle was maintained at 210 °C, as per the recommendations of the material supplier. The heating bed and chamber were maintained at 65 °C and 45 °C, respectively. The air gap between the beads of the fabricated material was set to zero. The ultimate tensile strength was measured as the average value for at least four specimens and was used to investigate the effects of recycling on the initial mechanical properties of the material. The fracture surfaces produced via tensile testing were investigated using SEM at an accelerating voltage of 5 kV (S-4800, Hitachi High-Technologies Corp., Schaumburg, IL, USA).

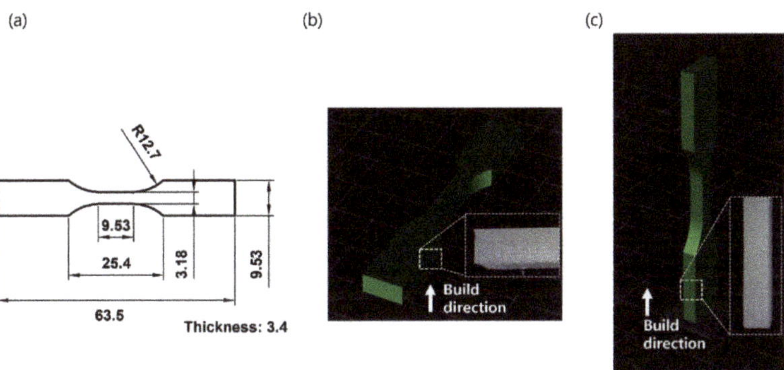

Figure 2. (**a**) Graphical representation of the tensile test specimen specified in the ASTM standard D638-10 (type 5); tensile test specimens printed (**b**) horizontally and (**c**) vertically. Magnified images of the specimens are also shown.

GPC analysis was performed using a WATERS GPC system with a Waters 410 differential refractometer and Shodex LF-804 (7.8 mm × 300 mm) column. As this type of size-exclusion chromatography separates analytes based on their size, it is a useful analytical tool for monitoring the molecular weight of a polymer. PLA was dissolved in chloroform, and the resulting mobile phase was passed through a column with a highly porous structure, where separation occurs based on the hydrodynamic volume (i.e., the radius of gyration) of each analyte. The molecular weights of pristine and recycled PLA were compared as they affect several characteristic physical properties of a polymer, such as its tensile strength, adhesive strength, brittleness, elastic modulus, and melt viscosity.

The differential scanning calorimetry(DSC) was carried out to investigate the change in the degree of crystallinity according to the number of recycling process(DSC Q20, TA Instrument, New Castle, DE, USA). A set of heating and cooling processes were carried out following three steps: PLA was heated to 200 °C at 5 °C/min with purged nitrogen environment to remove its thermal history. The PLA was then cooled down to 20 °C and heated back to 200 °C at the same rate. The degree of crystallinity(X_c) was calculated by using the below equation based on the second heating:

$$X_c = \frac{\Delta H_m}{\Delta H_0} \times 100 \tag{1}$$

where ΔH_m is the heat of melting and ΔH_m is the heat of meting for an infinitely large crystal, 93.6 g/J).

3. Results and Discussion

Printed and failed parts from 3D printing processes were used to extrude the recycled PLA filaments using a custom mechanical recycling system. The mechanical strength of recycled PLA specimens (which is an important parameter in 3D-printed parts) fabricated using recycled filaments was investigated via tensile tests. Figures 3 and 4 show the tensile stress and strain curves of the pristine and recycled PLA specimens fabricated in the horizontal and vertical directions, respectively. Several specimens were broken near the grip location during the tensile test, and the data from these specimens were excluded. This phenomenon was more frequently observed in the specimens fabricated using recycled filaments, which can be attributed to the shredded specimen being pulled into the filament without fully mixing during the recycling process. Among the two specimens, the ultimate tensile strength of the horizontally fabricated specimen was higher than that of the vertically fabricated specimen. This indicates that the strength of the fabricated PLA material is higher than the adhesion strength between the fabricated beads, resulting in the typical anisotropic property of the FDM-type 3D-printed product. These findings are consistent with those from previous studies [31–34]. In addition, in the case of the vertically fabricated pristine PLA specimen, no elongation was observed after the maximum stress before breaking, which implies that the fracture occurs between the adhesion interfaces and not in the material itself.

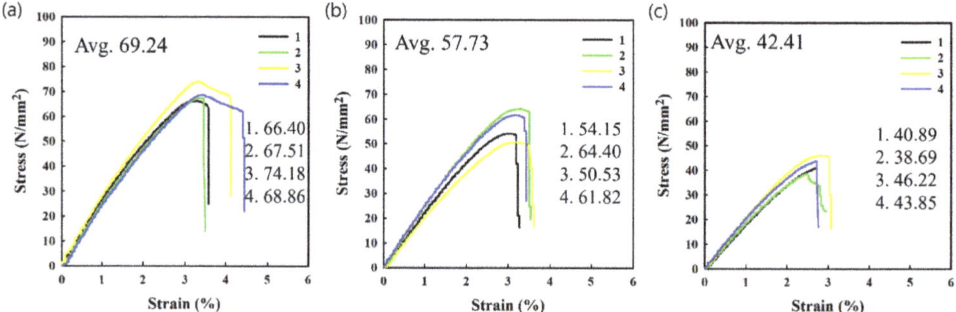

Figure 3. Stress–strain curves for (**a**) pristine, (**b**) one-time recycled, and (**c**) three-time recycled specimens fabricated horizontally.

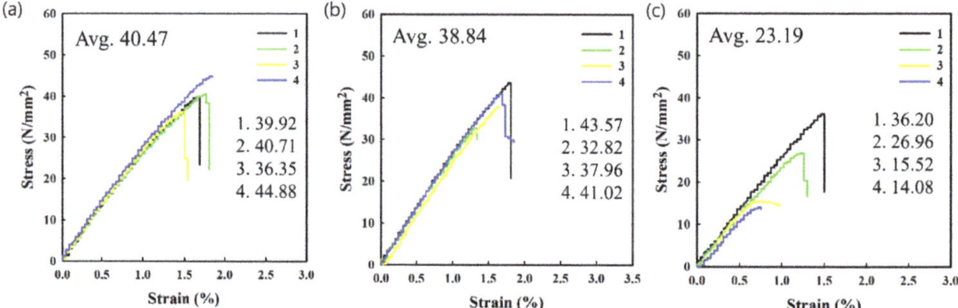

Figure 4. Stress–strain curves for (**a**) pristine, (**b**) one-time recycled, and (**c**) three-time recycled specimens fabricated vertically.

Considering the change in the mechanical strength of the horizontally fabricated specimens after recycling, both the maximum stress before break and the strain at break of the horizontally fabricated specimens decreased as the number of recycling steps increased, as shown in Figures 3 and 5. After three recycling steps, the maximum tensile strength and strain at break of the horizontally fabricated specimens reduced by 38.7% and 26.3%, respectively, on average. The specimens for the tensile strength test were also prepared by recycling the material up to five times; however, the tensile strength data of these samples were excluded owing to the large deviations in the measured values for these samples. The experimental results indicate that recycling led to a decrease in the mechanical strength and an increase in the brittleness of the material. The degradation of the mechanical strength of the polymeric material is due to a reduction in its molecular weight [4,27,34–37]. This is consistent with previous reports suggesting that mechanical recycling at elevated temperatures degrades the macromolecular structure, resulting in the chain scission of the polymer structure [4,27,35–37]. The resulting shorter chains increase the number of chain ends in the structure and the stress at which fracture occurs [38]. Similar observations regarding the degradation of PLA have shown that a correlation generally exists between the tensile strength and molecular weight of polymer materials, which can be approximated by the inverse relation [37,39,40]:

$$S = S_\infty - \frac{A}{M} \tag{2}$$

where S_∞ is the saturated tensile strength for an infinite molecular weight, and A is the correction factor for the material type.

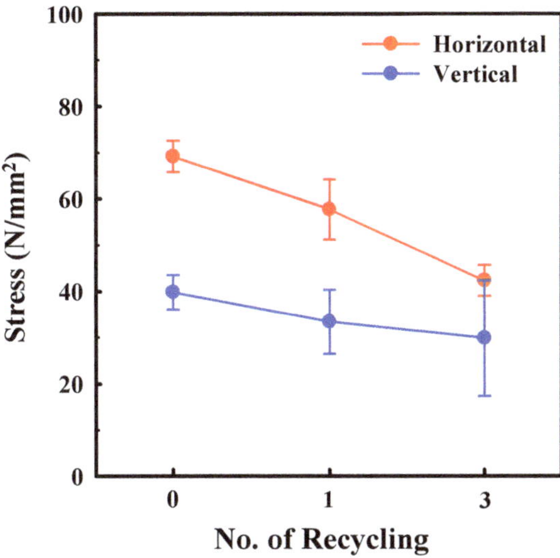

Figure 5. Averaged ultimate tensile strength as a function of the number of recycling steps for both horizontally and vertically fabricated specimens.

To determine the change in the molecular weight of PLA caused by recycling, the weight-averaged molecular weights of pristine and recycled PLA were determined using GPC with polystyrene as reference material. In Figure 6, the shift in the peak from left to right indicates a longer elution time, and therefore, a lower molecular weight. This elution time was correlated with that of the reference material to obtain the molecular weights, as shown in Figure 7, revealing that the molecular weight of the pristine material (175,888 g/mol) was reduced to 90,021 g/mol after five recycles. Recycling at elevated temperatures and repeated 3D printing processes are therefore considered conducive to chain scission, resulting in the degradation of the mechanical properties of PLA. However, other degradation mechanisms might also occur during recycling, such as the depolymerization of macromolecular chains due to residual catalyst, non-radical reactions, and mechanical degradation due to interactions between PLA and the equipment used in the fabrication process [4,41].

The degree of crystallinity was calculated using Equation (1), and results were summarized in Table 1. It was found that the degree of crystallization increased with the number of recycling. Based on previous studies, the change in crystallinity of polymer according to the degradation of materials can be explained in two different ways. The first one is that the degree of crystallinity decreased with increased number of recycling process. It was reported that the decrease in crystallinity is due to the radical reactions [42]. The radical reactions generated after recycling might cause the crosslinking of the polymer and reduce its crystallinity [43]. The second one is that the degree of crystallinity increased with the number of recycling of polymer. It was also reported that the increase in the degree of crystallinity with recycling is a kinetic effect due to the reduction of the molecular weight [15]. As provided in the previous paragraph of this work, the molecular weight of PLA decreases with recycling due to chain scissions. The shortened chain length due to chain scissions allows to have a better mobility resulting in a crystallite thickening process rather than new crystallization [44]. Therefore, it is expected that increased crystallinity induces the brittleness of PLA and resulting in the degradation of mechanical strength. In our study, it was confirmed that crystallinity increases with the number of recycling based on the results from GPC, DSC, and tensile testing experiments.

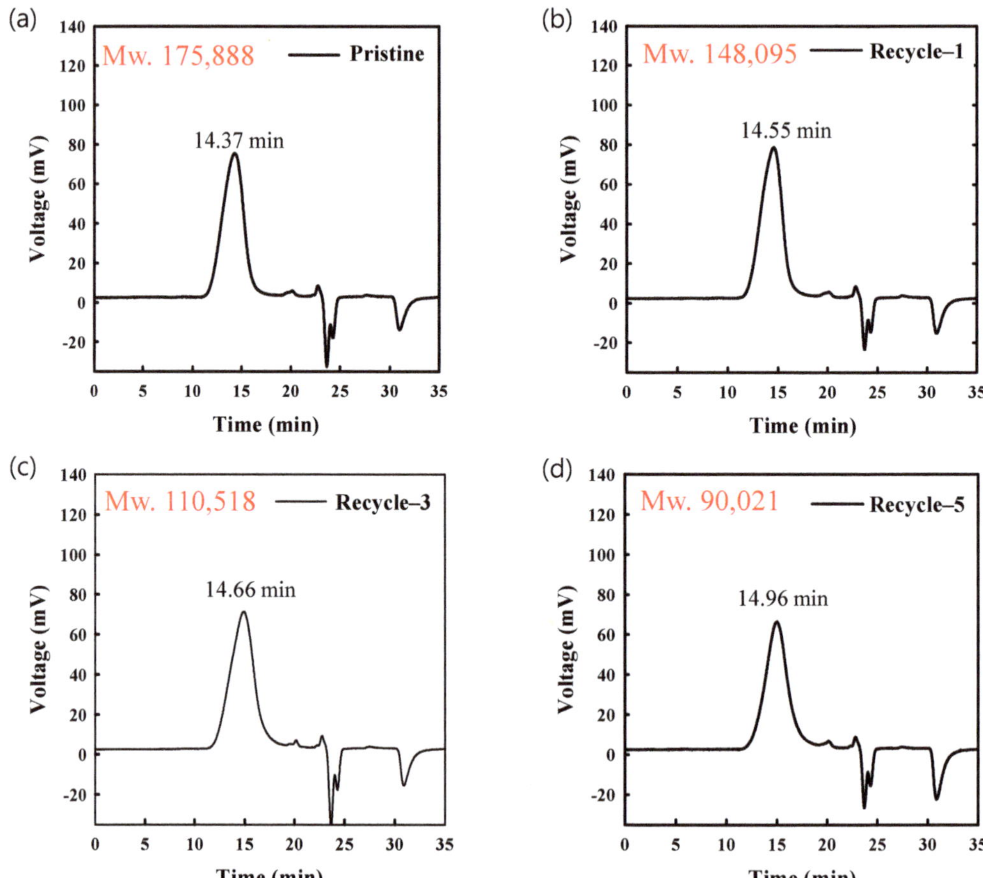

Figure 6. GPC results for pristine and recycled PLA filament.

Table 1. Evaluating the thermal properties of PLA as a function of the recycling process.

Sample (PLA)	ΔH_m	ΔH_0 (J/g)	X_c (%)
Pristine	30.89	93.6	33.0
Recycle 1	33.99	93.6	36.3
Recycle 3	38.66	93.6	41.3
Recycle 5	38.09	93.6	40.7

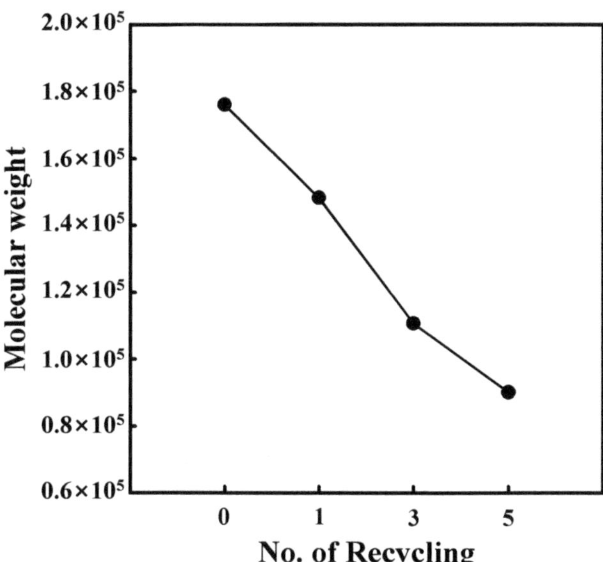

Figure 7. The molecular weight of PLA as a function of the number of recycling steps.

The maximum tensile strength and strain before breaking of the vertically fabricated PLA specimens also decreased with the number of recycling steps, as shown in Figures 4 and 5. After three cycles, the maximum tensile strength and strain at break of the horizontally fabricated specimens reduced to 42.5% and 34.2%, respectively, on average. Therefore, the deterioration rate of the sample fabricated in the vertical direction is higher than that of the sample fabricated in the horizontal direction. This indicates that the chain scission of the PLA matrix reduces not only the strength of the material but also the adhesion between the beads of the fabricated PLA, which in the case of a 3D-printed structure, acts as an adhesive material. Thus, heating during recycling and the 3D printing process causes thermally induced chain scission and a reduction in cohesive strength. These results suggest that the maximum tensile strength of the vertically fabricated specimens can be improved by adding an adhesion promoter. Studies on improving the mechanical properties of polymers by coating them with an adhesion promoter, such as polydopamine (which is an adhesive polymer derived from mussels), have been conducted. Based on the results of this study on the degradation mechanism, the mechanical properties of the recycled filament were noticeably improved [30]. The maximum tensile strength of the vertically fabricated pristine PLA was lower than that of the horizontally fabricated specimen because the strength of a vertically fabricated specimen is dependent on the adhesion between the beads rather than the strength of the PLA beads. This is in agreement with previous results [45] and indicates that vertically fabricated specimens exhibit a more brittle behavior than horizontally fabricated specimens, that is, they break abruptly when stress is induced during a tensile test; this behavior is not exhibited by the pristine material. In contrast, horizontally fabricated specimens exhibited elongation even after reaching their ultimate tensile strength, which is consistent with the intrinsic properties of the polymer (particularly for the samples produced via injection molding). However, this elongation after the ultimate tensile strength decreased as the number of recycling steps increased, owing to the embrittlement of recycled PLA, as shown in Figure 3.

The cross-sectional area after tensile testing was investigated via SEM imaging, and the fracture surfaces of the two different specimens are presented in Figure 8. These figures clearly show the difference in fracture morphology, and the characteristic fracture surface of the horizontally fabricated specimen indicates that the fabricated PLA beads are broken. In contrast, Figure 8b reveals that the cracks propagate entirely through the bonding interface between the beads in the vertically fabricated specimen. This confirms that the mechanical strength of the PLA beads determines the mechanical strength of the horizontally fabricated specimens, whereas the adhesion strength between the beads governs the mechanical properties of the vertically fabricated specimens. As the former is stronger than the latter, this imparts anisotropic mechanical properties to the 3D-printed objects. Hence, these factors must be considered during the design and manufacturing of products via 3D printing. As the surface morphologies of the fracture surfaces are closely correlated with the mechanical characteristics of printed parts, lots of research have been reported on the method of characterizing the surface topologies of the fracture surface such as optical microscope, scanning electron microscopy, X-ray computed tomography and the correlation with the mechanical properties [42–46]. Based on results in the previous reports, it could be confirmed that, as the void in the fracture surface decreases, the mechanical strength of the printed parts increases. Although not included in this study, it would be an important study to optimize the process conditions to reduce void and improve the mechanical properties when using recycled filaments [46]. This part was not included in this study and was left as a topic for the next research.

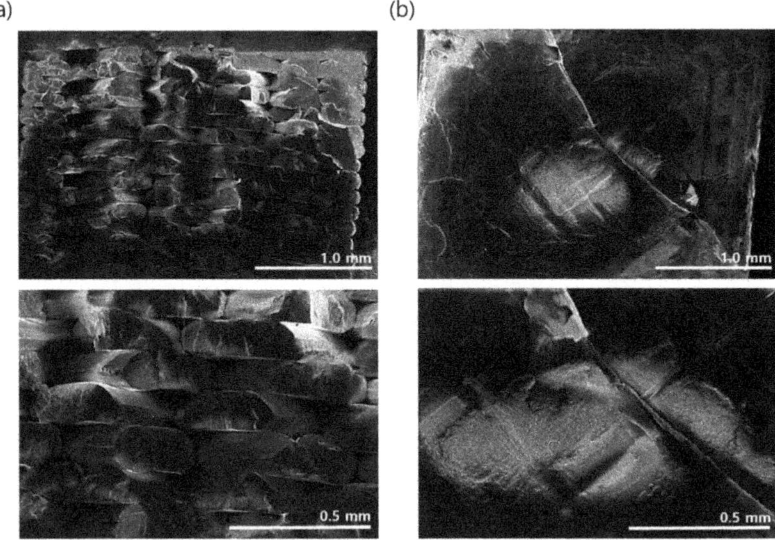

Figure 8. Fracture surface morphology of 3D-printed specimens after the tensile test: (**a**) horizontally and (**b**) vertically fabricated specimens.

4. Conclusions

The recycling of polymeric materials enables waste materials to be converted into filaments, which can then be used repeatedly in 3D printing systems. PLA, which has been widely used in 3D printing applications, was recycled via a distributed recycling system, and the recycled PLA filaments and printed specimens were characterized to investigate the effectiveness of the recycled filament. The characterization of the mechanical properties of the PLA filament, for use in 3D printers, reveals that mechanical recycling induces thermally activated degradation, thereby reducing the molecular weight of the polymer with each successive cycle. This results in the chain scission of PLA and degradation of the mechanical properties of the fabricated product. Using two different types of specimens (horizontally and vertically fabricated), mechanical recycling was shown to degrade the mechanical strength of the PLA bead as well as the adhesion between the beads. Thus, the mechanical strength of the PLA filaments was sensitive to the number of successive thermomechanical recycling processes. Therefore, recycled filaments are more suitable for fabricating models and prototypes than the parts operating under heavy loads because of the degradation of their mechanical strength. The investigation of the degradation mechanism revealed that this limitation can be overcome using chain extenders or adhesion promoters during recycling.

Author Contributions: Conceptualization and writing, N.K.; Experiment and visualization, D.L., Y.L., I.K. and K.H. All authors have read and agreed to the published version of the manuscript.

Funding: This study was supported by Konkuk University in 2018.

Institutional Review Board Statement: Not applicable.

Data Availability Statement: Not applicable.

Conflicts of Interest: The authors declare no conflict of interest. The funders had no role in the design of the study; in the collection, analyses, or interpretation of data; in the writing of the manuscript, or in the decision to publish the results.

References

1. Bassoli, E.; Gatto, A.; Iuliano, L.; Violante, M.G. 3D printing technique applied to rapid casting. *Rapid Prototyp. J.* **2007**, *13*, 148–155. [CrossRef]
2. Kruth, M.C.L.J.-P.; Nakagaw, T. Nakagaw Progress in Additive Manufacturing and Rapid Prototyping. *CIRP Ann. Manuf. Technol.* **1998**, *47*, 525–540. [CrossRef]
3. Crane, N.B.; Tuckerman, J.; Nielson, G.N. Self-assembly in additive manufacturing: Opportunities and obstacles. *Rapid Prototyp. J.* **2011**, *17*, 211–217. [CrossRef]
4. Kopinke, F.D.; R, M.; Mackenzie, K.; Möder, M.; Wachsen, O. Thermal decomposition of biodegradable polyesters-2. Poly(lactic acid). *Polym. Degrad. Stab.* **1996**, *53*, 329–342. [CrossRef]
5. Campbell, T.; Ivanova, C.W.O.; Garrett, B. *Could 3D Printing Change the World? Technologies, Potential, and Implications of Additive Manufacturing*; The Atlantic Council: Washington, DC, USA, 2011.
6. Hamad, K.; Kaseem, M.; Deri, F. Recycling of waste from polymer materials: An overview of the recent works. *Polym. Degrad. Stab.* **2013**, *98*, 2801–2812. [CrossRef]
7. Gomez-Gras, G.; Jerez-Mesa, R.; Travieso-Rodriguez, J.A.; Lluma-Fuentes, J. Fatigue performance of fused filament fabrication PLA specimens. *Mater. Des.* **2018**, *140*, 278–285. [CrossRef]
8. Song, Y.; Li, Y.; Song, W.; Yee, K.; Lee, K.Y.; Tagarielli, V.L. Measurements of the mechanical response of unidirectional 3D-printed PLA. *Mater. Des.* **2017**, *123*, 154–164. [CrossRef]
9. Liu, Q.; Zhao, M.; Zhou, Y.; Yang, Q.; Shen, Y.; Gong, R.H.; Zhou, F.; Li, Y.; Deng, B. Polylactide single-polymer composites with a wide melt-processing window based on core-sheath PLA fibers. *Mater. Des.* **2018**, *139*, 36–44. [CrossRef]
10. Foruzanmehr, M.; Vuillaume, P.Y.; Elkoun, S.; Robert, M. Physical and mechanical properties of PLA composites reinforced by TiO2 grafted flax fibers. *Mater. Des.* **2016**, *106*, 295–304. [CrossRef]
11. Nofar, M. Effects of nano-/micro-sized additives and the corresponding induced crystallinity on the extrusion foaming behavior of PLA using supercritical CO2. *Mater. Des.* **2016**, *101*, 24–34. [CrossRef]
12. Aghjeh, M.R.; Nazari, M.; Khonakdar, H.A.; Jafari, S.H.; Wagenknecht, U.; Heinrich, G. In depth analysis of micro-mechanism of mechanical property alternations in PLA/EVA/clay nanocomposites: A combined theoretical and experimental approach. *Mater. Des.* **2015**, *88*, 1277–1289. [CrossRef]
13. Santana, L.; Alves, J.L.; Netto, A.D.S. A study of parametric calibration for low cost 3D printing: Seeking improvement in dimensional quality. *Mater. Des.* **2017**, *135*, 159–172. [CrossRef]
14. Kashi, S.; Gupta, R.K.; Baum, T.; Kao, N.; Bhattacharya, S.N. Dielectric properties and electromagnetic interference shielding effectiveness of graphene-based biodegradable nanocomposites. *Mater. Des.* **2016**, *109*, 68–78. [CrossRef]
15. Pillin, I.; Montrelay, N.; Bourmaud, A.; Grohens, Y. Effect of thermo-mechanical cycles on the physico-chemical properties of poly(lactic acid). *Polym. Degrad. Stab.* **2008**, *93*, 321–328. [CrossRef]
16. Liang, J.Z.; Duan, D.R.; Tang, C.Y.; Tsui, C.P.; Chen, D.Z. Tensile properties of PLLA/PCL composites filled with nanometer calcium carbonate. *Polym. Test.* **2013**, *32*, 617–621. [CrossRef]
17. Yang, S.L.; Wu, Z.H.; Yang, W.; Yang, M.B. Thermal and mechanical properties of chemical crosslinked polylactide (PLA). *Polym. Test.* **2008**, *27*, 957–963. [CrossRef]
18. Kreiger, M.; Pearce, J.M. Environmental Life Cycle Analysis of Distributed Three-Dimensional Printing and Conventional Manufacturing of Polymer Products. *ACS Sustain. Chem. Eng.* **2013**, *1*, 1511–1519. [CrossRef]
19. Themelis, N.J.; Castaldi, M.J.; Bhatti, J.; Arsova, L. *Energy and Economic Value of Non-recycled Plastics (NRP) and Municipal Solid Wastes (MSW) that are Currently Landfilled in the Fifty States*; Columbia University: New York, NY, USA, 2011.

20. Al-Salem, S.M.; Lettieri, P.; Baeyens, J. Recycling and recovery routes of plastic solid waste (PSW): A review. *Waste Manag.* **2009**, *29*, 2625–2643. [CrossRef]
21. Pearce, J.; Blair, C.; Laciak, K.; Andrews, R.; Nosrat, A.; Zelenika-Zovko, I. 3-D Printing of Open Source Appropriate Technologies for Self-Directed Sustainable Development. *J. Sustain. Dev.* **2015**, *3*, 17–29. [CrossRef]
22. Kreiger, M.A.; Mulder, M.L.; Glover, A.G.; Pearce, J.M. Life cycle analysis of distributed recycling of post-consumer high density polyethylene for 3-D printing filament. *J. Clean. Prod.* **2014**, *70*, 90–96. [CrossRef]
23. Baechler, C.; DeVuono, M.; Pearce, J.M. Distributed recycling of waste polymer into RepRap feedstock. *Rapid Prototyp. J.* **2013**, *19*, 118–125. [CrossRef]
24. Available online: https://www.filabot.com/ (accessed on 3 September 2022).
25. Lee, D.; Lee, Y.; Lee, K.; Ko, Y.; Kim, N. Development and Evaluation of a Distributed Recycling System for Making Filaments Reused in Three-Dimensional Printers. *J. Manuf. Sci. Eng.* **2019**, *141*, 021007–021015. [CrossRef]
26. Signori, F.; Coltelli, M.B.; Bronco, S. Thermal degradation of poly(lactic acid) (PLA) and poly(butylene adipate-co-terephthalate) (PBAT) and their blends upon melt processing. *Polym. Degrad. Stab.* **2009**, *94*, 74–82. [CrossRef]
27. Le Duigou, A.; Pillin, I.; Bourmaud, A.; Davies, P.; Baley, C. Effect of recycling on mechanical behaviour of biocompostable flax/poly(L-lactide) composites. *Compos. Part A Appl. Sci. Manuf.* **2008**, *39*, 1471–1478. [CrossRef]
28. Taubner, V.; Shishoo, R. Influence of processing parameters on the degradation of poly(L-lactide) during extrusion. *J. Appl. Polym. Sci.* **2001**, *79*, 2128–2135. [CrossRef]
29. Lanzotti, A.; Grasso, M.; Staiano, G.; Martorelli, M. The impact of process parameters on mechanical properties of parts fabricated in PLA with an open-source 3-D printer. *Rapid Prototyp. J.* **2015**, *21*, 604–617. [CrossRef]
30. Zhao, X.G.; Hwang, K.J.; Lee, D.; Kim, T.; Kim, N. Enhanced mechanical properties of self-polymerized polydopamine-coated recycled PLA filament used in 3D printing. *Appl. Surf. Sci.* **2018**, *441*, 381–387. [CrossRef]
31. Carneiro, O.S.; Silva, A.F.; Gomes, R. Fused deposition modeling with polypropylene. *Mater. Des.* **2015**, *83*, 768–776. [CrossRef]
32. Chacon, J.M.; Caminero, M.A.; Garcia-Plaza, E.; Nunez, P.J. Additive manufacturing of PLA structures using fused deposition modelling: Effect of process parameters on mechanical properties and their optimal selection. *Mater. Des.* **2017**, *124*, 143–157. [CrossRef]
33. Zhuang, Y.; Song, W.T.; Ning, G.; Sun, X.Y.; Sun, Z.Z.; Xu, G.W.; Zhang, B.; Chen, Y.N.; Tao, S.Y. 3D-printing of materials with anisotropic heat distribution using conductive polylactic acid composites. *Mater. Des.* **2017**, *126*, 135–140. [CrossRef]
34. Casavola, C.; Cazzato, A.; Moramarco, V.; Pappalettere, C. Orthotropic mechanical properties of fused deposition modelling parts described by classical laminate theory. *Mater. Des.* **2016**, *90*, 453–458. [CrossRef]
35. Chrissafis, K.; Paraskevopoulos, K.M.; Bikiaris, D. Thermal degradation kinetics and decomposition mechanism of two new aliphatic biodegradable polyesters poly(propylene glutarate) and poly(propylene suberate). *Ther. Acta* **2010**, *505*, 59–68. [CrossRef]
36. Chrissafis, K.; Paraskevopoulos, K.M.; Papageorgiou, G.Z.; Bikiaris, D.N. Thermal decomposition of poly(propylene sebacate) and poly(propylene azelate) biodegradable polyesters: Evaluation of mechanisms using TGA, FTIR and GC/MS. *J. Anal. Appl. Pyrol.* **2011**, *92*, 123–130. [CrossRef]
37. Kim, N.; Kang, H.; Hwang, K.J.; Han, C.; Hong, W.S.; Kim, D.; Lyu, E.; Kim, H. Study on the degradation of different types of backsheets used in PV module under accelerated conditions. *Sol. Energy Mater. Sol. Cells* **2014**, *120*, 543–548. [CrossRef]
38. Abad, M.J.; Ares, A.; Barral, L.; Cano, J.; Diez, F.J.; Garcia-Garabal, S.; Lopez, J.; Ramirez, C. Effects of a mixture of stabilizers on the structure and mechanical properties of polyethylene during reprocessing. *J. Appl. Polym. Sci.* **2004**, *92*, 3910–3916. [CrossRef]
39. Smith, P.; Lemstra, P.J.; Pijpers, J.P. Tensile strength of highly oriented polyethylene. II. Effect of molecular weight distribution. *J. Polym. Sci. Part A Polym. Chem.* **1982**, *20*, 2229–2241. [CrossRef]
40. Dobkowski, Z. Determination of critical molecular weight for entangled macromolecules using the tensile strength data. *Rheol. Acta* **1995**, *34*, 578–585. [CrossRef]
41. Fan, Y.J.; Nishida, H.; Shirai, Y.; Endo, T. Thermal stability of poly (L-lactide): Influence of end protection by acetyl group. *Polym. Degrad. Stab.* **2004**, *84*, 143–149. [CrossRef]
42. Shojaeiarani, J.; Bajwa, D.S.; Rehovsky, C.; Bajwa, S.G.; Vahidi, G. Deterioration in the Physico-Mechanical and Thermal Properties of Biopolymers Due to Reprocessing. *Polymers* **2019**, *11*, 11010058. [CrossRef]
43. Pedroso, A.G.; Rosa, D.D.S. Mechanical, thermal and morphological characterization of recycled LDPE/corn starch blends. *Carbohydr. Polym.* **2005**, *59*, 1–9. [CrossRef]
44. Rasselet, D.; Ruellan, A.; Guinault, A.; Miquelard-Garnier, G.; Sollogoub, C.; Fayolle, B. Oxidative degradation of polylactide (PLA) and its effects on physical and mechanical properties. *Eur. Polym. J.* **2014**, *50*, 109–116. [CrossRef]
45. Angel, R.; Torrado, D.A.R. Failure Analysis and Anisotropy Evaluation of 3D-Printed Tensile Test Specimens of Different Geometries and Print Raster Patterns. *J. Fail. Anal. Prev.* **2016**, *16*, 154–164.
46. Tao, Y.; Kong, F.; Li, Z.; Zhang, J.; Zhao, X.; Yin, Q.; Xing, D.; Li, P. A review on voids of 3D printed parts by fused filament fabrication. *J. Mater. Res. Technol.* **2021**, *15*, 4860–4879. [CrossRef]

Article

Upcycling Different Particle Sizes and Contents of Pine Branches into Particleboard

Anita Wronka * and Grzegorz Kowaluk *

Institute of Wood Sciences and Furniture, Warsaw University of Life Sciences—SGGW, Nowoursynowska St. 159, 02-776 Warsaw, Poland
* Correspondence: anita_wronka@sggw.edu.pl (A.W.); grzegorz_kowaluk@sggw.edu.pl (G.K.); Tel.: +48-22-59-38-546 (G.K.)

Abstract: A growing world population means that demand for wood-based materials such as particleboard is constantly increasing. In recent years, wood prices have reached record highs, so a good alternative can be the utilization of branches, which can reduce the cost of raw materials for particleboard production. The goal of the study was to confirm the feasibility of using an alternative raw material in the form of *Pinus sylvestris* L. pine branches for the production of three-layer particleboard. Characterization of the alternative raw material was also carried out, and the bulk density was determined. As part of the research, six variants of particleboard, 0%, 5%, 10%, 25%, and 50%, w/w, and two variants where the first one had the face layer made of branch particles and the core layer made of industrial particles, and the reverse variant (all produced panels were three-layer) were produced and then their physical and mechanical properties were studied. The results show that even if the bulk density of branch particles is significantly higher than industrial material, the internal bond and water absorption rises as branch particle content increases. In the case of bending strength and modulus of elasticity, these were decreased with a branch particle content increase. The conducted tests confirmed the possibility of using the raw material, which was usually used as fuel or mulch, to produce particleboards even in 50% content. The present solution also contributes to the positive phenomenon of carbon storage, due to incorporating the branches' biomass into panels rather than burning it. Further research should be focused on the modification of particle production from branches to obtain lower bulk density and to reach fraction shares closer to industrial particles. Furthermore, the chemical characterization of the pine branch particles (cellulose and lignin content, extractives content, pH value) would provide valuable data about this potential alternative raw material.

Keywords: wood; upcycling; particleboard; mechanical properties; bulk density; physical properties; carbon storage; wood branches

Citation: Wronka, A.; Kowaluk, G. Upcycling Different Particle Sizes and Contents of Pine Branches into Particleboard. *Polymers* **2022**, *14*, 4559. https://doi.org/10.3390/polym14214559

Academic Editors: Sheila Devasahayam and Laurence Dyer

Received: 3 October 2022
Accepted: 24 October 2022
Published: 27 October 2022

Publisher's Note: MDPI stays neutral with regard to jurisdictional claims in published maps and institutional affiliations.

1. Introduction

In times of scarcity of wood, it is extremely important to make the best possible use of once harvested raw material. Sometimes this is not enough, so it is necessary to look for new alternative raw materials. Branches are very often not widely used in the wood industry due to their small diameter, irregular shape and size, and a high proportion of bark, among other reasons. In pine farms, pruning is carried out every few years, removing branches to obtain as much wood as possible in a short period. In pine forests, on the other hand, unwanted branches are also regularly removed to obtain better wood quality. In addition, stands that obstruct the growth of more mature trees are also removed. Each of these operations produces by-products that can be used as fuel, compost, or ballast for trees [1], or left in the forest during harvesting [2], so using this waste can have positive benefits for the environment and the production of wood-based panels. A growing world population means that demand for wood-based materials such as particleboard is constantly increasing.

The worldwide amount of particleboards produced in 2020 was 96.01 million m^3 [3]. In recent years, wood prices have reached record highs, so a good alternative is to use branches, which can reduce the cost of raw materials for particleboard production. This solution gives hope to wood materials producers to increase the profitability of production, but it also allows carbon dioxide to remain bound in the atmosphere for longer. This approach fits well with the general direction of EU policy concerning Carbon Capture and Storage [4]. The Scots pine tree accounts for 67% of Poland's forest area. It can be found in almost all environmental conditions, which proves its easy availability [5]. Branches, as well as high-grade wood, need to be mechanically processed. The research shows that the chipping of branches produces smaller particles than particles extracted from pine logs. The amount of dust was also higher in the particles produced from the branch material. The significantly higher amount of fine particles in pine particles may be related to the presence of needles. The smaller particles used in particleboard production may contribute to lower strength parameters [6].

Research conducted so far does not confirm the high popularity of using branches in the aspect of wood-based panel production, so it is difficult to find examples strictly related only to pine branches. In Iran, due to the scarcity of conventional wood raw materials, agricultural waste started to be of interest in the context of obtaining raw materials for particleboard production, thus date palm (*Phoenix dactylifera*) branches were used, which were considered to be easily available. Research has confirmed that date palms can be used as a substitute for conventional raw material and that the boards produced can serve as a sound and heat insulating material [7]. A similar reason—the search for alternative raw materials for particleboard—has been a starting point to produce the particleboards from sorghum bagasse bonded with maleic acid [8] and agro-industrial residues (cassava stem, sengon wood waste, and rice husk) and different contents of natural rubber latex adhesive [9]. Both papers confirmed the usefulness of the tested alternative raw materials in particleboard production.

Pine branches have so far been used in particleboard production as a finishing element on the board. This application was intended to explore the decorative possibilities and also to produce low-emission particleboard covered with pine branches. The pine branches were cut into thin slices so that the annual rings are visible. The slices have been overlapped on both sides of the board and there are many possible ways of arranging them to create interesting patterns. Thanks to the pressing of the pine slices, the surface of the board is smooth, allowing the sanding process to be omitted and the board to be varnished more quickly. For the middle layer of the particleboard, the waste from wood production was used, which additionally increases the ecological value of the board. Such boards can be used as decorative furniture fronts or as decorative panels for construction elements [10].

Another example of branches that are used in the wood industry is those of fruit trees grown in Greece and evergreen deciduous shrubs. Every year, 415,000 tonnes of green woody biomass from apple, apricot, peach, pear, and cherry trees are left in the fields or burnt. Studies have confirmed the possibility of using wood particles from fruit trees and evergreen shrubs in combination with Greek fir wood to produce particleboard that meets the requirements for EN 312:2010 class P4 (load-bearing boards for use in dry conditions). The content of the alternative raw material was above 50% [11].

Research has been conducted into the possibility of using wood particles from apple tree branches, which come from the annual maintenance of these trees. Most often, these branches are used as fuel, or if left in the field, they contribute to the development of diseases and fungi. Various variants of panels were produced as part of the tests, of which only variants that had less than 50% branch particles in their composition met the standards. A higher proportion of apple branches lowered the parameters of the panels produced. This may be due to the higher bulk density of apple wood (510–600 kg m^{-3}) [12]. To date, other unusual raw materials used in particleboard production include raspberry stems (*Rubus Idaeus* L.) [13], sunflowers [14], bamboo branches and wastes [15], particles of the trunk and branches of the bhadi tree (*Lannea coromandelica*) [16], grape trees pruning residues [17,18], dry branches of *Araucaria angustifolia* (*Bertol.*) [19], particles of *Phoenix Dactylifera*-L (Date Palm) [20], or chilli pepper steams [21].

Research confirms that tree branches have so far been successfully used to produce lightweight structural panels, whereas from a chemical point of view, they can also be used as biofuel, for example [22].

The present research was aimed at proving that it is possible to use pine branches for the production of particleboard, the properties of which meet standard requirements for use in the furniture industry. The research also included the characterization of the wood particles obtained by grinding the branches, and their bulk density and geometry were determined.

2. Materials and Methods

2.1. Materials

Pine (*Pinus sylvestris* L.) branches that were left as waste in the forest during felling were used in the study. The largest diameter of the collected branches was about 40 mm, whereas the lowest was about 10 mm. The collected branches were dried using a chamber dryer, where the temperature was 70 °C, and the drying resulted in a branch equivalent moisture content of approximately 10–12%. The dried branches were pre-crushed on a saw blade into 50 mm long chips, and the next step was to mill these chips using a laboratory hammer mill. As a reference material, the industrial particles consisting of over 95% of *Pinus sylvestris* L. have been taken from one of the commercial 3-layer particleboard production lines located in Poland.

An air gun was used to spray the glue over the particles, using a commercial urea-formaldehyde (UF) resin Silekol S-123 (Silekol Sp. z o.o., Kędzierzyn-Koźle, Poland) with a molar ratio of 0.89 and solid content of 66.5%. The resination was as follows: 12% for face layer particles and 10% for core layer particles of dry resin content referred to dry particles (w/w), where 2.0% of water solution of ammonium nitrate hardener was applied, and both were calculated for dry resin content. No hydrophobic agents were added.

2.2. Wood Raw Material Upcycling and Characterization

The chips made of pine branches were re-milled on a laboratory knife mill (laboratory prototype delivered by Research and Development Centre for Wood-Based Panels Sp. z o. o. in Czarna Woda, Poland), equipped with three knives, two contra-knives, and a 6 × 12 mm^{-2} mesh. Using the volumetric method, the bulk density of the particles obtained was tested. The measurement was repeated five times for each fraction. The particles obtained were sorted into a face layer (0.5 mm and 1 mm sieves) and a core layer (8 mm and 2 mm sieves). This procedure allowed the elimination of oversized particles. The fractions of particles were tested with an IMAL (Imal s.r.l., San Damaso (MO), Italy) vibrating laboratory sorter with seven sieves. The selected sieve sizes were 8, 4, 2, 1, 0.5, 0.25, and <0.25 mm. The amount of tested material for each fraction was about 100 g, and the set time of continuous vibrating was 15 min. As many as five repetitions were done for every tested material.

A brief characterization of *Pinus sylvestris* L. bark mechanical parameters (internal bond strength according to [23]), as well as the density was also conducted. The dimensions of the bark samples taken (as in Figure 1) were about 30 mm × 30 mm. As many as 25 samples have been used for every test. The bark content (w/w) for 50 mm long branch sections was also established. The bark was manually separated from 50 samples.

Figure 1. The samples of the *Pinus sylvestris* L. bark. (**a**) mature and (**b**) juvenile were taken for investigation.

2.3. Elaboration of Composites

Three-layer particleboard with different branch particle contents and in three different combinations was produced from pine branch particles. The particles used were dried to a moisture content of 5%. All particleboard variants produced had a density of 670 kg m^{-3}, 32% (w/w) face layers content, and a nominal thickness of 16 mm. The tests produced reference particleboards and particleboards with branch particle content sequentially: 0%, 5%, 10%, 25%, 50%, and 100% by weight, as well as particleboards in which the face layer was produced as in the reference particleboard (industrial particles) but 100% of branches particles in the core layer (called here 100 cl) and the opposite structure (100% branch particles in face and 100% of industrial particles in core) were called 100 fl. The manually formed mats of particleboards were pressed on a hydraulic press (ZUP-NYSA PH-1P125) at a maximum unit pressure of 2.5 MPa, a temperature of 200 °C, and a time factor of 20 s per one mm of nominal panel thickness. The conditioning of the panels before the tests took place at 20 °C and 65% humidity until a constant mass was obtained. The entire process of an attempt of upcycling the wood raw material from branches is presented in Figure 2. The composition of produced panels is presented in Table 1.

Figure 2. The general production process of three-layer particleboard from pine branches.

Table 1. Compositions of elaborated particleboards.

Panel Name	Pine Branches Particles Content [% by Weight]		Industrial Particles Content [% by Weight]	
	Face Layer	Core Layer	Face Layer	Core Layer
0	0	0	100	100
5	5	5	95	95
10	10	10	90	90
25	25	25	75	75
50	50	50	50	50
100	100	100	0	0
100 cl	0	100	100	0
100 fl	100	0	0	100

2.4. Characterization of the Elaborated PANELS

The test specimens were cut on a saw blade as required by European standards EN-326-2 [24] and EN-326-1 [25]. The modulus of rupture (MOR) and elasticity (MOE) were determined according to EN 310 [26], and the internal bond (IB) was determined according to EN 319 [23]. All the mechanical properties were examined with an INSTRON 3369 (Instron, Norwood, MA, USA) laboratory-testing machine, and, whenever applicable, the results were referred to as standards [27]. Board density was determined according to EN 323 [28], thickness swelling (TS), and water absorption (WA) due to EN 317 [29]. The screw withdrawal resistance (SWR) was measured according to [30]. The density profiles of the tested panels were measured on a GreCon DAX 5000 device (Fagus-GreCon Greten GmbH and Co. KG, Alfeld/Hannover, Germany).

2.5. Statistical Analysis

Analysis of variance (ANOVA) and t-tests calculations were used to test (α = 0.05) for significant differences between factors and levels, and where appropriate, using IBM SPSS statistic base (IBM, SPSS 20, Armonk, NY, USA). A comparison of the means was performed when the ANOVA indicated a significant difference by employing the Duncan test. The statistically significant differences in achieved results are given in the Results and Discussion paragraph whenever the data were evaluated.

3. Results and Discussion

3.1. Materials Characterization

The measured density of mature bark samples was 275 kg m^{-3} $\pm$ 21 kg m^{-3}, whereas the density of juvenile pine bark was 413 kg m^{-3} $\pm$ 18 kg m^{-3}. The measured internal bond of mature bark samples was 0.07 N mm^{-2} $\pm$ 0.01 N mm^{-2}, since, in the case of juvenile bark, the internal bond values were almost zero. These samples were delaminated during sample preparation and fixed in a testing machine. The measured bark content (w/w) on the branch sections was about 11%. According to the literature [31], the bark content in the mature pine log is 6.7%.

The results of the fraction share analysis of particles used in the research have been displayed in Figure 3. As it can be seen, in the case of face layer particles, a higher amount of smaller particles has been produced from pine branches. This could be caused by a higher amount of bark content in pine branches. In the case of core layer particles, a large amount of coarse particles was made from pine branches. This was especially visible for fractions 4 mm in size.

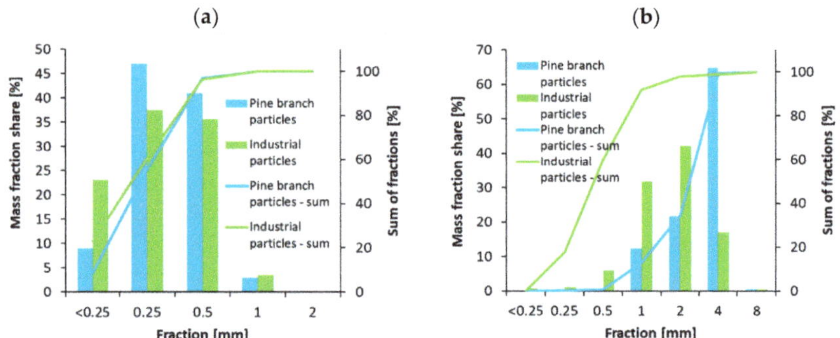

Figure 3. The fraction analysis results of the face (**a**) and core (**b**) layer particles.

The bulk density of the alternative raw material of pine (*Pinus sylvestris* L.) branch particles (Figure 4) was higher for both the outer layer (by 54 kg m^{-3}) and the inner layer (by 63 kg m^{-3}). This higher bulk density of branch particles can be influenced by the high content of bark, mentioned above. However, the bark density is lower than that of pine wood density, but the weak bark mechanical properties, including internal bond, lead to the production of a high amount of fine particles, including dust, which contributes to high bulk density. The higher bulk density of the raw material may adversely affect the strength values of the panels produced, as was the case with panels made from branches from pruning fruit trees [12]. Particles from branch shredding have a different geometry and a higher bark content, which also reduces strength parameters.

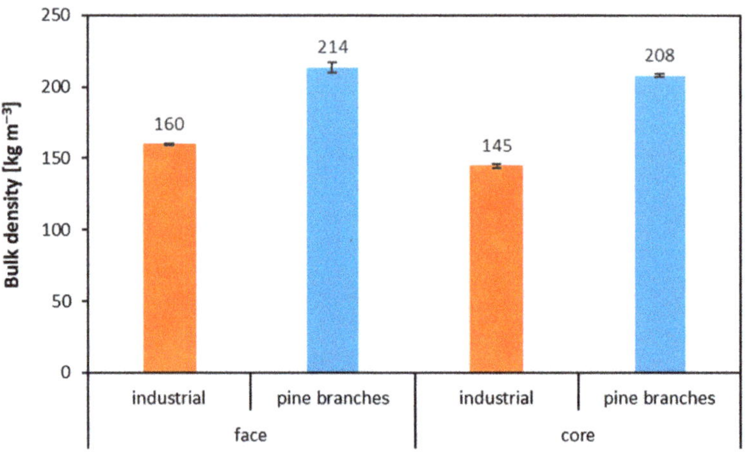

Figure 4. Bulk density of particles from pine branches used in the manufacture of three-layer particleboard.

3.2. Modulus of Rupture and Modulus of Elasticity

In the accompanying graph (Figure 5), it can be seen that as the pine branch particle content increases, the MOR and MOE values decrease. For MOR, the variant with an alternative particle content of 10% showed almost comparable results to the reference panels. The largest statistically significant differences were shown between the reference panels and the panels with an alternative raw material content of 100%. Furthermore, statistically significant differences have been found for panels 0 and 10, 25, 50, 100 cl, and 100 fl. The average values of MOR of the panels 100 against 100 cl and 100 fl were also statistically and significantly different. All the variants produced met the requirements of the EN standard.

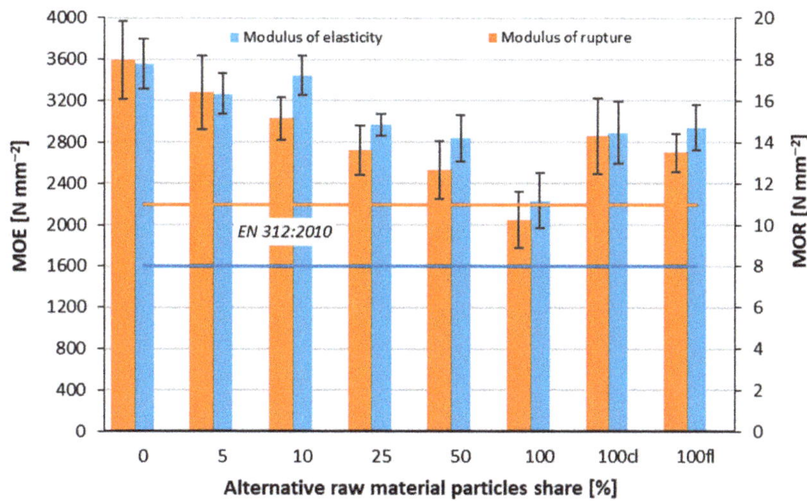

Figure 5. Modulus of rupture and modulus of elasticity of tested composites.

The MOE values obtained are lower, with a difference of 1320 N mm^{-2} between the highest and lowest parameters, so all variants, in this case, met the applicable standard. The lowest values were characterized by boards made entirely of alternative particles, where the MOE was 2235 N mm^{-2}. Concerning the statistical significance of the mean values of MOE, the relations found in the case of MOR can be applied.

Due to the nature of the alternative material, which is anatomically very similar to the pine wood commonly used for the production of wood-based composites, better strength parameters were expected, but due to the smaller diameters of the pine branches, they have lower strength and more bark, which has a lower density and therefore adversely affects the strength parameters. Branches with very small diameters may be too small after shredding, resulting in a large amount of dust, which is not desirable for the production of wood-based materials.

Similar relationships regarding MOE in their study were obtained by researchers investigating the strength of particleboards made with the addition of particles derived from sequoia branches [32].

3.3. Internal Bond and Screw Withdrawal Resistance

The results of the internal bond were presented in Figure 6. As the proportion of particles derived from pine branches increased, a favorable effect on the resistance values for internal bonds was noted. The results for the reference boards were 0.77 N mm^{-2} whereas the results for the boards made from 100% branch particles were 1.48 N mm^{-2}. The best results were obtained for the variant in which the core layer of the three-layer particleboard was made from 100% branch particles and the face layer was made from particles used as standard in the industry. The values obtained for this variant were at 1.69 N mm^{-2} and were thus more than double that of the variant where the middle layer was made industrially and the outer layer was made entirely of particles from branches. In addition, all variants meet the requirements of the standard, and statistical analyses for the alternative raw material used show no statistically significant differences between the average IB values of 5 and 10% and 25 and 50%, as well as for 100 and 100 cl. Alamsyah et al. [33] in their study indicate that IB results can be influenced by the mixing of particles with glue, sheet molding, and pressing and that the higher bulk density of the alternative raw material translates into higher IB results. In another study, the researchers used mulberry branches for particleboard, and the IB study obtained the following results sequentially for variants depending on particles size: 0.25–1 mm—1.43 N mm^{-2}, 1–2 mm—1.54 N mm^{-2}, and 2–4 mm—2.30 N mm^{-2}, at a particleboard density of 800–830 kg m^{-3} [34].

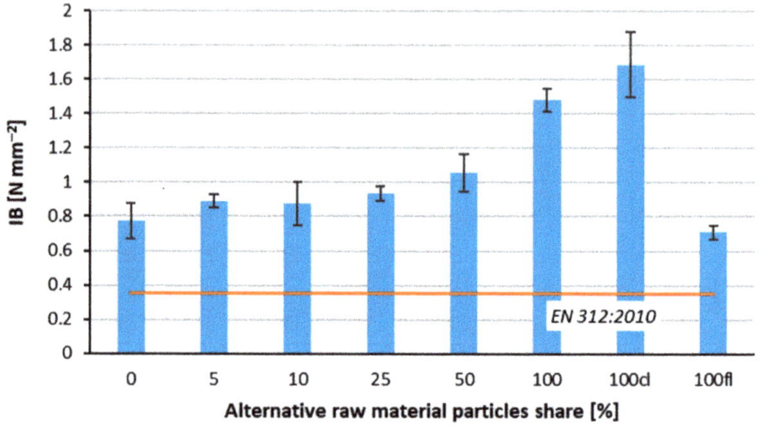

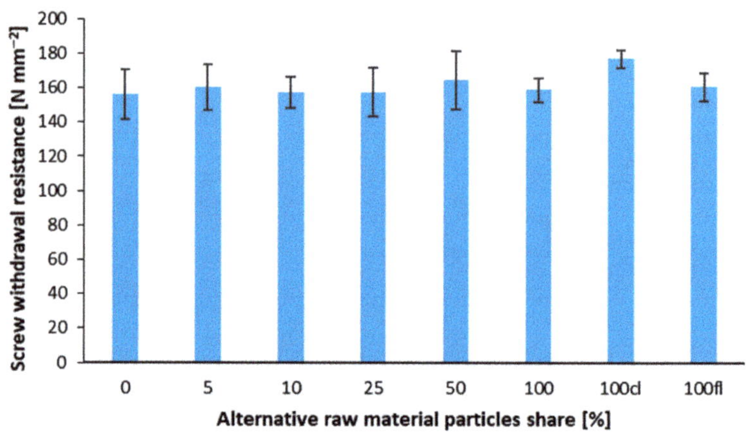

Figure 6. Internal bond of tested composites.

The only statistically significant differences among the average values of screw withdrawal resistance have been found between samples 100 and 100 cl (Figure 7). However, it is worth adding that the achieved values of SWR, between 156 N mm and 177 N mm were significantly higher than the SWR values found by [12] for the panels of the same purposes.

Figure 7. Screw withdrawal resistance of the panels with various branch particles share.

3.4. Thickness Swelling and Water Absorption

The accompanying graph (Figure 8) shows the results of swelling per thickness of the tested composites after 2 h and 24 h of soaking in water. When considering the two additional custom variants, after 2 h of soaking, the lowest TS was obtained for the variant in which the core layer was made entirely of alternative raw material, which was 22%. The highest value was obtained for the variant in which only the face layers were made of 100% alternative raw material; the TS was 33%. The rationale for this result can be found in the higher bark content of the alternative raw material used, which is more porous and has a lower specific density, which means it can absorb more water [35].

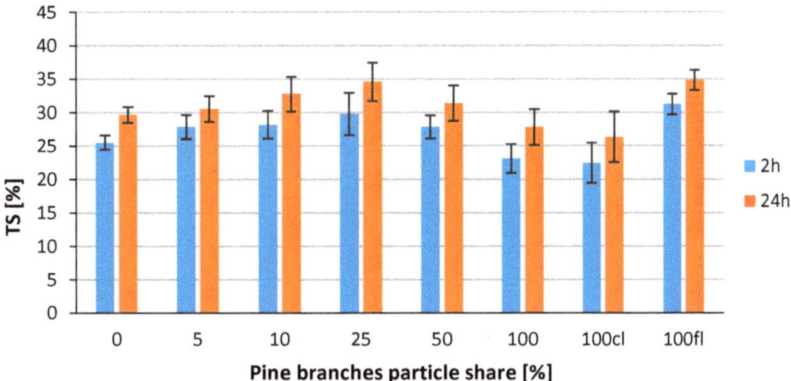

Figure 8. Thickness swelling of tested composites.

Analyzing the standard variants with the content of particles derived from branches from 5 to 100%, the lowest TS value after 2 h was recorded for the board made from 100% alternative raw material; TS was 23%, with the highest value after 2 h recorded for the 25% variant. With a longer soaking time, TS increases, and the proportion of dimensional change after 24 h is similar to that measured after 2 h.

WA results of the different variants of the manufactured branch particleboards are shown in Figure 9. After 2 h, water absorption was lowest for the reference variant—72%—whereas the highest values were recorded for variants 25 and 100% (80%). After 24 h, the discrepancies between the variants were slightly greater; still, the lowest WA value was characterized by the reference panels, where the WA was 82%. The highest WA value was 93% and was recorded for the variant made from 100% alternative material.

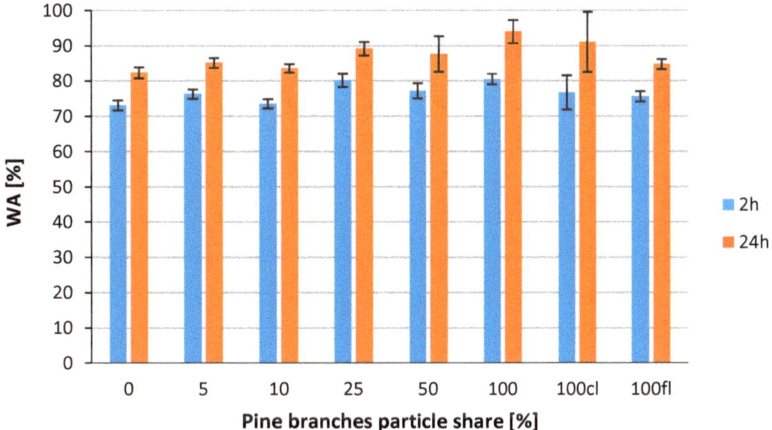

Figure 9. Water absorption of tested composites.

3.5. Density and Density Profiles

The obtained results of the density profile measurement are shown in Figure 10. The highest density in the panel was found in the outer layer, where 930 kg m^{-3} was recorded for industrial particles at a depth of about 1.5 mm measured from the surface. The lowest density in the outer layer was recorded for the variant that consisted of 10% particles from pine branches, the maximum recorded value, and in this case, was 900 kg m^{-3}, and was also at a depth of 1.5 mm. The lowest density of the entire panel was recorded for the variant that consisted of 100% pine branches. That lowest density was 520 kg m^{-3} at a depth of 7 mm. All the produced variants of panels have the same density of approx. 670 kg m^{-3}

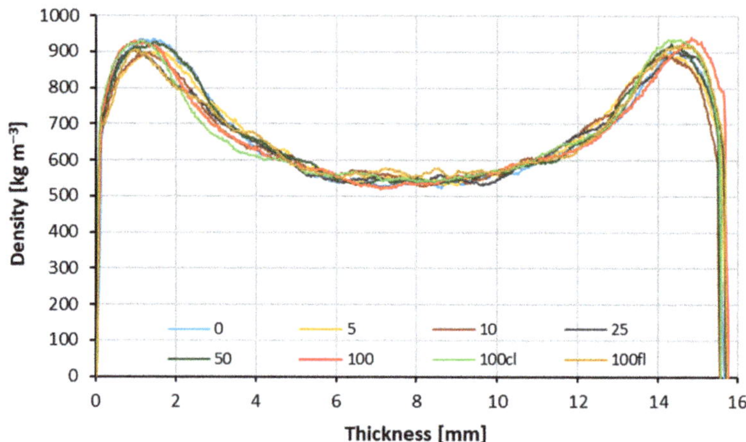

Figure 10. Density profiles of tested composites.

Higher density of the face layers of composites very often has a positive effect on strength parameters such as MOE and MOR. It is this part of the board that is responsible for transferring compressive and tensile stresses during bending. Due to the great similarity of the alternative material to the conventionally used raw material, it is difficult to notice significant differences in the density profile graph between the variants.

4. Conclusions

According to the conducted research and the analysis of the achieved results, the following conclusions and observations can be drawn:

1. It has been confirmed that particles of pine (*Pinus sylvestris* L.) branches can be successfully upcycled and used in the production of three-layer particleboard. This profitable opportunity contributes to the policy of carbon capture and storage.
2. The bulk density of particles derived from pine branches is characterized by a higher bulk density in both the face and core layer particles in comparison with industrial particles.
3. The content of pine branch particles was found to have a significant effect on the orthogonal tensile strength (IB): increasing the proportion of pine branch particles results in an increase in IB strength. Each of the produced variants meets the requirements of EN 312:2010 for P2-type panels.
4. There is no significant influence of the pine branch particles share in particleboards on their screw withdrawal resistance; however, the highest SWR has been found for the panel made of 100% of pine branch particles in the core layer and 100% of industrial particles in the face layers.
5. The highest thickness swelling of the tested panels has been found for those made of 25% of pine branch particles, whereas TS for panels made of 100% of pine branch particles in the core layer and 100% of industrial particles in the face layers were found in the lowest.
6. The water absorption test showed increasing dynamics in soaking time as the proportion of particles from pine branches increased. In addition, the general rule is that the water absorption rises with the pine branch particles' content.
7. The density of the face layers of the produced composites is comparable for the different variants; however, the slight tendency of the higher densification of the face layers has been found with the increase of the pine branch particles share.

8. Since several features of produced particleboards have been significantly influenced by bulk density and size of branch particles, further research should be focused on the modification of particle production from branches to obtain lower bulk density and to reach fraction shares closer to industrial particles. Furthermore, the chemical characterization of the pine branch particles (cellulose and lignin content, extractives content, pH value) would provide valuable data about this potential alternative raw material.

Author Contributions: A.W. took part in designing the experiments and performed the experiments, analyzed the data, wrote the first draft of the paper, and gained the funding; G.K. statistically analyzed data, wrote the final version of the paper, designed the experiments, and analyzed the data. Both authors assisted in the writing and improving of the paper. All authors have read and agreed to the published version of the manuscript.

Funding: This research received no external funding.

Institutional Review Board Statement: Not applicable.

Informed Consent Statement: Not applicable.

Data Availability Statement: The data presented in this study are available on request from the corresponding author.

Acknowledgments: Some of the mentioned tests have been completed and partially funded within the Student Furniture Scientific Group (Koło Naukowe Meblarstwa), Faculty of Wood Technology, Warsaw University of Life Sciences—SGGW.

Conflicts of Interest: The authors declare no conflict of interest.

References

1. Eker, M. Trends in woody biomass utilization in Turkish forestry. *Croat. J. For. Eng.* **2014**, *35*, 255–270.
2. Côté, W.A. *Defects and Abnormalities of Wood. Principles of Wood Science and Technology;* Springer: Berlin/Heidelberg, Germany, 1968; ISBN 9783642879302.
3. Lee, S.H.; Lum, W.C.; Boon, J.G.; Kristak, L.; Antov, P.; Pędzik, M.; Rogoziński, T.; Taghiyari, H.R.; Lubis, M.A.R.; Fatriasari, W.; et al. Particleboard from agricultural biomass and recycled wood waste: A review. *J. Mater. Res. Technol.* **2022**, *20*, 4630–4658. [CrossRef]
4. Bui, M.; Adjiman, C.S.; Bardow, A.; Anthony, E.J.; Boston, A.; Brown, S.; Fennell, P.S.; Fuss, S.; Galindo, A.; Hackett, L.A.; et al. Carbon capture and storage (CCS): The way forward. *Energy Environ. Sci.* **2018**, *11*, 1062–1176. [CrossRef]
5. Polish forests. Available online: https://www.lasy.gov.pl/pl/nasze-lasy/polskie-lasy (accessed on 24 September 2022).
6. Nati, C.; Spinelli, R.; Fabbri, P. Wood chips size distribution in relation to blade wear and screen use. *Biomass Bioenergy* **2010**, *34*, 583–587. [CrossRef]
7. Ghofrani, M.; Ashori, A.; Mehrabi, R. Mechanical and acoustical properties of particleboards made with date palm branches and vermiculite. *Polym. Test.* **2017**, *60*, 153–159. [CrossRef]
8. Sutiawan, J.; Hadi, Y.S.; Nawawi, D.S.; Abdillah, I.B.; Zulfiana, D.; Lubis, M.A.R.; Nugroho, S.; Astuti, D.; Zhao, Z.; Handayani, M.; et al. The properties of particleboard composites made from three sorghum *(Sorghum bicolor)* accessions using maleic acid adhesive. *Chemosphere* **2022**, *290*, 133163. [CrossRef]
9. Hidayat, W.; Aprilliana, N.; Asmara, S.; Bakri, S.; Hidayati, S.; Banuwa, I.S.; Lubis, M.A.R.; Iswanto, A.H. Performance of eco-friendly particleboard from agro-industrial residues bonded with formaldehyde-free natural rubber latex adhesive for interior applications. *Polym. Compos.* **2022**, *43*, 2222–2233. [CrossRef]
10. Pinchevska, O.; Šmidriaková, M. Wood particleboard covered with slices made of pine tree branches. *Acta Fac. Xylologiae* **2016**, *58*, 67–74. [CrossRef]
11. Lykidis, C.; Grigoriou, A.; Barboutis, I. Utilisation of wood biomass residues from fruit tree branches, evergreen hardwood shrubs and Greek fir wood as raw materials for particleboard production. Part A. Mechanical properties. *Wood Mater. Sci. Eng.* **2014**, *9*, 202–208. [CrossRef]
12. Kowaluk, G.; Szymanowski, K.; Kozlowski, P.; Kukula, W.; Sala, C.; Robles, E.; Czarniak, P. Functional Assessment of Particleboards Made of Apple and Plum Orchard Pruning. *Waste Biomass Valorization* **2019**, *11*, 2877–2886. [CrossRef]
13. Wronka, A.; Kowaluk, G. Selected properties of particleboard made of raspberry *Rubus idaeus* L. lignocellulosic particles. *Ann. WULS For. Wood Technol.* **2019**, *105*, 113–124. [CrossRef]
14. Mahieu, A.; Alix, S.; Leblanc, N. Properties of particleboards made of agricultural by-products with a classical binder or self-bound. *Ind. Crops Prod.* **2019**, *130*, 371–379. [CrossRef]

15. Alam, D.; Rahman, K.-S.; Ratul, S.; Sharmin, A.; Islam, T.; Hasan, M.; Islam, M. Properties of Particleboard Manufactured from Commonly Used Bamboo (*Bambusa vulgaris*) Wastes in Bangladesh. *Adv. Res.* **2015**, *4*, 203–211. [CrossRef]
16. Rahman, K.; Shaikh, A.A.; Rahman, M.; Alam, D.M.N.; Alam, R. The Potential for Using Stem and Branch of Bhadi (*Lannea Coromandelica*) As a Lignocellulosic Raw Material for Particleboard. *Int. Res. J. Biol. Sci.* **2013**, *2*, 8–12.
17. Spinelli, R. Woody Utilization of Grape Trees Pruning Residues for use in Particleboard Industry. *Ecol. Iran.* **2020**, *10*, 164–170.
18. Guruler, H.; Balli, S.; Yeniocak, M.; Goktas, O. Estimation the properties of particleboards manufactured from vine prunings stalks using artificial neural networks. *Mugla J. Sci. Technol.* **2015**, *1*, 24–33.
19. Rios, P.D.; Vieira, H.C.; Stupp, Â.M.; Del Castanhel Kniess, D.; Borba, M.H.; Da Cunha, A.B. Physical and mechanical review of particleboard composed of dry particles of branches of Araucaria angustifolia (Bertol.) Kuntze and wood of Eucalyptus grandis Hill ex Maiden. *Sci. For. Sci.* **2015**, *43*, 283–289.
20. Iskanderani, F.I. Physical properties of particleboard panels manufactured from Phoenix dactylifera-L (date palm) mid-rib chips using ureaformaldehyde binder. *Int. J. Polym. Mater. Polym. Biomater.* **2008**, *57*, 979–995. [CrossRef]
21. Oh, Y.; Yoo, J. Properties of particleboard made from chili pepper stalks. *J. Trop. For. Sci.* **2011**, *23*, 473–477.
22. Suansa, N.I.; Al-Mefarrej, H.A. Branch wood properties and potential utilization of this variable resource. *BioResources* **2020**, *15*, 479–491. [CrossRef]
23. *EN 319*; Particleboards and Fibreboards—Determination of Tensile Strength Perpendicular to the Plane of the Board. European Committee for Standardization: Brussels, Belgium, 1993.
24. *EN 326-2:2010+A1*; Wood-Based Panels. Sampling, Cutting and Inspection. Initial Type Testing and Factory Production Control. European Committee for Standardization: Brussels, Belgium, 2014.
25. *EN 326-1*; Wood-Based Panels. Sampling, Cutting and Inspection. Sampling and Cutting of Test Pieces and Expression of Test Results. European Committee for Standardization: Brussels, Belgium, 1993.
26. *EN 310*; Wood-Based Panels—Determination of Modulus of Elasticity in Bending and of Bending Strength. European Committee for Standardization: Brussels, Belgium, 1993.
27. *EN 312*; Particleboards—Specifications. European Committee for Standardization:: Brussels, Belgium, 2010.
28. *EN 323*; Wood-Based Panels—Determination of Density. European Committee for Standardization: Brussels, Belgium, 1993.
29. *EN 317*; Particleboards and Fibreboards—Determination of Swelling in Thickness after Immersion in Water. European Committee for Standardization: Brussels, Belgium, 1993.
30. *EN 320*; Particleboards and Fibreboards—Determination of Resistance to Axial Withdrawal of Screws. European Committee for Standardization: Brussels, Belgium, 2011.
31. Krzysik, F. *Nauka O Drewnie*; Państwowe Wydawnictwo Naukow: Warszawa, Poland, 1975.
32. Colak, S.; Birinci, A.; Colakoglu, G. Comparison of technological properties of particleborads produced from branch and stem wood of sequoia Semra. *Sigma J. Eng. Nat. Sci.* **2019**, *10*, 55–58.
33. Alamsyah, E.M.; Sutrisno; Sumardi, I.; Darwis, A.; Suhaya, Y.; Hidayat, Y. The possible use of surian tree (*Toona sinensis* Roem) branches as an alternative raw material in the production of composite boards. *J. Wood Sci.* **2020**, *66*, 25. [CrossRef]
34. Ferrandez-Villena, M.; Ferrandez-Garcia, A.; Ferrandez-Garcia, M.T.; Garcia-Ortuño, T. Acoustic and Thermal Properties of Particleboards Made from Mulberry Wood (*Morus alba* L.) Pruning Residues. *Agronomy* **2022**, *12*, 1803. [CrossRef]
35. Kelly, M.W. Critical literature review of relationship between processing parameters and physical properties of particleboard. *For. Prod. Lab.* **1977**, 1–57.

Review

Current Prospects for Plastic Waste Treatment

Damayanti Damayanti [1,2], Desi Riana Saputri [2], David Septian Sumanto Marpaung [3,4], Fauzi Yusupandi [2], Andri Sanjaya [2], Yusril Mahendra Simbolon [2], Wulan Asmarani [2], Maria Ulfa [2] and Ho-Shing Wu [1,*]

1. Department of Chemical Engineering and Materials Science, Yuan Ze University, 135 Yuan-Tung Road, Chung-Li, Taoyuan 32003, Taiwan; damayanti@tk.itera.ac.id
2. Department of Chemical Engineering, Institut Teknologi Sumatera, Jl. Terusan Ryacudu, Way Huwi, Kec. Jati Agung, Lampung Selatan 35365, Indonesia; riana.saputri@tk.itera.ac.id (D.R.S.); fauzi.yusupandi@tk.itera.ac.id (F.Y.); andri.sanjaya@tk.itera.ac.id (A.S.); yusril.119280082@student.itera.ac.id (Y.M.S.); wulan.119280009@student.itera.ac.id (W.A.); maria.119280081@student.itera.ac.id (M.U.)
3. Department of Biosystems Engineering, Institut Teknologi Sumatera, Jl. Terusan Ryacudu, Way Huwi, Kec. Jati Agung, Lampung Selatan 35365, Indonesia; david.marpaung@tbs.itera.ac.id
4. Graduate School of Biotechnology and Bioengineering, Yuan Ze University, 135 Yuan-Tung Road, Chung-Li, Taoyuan 32003, Taiwan
* Correspondence: cehswu@saturn.yzu.edu.tw

Abstract: The excessive amount of global plastic produced over the past century, together with poor waste management, has raised concerns about environmental sustainability. Plastic recycling has become a practical approach for diminishing plastic waste and maintaining sustainability among plastic waste management methods. Chemical and mechanical recycling are the typical approaches to recycling plastic waste, with a simple process, low cost, environmentally friendly process, and potential profitability. Several plastic materials, such as polypropylene, polystyrene, polyvinyl chloride, high-density polyethylene, low-density polyethylene, and polyurethanes, can be recycled with chemical and mechanical recycling approaches. Nevertheless, due to plastic waste's varying physical and chemical properties, plastic waste separation becomes a challenge. Hence, a reliable and effective plastic waste separation technology is critical for increasing plastic waste's value and recycling rate. Integrating recycling and plastic waste separation technologies would be an efficient method for reducing the accumulation of environmental contaminants produced by plastic waste, especially in industrial uses. This review addresses recent advances in plastic waste recycling technology, mainly with chemical recycling. The article also discusses the current recycling technology for various plastic materials.

Keywords: polypropylene; polystyrene; polyvinyl chloride; high-density polyethylene; low-density polyethylene; polyurethanes; chemical–mechanical recycling

Citation: Damayanti, D.; Saputri, D.R.; Marpaung, D.S.S.; Yusupandi, F.; Sanjaya, A.; Simbolon, Y.M.; Asmarani, W.; Ulfa, M.; Wu, H.-S. Current Prospects for Plastic Waste Treatment. *Polymers* **2022**, *14*, 3133. https://doi.org/10.3390/polym14153133

Academic Editors: Sheila Devasahayam and Laurence Dyer

Received: 2 July 2022
Accepted: 28 July 2022
Published: 31 July 2022

Publisher's Note: MDPI stays neutral with regard to jurisdictional claims in published maps and institutional affiliations.

1. Introduction

Plastics are an integral component of our modern lives due to their wide range of applications in households and industry [1]. Worldwide plastic production is estimated at around 1.1 billion tons of plastic in 2050 [2]. The increase in interest in plastics as raw materials in various sectors comes from its ease of handling, transparency, and cost-effectiveness [3]. Plastics have shown extraordinary packaging performance for food, confectioneries, chemical products, and medicinal products [1]. Around 40% of plastic materials worldwide are used to store and package completed items from various factories. Nevertheless, massive plastic waste is generated, due to its mass consumption. Packaging is the most significant contributor to worldwide plastic waste, contributing to about 50% of the total weight [4]. Plastic waste from thermoplastic, thermoset, and elastomers of polymeric materials are not easily degraded [5] and could become abundant by producing primary environmental contamination. Moreover, the excessive amount of plastics generated over the last century, and poor waste management, have raised concerns about the depletion of fossil resources, the destruction of marine and terrestrial ecosystems, and climate change [6]. Therefore, the application of proper plastic waste management is critical to solving sustainability and environmental issues.

To date, plastic waste management has gained more attention worldwide due to its impact on human life sustainability. Typical plastic waste management strategies include landfills, incineration, microbial decomposition, thermal decomposition, mechanical pulverization, and recycling. Rapid and effective identification and classification of separate mixtures of waste plastic is challenging and this will be a crucial component in the waste plastic industry [7]. Therefore, plastic waste is mainly disposed of in landfills and discharged into the environment. These wastes, particularly plastic packaging, end up in rivers and seas, posing a significant hazard to aquatic habitats [5]. Landfill is becoming increasingly costly, as the volume of waste increases and landfill capacity decreases. More importantly, dumping plastic waste in a landfill could waste valuable resources and causes a series of problems, such as additive leaching and land occupation [8]. Meanwhile, incineration is commonly used in the energy recycling of plastic waste, since a significant amount of energy can be recovered, and the energy can be utilized to generate electricity, combined heat, power, or for other operations [9]. However, recycling waste plastics by incineration can be harmful, because various toxic components, which may cause carcinogenesis, teratogenesis, and mutagenesis, are detected in fly ash and residues, at concentrations exceeding the allowable limits [10]. Among these methods, plastic waste recycling simultaneously offers an acceptable and environmentally friendly approach.

Plastic waste recycling refers to the waste management process that collects plastic waste materials and turns them into raw materials reused to produce other valuable products. Recycling is not only a method for disposing of plastic waste, but it is also an effective process to minimize the need for virgin plastics, which can help lessen global warming [9]. According to the ASTM Standard D5033, plastic recycling can be categorized as primary, secondary, tertiary, and quaternary recycling [11]. Based on the mechanism of the methods, plastic waste recycling can be classified as mechanical, chemical, and biological recycling [5]. Chemical recycling, such as catalytic and thermal processes, can convert plastic waste into value-added chemicals/fuels. This process is a potential method to reduce plastic waste as a primary source of environmental issues. Due to the plastic manufacturing industry consuming almost 6% of all petroleum produced globally, extracting fuel oils from waste plastics can help reduce the global dependence on oil [12]. The activation energy of catalytic pyrolysis with the presence of a catalyst is decreased, and catalytic pyrolysis can be performed at a lower temperature, increasing the polymer conversion rate. The catalytic pyrolysis of waste plastics and petroleum sludge was performed. The catalysts include molecular sieves, transition metals, metal oxides, clays, and activated carbons used to recycle plastic, and molecular sieves and M-series catalyst (M = Al, Fe, Ca, Na, K) for treating petroleum sludge [13].

Mechanical recycling is often classified as primary or secondary recycling, and chemical and biological recycling is commonly classified as tertiary and quaternary recycling. Each method has its advantages and disadvantages, depending on the user needs. Another aspect required in the recycling of plastic waste is the separation of the different materials. For instance, PVC in the PET extrusion process damages the equipment, due to chlorine, decreasing product qualities, such as color and viscosity [14].

There are various separation methods for plastic waste recycling, including optical sorting (colors and peaks based), density separation (densities based), flotation (surface properties based), and Tribo electrostatic separation (effective surface work function based) [10]. Knowing the most acceptable combination of recycling and separation methods of plastic waste would be a powerful way to diminish the accumulation of pollutants in the environment caused by plastic waste.

This review aims to discuss the current technology of mechanical and chemical plastic waste recycling, to reduce plastic waste accumulation in the environment. Among several recycling methods, mechanical and chemical are typical approaches used for plastic waste recycling. Several research studies have developed mechanical and chemical plastic recycling methods to replace landfill and incineration methods. The chemical recycling of plastic waste depends on the degradation of the polymer chains. Meanwhile, mechanical recycling of plastic waste typically leads to re-granulation. Furthermore, this review focuses on the recycling method in general and the suitability of each recycling method for various types of plastic waste. In addition, the identification and separation methods of fresh plastic waste from the environment, until ready to be recycled, will also be discussed. The separation of various materials needs to occur before the actual recycling process. A better understanding of each plastic waste recycling method is necessary for policymakers to be able to determine the proper methods to solve the significant plastic waste issue.

2. Waste Plastic Recycling and Technology

The recycling process of plastic can be divided into various types: primary, secondary, tertiary, and quaternary recycling [15]. Primary recycling is the processing of a specified and uncontaminated material, commonly scrap, from an industrial process. Furthermore, to provide a good product quality, recycled scrap or waste plastics can be mixed with new materials [16]. Nevertheless, the primary recycling process needs homogeneous, clean, and non-degraded materials, such as packaging, bottles, and pre-consumer products, with the product of primary recycling being quite similar to a virgin one [17].

The mechanical recycling of plastic waste is secondary recycling; the most common method for recycling plastic waste. Mechanical recycling processes post-consumer plastics, to produce the raw materials for various plastic products [18]. In comparison, the recycling process depends on the chemical and physical properties of the waste plastic feed, in terms of its origin, composition, and form [19,20]. Figure 1 is related to the technology of recycling waste plastics by the mechanical method. Mechanical recycling includes several techniques, for instance, collection, separation, sorting, and washing [21]. The main objective of the waste plastic sorting process is to obtain high-quality recycled plastic goods, especially from a single polymer stream. Waste sorting technologies are based on various chemical-physical properties of the plastic, for instance, chemical compounds, size, color, and shape. Furthermore, the materials from post-consumer waste contain various polymeric materials and organic substances [17,22]. The subsequent process is size reduction. The typical process for size reduction involves cutting or shredding; nevertheless, this process depends on the type of plastic waste stream and plant layout. These processes may occur before or after the sorting stage [23].

The other processes include size reduction, extrusion, and granulation. These may occur in different sequences and at different times [19]. The extrusion and granulation processes are required to create a granulation that is possible to convert into flakes. Furthermore, the polymer flakes are typically loaded into an extruder, heated, and pressed through a die, to form a continuous solid polymer product (strand). This can be chilled in a water bath before the pelletized process. The granulation method is often utilized to convert the strands into pellets, which can then be used to produce new products [23]. To consider the full life cycle of polyurethane foams (PUFs), PUFs were upcycled and reshaped to bulk polyurethanes (PUs) using a transcarbamoylation reaction of up to five cycles. Moreover, four PUFs were prepared and reshaped by compression molding at 160 °C for 30 min, demonstrating the potential of this recycling pathway for PUFs from different origins [24].

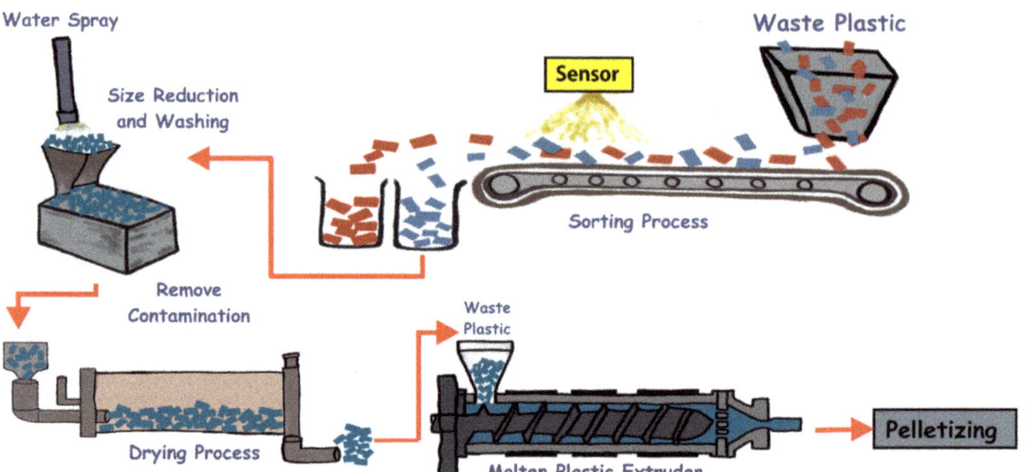

Figure 1. The technology of recycling waste plastic by a mechanical method: sorting process, size reduction, drying process, and molten plastic.

Tertiary or chemical recycling refers to the degradation of polymer bonds. As a result, the recovery of the oligomers monomers produces a smaller molecular weight. Hence, thermoplastic can be obtained with this method [25]. Some technologies can be applied in the following manner, as shown in Figure 2:

(a) Gasification: the polymer is utilized as a refuse-derived fuel using high temperature. It is converted to syngas with an H_2/CO molar ratio of 2:1 in a gasifier; the syngas produced depend on the various polymers.

(b) Pyrolysis: the plastic waste is converted to pyrolytic oil, which is equivalent to diesel oil. In this chemical recycling, the calorific value of the polymer affects the energy content of the diesel [26].

(c) Glycolysis: the ethylene glycol and waste plastic are added in the presence of a catalyst. The long polymer chain is degraded into building blocks, which can be recycled to produce new polymers.

(d) Hydrolysis: when biopolymers (e.g., PLA) are heated and broken down to their monomer building blocks, they can be dissolved in water. These monomers can be recycled and utilized to make new products [27,28].

Quaternary recycling is a method of recovering energy using a combustion process of the waste polymer [29]. The plastic waste is incinerated. Nevertheless, the released energy is captured and replaced with heat and power. These strategies present a hierarchy of choice, in ascending order, from primary to quaternary, for managing resources and minimizing the processing costs of converters. Table 1 lists the advantages and drawbacks of various methods of recycling waste plastic. Plastic waste is commonly subjected to mechanical conversion in a closed-loop recycling method. Nevertheless, in metropolitan areas, this strategy cannot alleviate the accumulated plastic waste [30].

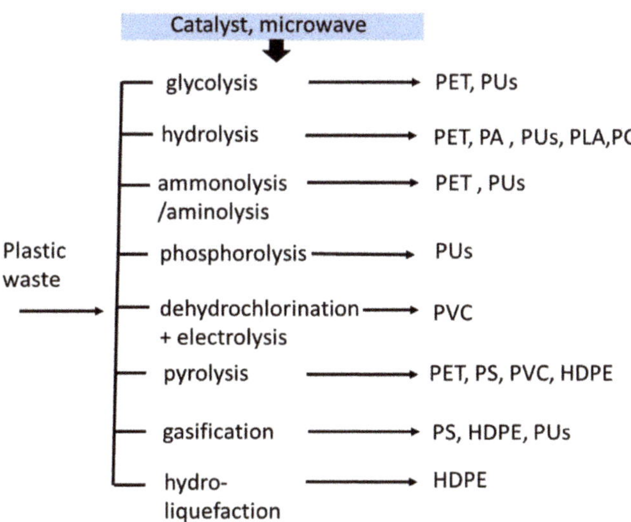

Figure 2. The chemical recycling route of plastic waste.

Table 1. The advantages and drawbacks of various methods of recycling plastic waste.

Methods	Advantages	Limitations	Ref.
Mechanical	- It is the easiest process for recycling metal matrix composites, and it is especially well-suited for fiber-reinforced polymers (FRP), where fiber breaking is accomplished through shredding.	- The decreased melt viscosity due to hydrolytic and thermal depolymerization.	[31,32]
	- The recycling facilities are simple and economical, and they use less energy and resources than chemical or physical recycling processes.	- The generation of cyclic and linear oligomers affects the printability and dyeability of the final product.	[32]
Chemical	- The greater rates of the monomer with a shorter reaction time.	- The expensive investment in developing technical infrastructure/processes.	[16,33]
	- Higher potential for profitability through new materials application.	- The feasibility of an industrial scale has not yet been completely established.	[34,35]
	- The most cost-effective approaches for high-performance recycling composites.	- High temperature and much energy are needed.	[36,37]
Biological	- The procedure is easy to follow and is also environmentally friendly.	- The depolymerization is extremely slow for the high molecular weight of hydrophobic plastic polymers.	[5]
	- Cost-effective process.		

Currently, certain technologies have become the most used techniques for identifying and analyzing plastics. These techniques can be divided into elemental/atomic spectroscopies and molecular spectroscopies. Nevertheless, these methods utilize different types of physical phenomena [38]. Table 2 lists a summary of the sensor identification and sorting of waste plastic. The molecular spectroscopies produce information about the sample's molecular identity and its molecular structure or conformation (the spectral signature of nearly any substance), allowing for accurate identification and characterization. Fourier transform infrared spectroscopy (FTIR), terahertz spectroscopy, Raman spectroscopy (RS), and near-infrared spectroscopy (NIR) are essential techniques in molecular spectroscopies. Fan et al. studied the admixture of microplastic containing polyethylene terephthalate (PET), polyethylene (PE), nylon (NY), polyvinyl chloride (PVC), and polypropylene (PP) by using Fourier transform infrared (FTIR) spectroscopy. The characteristic of wavelength numbers were PE (1472 and 1462 cm^{-1}), PVC (712 cm^{-1}), NY (3295 cm^{-1}), PET (793 cm^{-1}), and PP (841 cm^{-1}) [39].

During transportation and storage of plastic cups and bottles used for mineral water packaging, the plastic particles can be released due to long-term exposure to light, heat, and unfavorable chemical environments. These types of plastic particles can be identified using quantitative surface-enhanced Raman spectroscopy (SERS) [40]. On the contrary, with FTIR, Raman spectroscopy is far more responsive to the molecular structure (electronic bonds) than the functional groups. Consequently, FTIR and RS have already been considered as complementary methods [38,41]. Molecules can be stimulated to a greater energy level when photons from a laser interact with the molecular vibrations. The majority of this energy will be dispersed via elastic scattering (or Rayleigh scattering), in which the energy of the released photons is equal to that of the laser photon. Raman spectroscopy determines the wavelength of inelastically scattered photons. The released photon has a higher or lower energy than the photon emitted by the laser, which can be seen as the spectrum of intensity over wavelength [42]. The chemometric characterization of Raman spectra was demonstrated to be successful for classifying PE with various densities, due to the intensities of CH_2 with wagging and scissoring [43].

Laser-induced breakdown spectroscopy (LIBS) is a technology for elemental analysis, which has already been called "a future superstar" [44]. LIBS is already widely utilized for plastic sorting analyzers. The classification and identification of various types of waste plastics is the most common usage, consisting of different plastic objects; for instance, household applications, toys, electrical cables, containers, landmine casings, and various types of e-waste [45]. Furthermore, LIBS has received much attention for several plastic compounds such as plastic-based films, plastic-bonded explosives, and bio-plastic [46].

Several chemometric technologies can identify plastic polymers, and the data is collected using spectroscopic approaches. The broad category of chemometric technologies includes partial least squares regression (PLS), principal component analysis (PCA), and linear discrimination analysis (LDA) [42]. Henriksen et al. studied plastic identification using unsupervised machine learning models such as K-means clustering and PCA. In the thirteen types of plastics examined using PCA, the hyperspectral imaging with wavelengths ranging from 955 to 1700 nm proved that the spectral range was sufficient to identify plastics [47]. Furthermore, the most extensively used method of sorting waste plastic by multivariate analysis is with partial least squares discrimination analysis (PLS-DA), which is a very stable and straightforward approach for spectra data [48].

Near-infrared hyperspectral imaging (HIR-NIR) is a well-established technology for separating larger plastics in recycling plants and waste management. This sorting method is non-destructive, and the spectral range is a polymer fingerprint region with easily detectable C-N, N-H, and C-C absorption bands [49]. Vidal et al. investigated extensive microplastics identification using HIR-NIR. The microplastics had small particle sizes (<600 μm) and could be easily recognized, even though they were invisible by visual inspection or during handling. The features HIR-NIR are advantageous compared to traditional infrared (IR) spectrometers. A pixel size of 156 × 156 μm with a 75 cm^2 scan area was probed in under 1 min. The specificity and sensitivity of waste plastics such as PE, PP, PA, PET, and PS were over 99% [50].

Table 2. Summary sensor for identification and sorting of waste plastic recycling.

Waste Plastics	Analyzer	Chemometric Tool	Wavenumber, nm	Accuracy, %	Ref.
Waste electrical and electronic equipment plastic (PP, PS, ABS, ABS/PC)	NIR512 by Ocean Optics	PLS-DA PCA-LDA	900–1700	99	[51]
Household waste (PE, PP)	Specim ImSpector N17	PLS-DA	1000–1700	100	[52]
Standard plastic samples (PE, PET, ABS, PS, PC, PP, and PVC)	NIR	PCA-SVM PCA-KNN PCA-ANN	900–1700	100	[53]
PS, PP, and ABS	RS	PCA-SVM	100–3300 cm^{-1}	95	[54]
Waste plastics (PS, PP, PET, PVC, LDPE, HDPE)	MicroNIR	PLS-DA	900–1700	100	[55]
Black waste plastics (PS, PET, PP)	ATR FT-IR	FRBFNN	695–1376 cm^{-1}	99	[56]
Black waste plastics (PS, PET, PP)	RS	FRBFNN	410–2871 cm^{-1}	95	[56]
Waste plastics (PE, PP, PET, PVC, PS)	HSI-NIR	PLS-DA	1000–1700	100	[57]
Plastic solid waste (PET, PMMA, PP, PE, PS)	NIR	PCA-SVM	1000–1700	97.5	[58]
Black waste plastics (PET, PP, and PS)	RS	FRBFNN	200–3000 cm^{-1}	95	[59]
Electronic household appliances (PP, ABS, PS)	RS	NA	1000 cm^{-1}	94	[60]

2.1. Recycling Polypropylene

PP has a linear hydrocarbon chain, with a melting temperature of 160 °C [61]. PP is produced from propylene with a catalyst of metallocene or Ziegler-Natta. Furthermore, PP has excellent thermal, physical, and mechanical properties at ambient temperature. PP is an extensive application in plastics, stationery, furniture, food containers, and automotive industries [62]. PP recycling can be approached by chemical and mechanical methods, such as pyrolysis and hydrogenolysis. In addition, chemical recycling provides a chance to recycle waste plastic and convert it into higher value-added chemicals, especially for fuel additives. PP recycling approaches can be integrated with chemical refineries and generate a new generation of recyclable-by-design polymers [63,64].

Upcycling PP waste by hydrogenation has recently received much attention. Figure 3A illustrates the pathway of polypropylene degradation by hydrogenolysis. Chen et al. found a depolymerization mechanism of polypropylene under a catalyst by hydrogenolysis. The initial step of the reaction of hydrogenation is dehydration and adsorption. Hence, a hydrogen-depleted intermediate is formed. The hydrogenation of the fragments occurs during C-C cleavage, mainly resulting in new products. The lighter product with a R'-group will desorb more quickly [65].

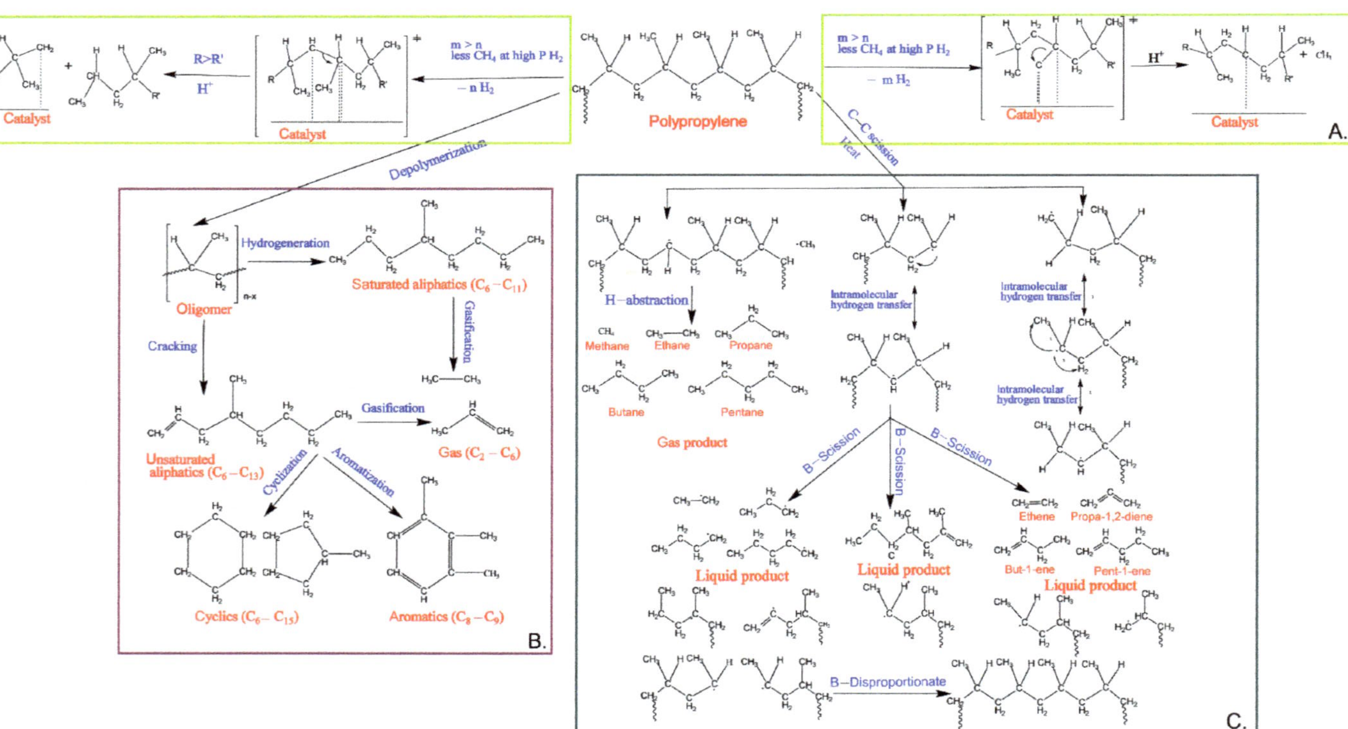

Figure 3. The depolymerization mechanism of polypropylene reaction: (**A**) hydrogenolysis into gas and liquid alkanes, (**B**) supercritical water depolymerization process, and (**C**) pyrolysis in a semi-batch reactor under atmospheric pressure [65–67].

Furthermore, Kane et al. investigated the detailed mechanism of depolymerization of waste polyolefin under hydrogenolysis. Generally, the hydrogenolysis process generates liquid alkanes over methane. There are several possible reactions during hydrogenolysis: (1) σ-Bonds metathesis is produced by primary carbons (1–3 kcal mol^{-1}); (2) the long-chain moves into β-H by an elimination process and reduces the yield of E-alkenes, with substitution by Z-alkenes, (3) the elimination of β-Alkyl and methyl (1 kcal mol^{-1}) following the majority Zr–C bond formation (2 kcal mol^{-1}); and (4) the stabilization process, by eliminating β-alky. A catalyst can promote the removal of the β-alkyl compound, degrading the long chain of PP more easily than LDPE [68]. The upcycling of plastics becomes exothermic via hydrogenolysis/hydrocracking, decreasing the reaction temperature to around 300 °C [69]. Noble metal nanoparticles placed on silica, alumina, ceria, or carbon in the presence of hydrogen can easily break C–C bonds in the polyolefin backbone through hydrogenolysis, resulting in low molecular-weight wax or small molecules of hydrocarbons [70,71]. Rorrer et al. investigated the hydrogenolysis of PP, and the process was under mild temperature (225–250 °C), with a selected hydrogen pressure (20–50 bar). In addition, the ruthenium nanoparticle supported on carbon (5 wt% Ru/C) was effective as a heterogeneous catalyst for the degradation of PP via hydrogenolysis. The hydrogenolysis of PP produces a liquid product with a yield of liquid over 68%, with the range of chemical compounds being liquid (C_5–C_{32}) and gas (C_1–C_5) [72].

The depolymerization of PP waste with supercritical water for liquefaction is illustrated in Figure 3B. In the first stage, PP waste degrades into oligomers in a short time (<0.5 h). Then, the bulk of unsaturated aliphatic is changed into cyclic via cyclization, when the reaction time is increased slightly. Most unsaturated aliphatic is converted into cyclic by cyclization. At the same time, small amounts of unsaturated aliphatic (olefin) may become saturated aliphatic (paraffin) and aromatics. Aromatization can theoretically occur through cyclic dehydrogenation or unsaturated aliphatic cyclotrimerization (olefins) [73]. Degradation of polymers takes place in the viscous polymer phase during pyrolysis. On the other hand, the supercritical water for liquefaction is produced by partial dissolution of the molten polymer phase. Furthermore, the dissolution of the polymer phase enhances unimolecular the reactions and polymer dissociation, for instance, β-scission. Consequently, coke formation, polycondensation, and gas creation processes are prevented [74].

Pyrolysis is a thermochemical degradation process; it can degrade the polymer compound at high temperatures without oxygen [75–77]. Non–catalytic pyrolysis is a standard technique for recycling large molecules of PP. Nevertheless, it requires a high temperature of up to (573–1173 K) and a long reaction time, and it produces a wide range of chemicals, for instance, alkanes, alkenes, aromatics, and gases [78]. Figure 3C shows the possible reaction during pyrolysis of PP. The mechanism of PP by pyrolysis starts with chain fission, radical recombination, allyl chain fission, intermolecular hydrogen abstraction, midchain β-scission, disproportionation, 1,3-end-hydrogen transfer, 1,4-end-hydrogen transfer, 1,5-end-hydrogen transfer, 1,6-end-hydrogen transfer, 1,3-mid-hydrogen transfer, 1,4-mid-hydrogen transfer, and 1,5-mid-hydrogen transfer [79]. Singh et al. studied waste tube tires (WTT) and waste polypropylene (WPP) for conversion into diesel fuel via catalytic pyrolysis and base $SrCO_3$. The major products were aromatics, naphthenes, monohydric alcohols, esters, amides, and halides. Diesel fuel from WTT and WPP has research octane numbers of 89.65 and 87.32, respectively [80].

Catalysts are frequently added to minimize the reaction time and increase product distribution. Several catalysts have shown increased yield products in PP processing with certain reactor conditions (Table 3). Acid catalysts, such as FCC catalyst, mesoporous silica, and zeolites, have been used in PP depolymerization [81,82]. The catalytic pyrolysis of medical waste PP under CO_2 was studied using Ni/SiO_2 catalyst. The chemicals, saturated hydrocarbons, olefins, and alcohols were produced in greater quantities when the H_2/CO ratio was controlled. Nevertheless, the catalytic pyrolysis process produced hydrocarbons ($\geq C_2$) and CH_4 and H_2 [83]. The depolymerization of PP pyrolysis was carried out in a stainless-steel batch reactor by Dutta et al.; the reaction temperature was 470 °C, with a maximum yield of 65%, and with the absence of a catalyst. On the other hand, a silica-alumina catalyst was added to the pyrolysis process. The maximum yield of liquid product was up to 75%, due to silica-alumina having better porosity and acidic properties. The caloric value of the liquid product was 41.1 MJ/Kg, with the major products being benzene, toluene, and xylene [84]. Harmon et al. studied the kinetic depolymerization of PP, and there were various methods used to investigate the activation energy (Ea), such as the Kissinger and Friedman methods. The Ea with the Friedman method was 235 kJ/mol, with a range of conversion up to 30–70%. Moreover, Ea = 236 kJ/mol and A = 5.60×1014 s^{-1} were determined by the Kissinger methods. The reaction profile was only 63% as wide as a first-order reaction [85].

Table 3. The products distribution catalytic and co-pyrolysis of PP.

Feedstocks	Catalyst	Condition			Yield of Product, %			Calorific Values of Liquid Product, kJ/kg	Major Product	Ref.
		Reactor	T, °C	t, min	Solid	Liquid	Gas			
PP	Spent FCC	Quartz Tube	510	60	NA	62	38	NA	Olefins, alkane	[86]
PP	Calcium bentonite clay	Batch	500	NA	0	88.5	~11.5	44,370	Alkene	[87]
PP + Ligno-cellulosic biomass; (1:2)	Spent FCC	Quartz Tube	510	60	12	52	36	NA	Aromatics, olefins, alkanes, oxygenates	[86]
PP	Bentonite clay	Fixed bed	500	10	NA	90.5	NA	44,763	Aromatics, alkanes, alkenes	[88]
PP	FCC	Stirred semi-batch	450	NA	3.6	92.3	4.1	NA	Olefins, paraffins, naphthene, aromatics	[89]
PP	Fe-SBA-15	Batch	540	300	2–0.8	73–77	24–21	NA	CH_4, C_2H_6, C_3H_6, and C_4	[90]
PP	Spent FCC	Batch	300	NA	2.3	72.4	23.7	43,435	Paraffin, olefins, naphthene, aromatic	[91]
PP	10% dolomite	Batch	400–500	90	NA	85.2	NA	43,000–46,000	Alkanes, alkenes	[92]
PP	Spent FCC	Stirred semi-batch reactor	400	NA	2	85	13	NA	Olefin, paraffin, naphthene, aromatic	[93]
PP	USY	Batch	450	45	1.2	82	16.8	NA	C_9, C_{12}, C_{15}, C_{18} and C_{21}	[94]
PP	Sulfated zirconium hydroxide	Batch	500	NA	<1	84.1	15	193.8	Paraffin, olefins	[95]
PP	Kaolin clay	Batch	450	30	23.67	67.5	8.85	46,470	Aromatics, olefins, amines, sulfide, hydroxyl	[96]

Compared to pyrolysis, the gasification process is conducted at higher temperatures, with a range of 700–1000 °C. The gasification agents, such as CO_2, air, O_2, and steam, are added to partially oxidize carbonaceous materials, producing a syngas consisting of CO, H_2, CO_2, and CH_4. Small amounts of other hydrocarbons can also be found. Syngas of high quality is needed for both chemical synthesis and fuel. Gasification with O_2-enriched air can generate syngas with a high calorific value and gas concentration in term of minimizing the N_2 dilution. Therefore, the air separation process can be expensive. To produce pure oxygen for partial oxidation, gasification can be combined with chemical looping processes and using transition metal-based oxygen carriers instead of gaseous O_2 [97,98]. Xiao et al. investigated the recycling PP plastic by air gasification in a fluidized bed gasifier, to produce a low tar content and fuel gas with a calorific value of 5.2–11.4 MJ/Nm^3. Approximately 250 mg/N m^3 of tar was found in the gas products, with a yield of fuel gas up to 3.9 N m^3/kg [99]. In addition, the dissolution process is one of the methods to depolymerize large compounds of PP into small molecules. The dissolution process can be affected by molecular weight, polymer size, dissolution time, temperature, and concentration [100]. During polymer breakdown, mechanisms such as solvent diffusion and chain disentanglement are implicated thermodynamically. Furthermore, polymer self-diffusion is critical during chain disentanglement [101].

2.2. Recycling Polystyrene

PS is an aromatic polymer produced by polymerizing a styrene monomer. PS is a popular material because it has excellent physical properties, such as strength, durability, versatility, and low cost [102]. PS is extensively used in the form of expanded PS foam, which has a low thermal conductivity, good resistance to many corrosives, and is nearly impervious to moisture [103]. Styrene polymerization can be achieved through various intermediates and/or active species, such as cationic, coordination polymerization, anionic, and radicals. Free-radical polymerization is mostly used in the commercial production of atactic PS with a higher molecular weight, up to 200,000–300,000 g/mol. It can produce an amorphous polymer with a comparatively high glass transition temperature of Tg = ~100 °C [104].

PS is difficult to degrade in natural environments. Nevertheless, chemical, mechanical, and thermal recycling are used to degrade the large molecule of PS. One of the methods for chemical recycling is dissolution. Figure 4 illustrates PS degradation by dissolution and a scheme of liquid products valuable in PS depolymerization, such as catalytic pyrolysis, dilute acid, and pyrolysis [105,106]. The polymeric components are first dissolved, then various processes are used to recover the solvent and polymer [107]. The recycling of foamed polymers by using solvents has several advantages. Filtration can be used to eliminate any insoluble impurities, leaving the polymer clean for any further treatment. Additionally, the dissolution process enables the separation of plastics from other waste and insoluble polymers, according to their chemical structure, a process known as selective dissolution. For expanded materials, dissolving the foam in a suitable solvent results in a significant volume decrease (over 100 times), lowering transportation expenses [108].

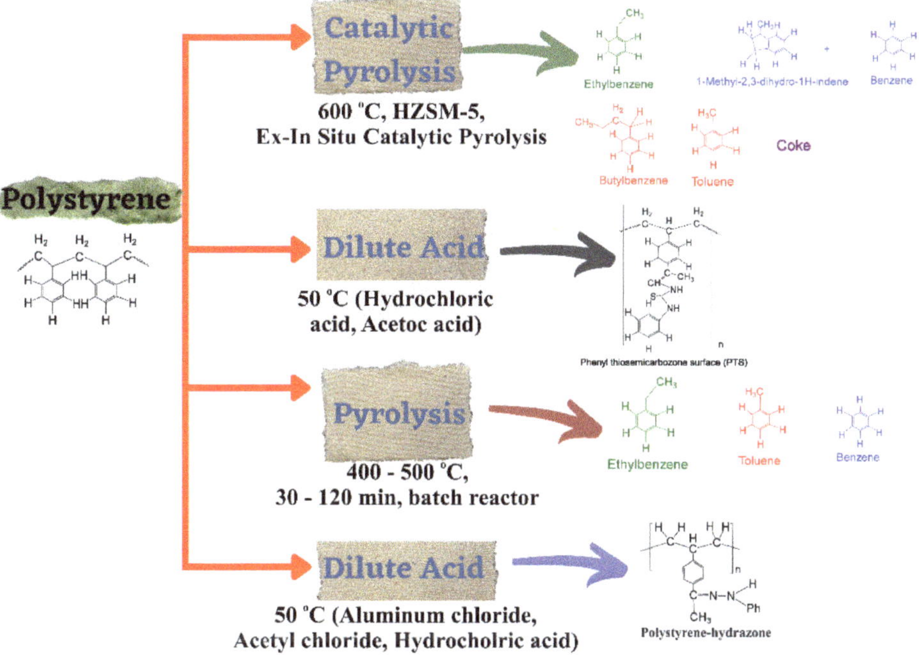

Figure 4. Scheme of valuable liquid products for PS depolymerization for different reaction paths.

Cymene, terpinene, phellandrene, and limonene are used for recycling expanded polystyrene. Furthermore, limonene is an excellent solvent and antioxidant for recycling expanded polystyrene throughout the heating process. A prior work discovered that using d-limonene as a diene compound in the thiolene reaction dissolves PS, allowing it to be recovered from the solution [101]. Furthermore, p-xylene/n-heptane, methyl ethyl ketone (MEK)/methanol, and MEK/n-hexane are suitable for recycling PS foam. Conventional solvents such as methanol/xylene can dissolve PS at various temperatures. Within a specific temperature range, a rising temperature will contribute to the rapid recovery of PS. Due to environmental damage, a more environmentally-friendly solvent (D-limonene, produced from citrus fruit rinds) was utilized to obtain 100% recycling of expanded polystyrene [109]. Gil-Jasso studied the application of essential oils to dissolve and recover PS waste. Various solvents were applied, such as chamomile, thyme, star anise, and eucalyptus oil, with a total percentage of PS recovery of more than 95%, and with a reaction time up to 833 s [110]. Polarity influences the solubility of polymers in solvents. A polymer is naturally inclined to dissolve more readily in the non-polar solvents that are chemically and physically closest to the XPS. Nevertheless, polar solvents can also be utilized in the recycling process if they do not have a strong tendency for hydrogen bond formation [108].

Pyrolysis and catalytic degradation of PS is a waste treatment process (thermal recycling) that can be utilized as a substitute for landfill disposal [111]. Simple thermal cracking at low temperatures can convert PS to styrene, without catalysts. PS pyrolysis is primarily influenced by catalyst presence, reaction time, temperature, and reactor type. Products consist primarily of liquid chemicals at low temperatures (mono aromatic). Coke and gas will be increased slightly at higher temperatures, and the liquid fraction includes many aromatics (dimer, trimer) [112]. Furthermore, adding a catalyst reduces the residence time of polymer degradation in the reactor and decreases the process temperature, by lowering the activation energy by breaking the chain of C–C bonds. PS catalytic depolymerization can be split into acid and alkaline [113]. Numerous research works on the catalytic pyrolysis of PS have been performed, including metallic oxides (alumina, alumina-silica, CuO/Al_2O_3, BaO, Al_2O_3, SiO_2, K_2O, CaO, or silica), assisted transition metals, mesoporous materials ($K_2O-BaO/MCM-4$, $K_2O/Si-MCM-41$, $MCM-41$ sepiolite derived from nature), as well as clay (pyrophyllite, albite, halloysite, montmorillonite) [112,114].

The degradation of PS waste with various basic and acidic catalysts was studied by Anwar et al. The catalytic pyrolysis was conducted with calcium oxide at temperatures ranging from 300 to 350 °C and at atmospheric pressure. The total distillate recovery was up to 77%. On the other hand, a metal carbonate catalyst generated pure styrene. Nevertheless, the yield of styrene was low [115]. The mechanism of PS catalytic pyrolysis with montmorillonite and albite as a catalyst was investigated before. The first stage of the reaction process was β-scission, followed by intermolecular H transfer, with major products being ethylbenzene and styrene [114].

Zayoud et al. studied pilot-scale pyrolysis of PS via a CSTR reactor and extruder under vacuum conditions. In the temperature operation increased to become 450 and 550 °C, the styrene yield rose 36 and 56%, respectively. In addition, at 450 °C and 0.02 bar, the yield of benzene, toluene, ethylbenzene, and xylene was enhanced from 4 wt% to 17 wt% at 450 °C and 1.0 bar [116]. Furthermore, Amjad et al. investigated PS catalytic cracking with Nb_2O_5 and NiO/Nb_2O_5 as catalysts. The Nb_2O_5 catalyst showed the highest catalytic cracking activity under a semi-batch reactor at 400 °C. The yield of ethylbenzene, toluene, α-methyl styrene, styrene, and dimers was 6%, 4%, 13%, 50%, and 6%, respectively [117]. PS thermal depolymerization can be applied to produce a styrene monomer. Nevertheless, this method has certain drawbacks, including equipment blockages and high-temperature requirements [113].

A microwave can be utilized in the pyrolysis process of PS. Microwave radiation interacts well with PS waste via the medium's dielectric constant and generates a quick heating process. Microwave processing features contain challenging characteristics: (a) rapid heating, (b) reduced energy consumption, (c) poor thermal inertia, and (d) high conversion efficiency of power [118]. The reaction was made possible by adjusting the microwave, temperature, powder, microwave design, catalyst, and absorber. The optimal temperature range for liquid products is between 600–500 °C, whereas a temperature beyond 700 °C produces more gas products. Silicon carbide and carbon are common absorbers used to increase microwave absorption. The synergistic interaction between reaction time, catalyst, and temperature enables the breakdown of long-chain hydrocarbon molecules [119]. PS pyrolysis through the microwave–metal interaction was studied by

Hussain et al. To induce rapid pyrolysis at elevated temperatures, an iron mesh of different shapes (cylindrical mesh, strips, and iron cylinder) was added. The cylindrical mesh produced heat within the temperature range of 1100–1200 °C and had a higher conversion, with a total liquid, solid, and gas of 80%, 5%, and 15%, respectively [120]. Rex et al. investigated the pyrolysis of a mixture of PS and polypropylene using a microwave-assisted method, with various types of activated carbon biomass. The operating condition of microwave pyrolysis was 900 W, with a reaction time of 10 min. The polymer and absorbent ratio was 10:1, and the maximum oil yield was 84.3 wt% [121].

2.3. Recycling of Polyvinyl Chloride

PVC is widely used in extensive applications, such as packaging, construction, electronic industries, and automotive products. Furthermore, PVC has excellent electrical, thermal, mechanical, and chemical resistance properties [122]. PVC can be degraded with a lower temperature reaction than other plastics. The detailed mechanism of PVC breakdown has been investigated with various models, which were modeled as three processes: (1) converse PVC through several intermediates compounds and HCl; (2) intermediate compounds are degraded into volatile compounds and polyene chains; and (3) polyene breakdown into toluene (and also other aromatics) [105]. Figure 5 shows how dehydrochlorination and electrodialysis are applied to recycle PVC waste. The recycling of PVC is divided into several parts. The dechlorinating process starts with a mixture of PVC waste and NaOH/ethylene glycol. The PVC waste is dechlorinated and transferred to the EG solution in the form of Cl. Then, the next process is NaCl recovery by electrodialysis with EG solution containing Cl^- and Na^+ through cation and anion exchange membranes [123]. Furthermore, Kameda et al. studied the electrodialysis of a NaCl/EG solution mixture through ion-exchange membranes. After 5 h, a high desalting ratio was obtained up to 98%. Nevertheless, the Donan effect was decreased by 0.5 wt.% of the efficiency NaCl, with total voltages greater than 4 V [123].

The other process for recycling PVC is pyrolysis. Nevertheless, PVC pyrolysis has some issues, because this process can produce fuel oil containing a large amount of chlorine [124]. The Cl^- in fuel oil products of pyrolysis can cause serious corrosion to parts of the machine and transfer toxic chemicals into the environment. As a result, the dichlorination process should be conducted before converting PVC into a high-quality fuel via pyrolysis [125]. On the other hand, catalytic and noncatalytic pyrolytic processes are used for PVC waste. Catalytic pyrolysis adds a catalyst, to increase the dichlorination process, while adding a sorbent reduces the product's chlorine compound.

Zakharyan et al. reported treatments of virgin and mixture PVC (multicomponent and binary PVC mixture, chlorine- and bromine-compounds mixtures, biomass, and municipal plastics waste) [126]. Pan et al. studied chlorine transformation and migration during PVC pyrolysis using TG-FTIR-MS methods. This showed that pyrolysis occurs in two primary steps: the first step of the reaction is using a temperature between 200–360 °C. This phase includes the dichlorination of PVC, which produces a massive amount of benzene and hydrogen chloride. The second step is a reaction temperature of 360–550 °C. The polyethylene chain is broken in the second step, due to a large amount of aromatic organic substances and chlorine-containing compounds [127]. In addition, the flash pyrolysis of PVC was conducted with a temperature reaction up to 500 °C. The major products were HCl, alkenes, monocyclic aromatics, and PAHs at 3.02%, 2.86%, 33.5%, and 48.3%, respectively [128].

Furthermore, the thermodynamic and kinetic parameters of a PVC cable sheath were investigated by Liu et al. The range of activation energy in the first stage was 132–149 kJ/mol, with the average activation energy being 141 kJ/mol. On the other hand, the activation energy increased in the second stage to 193.8–266.4 kJ/mol, and the median activation energy was 235.3 kJ/mol using the Flynn–Wall–Ozawa method [129]. Zhou et al. studied the upcycling of PVC waste into carbon compounds, chlorides, and pyrolysis gas using a one-pot dichlorination–carbonization-modification approach. The total solid yield of dechlorinated PVC was up to 80.8 wt% at 700 °C [130].

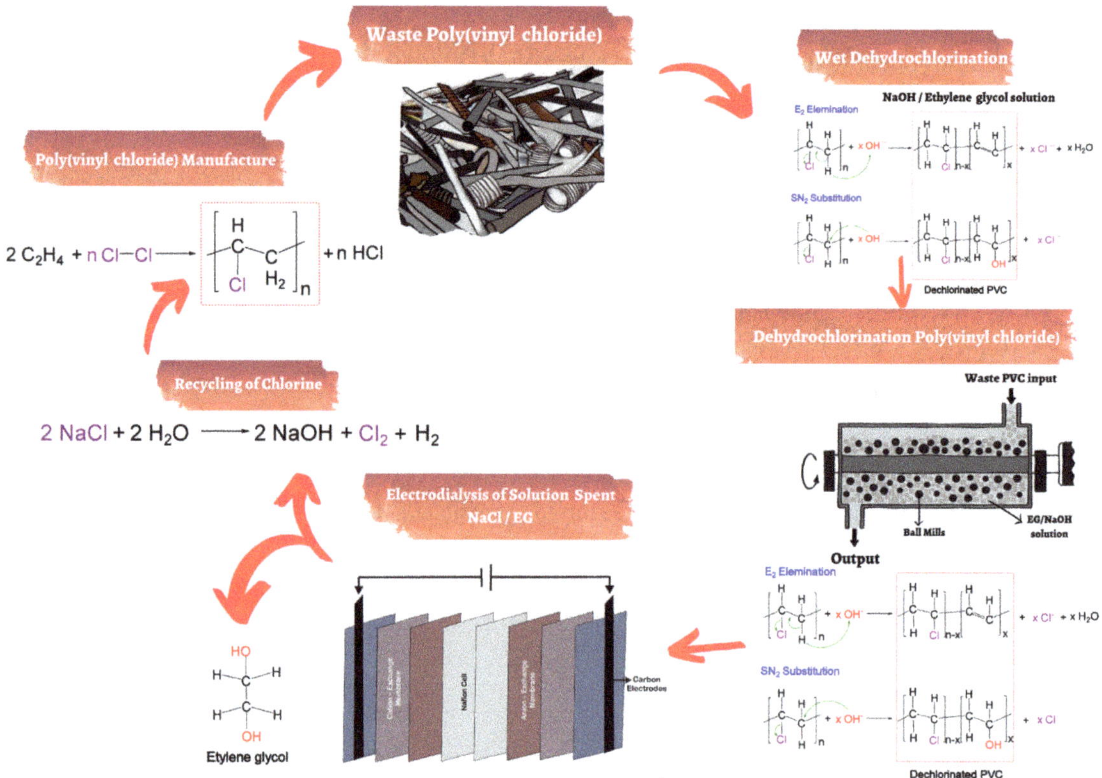

Figure 5. Technology for recycling of PVC waste by dehydrochlorination and electrodialysis [131].

Hydrothermal treatment is an effective method for removing chlorine from PVC [125]. A hydrothermal supercritical water treatment was evaluated to depolymerize PVC waste into organic compounds, such as gas and liquid products. The chlorine atoms were dissolved in water and did not cause the formation of organochlorine compounds. The sequential occurrence of three reactions was then postulated as a mechanism, as follows: (1) the zipper dehydrochlorination method was used to remove HCl from PVC, causing conjugate double bonds to form in the polymer chain; (2) polymer chain breakage; (3) aromatic molecules were synthesized by combining broken chains [132]. Enomoto et al. studied PVC depolymerization under high pressure in hot water [133]. Hydrothermal depolymerization under supercritical and subcritical regions was investigated by Takeshita et al. After the 300 °C degradation process, the chlorine compound in PVC was dissolved in water, and hazardous chlorinated organic compounds were detected in the gas and liquid fractions. The primary product with a temperature reaction of 250–350 °C was polyene as a residual solid and an aliphatic–aromatic compound in the gas and liquid fractions [132]. Zhao et al. studied the hydrothermal dichlorination of PVC wastes with an alkaline additive. Numerous chemical compounds were added, such as NH_3H_2O, KOH, Na_2CO_3, $NaHCO_3$, and NaOH, which took place in subcritical Ni_2^+ and involved water at 220 °C for 30 min. In addition, Na_2CO_3 is the most promising additive, due to its high dichlorination efficiency up to 65.1% [134].

2.4. Recycling of High-Density Polyethylene and Low-Density Polyethylene

High-density polyethylene (HDPE) and low-density polyethylene (LDPE) are the most popular forms of polyethylene. The manufacturing of HDPE adds the organometallic catalyst to polymerize ethylene. HDPE contains a higher proportion of crystalline regions than LDPE and is, hence, opaque and harder. The polymer chain in HDPE can be 500,000 to 1,000,000 carbon units long, with little branching. HDPE offers a wide range of applications in numerous plastic products, such as food containers, cleaning products, pipes, cables, tubes, and thin-film coating [135]. The recycling of HDPE waste is conducted using fluid catalytic cracking, pyrolysis, and gasification; it follows the open-loop recycling of HDPE waste [136].

The possible routes of using depolymerized HDPE waste for aromatic hydrocarbon formation are shown in Figure 6. The pyrolysis pattern of HDPE and LDPE is more intricate, and obtaining a large ethylene yield is challenging. The use of catalysts is an attractive solution to generate the desired product, especially with ethylene as the monomer of polyethylene. This process is called catalytic cracking. It can reduce energy consumption, due to lower temperatures than in thermal pyrolysis [137,138]. Singh studied the high yield liquid product of waste polyolefins pyrolysis, with a result of up to 92%, by applying $MgCO_3$ as a catalyst, with the primary products being aliphatic, alcohol hydrocarbons, ester, acetate, and aromatic [139]. Furthermore, Zeolite, e.g., HZSM-5, is frequently utilized in major catalytic reactions, due to its larger specific area, high selectivity, pore structure, and acid group, which provides a hydrogen transfer reaction [140,141]. Integration of a fluidized bed reactor filled with HZSM-5 catalysts for the pyrolysis process, and pressure swing adsorption (PSA) for light components, as well as an inert gas, such as nitrogen separation, is one of the current technologies for polyethylene waste recycling process. Hernandez et al. found that the optimum temperature for FBR, and which produced high-yield gaseous compounds, was 500 °C [142]. The presence of HZSM-5 favors C3–C5 hydrocarbons in large proportions. Meanwhile, Hernandez et al. investigated the HUSY catalyst with a lower ratio of silica/alumina and a larger surface area than the HZSM-5 catalyst. The results showed that C_{5+} components were the primary product. The gaseous compounds from the pyrolysis reactor were condensed to split light and heavy components in the separator [143].

Five distillation columns were used in light component separation to obtain ethylene, propane, and propylene. In the first distillation column, known as a demethanizer, methane was removed at 6 °C and 20 bar. The remaining light compounds were the bottom product of the demethanizer, and they then entered a second fractionation called a deethanizer, to split C_2 (ethane and ethylene) and C_{3+} as the bottom product [144]. Pure ethylene was obtained from the C_2 stream entering to deethylenizer, and a third distillation column was used with an operating temperature and pressure of -26 °C and 20 bar, respectively. The bottom product from the deethanizer was distilled in the fourth fractionation to produce C_3 (propane and propylene) and C_{4+} products mixed with heavy components from the separator in the pyrolysis unit. The last distillation yielded high propylene purity in the top column and propane in the bottom column [145].

Second step C_4 separation, is a technology used to recover n-butane, i-butane, butene mixture, and C_5 mixture. N-butane and i-butane are the raw materials for liquid petroleum gas (LPG) production, while butene mixture (trans-butene, 1-butene, isobutene, cis-2-butene, and 1,3-butadiene) and C_5 mixture (n-pentane and i-pentane) are frequently used as copolymers and solvents [143,146]. A mixture of heavy components from a light separation and pyrolysis unit was heated and split in a flash drum to remove the liquid aromatic mixture. The top product of the flash drum was condensed before entering the first fractionation, called the C_4 splitter. The column's overhead liquid was sent to the n-butane separator (4 bar, 35 °C), where the n-butane product was placed in the bottom column. The liquid in the top n-butane column contained an i-butane and butene mixture, refined in the butene column (4 bar, 35 °C), whereas the overhead liquid was an i-butane stream at the bottom column, producing a butene mixture. Moreover, the bottom product of the C_4

splitter flowed, to the C_5 splitter to generate a C_5 mixture in the top section and the rest of the heavier hydrocarbon was in the bottom column, which was a mixed aromatic mixture stream from the flash drum [147,148].

Aromatic components contain a mixture of benzene, toluene, and xylene (BTX), compounds that are regularly utilized as a feedstock in chemical industries. Liquid–liquid extraction is a popular method for recovering aromatic mixtures. There are many types of solvents, such as triethylene glycol (TEG), sulfolane, N-formylmorpholine (NFM), and N -methyl-2-pyrrolidinone (NMP), utilized to extract aromatic compounds [149,150]. The sulfolane process patented by UOP is commonly used in commercial plants, due to its high selectivity and boiling point, but low dissolvability. Therefore, to solve these drawbacks, a mixture of two solvents is the best option to increase the selectivity and lower the recycling rate and ratio of extractant [151,152]. Extraction of aromatic compounds using a co-solvent of TEG and sulfolane was studied by Galie et al. The significant selectivity of xylene was increased.

Meanwhile, the solvent and recycle feed ratios were reduced by 20% [151]. Other works showed that adding mixed solvents of sulfolane-NMP could extract 99% of benzene from reformate, and the distillate could be directly utilized as automobile gasoline [153]. However, liquid extraction process units have some drawbacks, due to their high investment cost. Conventional distillation is widely used for purifying the mixture of the components. Nevertheless, due to binary azeotrope conditions, the aromatic mixtures cannot be separated by traditional distillation. Extractive distillation (ED) is an alternative method to extract aromatic hydrocarbons with a high energy efficiency, low equipment investment, and modest process units. ED requires a particular component to raise the volatility to near boiling point [154,155].

The solvent in liquid–liquid extraction can be applied as a third compound in ED. Wang et al. reported that co-solvents of NMP and sulfolane were used to extract aromatic components with the ED technique. A feed with aromatic and non-aromatic substances flowed to the ED column (2.5 bar, 120 °C). A non-aromatic product was fed to the rectifying column in the top column, to produce non-aromatic compounds, and solvents were carried over to the overhead ED column. The bottom stream of the ED column entered the solvent recovery and regeneration column (1.01 bar, 101.4 °C), to obtain a high purity of aromatic mixtures and regenerated solvents, to reuse in the ED column. The recovery of aromatics with the ED method reached 99.92% [156].

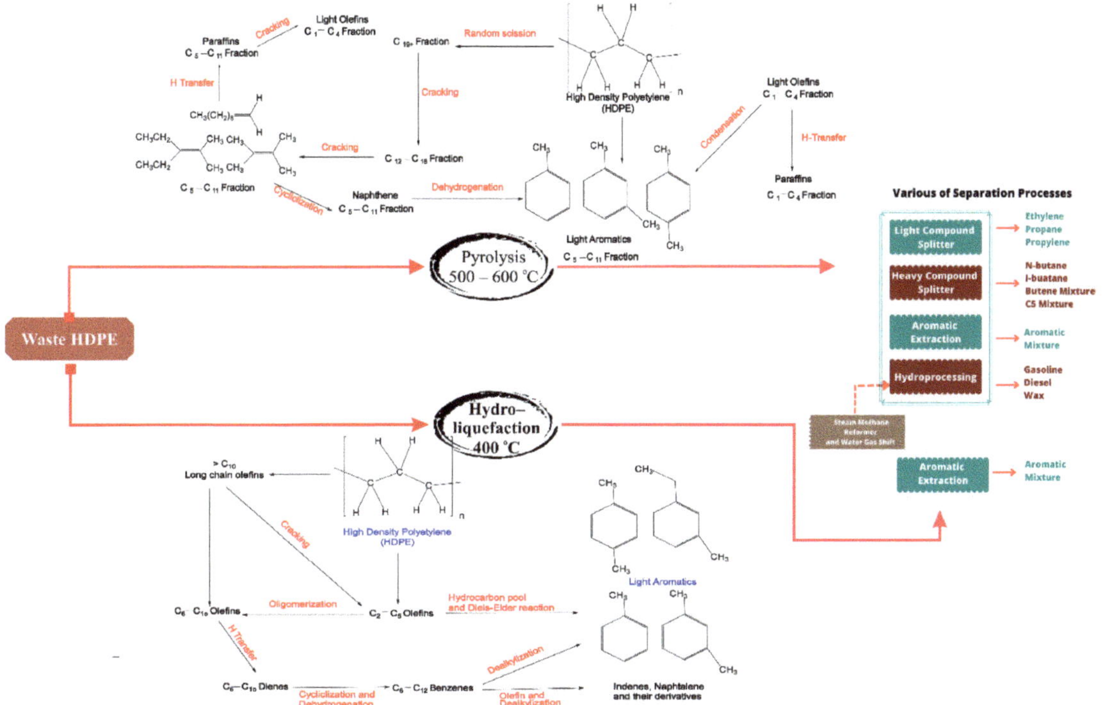

Figure 6. Possible routes of using depolymerized HDPE waste for aromatic hydrocarbon formation (**top**) by fast pyrolysis, using FCC spent catalysts in a fountain-confined conical spouted bed reactor, (**bottom**) by hydro-liquefaction over Ni/HZSM-5 [157–159].

On the other hand, adding hydrogen compounds is required to convert non-aromatic hydrocarbons into valuable products in a hydrogenator. Steam methane reforming, combined with a water gas shift reaction, is a mature process to produce hydrogen. A methane feed is mixed with steam, to carry out a methane reforming reaction. The products are hydrogen, carbon dioxide, carbon monoxide, and unconverted methane [160]. In addition, the water gas shift process is needed to boost hydrogen levels. Then, a PSA is installed to purify hydrogen from carbon dioxide, carbon monoxide, and unconverted methane. Afterward, the high-purity hydrogen and non-aromatic hydrocarbons are first heated to 200–450 °C, before flowing into the hydrogenation reactor. Gasoline (C_6–C_7), diesel (C_8–C_{16}), and wax (>C_{16}) are produced in a hydrogenator using NiMo catalyst and can be applied as a fuel and chemicals [161,162].

Table 4 shows the product distribution of catalytic pyrolysis of LDPE and HDPE waste. In the catalytic pyrolysis of PE waste, a critical issue is catalyst deactivation, which occurs when many tiny molecules enter the pores of the catalyst, and macromolecular hydrocarbons that plug the pores are formed. Consequently, the lifetime of the catalyst is short [163]. A hydro-liquefaction process combining cracking and hydrogenation is proposed as an alternative technique to convert PE waste directly to liquid fuels or aromatic compounds. In a hydro-liquefaction reactor, Ding et al. tested three catalysts (Ni/HSiAl, NiMo/HSiAl, and KC-2600) to convert HDPE waste to liquid fuels. The results showed that Ni/HSiAl produced high yields of light hydrocarbons ($\leq C_{13}$) and the properties of the product were better than commercial gasoline, owing to a lower content of aromatics and high isoparaffin components [164].

Table 4. Product distribution of HDPE and LDPE after catalytic pyrolysis.

Feedstock	Catalyst	Condition Operation			Yield of Product, %			Major Product	Ref.
		Reactor	T, °C	t, min	Solid	Liquid	Gas		
HDPE	HUSY	Batch	550	NA	1.9	41	39.5	C_3–C_7 Hydrocarbons	[165]
LDPE	Sulfated zirconium hydroxide	Batch	500	70	2	82	16	C_{10}–C_{24} hydrocarbons	[95]
HDPE	HZSM-5	Batch	550	NA	0.7	17.3	72.6	C_3–C_6 Hydrocarbons	[165]
LDPE	HUSY	Batch	550	NA	1.9	61.6	34.5	C_4–C_9 hydrocarbons	[165]
HDPE	Conventional Beta zeolite	Batch	380	120	45.7	45	9.3	C_1–C_4; C_5–C_{12}; >C_{13} hydrocarbon	[166]
HDPE	Hierarchical Beta (CTAB)	Batch	380	120	32.7	50.3	17	C_1–C_4; C_5–C_{12}; >C_{13} hydrocarbon	[166]
HDPE	Hierarchical Beta (PHAPTMS)	Batch	380	120	3	81.9	15.1	C_1–C_4; C_5–C_{12}; >C_{13} hydrocarbon	[166]
LDPE	HZSM-5	Batch	550	NA	0.5	18.3	70.7	C_3–C_7 hydrocarbons	[165]
LDPE	Bentonite	Fixed bed	700	NA	NA	86.6	NA	C_5–C_9; C_{10}–C_{13}; >C_{13}	[88]
HDPE	Bentonite	Fixed bed	700	NA	NA	88.7	NA	C_5–C_9; C_{10}–C_{13}; >C_{13}	[88]
HDPE	Sulfated zirconium hydroxide	Batch	500	70	<1	79.5	20.1	C_{10}–C_{24} hydrocarbons	[95]
HDPE	FCC	Semi-batch	420	60	4.2	89.1	6.7	C_4–C_9 Hydrocarbons	[167]
HDPE	MFI Zeolite—Syn	Flask	380	60	-	51	49	C_5–C_7 hydrocarbons	[168]
HDPE	Silica/NaOH	Packed bed	500	70	-	82	18	C_{10}–C_{28} hydrocarbons	[169]

Furthermore, Pan et al. investigated aromatic production from HDPE waste via the hydro-liquefaction method, using HZSM-5 with Ni to stabilize the aromatic product. Xylene was the dominant aromatic product, with a maximum yield of 28.9% at 400 °C. The aromatic selectivity reached 64.8%, with reaction time up to 4 h and a loading of Ni up to 15 wt% [158]. The patterning process of hydro-liquefaction is typical of direct coal liquefaction. Therefore, constructing a hydro-liquefaction plant can implement a straightforward coal liquefaction process.

2.5. Recycling of Polyurethanes Waste

Polyurethanes (PUs) are essential materials, due to the thermoset and thermoplastic that can modify their chemical, thermal, and mechanical properties by reacting with polyisocyanates and polyols. The major polymers with urethane groups (–HN–COO–) are classified as PUs, regardless of the rest of the molecule [170]. In addition, the polyether polyols based on polyethylene oxide, PP, aliphatic polyester polyols, tetrahydrofuran, polycarbonate polyols, aromatic polyester polyols, polybutadiene polyols, and acrylic polyols are the most often used polyols in the manufacturing of PUs [171]. The thermochemical recycling of PUs includes alcoholysis, glycolysis, ammonolysis, and hydrolysis [172]. Figure 7 shows the alternative depolymerization routes for polyurethane waste.

Furthermore, numerous studies have considered recycling PUs by hydrolysis. The recycling products of PUs are a high-quality yield of polyol, isomeric toluene diamines, and CO_2. This was achieved by dry atmospheric pressure steam under a temperature range of 190–230 °C [173]. Nevertheless, urethane linkages are quite stable, and protective groups such as the benzoxycarbony group are often utilized to cover their amino functions. As a result, PU hydrolysis must be performed using a strong acid, base, and quaternary compounds that can be added to generate the active hydrogen that contains polyethers and polyamines [174]. The yield of toluenediamine vs. time shows the presence of a parallel first-order reaction equation, in which the urethane chain reacted up to 50 times quicker than urea. Urethane bonds were broken by direct hydrolysis, whereas urea bonds were broken via thermal fragmentation, to parent isocyanate and amine [175]. The recycling of PU waste with 91% of I-PU and 98% of H-PU was hydrolyzed successfully by Motokucho et al. The operating conditions of this process, the CO_2 pressure and temperature of reaction, were up to 8.0 MPa and 190 °C, respectively. The water-soluble components were evaporated, to isolate the final products with a high yield [176]. The major drawback of the hydrolysis process is that it consumes a lot of energy to heat the batch and provide high pressure in the reactor, resulting in an uneconomical process. As a result, hydrolysis has yet to be commercialized [177,178].

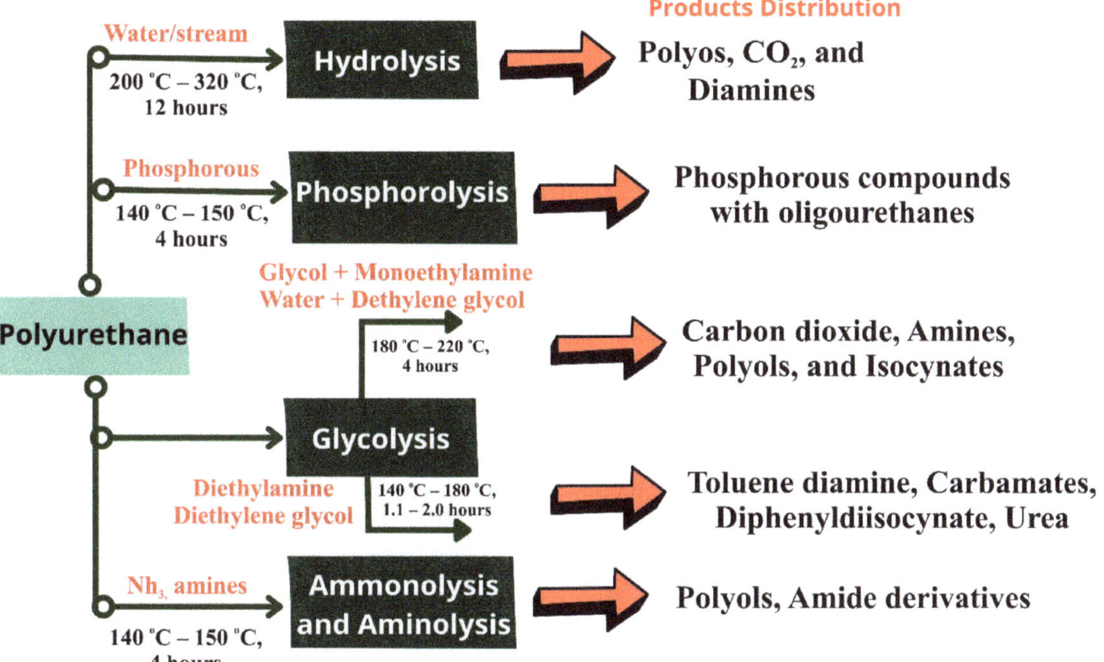

Figure 7. Alternative depolymerization routes for polyurethane waste [173].

Currently, glycolysis seems the most popular approach for chemical recycling of flexible and stiff polyurethane. Glycolysis is conducted by a transesterification reaction, where the hydroxyl group from glycol substitutes the ester group, by containing a carbonyl carbon of the urethane bond [179]. The common glycols applied in glycolysis are polyethylene glycol, diethylene glycol, and ethylene glycol, which are at high temperatures. In addition, the reactant and solvent should be glycol, and it is necessary to maintain the optimal ratio of PUs to glycol. Due to the long procedure time, a mixture of layers can be obtained. In addition, the recovered polyol is usually found in the top layer. It can also be used to produce new materials for PUs [180,181].

Furthermore, organometals, hydroxides, and acetates are various catalysts applied in the glycolysis of PU waste. The glycolysis can be performed with a co-reagent such as diethanol amine or amine. A reaction temperature below 180 °C results in a lack of catalytic activity, whereas a temperature above 220 °C results in unwanted amine side reactions. The major disadvantage of the chemically utilized reagents is implementing amine as a co-reactant, which causes issues with the recovered materials and is final-processed. Amines function as catalysts and accelerate the reaction between polyol and polyisocyanate by enhancing the electrophilicity of the isocyanate functional group. They also facilitate the polyols in generating PUs under uncontrolled and short-term conditions [175]. Ethylene glycol and diethylene glycol as reagents provided the best quality regarding the recovery of polyol, viscosity, and reaction time. Metal salts were used as catalysts, and lithium and $Zn (Ac)_2$ showed an excellent catalytic activity, by obtaining highly pure polyol and with a short reaction time. For instance, the alkaline metal hydroxides, KOH, appeared to dissolve stiff PUs more successfully than the other catalysts [181].

An alcoholysis reaction can be conducted by combining a sequence of metal hydroxide and alcohols, such as potassium hydroxide and sodium hydroxides, under a high pressure and temperature. The reaction of alcoholysis is identical to hydrolysis reactions. Polyols and urethane compounds are produced in this process. Furthermore, to convert the waste of PUs foam into liquid, solid, and gas products, the alcoholysis process requires a high temperature and oxygen-free environment [182]. Alcoholysis with 1,2-propanediol was used to recycle polyurethane foam waste. The compounds polyol and amine were discovered [183]. Gu et al. studied the alcoholysis agent with propylene glycol and ethylene glycol, while a catalyst was applied, such as $Zn_3[Co(CN)_6]_2$ (DMC) and KOH for PU rigid foam waste. A DMC catalyst was more effective than a KOH catalyst, according to the results of the experiments. The alcoholysis product using DMC had a higher hydroxyl content and lower viscosity, which is ideal for regenerating PU rigid foam [184].

For PU materials, gasification is a fascinating thermo-chemical feedstock recycling technique [185]. The gasification of plastic waste can be produce hydrogen compounds, generating a lot of attention for hydrogen as a future energy resource. Plastic gasification produces a variety of hydrocarbons. Nevertheless, a two-stage pyrolysis gasification method has been demonstrated to produce high-yield hydrogen from plastics. Furthermore, the pyrolysis of plastic waste is followed by steam gasification of the product in the presence of a catalyst, to produce hydrogen in a two-stage pyrolysis gasification reaction system [186]. Commercial gasification facilities have been operating for a range of feedstocks worldwide. SVZ GmbH's pilot plant was utilized to gasify waste plastics, pelletized shredder residue, multi-solid waste, and polyurethane foam using BG-Lurgi slagging-bed gasification under a high-temperature operation [185]. Guo et al. studied the gasification and pyrolysis of waste rigid polyurethane foam (WRPUF) using a fixed-bed reactor. The yield of gaseous products from WRPUF gasification was significantly greater than that from pyrolysis. The gasification yielded more volatile nitrogen than pyrolysis. The final product of the catalytic pyrolysis and gasification of WRPUF was dramatically influenced by metallic and metal compounds as catalysts [187].

2.6. Recycling of Polyethylene Terephthalate

The chemical recycling of PET was reviewed, such as pyrolysis, hydrolysis, methanolysis, glycolysis, ionic liquid, phase-transfer catalysis, and combinations of glycolysis and hydrolysis, glycolysis and methanolysis, and methanolysis and hydrolysis in our previous study [20]. Furthermore, reaction kinetics and conditions were investigated theoretically and experimentally. The recycling of PET is used to solve environmental problems and find another source of raw materials for petrochemical products and energy [20]. The hydrocracking of waste plastic was pyrolyzed into high-quality liquid fuel using various catalysts, e.g., zeolite [188]. On the other hand, PET can be reprocessed by mechanical recycling to develop wooden construction bricks for building purposes, with a composition of 75 wt% wood fiber and 25 wt% plastic waste, with a total hardness up to 21.270 HRR [189]. Therefore, the mechanical recycling of plastics can be applied in large-scale industries, to solve environmental issues [190].

3. Down Stream Problem

Decolorization technology is a critical problem for high-quality chemical recycling and recovery of plastic wastes. Only 1% of textile wastes, mainly whites, are recycled, so the final color quality of regenerated fibers is uncontrollable. Color removal is required for large-scale circulation of non-thermoplastic fibers. Technologies for color removal from plastic wastes include dye destruction or extraction for the pre-recycling process. Mu and Yang studied the minimization of fiber density using various solvents and temperatures, completely removing dispersed dyes, acid dyes, and direct dyes from PET, nylon, and cotton fibers [191]. The plastic waste was HDPE with blue and orange colorants (pigment and/or dye). Ferreira et al. reported that biosolvents derived from renewable sources were shown to be efficient in removing colorants from HDPE packaging waste by solvent extraction (dissolution-precipitation) [192].

Plastic waste must be subjected to a series of steps (washing, grinding, density separation, sensors-driven exclusion) to eliminate further contaminants (textiles, food, glass, other polymers). The sorting step is crucial to reduce contaminants and ensure that the contaminant threshold is not reached [193]. Increased recycling rates are a proposed solution to the current health and environmental crisis that is caused by the massive overproduction of plastics. However, almost all plastics contain toxic chemicals that are not removed during recycling, but which are carried over to the new products, and the recycling process can even generate new toxic substances such as dioxins. The increased recycling is intended to contribute to a so-called circular economy, but plastics containing toxic chemicals should not be recycled. Instead, they should be considered non-circular materials. Brosché et al. reported on the increasing amount of information about toxic chemicals transferred from plastic waste into recycled plastic pellets globally [194].

The life cycle assessment (LCA) methodology and applications have been developed, and LCA continues to be an important tool for understanding the environmental impacts of materials and processes. LCA of chemical recycling processes is a growing area. A critical analysis of nine chemical recycling LCA papers found that there are two approaches to modeling: a comparison of chemical recycling methods to other plastic waste management techniques, for example, mechanical recycling or the modeling of chemical recycling methods in combination with other plastic waste management methods, to treat mixed plastic waste [15]. Marson et al. presented a life cycle assessment of PUs foams with different recycled polyol contents [195].

4. Conclusions

The increase of plastic usage as packaging results in more waste in the environment, resulting in significant waste pollution issues. Among plastic waste management methods, plastic recycling offers a promising approach to reducing plastic waste, while maintaining sustainability. There are various plastic recycling technologies, each of which offers various advantages and disadvantages to the user and depending on the type of plastic waste. Chemical and mechanical recycling is the typical approach to recycling different plastic materials, such as PP, PS, PVC, HDPE, LDPE, and PUs. Mechanical recycling is a convenient technique to preserve the intrinsic value of plastic, while avoiding the waste of nonrenewable resources. Chemical recycling relies on the degradation of the polymer chains to obtain a low degree of pollution. The chemical and mechanical recycling approaches facilitate plastic waste recycling with a simple process, a low cost, environmentally friendly process, and potential profitability. However, plastic waste separation before the recycling process becomes a challenge, due to the different physical and chemical characteristics of plastic waste. Since each plastic type has various properties, such as melting temperature, density, and hardness, mixed polymers do not allow retaining their original properties and practical usefulness. Therefore, to increase the value and recycling rate of plastic waste, a reliable and effective separation method for plastic waste separation is very important. Integrating recycling and plastic waste separation technologies would be an effective strategy to reduce the accumulation of environmental pollutants caused by plastic waste, particularly for industrial applications.

The explosion of plastic usage in many industries has led to environmental damage issues. The increased focus on the ecological consequences of human activities and the rising need for energy and resources has resulted in a new perspective on plastic waste streams. All stakeholders must achieve good sustainable waste management practices, to maintain sustainability. Better knowledge of plastic waste recycling will lead the policy-makers to make proper rules to overcome the environmental problems caused by plastic waste. This review presents the plastic waste recycling approaches for the properties of each material, as waste management from an environmental sustainability perspective.

Author Contributions: Conceptualization, H.-S.W. and D.D.; validation, H.-S.W. and D.D.; investigation, D.D. and F.Y.; writing—original draft preparation, D.D., D.R.S., D.S.S.M., F.Y., A.S., Y.M.S., W.A., M.U. and H.-S.W.; writing—review and editing, D.D. and H.-S.W.; visualization, D.D.; supervision, H.-S.W. All authors have read and agreed to the published version of the manuscript.

Funding: This research received no external funding.

Institutional Review Board Statement: No applicable.

Informed Consent Statement: No applicable.

Data Availability Statement: No applicable.

Acknowledgments: We give thanks for the information from the Far Eastern New Century Corporation, Taiwan.

Conflicts of Interest: The authors declare no conflict of interest.

Abbreviations

ABS	Acrylonitrile butadiene styrene
EG	Ethylene glycol
FRP	Fiber-reinforced polymers
HDPE	High-density polyethylene
HRR	Rockwell R-scale hardness
LDA	Linear discrimination analysis
LDPE	Low-density polyethylene
LIBS	Laser-induced breakdown spectroscopy
PA	Polycaprolactam
PAHs	Polycyclic aromatic hydrocarbons
PBX	Plastic bonded explosive
PC	Polycarbonate
PCA-SVM	Principal component analysis
PE	Polyethylene
PLS	Partial least squares regression
PLS-DA	Partial least squares discrimination analysis
PET	Polyethylene terephthalate
PMMA	Poly(methyl methacrylate)
PP	Polypropylene
PVC	Polyvinyl chloride
PS	Polystyrene
PU	polyurethane
SERS	Surface-enhanced Raman spectroscopy

References

1. Evode, N.; Qamar, S.A.; Bilal, M.; Barceló, D.; Iqbal, H.M. Plastic waste and its management strategies for environmental sustainability. *Case Stud. Chem. Environ. Eng.* **2021**, *4*, 100142. [CrossRef]
2. Lai, Y.-Y.; Lee, Y.-M. Management strategy of plastic wastes in Taiwan. *Sustain. Environ. Res.* **2022**, *32*, 11. [CrossRef]
3. Joseph, B.; James, J.; Kalarikkal, N.; Thomas, S. Recycling of medical plastics. *Adv. Ind. Eng. Polym. Res.* **2021**, *4*, 199–208. [CrossRef]
4. Oberoi, G.; Garg, A. Single-Use Plastics: A Roadmap for Sustainability? *Supremo Amic.* **2021**, *24*, 585.
5. Prajapati, R.; Kohli, K.; Maity, S.K.; Sharma, B.K. Recovery and Recycling of Polymeric and Plastic Materials. In *Recent Developments in Plastic Recycling*; Springer: Singapore, 2021; pp. 15–41.
6. Abbate, E.; Rovelli, D.; Andreotti, M.; Brondi, C.; Ballarino, A. Plastic packaging substitution in industry: Variability of LCA due to manufacturing countries. *Procedia CIRP* **2022**, *105*, 392–397. [CrossRef]
7. Pan, D.; Su, F.; Liu, C.; Guo, Z. Research progress for plastic waste management and manufacture of value-added products. *Adv. Compos. Hybrid Mater.* **2020**, *3*, 443–461. [CrossRef]
8. Oehlmann, J.; Schulte-Oehlmann, U.; Kloas, W.; Jagnytsch, O.; Lutz, I.; Kusk, K.O.; Wollenberger, L.; Santos, E.M.; Paull, G.C.; Van Look, K.J. A critical analysis of the biological impacts of plasticizers on wildlife. *Philos. Trans. R. Soc. B Biol. Sci.* **2009**, *364*, 2047–2062. [CrossRef]

9. Astrup, T.; Møller, J.; Fruergaard, T. Incineration and co-combustion of waste: Accounting of greenhouse gases and global warming contributions. *Waste Manag. Res.* **2009**, *27*, 789–799. [CrossRef] [PubMed]

10. Wu, G.; Li, J.; Xu, Z. Triboelectrostatic separation for granular plastic waste recycling: A review. *Waste Manag.* **2013**, *33*, 585–597. [CrossRef]

11. *Standard D5033*; Standard Guide to Development of ASTM Standards Relating to Recycling and Use of Recycled Plastics. American Society for Testing and Materials (ASTM) International: West Conshohocken, PA, USA, 2000. [CrossRef]

12. Al-Salem, S. Energy production from plastic solid waste (PSW). In *Plastics to Energy*; Elsevier: Amsterdam, The Netherlands, 2019; pp. 45–64.

13. Pan, D.; Su, F.; Liu, H.; Liu, C.; Umar, A.; CastaÃ±eda, L.C.; Algadi, H.; Wang, C.; Guo, Z. Research Progress on Catalytic Pyrolysis and Reuse of Waste Plastics and Petroleum Sludge. *ES Mater. Manuf.* **2021**, *11*, 3–15. [CrossRef]

14. Awaja, F.; Pavel, D. Recycling of PET. *Eur. Polym. J.* **2005**, *41*, 1453–1477. [CrossRef]

15. Davidson, M.G.; Furlong, R.A.; McManus, M.C. Developments in the life cycle assessment of chemical recycling of plastic waste—A review. *J. Clean. Prod.* **2021**, *293*, 126163. [CrossRef]

16. Al-Sabagh, A.; Yehia, F.; Eshaq, G.; Rabie, A.; ElMetwally, A. Greener routes for recycling of polyethylene terephthalate. *Egypt. J. Pet.* **2016**, *25*, 53–64. [CrossRef]

17. Lerici, L.C.; Renzini, M.S.; Pierella, L.B. Chemical catalyzed recycling of polymers: Catalytic conversion of PE, PP and PS into fuels and chemicals over HY. *Procedia Mater. Sci.* **2015**, *8*, 297–303. [CrossRef]

18. Mazhandu, Z.S.; Muzenda, E.; Mamvura, T.A.; Belaid, M. Integrated and consolidated review of plastic waste management and bio-based biodegradable plastics: Challenges and opportunities. *Sustainability* **2020**, *12*, 8360. [CrossRef]

19. Ragaert, K.; Delva, L.; Van Geem, K. Mechanical and chemical recycling of solid plastic waste. *Waste Manag.* **2017**, *69*, 24–58. [CrossRef] [PubMed]

20. Damayanti; Wu, H.S. Strategic possibility routes of recycled PET. *Polymers* **2021**, *13*, 1475. [CrossRef]

21. Geyer, R.; Jambeck, J.R.; Law, K.L. Production, use, and fate of all plastics ever made. *Sci. Adv.* **2017**, *3*, e1700782. [CrossRef] [PubMed]

22. de Camargo, R.V.; Saron, C. Mechanical–chemical recycling of low-density polyethylene waste with polypropylene. *J. Polym. Environ.* **2020**, *28*, 794–802. [CrossRef]

23. Serranti, S.; Bonifazi, G. Techniques for separation of plastic wastes. In *Use of Recycled Plastics in Eco-Efficient Concrete*; Elsevier: Amsterdam, The Netherlands, 2019; pp. 9–37.

24. Quienne, B.; Cuminet, F.; Pinaud, J.; Semsarilar, M.; Cot, D.; Ladmiral, V.; Caillol, S. Upcycling Biobased Polyurethane Foams into Thermosets: Toward the Closing of the Loop. *ACS Sustain. Chem. Eng.* **2022**, *10*, 7041–7049. [CrossRef]

25. Garcia, J.M.; Robertson, M.L. The future of plastics recycling. *Science* **2017**, *358*, 870–872. [CrossRef]

26. Fulgencio-Medrano, L.; Garcia-Fernandez, S.; Asueta, A.; Lopez-Urionabarrenechea, A.; Perez-Martinez, B.B.; Arandes, J.M. Oil Production by Pyrolysis of Real Plastic Waste. *Polymers* **2022**, *14*, 553. [CrossRef] [PubMed]

27. Vollmer, I.; Jenks, M.J.; Roelands, M.C.; White, R.J.; van Harmelen, T.; de Wild, P.; van Der Laan, G.P.; Meirer, F.; Keurentjes, J.T.; Weckhuysen, B.M. Beyond mechanical recycling: Giving new life to plastic waste. *Angew. Chem. Int. Ed.* **2020**, *59*, 15402–15423. [CrossRef]

28. Schwarz, A.; Ligthart, T.; Bizarro, D.G.; De Wild, P.; Vreugdenhil, B.; van Harmelen, T. Plastic recycling in a circular economy; determining environmental performance through an LCA matrix model approach. *Waste Manag.* **2021**, *121*, 331–342. [CrossRef]

29. Siddiqui, M.N. Conversion of hazardous plastic wastes into useful chemical products. *J. Hazard. Mater.* **2009**, *167*, 728–735. [CrossRef]

30. Al-Salem, S.; Antelava, A.; Constantinou, A.; Manos, G.; Dutta, A. A review on thermal and catalytic pyrolysis of plastic solid waste (PSW). *J. Environ. Manag.* **2017**, *197*, 177–198. [CrossRef]

31. Khalid, M.Y.; Arif, Z.U.; Ahmed, W.; Arshad, H. Recent trends in recycling and reusing techniques of different plastic polymers and their composite materials. *Sustain. Mater. Technol.* **2021**, *31*, e00382. [CrossRef]

32. Das, S.K.; Eshkalak, S.K.; Chinnappan, A.; Ghosh, R.; Jayathilaka, W.; Baskar, C.; Ramakrishna, S. Plastic recycling of polyethylene terephthalate (PET) and polyhydroxybutyrate (PHB)—A comprehensive review. *Mater. Circ. Econ.* **2021**, *3*, 9. [CrossRef]

33. Shamsuyeva, M.; Endres, H.-J. Plastics in the context of the circular economy and sustainable plastics recycling: Comprehensive review on research development, standardization and market. *Compos. Part C Open Access* **2021**, *6*, 100168. [CrossRef]

34. Lee, A.; Liew, M.S. Tertiary recycling of plastics waste: An analysis of feedstock, chemical and biological degradation methods. *J. Mater. Cycles Waste Manag.* **2021**, *23*, 32–43. [CrossRef]

35. Francis, R. *Recycling of Polymers: Methods, Characterization and Applications*; John Wiley & Sons: Hoboken, NJ, USA, 2016.

36. Bhadra, J.; Al-Thani, N.; Abdulkareem, A. Recycling of polymer-polymer composites. In *Micro and Nano Fibrillar Composites (MFCs and NFCs) from Polymer Blends*; Elsevier: Amsterdam, The Netherlands, 2017; pp. 263–277.

37. Yang, L.; Gao, J.; Liu, Y.; Zhuang, G.; Peng, X.; Wu, W.-M.; Zhuang, X. Biodegradation of expanded polystyrene and low-density polyethylene foams in larvae of Tenebrio molitor Linnaeus (Coleoptera: Tenebrionidae): Broad versus limited extent depolymerization and microbe-dependence versus independence. *Chemosphere* **2021**, *262*, 127818. [CrossRef] [PubMed]

38. Araujo-Andrade, C.; Bugnicourt, E.; Philippet, L.; Rodriguez-Turienzo, L.; Nettleton, D.; Hoffmann, L.; Schlummer, M. Review on the photonic techniques suitable for automatic monitoring of the composition of multi-materials wastes in view of their posterior recycling. *Waste Manag. Res.* **2021**, *39*, 631–651. [CrossRef] [PubMed]

39. Fan, C.; Huang, Y.-Z.; Lin, J.-N.; Li, J. Microplastic constituent identification from admixtures by Fourier-transform infrared (FTIR) spectroscopy: The use of polyethylene terephthalate (PET), polyethylene (PE), polypropylene (PP), polyvinyl chloride (PVC) and nylon (NY) as the model constituents. *Environ. Technol. Innov.* **2021**, *23*, 101798. [CrossRef]
40. Lin, P.-Y.; Wu, I.-H.; Tsai, C.-Y.; Kirankumar, R.; Hsieh, S. Detecting the release of plastic particles in packaged drinking water under simulated light irradiation using surface-enhanced Raman spectroscopy. *Anal. Chim. Acta* **2022**, *1198*, 339516. [CrossRef]
41. Damayanti, D.; Wulandari, L.A.; Bagaskoro, A.; Rianjanu, A.; Wu, H.S. Possibility Routes for Textile Recycling Technology. *Polymers* **2021**, *13*, 3834. [CrossRef]
42. Neo, E.R.K.; Yeo, Z.; Low, J.S.C.; Goodship, V.; Debattista, K. A review on chemometric techniques with infrared, Raman and laser-induced breakdown spectroscopy for sorting plastic waste in the recycling industry. *Resour. Conserv. Recycl.* **2022**, *180*, 106217. [CrossRef]
43. Sato, H.; Shimoyama, M.; Kamiya, T.; Amari, T.; Šašic, S.; Ninomiya, T.; Siesler, H.W.; Ozaki, Y. Raman spectra of high-density, low-density, and linear low-density polyethylene pellets and prediction of their physical properties by multivariate data analysis. *J. Appl. Polym. Sci.* **2002**, *86*, 443–448. [CrossRef]
44. Winefordner, J.D.; Gornushkin, I.B.; Correll, T.; Gibb, E.; Smith, B.W.; Omenetto, N. Comparing several atomic spectrometric methods to the super stars: Special emphasis on laser induced breakdown spectrometry, LIBS, a future super star. *J. Anal. At. Spectrom.* **2004**, *19*, 1061–1083. [CrossRef]
45. Jull, H.; Bier, J.; Künnemeyer, R.; Schaare, P. Classification of recyclables using laser-induced breakdown spectroscopy for waste management. *Spectrosc. Lett.* **2018**, *51*, 257–265. [CrossRef]
46. Zeng, Q.; Sirven, J.-B.; Gabriel, J.-C.P.; Tay, C.Y.; Lee, J.-M. Laser induced breakdown spectroscopy for plastic analysis. *TrAC Trends Anal. Chem.* **2021**, *140*, 116280. [CrossRef]
47. Henriksen, M.L.; Karlsen, C.B.; Klarskov, P.; Hinge, M. Plastic classification via in-line hyperspectral camera analysis and unsupervised machine learning. *Vib. Spectrosc.* **2022**, *118*, 103329. [CrossRef]
48. Liu, K.; Tian, D.; Wang, H.; Yang, G. Rapid classification of plastics by laser-induced breakdown spectroscopy (LIBS) coupled with partial least squares discrimination analysis based on variable importance (VI-PLS-DA). *Anal. Methods* **2019**, *11*, 1174–1179. [CrossRef]
49. Serranti, S.; Palmieri, R.; Bonifazi, G.; Cózar, A. Characterization of microplastic litter from oceans by an innovative approach based on hyperspectral imaging. *Waste Manag.* **2018**, *76*, 117–125. [CrossRef]
50. Vidal, C.; Pasquini, C. A comprehensive and fast microplastics identification based on near-infrared hyperspectral imaging (HSI-NIR) and chemometrics. *Environ. Pollut.* **2021**, *285*, 117251. [CrossRef]
51. Wu, X.; Li, J.; Yao, L.; Xu, Z. Auto-sorting commonly recovered plastics from waste household appliances and electronics using near-infrared spectroscopy. *J. Clean. Prod.* **2020**, *246*, 118732. [CrossRef]
52. Serranti, S.; Luciani, V.; Bonifazi, G.; Hu, B.; Rem, P.C. An innovative recycling process to obtain pure polyethylene and polypropylene from household waste. *Waste Manag.* **2015**, *35*, 12–20. [CrossRef] [PubMed]
53. Yang, Y.; Zhang, X.; Yin, J.; Yu, X. Rapid and nondestructive on-site classification method for consumer-grade plastics based on portable NIR spectrometer and machine learning. *J. Spectrosc.* **2020**, *2020*, 6631234. [CrossRef]
54. Musu, W.; Tsuchida, A.; Kawazumi, H.; Oka, N. Application of PCA-SVM and ANN Techniques for Plastic Identification by Raman Spectroscopy. In Proceedings of the 2019 1st International Conference on Cybernetics and Intelligent System (ICORIS), Denpasar, Indonesia, 22–23 August 2019; pp. 114–118.
55. Rani, M.; Marchesi, C.; Federici, S.; Rovelli, G.; Alessandri, I.; Vassalini, I.; Ducoli, S.; Borgese, L.; Zacco, A.; Bilo, F. Miniaturized near-infrared (MicroNIR) spectrometer in plastic waste sorting. *Materials* **2019**, *12*, 2740. [CrossRef]
56. Bae, J.-S.; Oh, S.-K.; Pedrycz, W.; Fu, Z. Design of fuzzy radial basis function neural network classifier based on information data preprocessing for recycling black plastic wastes: Comparative studies of ATR FT-IR and Raman spectroscopy. *Appl. Intell.* **2019**, *49*, 929–949. [CrossRef]
57. Calvini, R.; Orlandi, G.; Foca, G.; Ulrici, A. Developmentof a classification algorithm for efficient handling of multiple classes in sorting systems based on hyperspectral imaging. *J. Spectr. Imaging* **2018**, *7*, a13. [CrossRef]
58. Zhu, S.; Chen, H.; Wang, M.; Guo, X.; Lei, Y.; Jin, G. Plastic solid waste identification system based on near infrared spectroscopy in combination with support vector machine. *Adv. Ind. Eng. Polym. Res.* **2019**, *2*, 77–81. [CrossRef]
59. Roh, S.-B.; Oh, S.-K.; Park, E.-K.; Choi, W.Z. Identification of black plastics realized with the aid of Raman spectroscopy and fuzzy radial basis function neural networks classifier. *J. Mater. Cycles Waste Manag.* **2017**, *19*, 1093–1105. [CrossRef]
60. Tsuchida, A.; Kawazumi, H.; Kazuyoshi, A.; Yasuo, T. Identification of shredded plastics in milliseconds using Raman spectroscopy for recycling. In Proceedings of the SENSORS, 2009 IEEE, Christchurch, New Zealand, 25–28 October 2009; pp. 1473–1476.
61. Harussani, M.; Sapuan, S.; Rashid, U.; Khalina, A.; Ilyas, R. Pyrolysis of polypropylene plastic waste into carbonaceous char: Priority of plastic waste management amidst COVID-19 pandemic. *Sci. Total Environ.* **2021**, *803*, 149911. [CrossRef] [PubMed]
62. Thiounn, T.; Smith, R.C. Advances and approaches for chemical recycling of plastic waste. *J. Polym. Sci.* **2020**, *58*, 1347–1364. [CrossRef]
63. Mark, L.O.; Cendejas, M.C.; Hermans, I. Cover Feature: The Use of Heterogeneous Catalysis in the Chemical Valorization of Plastic Waste (ChemSusChem 22/2020). *ChemSusChem* **2020**, *13*, 5773. [CrossRef]
64. Ellis, L.D.; Rorrer, N.A.; Sullivan, K.P.; Otto, M.; McGeehan, J.E.; Román-Leshkov, Y.; Wierckx, N.; Beckham, G.T. Chemical and biological catalysis for plastics recycling and upcycling. *Nat. Catal.* **2021**, *4*, 539–556. [CrossRef]

65. Chen, L.; Zhu, Y.; Meyer, L.C.; Hale, L.V.; Le, T.T.; Karkamkar, A.; Lercher, J.A.; Gutiérrez, O.Y.; Szanyi, J. Effect of reaction conditions on the hydrogenolysis of polypropylene and polyethylene into gas and liquid alkanes. *React. Chem. Eng.* **2022**, *7*, 844–854. [CrossRef]

66. Miao, Y.; von Jouanne, A.; Yokochi, A. Current Technologies in Depolymerization Process and the Road Ahead. *Polymers* **2021**, *13*, 449. [CrossRef] [PubMed]

67. Yan, G.; Jing, X.; Wen, H.; Xiang, S. Thermal Cracking of Virgin and Waste Plastics of PP and LDPE in a Semibatch Reactor under Atmospheric Pressure. *Energy Fuels* **2015**, *29*, 2289–2298. [CrossRef]

68. Kane, A.Q.; Esper, A.M.; Searles, K.; Ehm, C.; Veige, A.S. Probing β-alkyl elimination and selectivity in polyolefin hydrogenolysis through DFT. *Catal. Sci. Technol.* **2021**, *11*, 6155–6162. [CrossRef]

69. Yao, L.; King, J.; Wu, D.; Chuang, S.S.; Peng, Z. Non-thermal plasma-assisted hydrogenolysis of polyethylene to light hydrocarbons. *Catal. Commun.* **2021**, *150*, 106274. [CrossRef]

70. Kots, P.A.; Liu, S.; Vance, B.C.; Wang, C.; Sheehan, J.D.; Vlachos, D.G. Polypropylene Plastic Waste Conversion to Lubricants over Ru/TiO$_2$ Catalysts. *ACS Catal.* **2021**, *11*, 8104–8115. [CrossRef]

71. Rahimi, A.; García, J.M. Chemical recycling of waste plastics for new materials production. *Nat. Rev. Chem.* **2017**, *1*, 46. [CrossRef]

72. Rorrer, J.E.; Troyano-Valls, C.; Beckham, G.T.; Román-Leshkov, Y. Hydrogenolysis of Polypropylene and Mixed Polyolefin Plastic Waste over Ru/C to Produce Liquid Alkanes. *ACS Sustain. Chem. Eng.* **2021**, *9*, 11661–11666. [CrossRef]

73. Chen, W.-T.; Jin, K.L.; Wang, N.H. Use of supercritical water for the liquefaction of polypropylene into oil. *ACS Sustain. Chem. Eng.* **2019**, *7*, 3749–3758. [CrossRef]

74. Akiya, N.; Savage, P.E. Roles of water for chemical reactions in high-temperature water. *Chem. Rev.* **2002**, *102*, 2725–2750. [CrossRef] [PubMed]

75. Verma, A.; Budiyal, L.; Sanjay, M.; Siengchin, S. Processing and characterization analysis of pyrolyzed oil rubber (from waste tires)-epoxy polymer blend composite for lightweight structures and coatings applications. *Polym. Eng. Sci.* **2019**, *59*, 2041–2051. [CrossRef]

76. Damayanti, D.; Wu, H.S. Pyrolysis kinetic of alkaline and dealkaline lignin using catalyst. *J. Polym. Res.* **2018**, *25*, 7. [CrossRef]

77. Damayanti, D.; Supriyadi, D.; Amelia, D.; Saputri, D.R.; Devi, Y.L.L.; Auriyani, W.A.; Wu, H.S. Conversion of Lignocellulose for bioethanol production, applied in bio-polyethylene terephthalate. *Polymers* **2021**, *13*, 2886. [CrossRef]

78. Kasar, P.; Sharma, D.; Ahmaruzzaman, M. Thermal and catalytic decomposition of waste plastics and its co-processing with petroleum residue through pyrolysis process. *J. Clean. Prod.* **2020**, *265*, 121639. [CrossRef]

79. Kruse, T.M.; Wong, H.-W.; Broadbelt, L.J. Mechanistic modeling of polymer pyrolysis: Polypropylene. *Macromolecules* **2003**, *36*, 9594–9607. [CrossRef]

80. Singh, M.V. Conversions of Waste Tube-Tyres (WTT) and Waste Polypropylene (WPP) into Diesel Fuel through Catalytic Pyrolysis Using Base SrCO$_3$. *Eng. Sci.* **2021**, *13*, 87–97. [CrossRef]

81. Munir, D.; Irfan, M.F.; Usman, M.R. Hydrocracking of virgin and waste plastics: A detailed review. *Renew. Sustain. Energy Rev.* **2018**, *90*, 490–515. [CrossRef]

82. Nakaji, Y.; Tamura, M.; Miyaoka, S.; Kumagai, S.; Tanji, M.; Nakagawa, Y.; Yoshioka, T.; Tomishige, K. Low-temperature catalytic upgrading of waste polyolefinic plastics into liquid fuels and waxes. *Appl. Catal. B Environ.* **2021**, *285*, 119805. [CrossRef]

83. Jung, S.; Lee, S.; Dou, X.; Kwon, E.E. Valorization of disposable COVID-19 mask through the thermo-chemical process. *Chem. Eng. J.* **2021**, *405*, 126658. [CrossRef] [PubMed]

84. Dutta, N.; Gupta, A. An experimental study on conversion of high-density polyethylene and polypropylene to liquid fuel. *Clean Technol. Environ. Policy* **2021**, *23*, 2213–2220. [CrossRef]

85. Harmon, R.E.; SriBala, G.; Broadbelt, L.J.; Burnham, A.K. Insight into polyethylene and polypropylene pyrolysis: Global and mechanistic models. *Energy Fuels* **2021**, *35*, 6765–6775. [CrossRef]

86. Praveen Kumar, K.; Srinivas, S. Catalytic co-pyrolysis of biomass and plastics (polypropylene and polystyrene) using spent FCC catalyst. *Energy Fuels* **2019**, *34*, 460–473. [CrossRef]

87. Panda, A.K. Thermo-catalytic degradation of different plastics to drop in liquid fuel using calcium bentonite catalyst. *Int. J. Ind. Chem.* **2018**, *9*, 167–176. [CrossRef]

88. Budsaereechai, S.; Hunt, A.J.; Ngernyen, Y. Catalytic pyrolysis of plastic waste for the production of liquid fuels for engines. *RSC Adv.* **2019**, *9*, 5844–5857. [CrossRef]

89. Abbas-Abadi, M.S.; Haghighi, M.N.; Yeganeh, H.; McDonald, A.G. Evaluation of pyrolysis process parameters on polypropylene degradation products. *J. Anal. Appl. Pyrolysis* **2014**, *109*, 272–277. [CrossRef]

90. Zhao, Y.; Wang, W.; Jing, X.; Gong, X.; Wen, H.; Deng, Y. Catalytic cracking of polypropylene by using Fe-SBA-15 synthesized in an acid-free medium for production of light hydrocarbon oils. *J. Anal. Appl. Pyrolysis* **2020**, *146*, 104755. [CrossRef]

91. Aisien, E.T.; Otuya, I.C.; Aisien, F.A. Thermal and catalytic pyrolysis of waste polypropylene plastic using spent FCC catalyst. *Environ. Technol. Innov.* **2021**, *22*, 101455. [CrossRef]

92. Sonawane, Y.; Shindikar, M.; Khaladkar, M. High calorific value fuel from household plastic waste by catalytic pyrolysis. *Nat. Environ. Pollut. Technol.* **2017**, *16*, 879.

93. Lee, K.-H.; Noh, N.-S.; Shin, D.-H.; Seo, Y. Comparison of plastic types for catalytic degradation of waste plastics into liquid product with spent FCC catalyst. *Polym. Degrad. Stab.* **2002**, *78*, 539–544. [CrossRef]

94. Kassargy, C.; Awad, S.; Burnens, G.; Kahine, K.; Tazerout, M. Experimental study of catalytic pyrolysis of polyethylene and polypropylene over USY zeolite and separation to gasoline and diesel-like fuels. *J. Anal. Appl. Pyrolysis* **2017**, *127*, 31–37. [CrossRef]
95. Panda, A.K.; Alotaibi, A.; Kozhevnikov, I.V.; Shiju, N.R. Pyrolysis of plastics to liquid fuel using sulphated zirconium hydroxide catalyst. *Waste Biomass Valoriz.* **2020**, *11*, 6337–6345. [CrossRef]
96. Hakeem, I.G.; Aberuagba, F.; Musa, U. Catalytic pyrolysis of waste polypropylene using Ahoko kaolin from Nigeria. *Appl. Petrochem. Res.* **2018**, *8*, 203–210. [CrossRef]
97. Haribal, V.P.; He, F.; Mishra, A.; Li, F. Iron-doped BaMnO3 for hybrid water splitting and syngas generation. *ChemSusChem* **2017**, *10*, 3402–3408. [CrossRef]
98. Huang, J.; Veksha, A.; Chan, W.P.; Giannis, A.; Lisak, G. Chemical recycling of plastic waste for sustainable material management: A prospective review on catalysts and processes. *Renew. Sustain. Energy Rev.* **2022**, *154*, 111866. [CrossRef]
99. Xiao, R.; Jin, B.; Zhou, H.; Zhong, Z.; Zhang, M. Air gasification of polypropylene plastic waste in fluidized bed gasifier. *Energy Convers. Manag.* **2007**, *48*, 778–786. [CrossRef]
100. Hadi, J.A.; Najmuldeen, F.G.; Ahmed, I. Quality restoration of waste polyolefin plastic material through the dissolution-reprecipitation technique. *Chem. Ind. Chem. Eng. Q.* **2014**, *20*, 163–170. [CrossRef]
101. Zhao, Y.-B.; Lv, X.-D.; Ni, H.-G. Solvent-based separation and recycling of waste plastics: A review. *Chemosphere* **2018**, *209*, 707–720. [CrossRef] [PubMed]
102. Chaudhary, A.K.; Vijayakumar, R. Studies on biological degradation of polystyrene by pure fungal cultures. *Environ. Dev. Sustain.* **2020**, *22*, 4495–4508. [CrossRef]
103. Chaudhary, A.; Dave, M.; Upadhyay, D.S. Value-added products from waste plastics using dissolution technique. *Mater. Today: Proc.* **2022**, *57*, 1730–1737. [CrossRef]
104. Terashima, T. Polystyrene (PSt). In *Encyclopedia of Polymeric Nanomaterials*; Kobayashi, S., Müllen, K., Eds.; Springer: Berlin/Heidelberg, Germany, 2015; pp. 2077–2091.
105. Gebre, S.H.; Sendeku, M.G.; Bahri, M. Recent Trends in the Pyrolysis of Non-Degradable Waste Plastics. *ChemistryOpen* **2021**, *10*, 1202–1226. [CrossRef]
106. Siyal, A.N.; Memon, S.Q.; Khuhawar, M. Recycling of styrofoam waste: Synthesis, characterization and application of novel phenyl thiosemicarbazone surface. *Pol. J. Chem. Technol.* **2012**, *14*, 11–18. [CrossRef]
107. Cella, R.F.; Mumbach, G.D.; Andrade, K.L.; Oliveira, P.; Marangoni, C.; Bolzan, A.; Bernard, S.; Machado, R.A.F. Polystyrene recycling processes by dissolution in ethyl acetate. *J. Appl. Polym. Sci.* **2018**, *135*, 46208. [CrossRef]
108. García, M.T.; Gracia, I.; Duque, G.; de Lucas, A.; Rodríguez, J.F. Study of the solubility and stability of polystyrene wastes in a dissolution recycling process. *Waste Manag.* **2009**, *29*, 1814–1818. [CrossRef] [PubMed]
109. Achilias, D.; Antonakou, E.; Koutsokosta, E.; Lappas, A. Chemical recycling of polymers from waste electric and electronic equipment. *J. Appl. Polym. Sci.* **2009**, *114*, 212–221. [CrossRef]
110. Gil-Jasso, N.D.; Segura-González, M.A.; Soriano-Giles, G.; Neri-Hipolito, J.; López, N.; Mas-Hernández, E.; Barrera-Díaz, C.E.; Varela-Guerrero, V.; Ballesteros-Rivas, M.F. Dissolution and recovery of waste expanded polystyrene using alternative essential oils. *Fuel* **2019**, *239*, 611–616. [CrossRef]
111. Mumbach, G.D.; Alves, J.L.F.; Da Silva, J.C.G.; De Sena, R.F.; Marangoni, C.; Machado, R.A.F.; Bolzan, A. Thermal investigation of plastic solid waste pyrolysis via the deconvolution technique using the asymmetric double sigmoidal function: Determination of the kinetic triplet, thermodynamic parameters, thermal lifetime and pyrolytic oil composition for clean energy recovery. *Energy Convers. Manag.* **2019**, *200*, 112031. [CrossRef]
112. Maafa, I.M. Pyrolysis of polystyrene waste: A review. *Polymers* **2021**, *13*, 225. [CrossRef] [PubMed]
113. Huang, J.; Cheng, X.; Meng, H.; Pan, G.; Wang, S.; Wang, D. Density functional theory study on the catalytic degradation mechanism of polystyrene. *AIP Adv.* **2020**, *10*, 085004. [CrossRef]
114. Cho, K.H.; Cho, D.R.; Kim, K.H.; Park, D.W. Catalytic degradation of polystyrene using albite and montmorillonite. *Korean J. Chem. Eng.* **2007**, *24*, 223–225. [CrossRef]
115. Anwar, J.; Munawar, M.A.; Waheed-uz-Zaman; Dar, A.; Tahira, U. Catalytic depolymerisation of polystyrene. *Prog. Rubber Plast. Recycl. Technol.* **2008**, *24*, 47–51. [CrossRef]
116. Zayoud, A.; Thi, H.D.; Kusenberg, M.; Eschenbacher, A.; Kresovic, U.; Alderweireldt, N.; Djokic, M.; Van Geem, K.M. Pyrolysis of end-of-life polystyrene in a pilot-scale reactor: Maximizing styrene production. *Waste Manag.* **2022**, *139*, 85–95. [CrossRef]
117. Tajjamal, A.; Ul-Hamid, A.; Faisal, A.; Zaidi, S.A.H.; Sherin, L.; Mir, A.; Mustafa, M.; Ahmad, N.; Hussain, M.; Park, Y.-K. Catalytic cracking of polystyrene pyrolysis oil: Effect of Nb2O5 and NiO/Nb2O5 catalyst on the liquid product composition. *Waste Manag.* **2022**, *141*, 240–250. [CrossRef]
118. Fan, S.; Zhang, Y.; Liu, T.; Fu, W.; Li, B. Microwave-assisted pyrolysis of polystyrene for aviation oil production. *J. Anal. Appl. Pyrolysis* **2022**, *162*, 105425. [CrossRef]
119. Putra, P.H.M.; Rozali, S.; Patah, M.F.A.; Idris, A. A review of microwave pyrolysis as a sustainable plastic waste management technique. *J. Environ. Manag.* **2022**, *303*, 114240. [CrossRef]
120. Hussain, Z.; Khan, K.M.; Hussain, K. Microwave–metal interaction pyrolysis of polystyrene. *J. Anal. Appl. Pyrolysis* **2010**, *89*, 39–43. [CrossRef]

121. Rex, P.; Masilamani, I.P.; Miranda, L.R. Microwave pyrolysis of polystyrene and polypropylene mixtures using different activated carbon from biomass. *J. Energy Inst.* **2020**, *93*, 1819–1832. [CrossRef]
122. Miliute-Plepiene, J.; Fråne, A.; Almasi, A.M. Overview of polyvinyl chloride (PVC) waste management practices in the Nordic countries. *Clean. Eng. Technol.* **2021**, *4*, 100246. [CrossRef]
123. Kameda, T.; Fukushima, S.; Shoji, C.; Grause, G.; Yoshioka, T. Electrodialysis for NaCl/EG solution using ion-exchange membranes. *J. Mater. Cycles Waste Manag.* **2013**, *15*, 111–114. [CrossRef]
124. Qi, Y.; He, J.; Li, Y.; Yu, X.; Xiu, F.-R.; Deng, Y.; Gao, X. A novel treatment method of PVC-medical waste by near-critical methanol: Dechlorination and additives recovery. *Waste Manag.* **2018**, *80*, 1–9. [CrossRef]
125. Yu, J.; Sun, L.; Ma, C.; Qiao, Y.; Yao, H. Thermal degradation of PVC: A review. *Waste Manag.* **2016**, *48*, 300–314. [CrossRef]
126. Zakharyan, E.; Petrukhina, N.; Maksimov, A. Pathways of Chemical Recycling of Polyvinyl Chloride: Part 1. *Russ. J. Appl. Chem.* **2020**, *93*, 1271–1313. [CrossRef]
127. Pan, J.; Jiang, H.; Qing, T.; Zhang, J.; Tian, K. Transformation and kinetics of chlorine-containing products during pyrolysis of plastic wastes. *Chemosphere* **2021**, *284*, 131348. [CrossRef]
128. Zhou, J.; Liu, G.; Wang, S.; Zhang, H.; Xu, F. TG-FTIR and Py-GC/MS study of the pyrolysis mechanism and composition of volatiles from flash pyrolysis of PVC. *J. Energy Inst.* **2020**, *93*, 2362–2370. [CrossRef]
129. Liu, H.; Wang, C.; Zhang, J.; Zhao, W.; Fan, M. Pyrolysis kinetics and thermodynamics of typical plastic waste. *Energy Fuels* **2020**, *34*, 2385–2390. [CrossRef]
130. Zhou, X.-L.; He, P.-J.; Peng, W.; Yi, S.-X.; Lü, F.; Shao, L.-M.; Zhang, H. Upcycling waste polyvinyl chloride: One-pot synthesis of valuable carbon materials and pipeline-quality syngas via pyrolysis in a closed reactor. *J. Hazard. Mater.* **2022**, *427*, 128210. [CrossRef] [PubMed]
131. Kumagai, S.; Lu, J.; Fukushima, Y.; Ohno, H.; Kameda, T.; Yoshioka, T. Diagnosing chlorine industrial metabolism by evaluating the potential of chlorine recovery from polyvinyl chloride wastes—A case study in Japan. *Resour. Conserv. Recycl.* **2018**, *133*, 354–361. [CrossRef]
132. Takeshita, Y.; Kato, K.; Takahashi, K.; Sato, Y.; Nishi, S. Basic study on treatment of waste polyvinyl chloride plastics by hydrothermal decomposition in subcritical and supercritical regions. *J. Supercrit. Fluids* **2004**, *31*, 185–193. [CrossRef]
133. Enomoto, H. Dechlorination Treatment of Poly (vinyl chloride). *J. Jpn. Soc. Waste Mgmt. Exp.* **1995**, *6*, 16–22. [CrossRef]
134. Zhao, P.; Li, T.; Yan, W.; Yuan, L. Dechlorination of PVC wastes by hydrothermal treatment using alkaline additives. *Environ. Technol.* **2018**, *39*, 977–985. [CrossRef]
135. Sogancioglu, M.; Yel, E.; Ahmetli, G. Pyrolysis of waste high density polyethylene (HDPE) and low density polyethylene (LDPE) plastics and production of epoxy composites with their pyrolysis chars. *J. Clean. Prod.* **2017**, *165*, 369–381. [CrossRef]
136. Lopez, G.; Artetxe, M.; Amutio, M.; Alvarez, J.; Bilbao, J.; Olazar, M. Recent advances in the gasification of waste plastics. A critical overview. *Renew. Sustain. Energy Rev.* **2018**, *82*, 576–596. [CrossRef]
137. Marcilla, A.; Gomez, A.; Reyes-Labarta, J.; Giner, A.; Hernández, F. Kinetic study of polypropylene pyrolysis using ZSM-5 and an equilibrium fluid catalytic cracking catalyst. *J. Anal. Appl. Pyrolysis* **2003**, *68*, 467–480. [CrossRef]
138. Miranda, R.; Pakdel, H.; Roy, C.; Vasile, C. Vacuum pyrolysis of commingled plastics containing PVC II. Product analysis. *Polym. Degrad. Stab.* **2001**, *73*, 47–67. [CrossRef]
139. Singh, M.V. Pyrolysis of Waste Polyolefins into Liquid Petrochemicals Using Metal Carbonate Catalyst. *Eng. Sci.* **2022**, *19*, 285–291. [CrossRef]
140. Marcilly, C. Evolution of refining and petrochemicals. What is the place of zeolites. In *Studies in Surface Science and Catalysis*; Elsevier: Amsterdam, The Netherlands, 2001; Volume 135, pp. 37–60.
141. Degnan Jr, T.F. The implications of the fundamentals of shape selectivity for the development of catalysts for the petroleum and petrochemical industries. *J. Catal.* **2003**, *216*, 32–46. [CrossRef]
142. del Remedio Hernández, M.; Gómez, A.; García, Á.N.; Agulló, J.; Marcilla, A. Effect of the temperature in the nature and extension of the primary and secondary reactions in the thermal and HZSM-5 catalytic pyrolysis of HDPE. *Appl. Catal. A Gen.* **2007**, *317*, 183–194. [CrossRef]
143. del Remedio Hernández, M.; García, Á.N.; Marcilla, A. Catalytic flash pyrolysis of HDPE in a fluidized bed reactor for recovery of fuel-like hydrocarbons. *J. Anal. Appl. Pyrolysis* **2007**, *78*, 272–281. [CrossRef]
144. Yang, M.; Tian, X.; You, F. Manufacturing ethylene from wet shale gas and biomass: Comparative technoeconomic analysis and environmental life cycle assessment. *Ind. Eng. Chem. Res.* **2018**, *57*, 5980–5998. [CrossRef]
145. Murali, A.; Berrouk, A.S.; Dara, S.; AlWahedi, Y.F.; Adegunju, S.; Abdulla, H.S.; Das, A.K.; Yousif, N.; Hosani, M.A. Efficiency enhancement of a commercial natural gas liquid recovery plant: A MINLP optimization analysis. *Sep. Sci. Technol.* **2020**, *55*, 955–966. [CrossRef]
146. Buekens, A.G.; Froment, G.F. Thermal cracking of propane. Kinetics and product distributions. *Ind. Eng. Chem. Process Des. Dev.* **1968**, *7*, 435–447. [CrossRef]
147. Haribal, V.P.; Chen, Y.; Neal, L.; Li, F. Intensification of ethylene production from naphtha via a redox oxy-cracking scheme: Process simulations and analysis. *Engineering* **2018**, *4*, 714–721. [CrossRef]
148. Qyyum, M.A.; Naquash, A.; Haider, J.; Al-Sobhi, S.A.; Lee, M. State-of-the-art assessment of natural gas liquids recovery processes: Techno-economic evaluation, policy implications, open issues, and the way forward. *Energy* **2022**, *238*, 121684. [CrossRef]

149. Ahmad, S.; Tanwar, R.; Gupta, R.; Khanna, A. Interaction parameters for multi-component aromatic extraction with sulfolane. *Fluid Phase Equilibria* **2004**, *220*, 189–198. [CrossRef]
150. Choi, Y.J.; Cho, K.W.; Cho, B.W.; Yeo, Y.-K. Optimization of the sulfolane extraction plant based on modeling and simulation. *Ind. Eng. Chem. Res.* **2002**, *41*, 5504–5509. [CrossRef]
151. Gaile, A.; Erzhenkov, A.; Semenov, L.; Varshavskii, O.; Zalishchevskii, G.; Somov, V.; Marusina, N. Extraction of aromatic hydrocarbons with triethylene glycol-sulfolane mixed extractant. *Russ. J. Appl. Chem.* **2001**, *74*, 1668–1671. [CrossRef]
152. Gaile, A.; Zalishchevskii, G.; Erzhenkov, A.; Kayfadzhyan, E.; Koldobskaya, L. Extraction of aromatic hydrocarbons from reformates with mixtures of triethylene glycol and sulfolane. *Russ. J. Appl. Chem.* **2007**, *80*, 591–594. [CrossRef]
153. Gaile, A.; Zalishchevskii, G.; Erzhenkov, A.; Koldobskaya, L. Benzene separation from the benzene fraction of reformer naphtha by extractive rectification with N-methylpyrrolidone-sulfolane mixtures. *Russ. J. Appl. Chem.* **2008**, *81*, 1375–1381. [CrossRef]
154. Lei, Z.; Li, C.; Li, J.; Chen, B. Suspension catalytic distillation of simultaneous alkylation and transalkylation for producing cumene. *Sep. Purif. Technol.* **2004**, *34*, 265–271. [CrossRef]
155. Kelly, M.F.; Uitti, K.D. Extractive Distillation of Aromatics with a Sulfolane Solvent. US3551327A, 12 March 1969.
156. Wang, Q.; Zhang, B.; He, C.; He, C.; Chen, Q. Optimal design of a new aromatic extractive distillation process aided by a co-solvent mixture. *Energy Procedia* **2017**, *105*, 4927–4934. [CrossRef]
157. Zhao, X.; You, F. Waste high-density polyethylene recycling process systems for mitigating plastic pollution through a sustainable design and synthesis paradigm. *AIChE J.* **2021**, *67*, e17127. [CrossRef]
158. Pan, Z.; Xue, X.; Zhang, C.; Wang, D.; Xie, Y.; Zhang, R. Production of aromatic hydrocarbons by hydro-liquefaction of high-density polyethylene (HDPE) over Ni/HZSM-5. *J. Anal. Appl. Pyrolysis* **2018**, *136*, 208–214. [CrossRef]
159. Orozco, S.; Artetxe, M.; Lopez, G.; Suarez, M.; Bilbao, J.; Olazar, M. Conversion of HDPE into Value Products by Fast Pyrolysis Using FCC Spent Catalysts in a Fountain Confined Conical Spouted Bed Reactor. *ChemSusChem* **2021**, *14*, 4291–4300. [CrossRef]
160. Farooqi, A.S.; Yusuf, M.; Zabidi, N.A.M.; Saidur, R.; Sanaullah, K.; Farooqi, A.S.; Khan, A.; Abdullah, B. A comprehensive review on improving the production of rich-hydrogen via combined steam and CO_2 reforming of methane over Ni-based catalysts. *Int. J. Hydrogen Energy* **2021**, *46*, 31024–31040. [CrossRef]
161. Wang, B.; Gebreslassie, B.H.; You, F. Sustainable design and synthesis of hydrocarbon biorefinery via gasification pathway: Integrated life cycle assessment and technoeconomic analysis with multiobjective superstructure optimization. *Comput. Chem. Eng.* **2013**, *52*, 55–76. [CrossRef]
162. Yang, H.; Zhang, C.; Gao, P.; Wang, H.; Li, X.; Zhong, L.; Wei, W.; Sun, Y. A review of the catalytic hydrogenation of carbon dioxide into value-added hydrocarbons. *Catal. Sci. Technol.* **2017**, *7*, 4580–4598. [CrossRef]
163. Butler, E.; Devlin, G.; McDonnell, K. Waste polyolefins to liquid fuels via pyrolysis: Review of commercial state-of-the-art and recent laboratory research. *Waste Biomass Valorization* **2011**, *2*, 227–255. [CrossRef]
164. Ding, W.; Liang, J.; Anderson, L.L. Hydrocracking and hydroisomerization of high-density polyethylene and waste plastic over zeolite and silica— alumina-supported Ni and Ni— Mo sulfides. *Energy Fuels* **1997**, *11*, 1219–1224. [CrossRef]
165. Marcilla, A.; Beltrán, M.; Navarro, R. Thermal and catalytic pyrolysis of polyethylene over HZSM5 and HUSY zeolites in a batch reactor under dynamic conditions. *Appl. Catal. B Environ.* **2009**, *86*, 78–86. [CrossRef]
166. Caldeira, V.P.; Peral, A.; Linares, M.; Araujo, A.S.; Garcia-Muñoz, R.A.; Serrano, D.P. Properties of hierarchical Beta zeolites prepared from protozeolitic nanounits for the catalytic cracking of high density polyethylene. *Appl. Catal. A Gen.* **2017**, *531*, 187–196. [CrossRef]
167. Abbas-Abadi, M.S.; Haghighi, M.N.; Yeganeh, H. Evaluation of pyrolysis product of virgin high density polyethylene degradation using different process parameters in a stirred reactor. *Fuel Process. Technol.* **2013**, *109*, 90–95. [CrossRef]
168. Lee, J.Y.; Park, S.M.; Saha, S.K.; Cho, S.J.; Seo, G. Liquid-phase degradation of polyethylene (PE) over MFI zeolites with mesopores: Effects of the structure of PE and the characteristics of mesopores. *Appl. Catal. B Environ.* **2011**, *108*, 61–71. [CrossRef]
169. Obeid, F.; Zeaiter, J.; Ala'a, H.; Bouhadir, K. Thermo-catalytic pyrolysis of waste polyethylene bottles in a packed bed reactor with different bed materials and catalysts. *Energy Convers. Manag.* **2014**, *85*, 1–6. [CrossRef]
170. Zia, K.M.; Bhatti, H.N.; Bhatti, I.A. Methods for polyurethane and polyurethane composites, recycling and recovery: A review. *React. Funct. Polym.* **2007**, *67*, 675–692. [CrossRef]
171. Zahedifar, P.; Pazdur, L.; Vande Velde, C.M.; Billen, P. Multistage chemical recycling of polyurethanes and dicarbamates: A glycolysis–hydrolysis demonstration. *Sustainability* **2021**, *13*, 3583. [CrossRef]
172. Gama, N.V.; Ferreira, A.; Barros-Timmons, A. Polyurethane foams: Past, present, and future. *Materials* **2018**, *11*, 1841. [CrossRef] [PubMed]
173. Sheel, A.; Pant, D. Chemical depolymerization of polyurethane foams via glycolysis and hydrolysis. In *Recycling of Polyurethane Foams*; Elsevier: Amsterdam, The Netherlands, 2018; pp. 67–75.
174. Greene, T.W.; Wuts, P.G. *Protective Groups in Organic Synthesis*; Wiley: Hoboken, NJ, USA, 1991.
175. Nikje, M.M.A.; Garmarudi, A.B.; Idris, A.B. Polyurethane waste reduction and recycling: From bench to pilot scales. *Des. Monomers Polym.* **2011**, *14*, 395–421. [CrossRef]
176. Motokucho, S.; Nakayama, Y.; Morikawa, H.; Nakatani, H. Environment-friendly chemical recycling of aliphatic polyurethanes by hydrolysis in a CO_2-water system. *J. Appl. Polym. Sci.* **2018**, *135*, 45897. [CrossRef]
177. Kemona, A.; Piotrowska, M. Polyurethane recycling and disposal: Methods and prospects. *Polymers* **2020**, *12*, 1752. [CrossRef] [PubMed]

178. Simón, D.; Borreguero, A.; De Lucas, A.; Rodríguez, J. Recycling of polyurethanes from laboratory to industry, a journey towards the sustainability. *Waste Manag.* **2018**, *76*, 147–171. [CrossRef]
179. Simón, D.; Borreguero, A.; De Lucas, A.; Rodríguez, J. Glycolysis of viscoelastic flexible polyurethane foam wastes. *Polym. Degrad. Stab.* **2015**, *116*, 23–35. [CrossRef]
180. Xu, S.; Li, X.; Sui, G.; Du, R.; Zhang, Q.; Fu, Q. Plasma modification of PU foam for piezoresistive sensor with high sensitivity, mechanical properties and long-term stability. *Chem. Eng. J.* **2020**, *381*, 122666. [CrossRef]
181. Heiran, R.; Ghaderian, A.; Reghunadhan, A.; Sedaghati, F.; Thomas, S. Glycolysis: An efficient route for recycling of end of life polyurethane foams. *J. Polym. Res.* **2021**, *28*, 22. [CrossRef]
182. Gadhave, R.V.; Srivastava, S.; Mahanwar, P.A.; Gadekar, P.T. Recycling and disposal methods for polyurethane wastes: A review. *Open J. Polym. Chem.* **2019**, *9*, 39–51. [CrossRef]
183. Yang, W.; Dong, Q.; Liu, S.; Xie, H.; Liu, L.; Li, J. Recycling and disposal methods for polyurethane foam wastes. *Procedia Environ. Sci.* **2012**, *16*, 167–175. [CrossRef]
184. Gu, X.; Lyu, S.; Cheng, W.; Liu, S. Effect of different catalysts on recovery and reuse of waste polyurethane rigid foam. *Mater. Res. Express* **2021**, *8*, 035105. [CrossRef]
185. Deng, Y.; Dewil, R.; Appels, L.; Ansart, R.; Baeyens, J.; Kang, Q. Reviewing the thermo-chemical recycling of waste polyurethane foam. *J. Environ. Manag.* **2021**, *278*, 111527. [CrossRef] [PubMed]
186. Chandrasekaran, S.R.; Sharma, B.K. From waste to resources: How to integrate recycling into the production cycle of plastics. In *Plastics to Energy*; Elsevier: Amsterdam, The Netherlands, 2019; pp. 345–364.
187. Guo, X.; Zhang, W.; Wang, L.; Hao, J. Comparative study of nitrogen migration among the products from catalytic pyrolysis and gasification of waste rigid polyurethane foam. *J. Anal. Appl. Pyrolysis* **2016**, *120*, 144–153. [CrossRef]
188. Akhmetova, F.; Aubakirov, Y.A.; Tashmukhambetova, Z.H.; Sassykova, L.R.; Arbag, H.; Kurmangaliyeva, A. Recycling of waste plastics to liquid fuel mixture over composite zeolites catalysts. *Chem. Bull. Kazakh Natl. Univ.* **2021**, *101*, 12–18. [CrossRef]
189. Maddodi, B.S.; Lathashri, U.A.; Devesh, S.; Rao, A.U.; Shenoy, G.B.; Wijerathne, H.T.; Sooriyaperkasam, N.; Kumar M, P. Repurposing Plastic Wastes in Non-conventional Engineered Wood Building Bricks for Constructional Application—A Mechanical Characterization using Experimental and Statistical Analysis. *Eng. Sci.* **2022**, *18*, 329–336. [CrossRef]
190. Martín-Lara, M.; Moreno, J.; Garcia-Garcia, G.; Arjandas, S.; Calero, M. Life cycle assessment of mechanical recycling of post-consumer polyethylene flexible films based on a real case in Spain. *J. Clean. Prod.* **2022**, *365*, 132625. [CrossRef]
191. Mu, B.; Yang, Y. Complete separation of colorants from polymeric materials for cost-effective recycling of waste textiles. *Chem. Eng. J.* **2022**, *427*, 131570. [CrossRef]
192. Ferreira, A.M.; Sucena, I.; Otero, V.; Angelin, E.M.; Melo, M.J.; Coutinho, J.A.P. Pretreatment of Plastic Waste: Removal of Colorants from HDPE Using Biosolvents. *Molecules* **2021**, *27*, 98. [CrossRef]
193. Tsochatzis, E.D.; Lopes, J.A.; Corredig, M. Chemical testing of mechanically recycled polyethylene terephthalate for food packaging in the European Union. *Resour. Conserv. Recycl.* **2022**, *179*, 106096. [CrossRef]
194. Brosché, S.; Strakova, J.; Bell, L.; Karlsson, T. *Widespread Chemical Contamination of Recycled Plastic Pellets Globally*; IPEN: Berkeley, CA, USA; Los Angeles, CA, USA, 2021.
195. Marson, A.; Masiero, M.; Modesti, M.; Scipioni, A.; Manzardo, A. Life Cycle Assessment of Polyurethane Foams from Polyols Obtained through Chemical Recycling. *ACS Omega* **2021**, *6*, 1718–1724. [CrossRef] [PubMed]

Article

Investigation of Recycled and Coextruded PLA Filament for Additive Manufacturing

Jana Sasse *,†, , Lukas Pelzer †, , Malte Schön †, , Tala Ghaddar and Christian Hopmann

Institute for Plastics Processing, RWTH Aachen University, 52074 Aachen, Germany;
lukas.pelzer@ikv.rwth-aachen.de (L.P.); malte.schoen@ikv.rwth-aachen.de (M.S.);
tala.ghaddar@alumni.fh-aachen.de (T.G.); office@ikv.rwth-aachen.de (C.H.)
* Correspondence: jana.sasse@ikv.rwth-aachen.de; Tel.: +49-241-80-27271
† These authors contributed equally to this work.

Abstract: Polylactide acid (PLA) is one of the most used plastics in extrusion-based additive manufacturing (AM). Although it is bio-based and in theory biodegradable, its recyclability for fused filament fabrication (FFF) is limited due to material degradation. To better understand the material's recyclability, blends with different contents of recycled PLA (rPLA) are investigated alongside a coextruded filament comprised of a core layer with high rPLA content and a skin layer from virgin PLA. The goal was to determine whether this coextrusion approach is more efficient than blending rPLA with virgin PLA. Different filaments were extruded and subsequently used to manufacture samples using FFF. While the strength of the individual strands did not decrease significantly, layer adhesion decreased by up to 67%. The coextruded filament was found to be more brittle than its monoextruded counterparts. Additionally, no continuous weld line could be formed between the layers of coextruded material, leading to a decreased tensile strength. However, the coextruded filament proved to be able to save on master batch and colorants, as the outer layer of the filament has the most impact on the part's coloring. Therefore, switching to a coextruded filament could provide economical savings on master batch material.

Keywords: additive manufacturing; coextrusion; polylactide acid; recycling; filament

Citation: Sasse, J.; Pelzer, L.; Schön, M.; Ghaddar, T.; Hopmann, C. Investigation of Recycled and Coextruded PLA Filament for Additive Manufacturing. *Polymers* **2022**, *14*, 2407. https://doi.org/10.3390/polym14122407

Academic Editors: Sheila Devasahayam and Chin-San Wu

Received: 25 March 2022
Accepted: 4 June 2022
Published: 14 June 2022

Publisher's Note: MDPI stays neutral with regard to jurisdictional claims in published maps and institutional affiliations.

1. Introduction

With benefits such as tool-less material processing, high geometric freedom, fast prototyping and cost-efficient small-scale production, additive manufacturing (AM) has the potential to revolutionize the manufacturing industry. This is reflected in the current value of the AM market, which is estimated at USD 12.6 billion in 2020 with a 21% year over year growth. The current projection estimates the AM market to reach a value of USD 37.2 billion in 2026 [1]. Material extrusion technologies, accounting for the largest share of the AM market [2], are expected to grow even faster at a rate of 27.43% between 2018 and 2024 [3]. To sustain such growth, users are demanding more sustainable technologies and materials [4]. This is particularly apparent when considering the plastics industry. With a global polymer use of 300 Mt in 2019 and an estimated 350 Mt in 2023, the consumption of resources is at an all-time high [5].

While the increase in extrusion-based systems, such as fused filament fabrication (FFF), also increases material use and waste, i.e., through failed parts or support structures [6], it can have a substancial impact on creating a more sustainable manufacturing environment. Based on the technology's high degree of freedom when creating parts, internal structures can be filled sparsely, resulting in reduced material use and lightweight parts [7,8]. Because volumetric elements are added rather than subtracted, material utilization is high [6]. AM also has the potential to avoid over-production by manufacturing on demand [7,8]. Finally, since extrusion-based systems process thermoplastics, excess material can be recycled [9].

At this point, however, an endless closed loop of reusing the same material indefinitely is not possible [10].

The durability of polylactide acid (PLA), one of the most used polymers in extrusion-based AM [11], is limited, as PLA degrades over time and with every processing step. Limiting factors on the recyclability of PLA include thermal decomposition, hydrolysis, photo-oxidation, natural weathering and thermo-oxidative degradation [12]. While hydrolytic degradation can be part of its desired properties as it is key to PLA's biodegradability, it can also limit the applications for this material. Thermal decomposition and thermo-oxidative degradation are the most dominant factors regarding recyclability. With each cycle of additive manufacturing, shredding and the production of new filament from PLA, the material is re-extruded at high temperatures, leading to random chain scission responsible for a reduction in molar mass, which in turn affects the glass transition temperature T_g and the degree of crystallinity [12]. Previous research has shown that PLA displays brittle behavior when the molar mass $\bar{M}_n$ drops below 40 kg/mol [12]. Amorin et al. [13] have shown a reduction in molar mass from an initial $\bar{M}_n = 70.7$ to $\bar{M}_n = 61.7$ and an increase of the melt flow index (MFI) of about 57.6% after five extrusion cycles.

These accumulated degradation mechanisms lead to decreased mechanical properties of recycled PLA (rPLA) compared to virgin PLA, which can be measured, e.g., using injection-molded samples of both virgin and recycled PLA. Ženkiewicz et al. [14] found a reduction in tensile strength by 5.2% and a reduction in impact strength by 20.02% after ten extrusion cycles. Another study conducted by Budin et al. [15] found a decrease in tensile strength by 11%, along with a 5% decrease in transverse rupture strength, a 50% decrease in impact strength and a 4% decrease in hardness for the rPLA samples.

In addition to the mechanical degradation mechanisms, the process of additive manufacturing introduces new challenges for the use of rPLA due to the delicacy of the filament and additional requirements on the weld lines between layers. An investigation by Anderson [16] analyzed the mechanical properties of additively manufactured samples using both virgin and recycled PLA and a decrease in tensile strength by 10.9%, an increase in shear strength by 6.8% and a decrease in hardness by 2.4 % were found, with increased variability in the results and occasional nozzle blockage with the recycled filament. An analysis by Cruz et al. [17] found no significant decrease in tensile strength at break across five reprocessing cycles, although a reduction in the strain at break of 10.63% was observed. In addition, a decrease in molecular weight by 46.91% after five reprocessing cycles and a six-fold increase in the MFI were found. Breški et al. [18] conducted a study investigating the suitability of recycled PLA filaments for additive manufacturing processes. The authors found inconsistent filament diameters and subsequent potential nozzle blockage for filament made from recycled PLA. Another study conducted by Babagowda et al. [19] investigating the mechanical properties of blends with virgin and recycled PLA found that apart from the recycled PLA content, the layer thickness in the printed test samples also played an important role in the tensile and flexural strength.

A review by Pakkanen et al. [20] found that while relevant, the field of recycling PLA for additive manufacturing purposes is still insufficiently studied. The authors suggest the use of a blend of virgin and recycled material to find an acceptable trade-off between mechanical properties and environmental concerns. They also point out that improvements in waste management are critical for this endeavor, as the contamination of material from post-consumer waste can pose a problem for the recyclability of PLA. A review by Shanmugam et al. [21] found that existing research to justify the use of recycled plastics in additive manufacturing was still lacking with regard to bending characteristics, the influence of FFF process parameters and the bonding between the layers of printed recycled parts.

The weakened mechanical properties of additively manufactured parts produced using rPLA largely stem from the weakened weld line between layers. Due to the laminar flow of the heated filament in the hotend, the material on the outer layer of the filament is also the material comprising the weld lines and outer layer visible to the consumer [22,23].

Therefore, the 'ideal filament' would have an outer layer with good optical properties and the capability to form strong weld lines, while the inner core of the filament could be comprised of recycled material in order to save cost and improve the ecological impact. While this method has already been used in other applications in plastics extrusion since the 1990s [24] and has applications in various fields such as profile extrusion [25], blow molding [26] and in food packaging [27], it recently has gained traction due to an increase in the demand of more sustainable packaging.

Recent research has already shown the potential of coextrusion in the production of filament for additive manufacturing applications. Ruckdashel at al. [23] have demonstrated how coextrusion can be used to enhance the content of carbon-based or inorganic fillers in 3D filament without compromising its mechanical properties. In another paper by Hart et al. [28], a method for the coextrusion of dual material filament was presented, where a star-shaped polycarbonate (PC) core is used to enhance the mechanical properties of acrylonitrile butadiene styrene (ABS) filament. However, to the authors' knowledge, coextrusion has not been investigated as a tool to aid the processing of rPLA.

In this study, PLA filament is manufactured using varying proportions of recycled material, both in a monoextrusion and for the first time a coextrusion process. The parts manufactured from this filament are tested mechanically and optically to investigate the influence of various recycling strategies on part quality. The authors' work intends to verify that the well-known benefits of coextrusion in the processing of recycled materials also apply to FFF. More specifically, the authors hypothesize that coextrusion gives better mechanical properties compared to simply blending virgin and recycled plastics, and that coextrusion can save on master batch. The work presented is significant due to its implications regarding the sustainability in plastics production by significantly reducing use of virgin material while maintaining high part quality. This contributes to the increasingly relevant fields of AM as well as sustainable practices in production by finding ways to incorporate recycled materials in existing processes.

2. Materials and Methods

All materials used were originally purchased in filament form and manually shredded to granulate size. As virgin material, PLA Extrafill Natural (Fillamentum Manufacturing Czech s.r.o., Hulin, Czech Republic) with 5% master batch material in the form of 3DJake EcoPLA White (Niceshops GmbH, Paldau, Austria) was used. The recycled material was produced from aged samples of PLA Neutral (German Reprap GmbH, Feldkirchen, Germany) mixed with 5% CCTree ST-PLA Pro Black (CCTree, Chenzhen, China) as a master batch material.

A rheological characterization of both materials was performed via plate–plate rheometry with a gap of 1 mm, a frequency of 1 Hz and an amplitude of 0.1% at 270 °C. The molecular weight can be deferred using Relation (1) described in [29].

$$\eta_0 = K \cdot M_W^{3.7}, \quad \text{with} \quad lgK = -16.1 \tag{1}$$

For the application of Relation (1), which is valid at 180 °C, a temperature correction using a WLF approach was used, with a glass transition temperature of $T_g = 332.65\,K$ [14], resulting in a standard temperature $T_s = 382.65\,K$.

2.1. Extrusion

Prior to extrusion, all materials were dried in a dry air dryer at a temperature of 60 °C for two hours.

Extrusion trials were carried out using single-screw extruders with a 19 mm screw diameter and a length of 25 D. For monoextrusion, the screw was revolving at 15 rpm, corresponding to a throughput of 0.29 kg/h, and the temperatures in the barrel zones were set up as a rising temperature profile (170 °C, 180 °C, 190 °C) with a nominal temperature of 145 °C in the extrusion die. Downstream of the die, the haul off of a 3 m hot air shock canal was used as a cooling section for the extrudate with subsequent manual spooling. The haul-off speed was adjusted until a consistent filament diameter of 1.75 mm was reached and ranged from 1.5 to 1.8 m/min, depending on the rPLA content.

First, monoextruded filaments with varying degrees of rPLA content were produced. Continuous extrusion of filament was only achieved for material mixtures with up to 60% rPLA content. At proportions exceeding 60% rPLA content, the filament produced was too brittle for further processing. Therefore, monoextrusion filament samples were limited to 0%, 20%, 40% and 60% rPLA content, respectively. For the extrusion of 60% rPLA filament, the temperatures of the barrel zones on the extruder had to be reduced by 10 °C to achieve good strechability. With increasing rPLA content, higher variations in the filament diameter were observed, with diameters ranging between 1.6 and 1.8 mm.

The coextruded filament was produced in an extrusion line setup visualized in Figure 1. In this setup, two single-screw extruders were used to feed the coextrusion die with both a core layer of blended material and a skin layer of virgin material. The coextrusion die (Figure 2) is based on a spiral mandrel die design for the skin layer, which is coating the core layer fed through the middle section. The coextruded filament was produced with a core layer with 60% rPLA content and a skin layer of virgin PLA. The ratio of the core and skin layer was controlled by the respective throughputs of the two extruders. For a 20% skin layer, the main extruder was set to 8 rpm, while the side extruder revolved at 2 rpm. This resulted in an overall rPLA content of 48% for the coextruded filament.

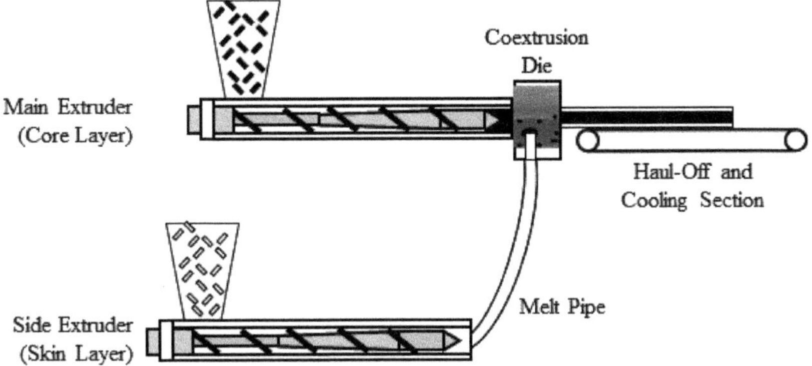

Figure 1. Schematic extrusion line configuration for coextrusion of PLA filament.

2.2. Additive Manufacturing

To evaluate the effect of using recycled material in filament production on the mechanical properties of manufactured parts, tensile tests and impact tests are conducted. The tensile tests are carried out according to DIN EN ISO 527. However, the test specimen geometry is adapted from the typically used 1BA sample and modified to be better suited to the FFF process. Specifically, the testing zone's width is increased to 8 mm while the thickness is increased to 6 mm, allowing for enough volume to also include sparse infill. Additionally, the total length is increased to 90 mm while the testing zone's length is decreased to 20 mm, allowing for a larger radius in the transition between the testing zone and the clamping zone. This decreases the chance for failure of the test specimen outside the testing zone, resulting in a noticeably increased number of valid tests. The charpy impact tests are conducted according to DIN EN ISO 791, with test specimen dimensions of 80 mm by 10 mm by 4 mm and a v-shaped notch, which is included in the specimen design and therefore manufactured during the AM process.

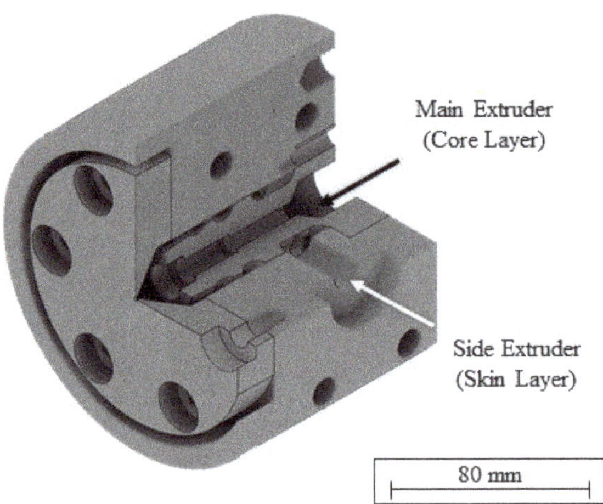

Figure 2. Coextrusion die based on a spiral mandrel die.

To evaluate the impact of various amounts of recycled filament on the strength of individual strands as well as inter-layer adhesion, all test specimens are manufactured in a horizontal and a vertical orientation (Figure 3). For this, a CR-10S 3D-printer is used. The machine is equipped with a 0.6 mm nozzle to prevent clogging from potential impurities in the filament. The G-Code is prepared using Slic3r version 1.3.0, which is an open source slicing software. For both build orientations, five specimens are manufactured from each monoextruded material (100% virgin PLA, 20% rPLA, 40% rPLA, 60% rPLA) and the coextruded material (48% rPLA). Additionally, five reference samples for both orientations are manufactured from virgin PLA filament. All additional process parameters are kept constant according to Table 1.

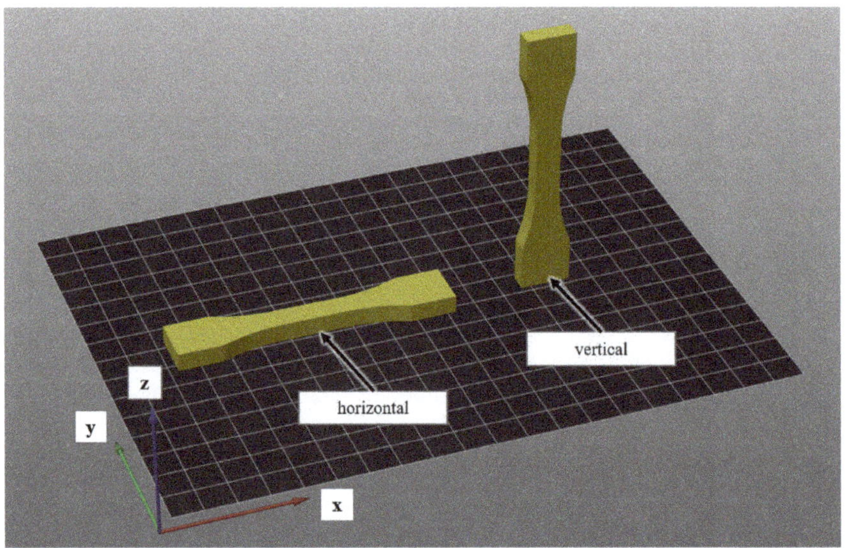

Figure 3. Horizontal manufacturing orientation (**left**) and vertical manufacturing orientation (**right**).

Table 1. Process parameters for the manufacturing of test specimens.

Process Parameter	Value	Unit
Perimeters	2	[-]
Top/bottom layers	3	[-]
Layer height	0.3	[mm]
Manufacturing speed	49	[mm/s]
Top/bottom infill pattern	Rectilinear	[-]
Top/bottom infill angle	45°	[-]
Interior infill pattern	Gyroid	[-]
Interior infill density	40%	[-]
Extrusion width	0.84	[mm]
Temperature nozzle	230	[°C]
Temperature heat bed	55	[°C]
Part cooling	100%	[-]

Figure 4 shows one specimen from each category after testing. While the reference sample is somewhat translucent, based on the natural, uncolored reference material, the samples from recycled material appear in various shades of gray, which are caused by the varying mixtures of white-colored virgin PLA and black-colored rPLA used during filament production.

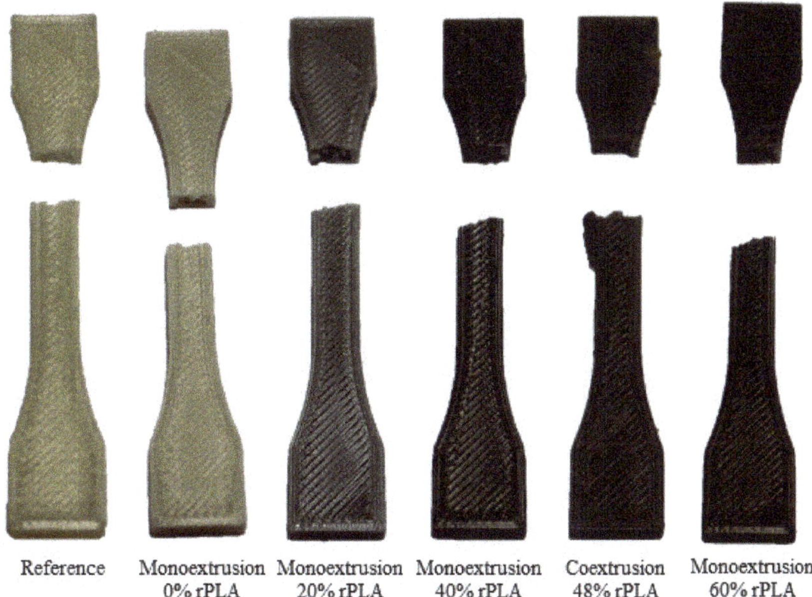

Figure 4. Tensile testing samples manufactured in horizontal position.

3. Results

In the initial rheological characterization using plate–plate rheometry, the zero shear viscosity at 270 °C of the material with 100% virgin PLA was measured at 358.265 Pa s, while the zero shear viscosity at 270 °C of the recycled material was reduced by 71.7% with 101.322 Pa s. Using Relation (1), the molecular mass of the recycled material was approximated to be only 70.6% of the virgin material.

3.1. Mechanical Analysis

The manufactured tensile specimens were tested using a Z150 universal test machine which is equipped with a contact multiXtens extensometer for strain measurement. To ensure a constant load on the specimen in the clamping, the jaws were tightened to 15 Nm. Young's modulus was measured at a speed of 1 mm/min, while the remaining test was conducted at 5 mm/min. The impact specimens were tested using a small impact pendulum Nr. 5102 with a 0.5 J impact hammer and a pendulum length of 0.225 m. All tests were conducted at room temperature.

The tensile strength tests were conducted with samples manufactured in both the horizontal and vertical position. While the tensile strength of the samples printed in the horizontal position mainly indicates the strength of the individual strands, the tensile strength of the samples printed in the vertical position is a measure for the strength of the weld line between the layers. For each material and printing orientation, five samples were analyzed, and ANOVA was performed to indicate the significance of the observed differences ($p < 0.05$).

In the results section, all significant differences are marked in the diagrams using a Tukey's honestly significant difference (HSD) test.

The tensile strength of the monoextruded filament decreases with increasing rPLA content. Figure 5, on the left, shows the tensile strength of the samples manufactured in the horizontal position. The filament with 100% virgin PLA content had an average tensile strength of 34.45 MPa, and while it decreased with increasing rPLA content, the differences were not significant. The coextruded filament, on the other hand, was brittle, and the tensile strength was significantly lower than for all the monoextruded filaments. With an average of 21.17 MPa, the tensile strength was reduced by about 39% compared to the 100% virgin PLA filament.

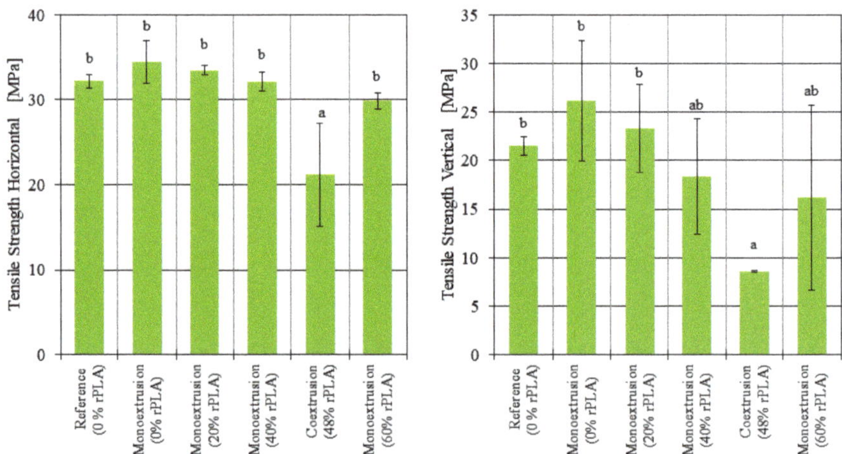

Figure 5. Tensile strength of additively manufactured samples manufactured in horizontal and vertical position.

The tensile strength of the specimens printed in the vertical position (Figure 5 right) was more inconsistent. Since a small defect or inconsistency in one layer is enough to initiate part failure, and the number of layers in vertical specimens is much larger than in horizontal specimens, while the area of each layer is much smaller for vertical samples, a larger variation in measured results is to be expected. With increasing rPLA content, the number of failed tests due to failure in the clamping zone also increased. As a result, only two out of five specimens at 60% rPLA content could be correctly analyzed. The two remaining specimens display divergent behavior, where one specimen's tensile strength was measured at 6.63 MPa, while the other one's was measured at 25.69 MPa. This makes an interpretation of the results difficult. For the vertical samples, the tensile strength of the monoextruded filaments decreased with increasing rPLA content. With 8.61 MPa, the tensile strength of the coextruded samples in the vertical position was reduced by 67% compared to the 100% virgin PLA content samples. However, only the difference in tensile strength between the samples with 100% virgin PLA content and the coextruded samples was found to be statistically significant ($p = 0.024$). There was barely any variation in the tensile strength of the coextruded samples, and there were less failed tensile tests. This might suggest that the coextruded filament produced more reliable weld lines than the monoextruded filaments with 40% and 60% rPLA content.

In addition, the Young's modulus of the samples was evaluated. For the samples produced from monoextruded filament and printed in the horizontal position (Figure 6 left), no significant changes were observed. On the other hand, the Young's modulus of the coextruded samples printed in the horizontal position was reduced by 55% compared to the samples with 100% virgin PLA ($p = 0.047$). The results for the Young's modulus of the samples printed in the vertical position are depicted in Figure 6 right. Here, no statistically significant differences between the monoextruded or coextruded samples were found.

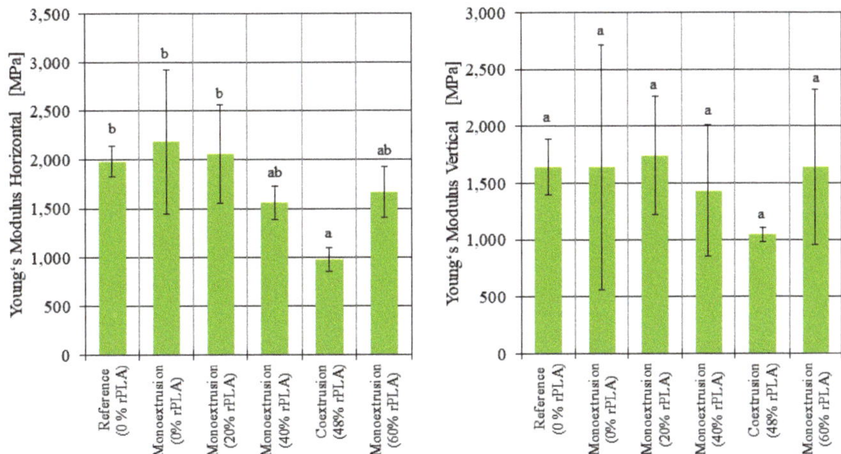

Figure 6. Young's modulus of additively manufactured samples manufactured in horizontal and vertical position.

The Charpy impact strength measured for the samples printed in the horizontal position (Figure 7 left) ranged between 2.32 and 3.24 kJ/m², with no significant dependency from the rPLA content. The samples printed in the vertical position (Figure 7 right) had an impact strength between 1.11 and 2.00 kJ/m². Here, the only difference found to be statistically significant ($p = 0.041$) was between the samples printed from 100% virgin PLA filament and the samples printed from filament with 40% rPLA content.

3.2. Optical Analysis

To investigate a possible cause for the comparatively low mechanical performance of samples manufactured from coextruded filament and for the high standard deviation of some samples, optical analyses of the extruded filament and the manufactured test samples were performed.

The coextruded filament was analyzed in bright-field microscopy as well as optical microscopy with differential interference contrast (DIC). A comparison of the cross-sections of different samples of the same coextruded filament shows high deviations from an idealized circular shape with a 1.75 mm diameter (Figure 8). While some coextruded samples were of roughly circular shape, others were much more inconsistent. Spiral patterns from the screw were apparent in all samples, suggesting a bad mixing behavior of material and master batch. Not all samples display clearly visible coextrusion with a clear interface between the core layer and skin layer. Sample a shows a clear interface between the two layers and is also of a roughly circular shape. Sample b, on the other hand, might show an interface, but the interface is much less pronounced. Samples c and d show no apparent signs of coextrusion. While sample e shows a roughly circular interface between the layers, the filament's shape has a lot of kinks, suggesting an uneven extrusion of the skin layer. This is even more pronounced in sample f, where both the outer and the inner interface of the skin layer are noticeably uneven.

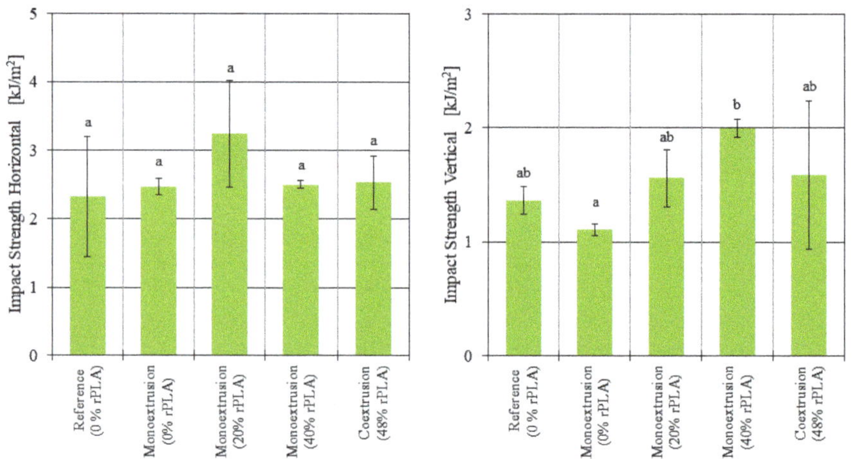

Figure 7. Charpy impact strength of additively manufactured samples manufactured in horizontal and vertical position.

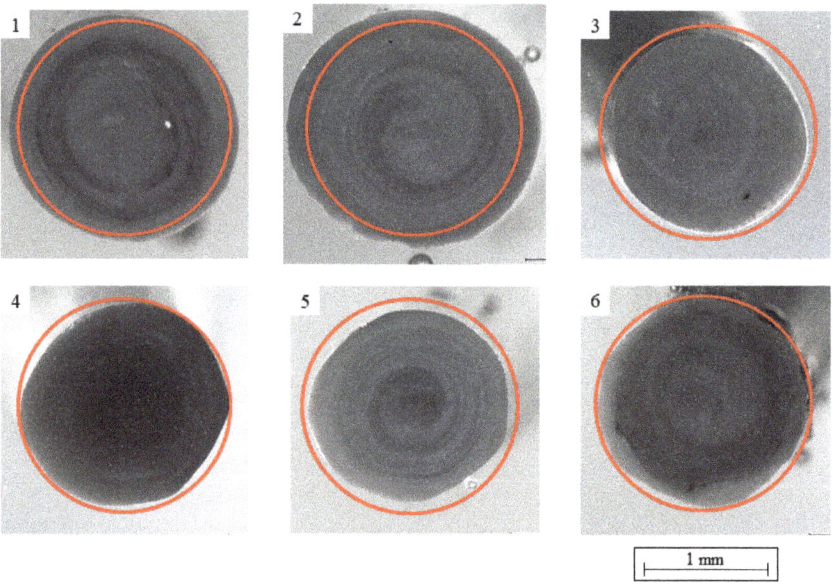

Figure 8. Microscopic analysis of six samples of the coextruded filament with a 20% skin layer, where the circle marks an ideal cross-section with d = 1.75 mm.

One interpretation of these results might be that the skin layer was extruded at such a low throughput that no consistent material output was achieved. However, an increase in throughput was not feasible due to the limitations in haul-off speed and cooling line length in the lab setup used. In addition, many coextruded samples displayed a significantly smaller cross section area. This is also apparent in the mass of the printed samples, which is visualized alongside the tensile strengths in Figure 9. It was observed that both the tensile strength and the mass of the individual tensile testing specimens decreased with increasing rPLA content. The coextruded samples were on average 17% lighter than the 100% virgin PLA monoextruded samples ($p = 1 \cdot 10^{-4}$). The monoextruded filaments with 60% rPLA content show similar problems with respect to fluctuations in diameter and a 16% reduction in sample mass compared to the 100% virgin PLA samples ($p = 0.047$). Since filament-based AM machines feed material based on length rather than based on weight or volume, every deviation in filament diameter and filament roundness influences the resulting parts. Based on the smaller filament diameter (Figure 8) and the lower measured weight of the samples, it can be concluded that less material than specified was introduced to the part. This, in turn, means that the layer cross-sections are smaller, resulting in lower mechanical properties when loaded in the strand direction. Additionally, layers are not compressed as much, and the interface between the layers is smaller, resulting in lower mechanical properties of the part when loaded in the build direction.

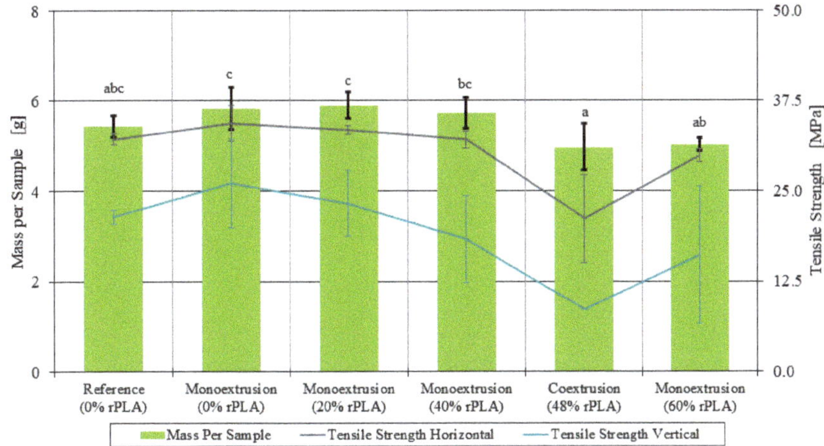

Figure 9. Average mass of tensile testing samples printed in both horizontal and vertical position.

Another aspect to consider is failure initiation due to the notch effect. To investigate whether the roundness deviation in the extruded filament results in increased roughness of the manufactured samples, an analysis of the surface roughness of the printed horizontal tensile testing samples was performed using a laser scanning microscope (LSM). This method enables the three-dimensional reconstruction of the part's surface (Figure 10 bottom), allows for tilt and form correction (Figure 10 top), and calculates all relevant surface roughness metrics. To obtain an accurate comparison of the roughness caused by inconsistent filament diameter and roundness, the tilt and form correction was used to correct for bulging of the samples toward the heat bed. Known as elephant's foot, this bulging is a typical artifact in FFF parts and can be a result of excess heat from the heat bed during the manufacturing process, preventing the lower layers from solidifying completely. As a result, those lower layers are compressed by the weight of subsequent layers. The bulging may also be attributed to an undersized gap between the nozzle and the bed during the first layer. By choosing a curved profile for correction, the bulge is accounted for, and surface roughness caused by the individual strands can be measured.

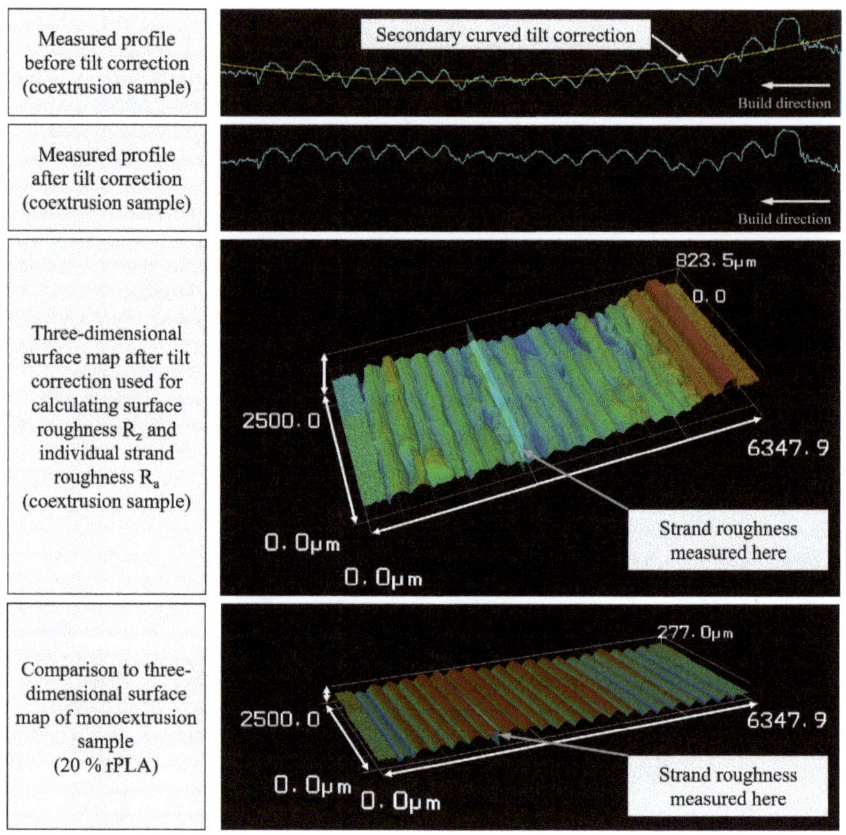

Figure 10. Surface tilt correction and three-dimensional surface map exemplified for the LSM measurement of a sample manufactured from coextruded filament. Every peak represents one individual layer.

Using the corrected surfaces, the peak-to-valley height R_z [30] over the sample's complete depth was calculated. Figure 11 shows the distribution of roughness values for the parts manufactured from mono- and coextruded filaments. While the values for the monoextruded filament ranged from $R_z = 206.62$ µm at 20% rPLA content to $R_z = 328.47$ µm at 40% rPLA content, the surface roughness for the coextruded material was significantly higher at $R_z = 921.01$ µm.

Similar results can be observed when evaluating single strands of material. Here, the arithmetic mean roughness value R_a [30] was used to quantify deviation inside a single strand (right axis data in Figure 11). With values between $R_a = 1.77$ µm at 40% rPLA content and $R_a = 9.05$ µm at 20% rPLA content, individual strands manufactured from monoextruded filament showed only a low surface roughness. In comparison, the single-strand roughness of coextruded filament after the AM process was $R_a = 25.17$ µm.

These investigations show the significantly higher roughness of the parts produced from coextruded filament, which can be attributed to the diameter and roundness deviation caused by the extrusion process. In addition to reducing the surface quality in manufactured parts, this can be a possible explanation for the coextruded sample's lower mechanical performance.

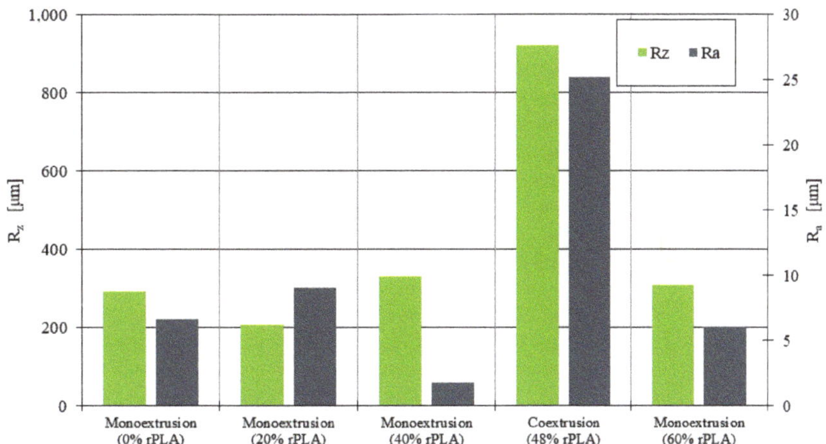

Figure 11. Surface roughness R_z and individual strand roughness R_a of additively manufactured tensile testing samples.

The fluctuations in diameter of the filament with 40% and 60% rPLA content did not show in the surface roughness of the samples. On the other hand, the coextruded samples displayed a highly increased surface roughness, suggesting that the roundness of the filament cross-section is a much higher influence factor than fluctuations in the filament diameter over time.

In addition, measurements of the reflectance of printed samples were performed using a spectral photometer. Due to the use of black and white master batch material, the filaments also display differences in their color, and a high reflectance should correspond to a low rPLA content. Figure 12 shows the results for the printed samples. For the monoextruded filaments, the virgin material displayed a reflectance of 75%, while monoextruded filaments with 20%, 40% and 60% rPLA content had a decreased reflectance of 55%, 26% and 29%, respectively. On the contrary, the coextruded filament with a 20% skin layer of virgin material had a reflectance of 42%. All differences observed were statistically significant, with $p < 1 \cdot 10^{-8}$ for all comparisons but 40% rPLA v. 60% rPLA where $p = 0.046$. This suggests that the skin layer was largely contributing to the overall reflectance of the samples.

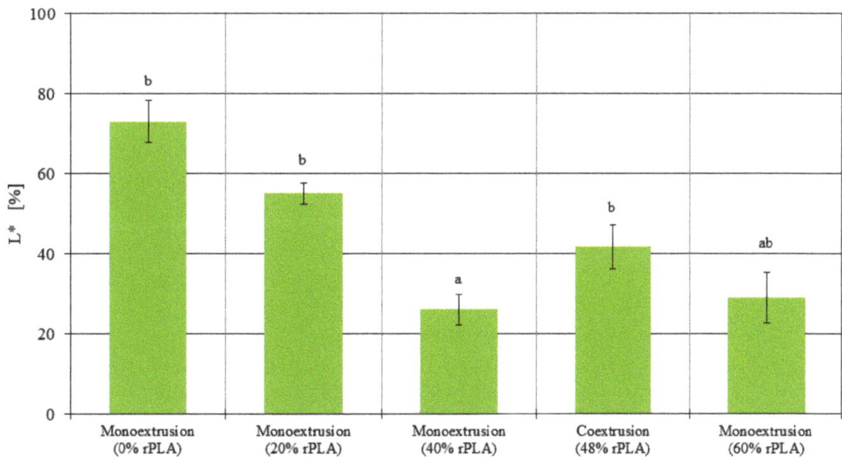

Figure 12. Reflectance of all additively manufactured tensile testing samples.

4. Discussion

For the samples manufactured in the horizontal position, where the majority of the tensile strength comes from the strength of the individual strands, a recycled PLA content did not significantly affect the tensile strength. The samples manufactured in the vertical position, where the tensile strength is largely affected by the strength of the weld lines between the layers, there was a tensile strength comparable to the virgin and reference samples only for up to 40% recycled PLA content. The samples printed with 60% rPLA content had so much variation between the individual specimens, and no reliable conclusion can be reached. The impact strength was observed to decrease with increasing rPLA content, although the variation in the results does not allow for further conclusions on this matter.

While previous research regarding the mechanical properties of additively manufactured parts using recycled PLA focused on the degradation caused by the re-extrusion process [13,17], the material used in this study also experienced degradation through aging. This could serve as an explanation for why a larger decrease in tensile strength was observed. These results are in agreement with previous research by Ong et al. [31], where a 50% reduction in tensile strength after one recycling step was found, and after two cycles, only few specimen were actually able to be analyzed. Due to their setup using post-consumer PLA from a university 3D printing laboratory, their material is assumed to be fairly similar to the recycled material present in this study.

With increasing rPLA content, more instabilities in the extrusion process were observed in our experiments, leading to increased variation in the filament diameter and eventually to potential clogging of the nozzle during the AM process, which is consistent with previous findings [16,18,31,32].

The tensile tests performed highlighted a number of apparent differences between the monoextruded and the coextruded filaments. While all tensile tests performed on the monoextruded samples manufactured in the horizontal position showed some degree of ductile behavior and were in a similar tensile strength range, the coextruded samples were brittle with a tensile strength up to 38.5% lower than the monoextruded counterparts. The strength of the weld lines also did not improve through the coextrusion process. The tensile strength of the coextruded samples printed in the vertical position was 67% lower than the baseline samples made from virgin PLA.

One reason for this might be detectable in the microscopic analysis of the coextruded filaments. The filaments produced using coextrusion had an inconsistent quality. It is apparent that the outer layer of the virgin material, which was supposed to improve the strength of the weld line, has large variations in thickness and is not continuous. This leads to inconsistencies in the roundness and diameter of the filament and subsequently to variations in weight. The subsequent effects on the surface quality of the filament and thus the quality of the weld lines might be the underlying cause for the consistently low tensile strength. The authors' hypothesis of coextrusion leading to better mechanical properties could therefore not be proven.

On the other hand, the reflectance measurements suggest that the coextrusion method presented is a viable option to reduce the amount of master batch and colorants needed for filament production. It has been confirmed that due to the laminar flow in the 3D printer's hot-end, the color of the outer layer of the filament has the largest influence on the color of the additively manufactured part. Subsequently, the coextrusion method as presented here could be a useful tool to save production cost in the long-term, provided that further research increases the quality of filament produced by means of PLA–rPLA coextrusion.

5. Conclusions

In this study, monoextruded filaments for FFF with varying recycled PLA content have been produced and tested. It was found that the strength of the strand itself, tested by means of samples printed in the horizontal position, did not decrease significantly. The strength of the weld lines between the layers in samples printed in the vertical position on the other hand decreased noticeably, up to the point where no reliable results could be evaluated at 60% rPLA content.

To test the hypothesis that recycled filament with an outer layer of virgin material is able to form stronger weld lines than just recycled material, a method for the coextrusion of filament with a high rPLA content core layer and a 20% outer virgin PLA layer was developed. Coextruded filament was produced, and tensile as well as Charpy test samples in both horizontal and vertical positions were manufactured via extrusion-based AM. The parts manufactured from coextruded filament were found to be more brittle than their monoextruded counterparts. Due to inconsistencies in the coextrusion process, no continuous weld line could be formed between the layers, leading to a decreased tensile strength. However, the coextruded filament enabled the reduction of master batch and colorants as the outer layer of the filament appears to be mainly responsible for the manufactured part's coloring. Therefore, switching to a coextrusion line could provide economical savings on master batch material.

Additional research improving methods to increase the use of recycled material in FFF filament is necessary. Further improvements in the coextrusion process are needed to make this a viable method. An improvement in the optical results could be achieved by compounding of the material and master batch prior to extrusion or the use of additional mixing elements before the die. The irregularities in the skin layer might be resolved by increasing the throughput to avoid flow surges. To address the issue of inconsistent filament diameter, the main cause for decreased mechanical properties of the parts manufactured from coextruded material, a closed-loop material feed system can be implemented in the AM process. By continuously measuring and correcting the material throughput, the reliance on a precise filament diameter can be circumvented. In addition, the use of a gear pump can help to reduce the pulsatility in the extruder output.

Finally, the optical results encourage further research in the concept of coextrusion for the saving of master batch. To test this concept, additional coextrusion trials with differently colored virgin material skin and core layers are necessary to analyze the mechanical performance of the coextrusion samples independently from the recycled PLA content, provided that further research increases the quality of filaments produced by means of PLA–rPLA coextrusion.

Author Contributions: Conceptualization, M.S., J.S. and L.P.; methodology, M.S.; validation, T.G., L.P. and J.S.; formal analysis, J.S.; investigation, T.G.; resources, C.H.; data curation, J.S. and L.P.; writing—original draft preparation, J.S. and L.P.; writing—review and editing, M.S. and C.H.; visualization, J.S.; supervision, M.S. and C.H.; project administration, C.H.; funding acquisition, C.H. All authors have read and agreed to the published version of the manuscript.

Funding: Funded by the Deutsche Forschungsgemeinschaft (DFG, German Research Foundation) under Germany's Excellence Strategy—EXC-2023 Internet of Production—390621612.

Institutional Review Board Statement: Not applicable.

Informed Consent Statement: Not applicable.

Data Availability Statement: Data available on request.

Conflicts of Interest: The authors declare no conflict of interest.

Abbreviations

The following abbreviations are used in this manuscript:

PLA	Polylactide Acid
rPLA	Recycled PLA
FFF	Fused Filament Fabrication
AM	Additive Manufacturing
MFI	Melt Flow Index
PC	Polycarbonate
ABS	Acrylonitrile Butadiene Styrene
HSD	Honestly Significant Difference
DIC	Differential Interference Contrast
LSM	Laser Scanning Microscope
IKV	Institute for Plastics Processing

References

1. Roberts, T. *Additive Manufacturing Trend Report 2021*; Hubs B.V.: Amsterdam, The Netherlands, 2021.
2. Wohlers, T. *Wohlers Report 2019: 3D Printing and Additive Manufacturing State of the Industry*; ASTM International: Washington, DC, USA, 2019.
3. Fused Deposition Modeling 3D Printing Technology Market Size, Share, and Trends by Component (Hardware, Software, Services, Material), by End-User (Healthcare, Automotive, Aerospace and Defense, Construction, Others), and by Geography—Forecasts from 2019 to 2024. Available online: https://www.knowledge-sourcing.com/report/fused-deposition-modeling-3d-printing-technology-market (accessed on 23 March 2022).
4. Moreau, C. *The State of 3D Printing 2021*; Sculpteo: Villejuif, France, 2021.
5. AMI. *Summary of the Global Plastics Industry*; AMI: Bristol, UK, 2019.
6. Caviezel, C.; Grünwald, R.; Ehrenberg-Silies, S.; Kind, S.; Jetzke, T.; Bovenschulte, M. *Additive Fertigungsverfahren (3-D-Druck): Innovationsanalyse*; Büro für Technikfolgen-Abschätzung beim Deutschen Bundestag: Berlin, Germany, 2017.
7. Chen, S.-C.; Zhang, X.-M.; Liu, M.; Ma, J.-P.; Lu, W.-Y.; Chen, W.-X. Rheological Characterization and Thermal Stability of Different Intrinsic Viscosity Poly(ethylene terephthalate) in Air and Nitrogen. *Int. Polym. Process.* **2016**, *31*, 292–300. [CrossRef]
8. Mani, M.; Lyons, K.W.; Gupta, S.K. Sustainability Characterization for Additive Manufacturing. *J. Res. Natl. Inst. Stand. Technol.* **2014**, *119*, 419–428. [CrossRef] [PubMed]
9. Beltrán, F.R.; Arrieta, M.P.; Moreno, E.; Gaspar, G.; Muneta, L.M.; Carrasco-Gallego, R.; Yáñez, S.; Hidalgo-Carvajal, D.; de La Orden, M.U.; Martínez Urreaga, J. Evaluation of the Technical Viability of Distributed Mechanical Recycling of PLA 3D Printing Wastes. *Polymers* **2021**, *13*, 1247. [CrossRef] [PubMed]
10. Zhao, P.; Rao, C.; Gu, F.; Sharmin, N.; Fu, J. Close-looped recycling of polylactic acid used in 3D printing: An experimental investigation and life cycle assessment. *J. Clean. Prod.* **2018**, *197*, 1046–1055. [CrossRef]
11. Wickramasinghe, S.; Do, T.; Tran, P. FDM-Based 3D Printing of Polymer and Associated Composite: A Review on Mechanical Properties, Defects and Treatments. *Polymers* **2020**, *12*, 1529. [CrossRef] [PubMed]
12. Rasselet, D.; Ruellan, A.; Guinault, A.; Miquelard-Garnier, G.; Sollogoub, C.; Fayolle, B. Oxidative degradation of polylactide (PLA) and its effects on physical and mechanical properties. *Eur. Polym. J.* **2014**, *50*, 109–116. [CrossRef]
13. Amorin, N.; Rosa, G.; Alves, J.; Franchetti, S.; Gonçalves, S.; Fechine, G.J.M. Study of Thermodegradation and Thermostabilization of Poly(lactide acid) Using Subsequent Extrusion Cycles. *J. Appl. Polym. Sci.* **2014**, *131*. [CrossRef]
14. Żenkiewicz, M.; Richert, J.; Rytlewski, P.; Moraczewski, K.; Stepczyńska, M.; Karasiewicz, T. Characterisation of multi-extruded poly(lactic acid). *Polym. Test.* **2009**, *28*, 412–418. [CrossRef]
15. Budin, S.; Maideen, N.C.; Koay, M.H.; Ibrahim, D.; Yusoff, H. A comparison study on mechanical properties of virgin and recycled polylactic acid (PLA). *J. Phys. Conf. Ser.* **2019**, *1349*, 12002. [CrossRef]
16. Anderson, I. Mechanical Properties of Specimens 3D Printed with Virgin and Recycled Polylactic Acid. *3D Print. Addit. Manuf.* **2017**, *4*, 110–115. [CrossRef]
17. Cruz Sanchez, F.; Lanza, S.; Boudaoud, H.; Hoppe, S.; Camargo, M. Polymer Recycling and Additive Manufacturing in an Open Source context : Optimization of processes and methods. In Proceedings of the Annual International Solid Freeform Fabrication Symposium-An Additive Manufacturing Conference, Austin, TX, USA, 10–12 August 2015.
18. Breški, T.; Hentschel, L.; Godec, D.; Đuretek, I. Suitability of Recycled PLA Filament Application in Fused Filament Fabrication Process. *Teh. Glas.* **2021**, *15*, 491–497. [CrossRef]
19. Babagowda; Kadadevara Math, R.S.; Goutham, R.; Srinivas Prasad, K.R. Study of Effects on Mechanical Properties of PLA Filament which is blended with Recycled PLA Materials. *IOP Conf. Ser. Mater. Sci. Eng.* **2018**, *310*, 12103. [CrossRef]
20. Pakkanen, J.; Manfredi, D.; Minetola, P.; Iuliano, L. About the Use of Recycled or Biodegradable Filaments for Sustainability of 3D Printing. In Proceedings of the SDM: International Conference on Sustainable Design and Manufacturing, Bologna, Italy, 26–28 April 2017; pp. 776–785.
21. Shanmugam, V.; Das, O.; Neisiany, R.E.; Babu, K.; Singh, S.; Hedenqvist, M.S.; Berto, F.; Ramakrishna, S. Polymer Recycling in Additive Manufacturing: An Opportunity for the Circular Economy. *Mater. Circ. Econ.* **2020**, *2*, 11. [CrossRef]
22. Lu, B.; Zhang, H.; Maazouz, A.; Lamnawar, K. Interfacial Phenomena in Multi-Micro-/Nanolayered Polymer Coextrusion: A Review of Fundamental and Engineering Aspects. *Polymers* **2021**, *13*, 417. [CrossRef] [PubMed]

23. Ruckdashel, R.; Wang, S.; Ullrich, F.; Kisil, M.; Lam, A.; Mangkhalakhili, J.; Tang, S.; Park, J.; Vera-Sorroche, J. Design and Evaluation of Bicomponent Core-Sheath Die for 3D Printer Filament Feedstock Co-extrusion. In Proceedings of the SPE ANTEC 2020, Online, 5 May 2020.
24. DiRaddo, R.W.; Dube, F.A.; Garcia-Rejon, A. Multilayer annular flow of a recycled/virgin material combination. In Proceedings of the ENERCOMP 95: International Conference on Composite Materials and Energy, Montreal, QC, Canada, 8–10 May 1995.
25. Ryckebosh, K.; Gupta, M. Optimization of a Profile Coextrusion Die Using a Three-Dimensional Flow Simulation Software. *SPE ANTEC Tech. Pap.* **2015**. Available online: https://plasticflow.com/papers/Deceuninck_ANTEC15.pdf (accessed on 23 March 2022).
26. Dyadichev, A.V.; Dyadichev, V.V.; Kolesnikov, A.V.; Menyuk, S.G.; Dyadicheva, E.A.; Chornobay, S.Y. Model of preform deflected mode in the process of secondary polymer materials coextrusion processing. *J. Phys. Conf. Ser.* **2019**, *1260*, 062005 [CrossRef]
27. Radusin, T.; Nilsen, J.; Larsen, S.; Annfinsen, S.; Waag, C.; Eikeland, M.S.; Pettersen, M.K.; Fredriksen, S.B. Use of recycled materials as mid layer in three layered structures-new possibility in design for recycling. *J. Clean. Prod.* **2020**, *259*, 120876. [CrossRef]
28. Hart, K.R.; Dunn, R.M.; Wetzel, E.D. Tough, Additively Manufactured Structures Fabricated with Dual-Thermoplastic Filaments. *Adv. Eng. Mater.* **2020**, *22*, 2070013. [CrossRef]
29. Standau, T.; Long, H.; Murillo Castellón, S.; Brütting, C.; Bonten, C.; Altstädt, V. Evaluation of the Zero Shear Viscosity, the D-Content and Processing Conditions as Foam Relevant Parameters for Autoclave Foaming of Standard Polylactide (PLA). *Materials* **2020**, *13*, 1371. [CrossRef] [PubMed]
30. Mitutoyo America Corporation. *Quick Guide to Surface Roughness Measurement*; Mitutoyo America Corporation: Aurora, IL, USA, 2016.
31. Ong, T.; Choo, H.L.; Choo, W.; Koay, S.C.; Pang, M. Recycling of Polylactic Acid (PLA) Wastes from 3D Printing Laboratory. In *Advances in Manufacturing Engineering*; Emamian, S.S., Awang, M., Yusof, F., Eds.; Springer: Singapore, 2020; pp. 725–732. ISBN 978-981-15-5752-1.
32. Herianto; Atsani, S.I.; Mastrisiswadi, H. Recycled Polypropylene Filament for 3D Printer: Extrusion Process Parameter Optimization. *IOP Conf. Ser. Mater. Sci. Eng.* **2020**, *722*, 12022. [CrossRef]

Review

Thermocatalytic Conversion of Plastics into Liquid Fuels over Clays

Evgeniy S. Seliverstov [1], Lyubov V. Furda [2] and Olga E. Lebedeva [2,*]

1 Department of Biology, Institute of Pharmacy, Chemistry and Biology, Belgorod State National Research University, 308015 Belgorod, Russia; seliverstov.evgeniy.s@gmail.com
2 Department of General Chemistry, Institute of Pharmacy, Chemistry and Biology, Belgorod State National Research University, 308015 Belgorod, Russia; furda@bsu.edu.ru
* Correspondence: olebedeva@bsu.edu.ru

Abstract: Recycling polymer waste is a great challenge in the context of the growing use of plastics. Given the non-renewability of fossil fuels, the task of processing plastic waste into liquid fuels seems to be a promising one. Thermocatalytic conversion is one of the methods that allows obtaining liquid products of the required hydrocarbon range. Clays and clay minerals can be distinguished among possible environmentally friendly, cheap, and common catalysts. The moderate acidity and the presence of both Lewis and Brønsted acid sites on the surface of clays favor heavier hydrocarbons in liquid products of reactions occurring in their pores. Liquids produced with the use of clays are often reported as being in the gasoline and diesel range. In this review, the comprehensive information on the thermocatalytic conversion of plastics over clays obtained during the last two decades was summarized. The main experimental parameters for catalytic conversion of plastics according to the articles' analysis, were the reaction temperature, the acidity of modified catalysts, and the catalyst-to-plastic ratio. The best clay catalysts observed were the following: bentonite/spent fluid cracking catalyst for high-density polyethylene (HDPE); acid-restructured montmorillonite for medium-density polyethylene (MDPE); neat kaolin powder for low-density polyethylene (LDPE); Ni/acid-washed bentonite clay for polypropylene (PP); neat kaolin for polystyrene (PS); Fe-restructured natural clay for a mixture of polyethylene, PP, PS, polyvinyl chloride (PVC), and polyethylene terephthalate (PET). The main problem in using natural clays and clay minerals as catalysts is their heterogeneous composition, which can vary even within the same deposit. The serpentine group is of interest in studying its catalytic properties as fairly common clay minerals.

Keywords: secondary raw materials; plastics; fuel; catalysts; clays; clay minerals; thermocatalytic conversion

Citation: Seliverstov, E.S.; Furda, L.V.; Lebedeva, O.E. Thermocatalytic Conversion of Plastics into Liquid Fuels over Clays. *Polymers* **2022**, *14*, 2115. https://doi.org/10.3390/polym14102115

Academic Editors: Antonio Pizzi and Sheila Devasahayam

Received: 23 March 2022
Accepted: 20 May 2022
Published: 23 May 2022

Publisher's Note: MDPI stays neutral with regard to jurisdictional claims in published maps and institutional affiliations.

1. Introduction

The last few centuries have been marked by the rapid development of mankind. The obvious benefits that it brought were accompanied by new, serious anthropogenic challenges. One of them was the emergence in the 1950s of new synthetic materials—plastics. The main ingredient of plastic are polymers, such as polyolefins (with commercially dominant polyethylene and polypropylene) possessing the general formula $(CH_2CHR)_n$ where R is an alkyl group, polystyrene $((C_6H_5CH = CH_2)_n)$, polyvinyl chloride $((C_2H_3Cl)_n)$, etc. Disposable tableware, containers, packaging, and many other plastic products have firmly entered our everyday life, but their uncontrolled disposable use has created a huge threat to the environment. Nature was not ready for this amount of difficult-to-recycle material in a very short time, and despite recent reports of microorganisms across the globe adapting themselves to plastic degradation [1], it is still our urgent responsibility to resolve this problem.

One of the promising solutions is the conversion of plastic waste into liquid fuels. With a catalyst sufficiently selective to produce a mixture of hydrocarbons with an expected carbon number range, it would be possible to obtain liquid products with a composition similar to that of fuels such as gasoline and diesel. Since the production of various catalysts is often accompanied by environmental pollution, a complex preparation process, and, as a result, a high price of the final product, the catalysts must also comply with the principles of green chemistry and have a low cost.

Solid acid catalysts are among the most effective in the catalytic conversion of plastics. The process of the thermocatalytic transformation using these catalysts mostly depends on the presence of acid sites [2] and the number and size of catalyst pores [3]. Many works, including those of our group, are devoted to the use of synthetic aluminosilicates [4,5]. In particular, the synthesis and application of specific nanosponges of solid acids, named "acidic aluminosilicates", should be mentioned as one of the latest achievements [6].

Microporous zeolite catalysts have high acidity active sites, which makes them able to split carbon-carbon bonds [2]. However, the small pore size of zeolites restricts the access of large molecules to acid sites located inside the channels. The presence of this steric factor leads to a higher yield of gases and a relatively high concentration of branched hydrocarbons among the degradation products since the contact of the polymer chain occurs mainly with the outer surface of the zeolite. In addition, a significant number of solid degradation products are formed on the surface and inside the pores of the catalyst. This leads to zeolite pore closure and catalytic deactivation. In the process of polymer cracking, these catalysts provide high selectivity for gaseous products.

Clays are moderately acidic, so reactions occurring within their pores favor the transformation of heavier hydrocarbons into liquid products than those of zeolites [7].

Liquids produced from plastic waste with the use of clay catalysts are often reported as being in the gasoline and diesel range [8]. Moreover, the layered structure of clays allows the formation of a porous network by alternating plates with so-called pillars (three-dimensional species as interlayer cations), thus creating interconnected micropores larger than those of zeolites [7]. Such materials demonstrate high stability and the possibility of reuse during heating to high temperatures [9].

Several reviews have been published on the topic of catalytic pyrolysis of plastics and the search for low-cost catalysts recently [10,11]. However, the works devoted to using clay-based catalysts were covered there briefly, and they did not include all available research on the activity of clay minerals of different groups. Peng et al. mentioned only montmorillonites and their analogs [10], while Fadillah et al. considered a few articles on kaolin and bentonite activity [11].

The task of collecting comprehensive information on the thermocatalytic conversion of plastics over clays obtained by different authors during the last two decades was set during the literature analysis. Works published in the last five years (2017–2021, including 2022) have been highlighted in bold in the tables to focus the attention on the latest results.

Natural materials containing clay minerals (hydrous aluminum silicates with variable amounts of cations) originated from natural sites or synthesized are designated as «clays» in the context of the present review. The application of pristine clays is rather rare. Usually, the clays are modified by different treatments. Modified clays are also the subjects of this review.

The following types of plastic materials are designated merely by the abbreviations to simplify the perception in the following text below: high-density polyethylene (HDPE), medium-density polyethylene (MDPE), low-density polyethylene (LDPE), polypropylene (PP), polystyrene (PS), polyethylene terephthalate (PET), polyvinyl chloride (PVC), ethylene-vinyl acetate (EVA).

2. Nature of Catalytic Activity of Clays

Clays belong to solid acids. They have both Lewis and Brønsted acid sites (Figure 1) [7].

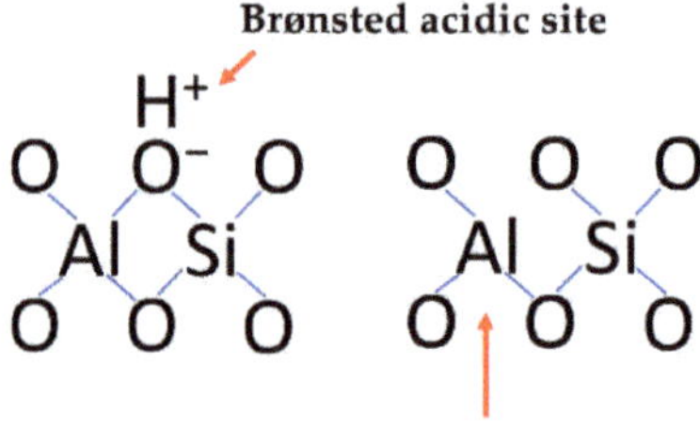

Figure 1. Acidic sites of clays.

The acidic sites are comparatively strong (H_0 typically quoted in the range from -5.6 to -8.2), though not as strong as the zeolite ones [7]. All the clays being aluminosilicates, the nature of the active sites is essentially the same for all types of clays. It is porosity that defines the specific features of different clays. Microporosity depends on the crystallographic structure of the material. There is an additional factor influencing the porosity of the clays. Their part-amorphous nature provides mesoporosity over a wide range of pore sizes.

Original clays in cationic forms usually contain an insufficient number of acidic sites since the sites involve protons (Figure 1). Only cationic deficient samples of clays demonstrate catalytic activity in the reactions of the acid-base type. Generally, acidic activation is necessary for obtaining catalytically active clays. The conditions of acidic treatment are often crucial for the efficiency of the clay catalysts.

3. Kaolin Group Catalytic Activity

The kaolin group is represented by layered phyllosilicate minerals with the chemical composition $Al_2Si_2O_5(OH)_4$. The layers of these clay minerals consist of corner-sharing tetrahedra and edge-sharing octahedra. Tetrahedra are formed by silicon atoms, and octahedrons are constructed from aluminum atoms. The way the layers are stacked and the nature of the material between the layers distinguishes the individual minerals (kaolinite, dickite, halloysite, and nacrite, sometimes also serpentine subgroup) in the group [12]. Rocks rich in kaolinite are thus called kaolin.

Kaolin-based catalysts are the most commonly mentioned among the articles on clay catalysts for the conversion of plastics into liquid fuels due to the abundant availability of natural kaolin. All results from work on kaolin clay catalysts are presented in Table 1. The symbol + is used when a mixture of polymers is described in the publication.

Table 1. Publications on the conversion of different plastics over clay minerals from kaolin group catalysts.

Catalyst	Plastic	Temperature, °C	Highest Liquid Yield, wt%	Specific Results	Reference
Kaolinite-containing natural clay	HDPE	478	16	Catalyst produced more alkanes than olefins in both gaseous and liquid oil products.	[13]
Kaolin and its modifications With CH_3COOH, HCl, H_3PO_4, HNO_3, and NaOH	HDPE	450	78.7	The liquid fuel consisted of petroleum products range hydrocarbons (C_{10}–C_{25}).	[14]
Kaolin	LDPE	450	79.5	The oil consists of paraffins and olefins with a predominance of C_{10}–C_{16} components.	[15]
Kaolin	LDPE	600	about 75	**The first addition of kaolin gives aliphatic compounds and C_6–C_{20} aromatics (90–95%).**	[16]
75% kaolinite with 25% bentonite	LDPE	580	74.45	**High yield of paraffins (70.62%). The percentage of aromatics was 5.27%.**	[17]

Table 1. *Cont.*

Catalyst	Plastic	Temperature, °C	Highest Liquid Yield, wt%	Specific Results	Reference
China clay (kaolinite)	LDPE	300	84	Components with a boiling point of 125–180°C were identified as alkanes, alkenes, and aromatics.	[18]
Kaolin	LDPE	450	99.82	The highest percentage component is heptane.	[19]
Al-substituted Keggin tungstoborate/kaolin composite	LDPE	295	84	During the catalytic cracking 70 mol.% of gasoline range hydrocarbons were produced.	[20]
tungstophosphoric acid/kaolin composite	LDPE	335	81	A high content of benzene-like hydrocarbons (C_{11}–C_{14}).	[21]
Ahoko kaolin	PP	450	79.85	Liquid products with properties comparable to conventional fuels (gasoline and diesel).	[22]
Hydrochloric acid/kaolin composite	PP	470	71.9	The condensable hydrocarbons contain dominantly alkanes and alkenes in the range C_6–C_{12}.	[23]
Commercial-grade kaolin clay	PP	450	89.5	Contains olefins, aliphatic, and aromatic hydrocarbons in the oil comparable with liquid fossil fuels.	[24]
Commercial-grade kaolin clay and kaolin treated with sulfuric acid	PP	500	92 (acid-treated), 87.5 (neat kaolin)	The oil from the neat kaolin—C_{10}–C_{18} products, from the acid-treated kaolin—mainly C_9–C_{13}.	[25]
Kaolin	PP	500	87.5	Fuel properties are identical to the different petroleum fuels.	[26]
Neat kaolin and kaolin treated with hydrochloric acid	PP	400–500	71.9	The highest yield of liquid hydrocarbons was achieved with kaolin clay treated with 3M HCl.	[27]
Kaolin	PP/vaseline (4.0 wt%)	520	52.5	The gasoline—32.77%, diesel—13.59%, residue—6.14%	[28]
CuO/kaolin and neat kaolin	PS	450	96.37 (neat kaolin), 92.48 (CuO/kaolin)	The oil contained aromatic hydrocarbons, but from CuO/kaolin—85% $C_{10}H_8$ and ~13% C_8H_8.	[29]
Zeolite-Y + metakaolin + aluminum hydroxide + sodium silicate all synthesized from kaolin	HDPE + LDPE + PP + PS + PET	350	46.7	Catalyzed fuel samples consist of 93% gasoline and 7% diesel fraction.	[30]
Kaolin	Virgin HDPE, HDPE waste and mixed plastic waste	425	79	The catalyst was the most selective in producing diesel, which yielded 63%.	[31]
Halloysite treated with hydrochloric acid	PS	450	90.2	Aromatic compounds of more than 99%. The main product is styrene (58.82%).	[32]

In the work of Liu et al., natural kaolinite-containing clay had no acidic sites and did not show any effect on the degradation temperature of HDPE [13]. However, it produced liquid oil with a yield of 16 wt%, a number of gaseous products much smaller than that of thermal degradation with a yield of 3.4 wt%, and a number of alkanes larger than that of olefins. The authors concluded that a clay catalyst was favorable for the enhancement of the intermolecular hydrogen transfer reaction and inhibition of the β-scission reaction of radicals compared to thermal degradation, which was related to the hydrogen bonds from the layer structure and large mesopores.

It appears that the best result for the high-density polyethylene degradation was obtained by Kumar and Singh using the response surface methodology (RSM) [14]. RSM allowed a reduction in the number of costly experiments by selecting the right experimental conditions. It can be considered a promising method for the evaluation of selected experimental variables in the planning step of such experiments. The optimized values of experimental variables were 450 °C, 0.341, and 1:4 for reaction temperature, the acidity of the catalyst, and the catalyst-to-waste HDPE ratio, respectively, to produce a maximum liquid fuel yield of 78.7%.

Luo et al. studied the possibility of reusing the kaolin catalyst. They found that the yield of aromatic hydrocarbons increases with the reuse of the catalyst, which is associated with an increase in particle size [16]. The optimal size of kaolin particles was investigated in the work of Erawati et al., where it was established as 7.5×6.5 cm [19].

The influence of non-inert reaction conditions on products of thermocatalytic conversion was shown by Uzair et al. [23]. Alcohols and ketones were formed due to oxidative cracking of PP.

The work of Auxilio et al. proved that catalyst surface Lewis acidity was critical for hydrocarbon fraction selectivity, and higher acidity favors gasoline formation, while low to mild acidity favors diesel formation [31]. They found that mesopore volume was a crucial factor in avoiding catalyst coking because the small mesopore volume favored high overall coke formation. In addition, the authors stated that using a pellet form catalyst was advantageous over powder form to avoid large pressure drops in the reactive distillation column.

In summarizing, the highest liquid yields described in the articles concerned with kaolin catalysts for different plastics are 78.7 wt% for HDPE over nitric acid-treated kaolin [14], 99.82 wt% for LDPE on neat kaolin powder [19], 92 wt% for PP on sulfuric acid-treated kaolin [25], and 96.37 wt% for PS over pristine kaolin [29]. Quite expectably, the most efficient catalysts were obtained using acidic treatment, which led to the generation of a sufficient number of acidic sites.

4. Smectite Group Catalytic Activity

Members of the smectite group include the dioctahedral minerals (montmorillonite, beidellite, and nontronite) and the trioctahedral minerals (hectorite, saponite, and sauconite). The basic structural unit of these clay minerals is a layer consisting of two inward-pointing tetrahedral sheets with a central alumina octahedral sheet [33]. The clay consisting mostly of montmorillonite is called bentonite, but in commerce, this term can be used in a more general way to refer to any swelling clay composed mostly of minerals from the smectite group.

The bentonite- and pure montmorillonite-based catalysts are the most commonly occurred besides smectite catalysts for plastic transformation. There are a few articles devoted to the use of saponite and beidellite. All results from the work on smectite clay catalysts are presented in Table 2.

Table 2. Publications on the conversion of different plastics over clay minerals from smectite group catalysts.

Catalyst	Plastic	Temperature, °C	Highest Liquid Yield, wt%	Specific Results	Reference
Bentonite (50 wt%)/spent fluid catalytic cracking catalyst (FCC)	HDPE	500	100	High yields of gasoline C_5–C_{11} (50 wt%) The yield of C_{12}–C_{20} hydrocarbons—8–10 wt%.	[34]
Pillared bentonite (PILC) intercalated with Fe or Al	**HDPE and heavy gas oil (HGO)**	500	>80	**The oil from the Fe-PILC-Fe-300 catalyst was more similar to the standard diesel.**	[35]
Bentonite (Gachi clay)	LDPE	300	77	Olefin and paraffin hydrocarbons.	[36]
South Asian clay classified as bentonite andmontmorillonite impregnated with nickel NPs	**LDPE and post-consumer polybags**	350	79.23 (LDPE), 76.01 (poly-bags)	**The final products are in the range of gasoline, kerosene, and diesel.**	[37]
Bentonite thin layer loaded with MnO$_2$ nanoparticles (NPs)	PP	750	**Parameters were designed to get off the liquid**	**The complete decomposition of plastics with the formation of gases (methane and hydrogen) and coke.**	[38]
Bentonite treated with 0.5M hydrochloric acid	PS	400	88.78	**The obtained liquid contains styrene. Toluene and benzene were the major components.**	[39]

Table 2. *Cont.*

Catalyst	Plastic	Temperature, °C	Highest Liquid Yield, wt%	Specific Results	Reference
Acid-washed bentonite clay (AWBC), Zn/AWBC, Ni/AWBC, Co/AWBC, Fe/AWBC, Mn/AWBC	PP, HDPE	300 for PP and 350 for HDPE	AWBC (PP 68.77, HDPE 70.19), Ni/AWBC (PP 92.76, HDPE 62.07), Co/AWBC (PP 82.8, HDPE 69.31), Fe/AWBC (PP 82.78, HDPE 71.34), Mn/AWBC (PP 80.4, HDPE 81.07), Zn/AWBC (PP 82.50, HDPE 91)	Co/AWBC/PP (mainly olefins and naphthenes) and Zn/AWBC/HDPE (mainly paraffins and olefins) were the most effective.	[40]
H_2SO_4-activated bentonite (synthesized)	PP + HDPE	328	79	The hydrocarbon oil.	[41]
A mixture of nature bentonite and zeolite (70:30)	PP, PET	400	78.42 (PP), 72.38 (PP + PET)	The number of C_3–C_{10} compounds increased.	[42]
Pelletized bentonite	PS, PP, LDPE, HDPE	500	88.5 (PS), 90.5 (PP), 87.6 (LDPE), 88.9 (HDPE)	PS—95% aromatic hydrocarbons; PP, LDPE, and HDPE—aliphatic hydrocarbons; LDPE, and HDPE—diesel fuel (96% similarity); PS—gasohol 91.	[8]
Calcium bentonite	PP, LDPE, HDPE, PP + LDPE + HDPE	500	88.5 (PP), 82 (LDPE), 82.5 (HDPE) 81 (PP + LDPE + HDPE)	The oil contained only a mixture of hydrocarbons and has matching fuel properties as that of fossil fuel. Mixed plastics—C_{10}-C_{28}.	[43]
Pillared bentonite (Al-PILC, Fe-PILC, Ti-PILC, Zr-PILC)	HDPE + PS + PP + PET	300–500	68.2 (Al-PILC), 79.3 (Fe-PILC), 62.8 (Ti-PILC), 62.1 (Zr-PILC)	80.5% diesel fraction was observed in presence of Fe-PILC.	[7]
Fe/Al pillared montmorillonite mixed with an acid Commercial bentonite as a binder	HDPE	600	About 40	The catalyst gave high yields of waxes, particularly rich in diesel hydrocarbon range (C_{11}–C_{21}).	[44]
commercial acid-restructured montmorillonite and Al- and Fe/Al-pillared derivative	MDPE	300	About 70	The clay-based catalysts gave higher yields of liquid products in the C_{15}–C_{20} range. Clay catalysts produce liquid hydrocarbons in the gasoline and diesel range.	[45]
Al_2O_3-pillared montmorillonite (calcium rich)	LDPE	430	70.2	Hydrocarbons from C_5 to C_{13}.	[46]
Montmorillonite (Zenith-N) and a pillared derivative	LDPE	427	68 (montmorillonite), 75 (pillared derivative)	Clays showed enhanced liquid formation due to their mild acidity.	[47]
Al-pillared montmorillonite (Al-PILC), and regenerated samples	LDPE	360	72 (Al-PILC), 68 (regenerated sample)	These products were in the boiling point range of motor engine fuels.	[48]
Montmorillonite (Zenith-N) and a pillared derivative	LDPE	360	75 (montmorillonite), 76 (pillared derivative)	These products were in the boiling point range of gasoline.	[49]
Ionically bonding macrocyclic Zr-Zr complex to montmorillonite	PP	300–400	-	A low molecular weight waxy product with paraffin wax characteristics was obtained.	[50]
Untreated and Al-pillared montmorillonite clay	PS	400	83.2 (untreated clay), 81.6 (Al-pillared clay)	Styrene was the major product, and ethylbenzene was the second most abundant one in the liquid product.	[51]
Four different types of montmorillonites: K5, K10, K20, K30	LDPE, PP, and the municipal waste plastics	begins at 250 for mK5 (LDPE), 210–435 for mK20 (PP)	Data not presented	The catalytic degradation products contain a relatively narrow distribution of light hydrocarbons.	[52]

Table 2. *Cont.*

Catalyst	Plastic	Temperature, °C	Highest Liquid Yield, wt%	Specific Results	Reference
Organically modified montmorillonite/Co_3O_4	PP + HDPE + PS	700	59.6	The catalyst promoted the degradation of mixed plastics into light hydrocarbons and aromatics.	[53]
cloisite 15 A as a natural montmorillonite modified with a quaternary ammonium salt	Industrial grade of HDPE, which was a copolymer with 1-hexene (1.5 wt%) as comonomer	473.7	Data not presented	**It was found that the nano clay reduces the temperature at a maximum degradation rate.**	[54]
Commercial acid-restructured saponite and Al- and Fe/Al-pillared derivatives	MDPE	300	About 70	The clay-based catalysts gave higher yields of liquid products in the C_{15}–C_{20} range. Clay catalysts produce liquid hydrocarbons in the gasoline and diesel range.	[45]
Saponite, with a small number of impurities, mainly sepiolite and a pillared derivative	LDPE	427	83 (saponite), 82 (coked pillaredderivative)	Clays showed enhanced liquid formation due to their mild acidity.	[47]
Al-pillared saponite and regenerated samples	LDPE	360	72 (pillared saponite), 67 (regenerated sample)	These products were in the boiling point range of motor engine fuels.	[48]
Saponite and a pillared derivative	LDPE	360	68 (saponite), 72 (pillared derivative)	These products were in the boiling point range of gasoline.	[49]
Commercial acid-restructured beidellite and Al- and Fe/Al-pillared derivatives	MDPE	300	About 70	The clay-based catalysts gave higher yields of liquid products in the C_{15}–C_{20} range. The catalysts produce liquid hydrocarbons in the gasoline and diesel range.	[45]

Pillared bentonite clays were selective to cracking HGO/HDPE in light hydrocarbons (C_{10}–C_{23}) and produced a light linear hydrocarbon content 63% higher than that produced with zeolite [35].

The work of Elordi et al. draws attention to the result obtained by the authors that pristine bentonite does not demonstrate catalytic activity at 500 °C in a conical spouted bed reactor in the continuous regime (1 g min^{-1} of HDPE is fed) [34]. However, agglomeration of 50 wt% bentonites with spent fluid catalytic cracking catalyst (FCC) allows the thermal cracking of the initial macromolecules in the mesopores of the clay until they reach the spent FCC particles.

Gobin and Manos noted that even if clays were less active than zeolites, they could fully degrade the polymer [47]. In this work, the authors used montmorillonite (Zenith-N), saponite (with a small number of impurities, mainly sepiolite), and their pillared derivatives. They showed enhanced liquid formation and lower coke formation. Regenerated pillared clays offered practically the same performance as fresh samples, but their original performance deteriorated after the removal of the formed coke.

Summing up, the highest liquid yields described in the articles for smectite catalysts for different plastics are 100 wt% for HDPE over bentonite (50 wt%)/spent fluid catalytic cracking catalyst (FCC) [34], about 70 wt% for MDPE over acid-restructured montmorillonite catalyst [45], 87.6 wt% for LDPE over pelletized bentonite [8], 92.76 wt% for PP over Ni/acid-washed bentonite clay [40], and 88.78 wt% for PS on acid-treated bentonite-based catalyst [39]. The full conversion of the HDPE in the case of using the FCC was achieved according to step-by-step reactions where on the first step, the thermal cracking of the initial macromolecules occurred in the mesopores of the bentonite until they reached the spent FCC catalyst particles.

5. Other Clay Minerals' Catalytic Activity

The variety of clay minerals is not limited to the above-mentioned two groups. Only a few examples of studying the catalytic activity of other clay minerals (sepiolite, vermiculite, talc, and pyrophyllite) in relation to plastics were found (Table 3).

Table 3. Publications on the conversion of different plastics over sepiolite, talc, pyrophyllite, and vermiculite catalysts.

Catalyst	Plastic	Temperature, °C	Highest Liquid Yield, wt%	Specific Results	Reference
Commercial sepiolite	PE, PP, PS, EVA	432.65 (PE), 401.65 (PP), 449.75 (PS), 459.85 (EVA)	Data not presented	Clay reduces the decomposition temperatures of PE and PP. However, steric effects associated with the PS and EVA substituents nullify this catalytic behavior.	[55]
Tetraethyl silicate modified vermiculite, Co, and Ni intercalated vermiculite	**PP + PE**	**300–480**	**80.6 (organic vermiculite), 73.2 (Co/verm), 70.7 (Ni/verm), 73.9 (Co/Ni/verm)**	**The obtained liquid is mainly composed of C_9–C_{12} and C_{13}–C_{20}.**	**[56]**
Talc (French chalk)	LDPE	300	91	Components with a boiling point of 125–180°C were identified as alkanes, alkenes, and aromatics.	[18]
Talc (plastic filler)	**PP**	**620**	**About 23**	**The liquid product contained a higher aromatic content (57.9%) and a lower n-alkene content (5.8%).**	**[57]**
Pyrophyllite treated with hydrochloric acid	PS	450	88.3	The catalysts showed selectivity to aromatics over 99%. Styrene (63.40%) is the major product, and ethylbenzene is the second-most abundant one (6.93%).	[32]

Interestingly, in the case of talc, its catalytic activity was revealed by chance [57]. Talc is often a filler in polypropylene that increases its stiffness. The product yields and compositions from pure PP and PP with fillers showed a significant difference, indicating a higher degree of degradation for PP with fillers, most likely resulting from the fillers acting as a catalyst. It produced a much higher gas yield (76.3%) and a negligible wax yield.

Khan and Hussain also reported the catalytic activity of talk (French chalk, as mentioned in the work) [18]. They indicated that the products of the pyrolysis of the French chalk catalyzed reactions contain no wax and give a greater proportion of the oil as well as gaseous products.

The results obtained for sepiolite show that, despite the low "nominal" catalytic activity of this clay, it has enough catalytic properties to decrease the temperature of decomposition of PE and PP [55]. However, the steric effects related to the substituents of PS and EVA cancel this catalytic behavior. Experiments performed in an oxidizing atmosphere showed that there was no noticeable decrease in the temperature that may be related to the presence of the clay.

The Co/Verm and Ni/Verm catalysts in the work of Chen et al. had higher selectivity for fractions with a carbon number greater than C_{13} [56]. Organic Verm and CoNi/Verm catalysts had higher selectivity for fractions with a carbon number less than C_{13}. Due to the interaction between acidity and texture properties, the modified catalyst could produce a large amount of diesel oil, a distillate from petroleum products, and H_2 in natural gas products.

The acid-treated pyrophyllite catalyst also showed good catalytic performance for the degradation of PS [32]. Compared to thermal degradation, catalysts showed much higher selectivity for ethylbenzene and much lower production of C_{16}–C_{21} (8.45%).

To sum it up, the most promising liquid yields were obtained by degradation of LDPE on talc (91 wt%) [18] and PS on pyrophyllite (88.3 wt%) [32]. Similar to the cases for the above-described kaolin acidic treatment allows reaching the highest efficiency in plastic conversion.

6. Catalytic Activity of Mixed Natural Clays

Some works were focused on uncharacterized mixed clays from different fields (Table 4).

Table 4. Publications on the conversion of different plastics over clays from different fields.

Catalyst	Plastic	Temperature, °C	Highest Liquid Yield, wt%	Specific Results	Reference
Acid-activated fire clay (Pradeep Enterprises, Ajmeri Gate, Delhi)	HDPE	450	41.4	The identified compounds were mainly paraffins and olefins with a carbon number range of C_6–C_{18}.	[58]
Indian Fuller's earth (Multan clay)	LDPE	300	58.33	The obtained liquid contained olefin, paraffin, and aromatic hydrocarbons. Light naphtha—15%, heavy naphtha—35%, middle distillate—60%.	[59]
Fuller's earth	LDPE	300	91	Components with a boiling point of 125-180°C were identified as alkanes, alkenes, and aromatics.	[18]
Natural clay mineral (Indonesia) with $LaFeO_3$ NPs	PP	460–480	88.8 (5th cycle)	The liquid fraction: alkanes (44.70%), alkenes (34.84%), cyclo-alkanes (9.87%), cyclo-alkenes (3.07), branched-chain alkanes (2.42%), branched-chain alkenes (0.88%).	[60]
natural clay with kaolinite, hematite, smectite, quartz	PS	410	86.68	Fuel properties of the liquid fraction obtained showed a good resemblance with gasoline and diesel oil.	[61]
Red clay (Auburn, Alabama, USA)	PS and LDPE (co-pyrolysis with a lignin)	500, 600, 700, 800	data not presented	The carbon yield of a lignin-derived compound, guaiacol, increased during co-pyrolysis of lignin with LDPE, and PS with red clay as a catalyst.	[62]
Shwedaung clay, Mabisan clay	HDPE + LDPE + PS + PP + PET	210–380	65.81 (Shwedaung clay), 67.06 (Mabisan clay)	Fuel can be used internal combustion engine after distillation. Char can be used as solid fuel.	[63]
Fe-restructured clay (Fe-RC)	PE + PP + PS + PVC + PET	450	83.73	High selectivity for the C_9–C_{12} and C_{13}–C_{19} oil fractions, which are the major constituents of kerosene and diesel fuel.	[64]
Romanian natural clays: Vadu Crişului clay and Lugoj clay	PS + PET + PVC	420	62.18 (Vadu Crişului clay), 54.98 (Lugoj clay)	The liquid products contained monoaromatic compounds such as styrene, toluene, ethylbenzene, or alpha-methylstyrene.	[65]

For example, Filip et al. investigated the thermal degradation processes at 420 °C of a plastic waste mixture (PS + PET + PVC) in the absence and presence of two types of natural Romanian clay catalysts [65]. The GC-MS results showed that the liquid fractions contained mainly monoaromatic compounds. The highest amounts of styrene come from thermal degradation of PS, which was the major component in the plastic mixture. The Vadu Crişului clay catalyst has been found as the most efficient catalyst for the thermal degradation of a plastic mixture.

In summarizing, the highest liquid yields were obtained for LDPE on Fuller's earth catalyst (91 wt%) [18]; PP on natural clay mineral from Indonesia impregnated with $LaFeO_3$ nanoparticles (88.8 wt% on the 5th cycle) [60]. It should be stressed that in the latter case, the yield growth was provided by the specific efficient promoter—lanthanum ferrite. This significant distinction of the catalyst attracts attention to the perspectives of non-acidic modification of clays.

7. Conclusions and Perspectives

The results of numerous researches give evidence that the main experimental parameters for thermocatalytic conversion of plastics were the reaction temperature, the acidity of modified catalysts, and the catalyst-to-plastic ratio. By varying the parameters, one can achieve an essential increase in the yield of the liquid hydrocarbons in the process of plastic conversion.

The best clay and clay-based catalysts with the highest liquid yields among works described in this review for each of the plastic were the following: bentonite/spent FCC for HDPE; acid-restructured montmorillonite for MDPE; neat kaolin powder for LDPE; Ni/acid-washed bentonite clay for PP; neat kaolin for PS; Fe-restructured natural clay for a mixture of PE, PP, PS, PVC, and PET. It can be seen that the modification of clay catalysts (acid-washing or pillaring) in some cases helps to achieve a higher yield of the liquid fraction. However, some pure clays and clay minerals are also showing excellent catalytic activity.

The principal problem in using natural clays and clay minerals as catalysts is their heterogeneous composition, which can vary even within the same deposit. Therefore, studies on their use should begin with a thorough characterization of the samples used—their elemental composition, particle size and porosity, acidity, etc. Otherwise, the main rule of reproducibility of scientific results is violated, and works using the same clays and clay minerals can obtain drastically different results, leading to confusion.

Despite the availability of well-studied catalysts based on kaolin, bentonite, and montmorillonite, many other clay minerals remain poorly studied as prospective catalysts. For instance, the serpentine group (often combined with kaolin in the kaolin-serpentine group) is a set of common rock-forming hydrous magnesium iron phyllosilicate $((Mg,Fe)_3Si_2O_5(OH)_4)$ minerals commonly found in serpentinite rocks. Serpentinite has not been used directly as a catalyst but has shown very interesting results as a precursor to producing active catalysts (i.e., the intercalation of serpentine with the alkaline metals gave rise to the basic catalysts for the production of biodiesel). Thus, the serpentine group is of interest in studying its catalytic properties as fairly common but not well-studied clay minerals. Minerals of the chlorite, illite, and halloysite groups also deserve a separate investigation.

Another promising direction of future studies is clay activation and modification. Various examples of modifications thus far applied by different authors cannot be considered a comprehensive list of possible treatments. Some well-known methods of clay activation, such as UV-irradiation, mechanical treatment, and especially chemical promotion, are still of interest.

Author Contributions: Conceptualization, E.S.S. and O.E.L.; investigation, E.S.S. and L.V.F.; writing—original draft preparation, E.S.S.; writing—review and editing, O.E.L. and L.V.F.; visualization, E.S.S.; supervision, O.E.L. All authors have read and agreed to the published version of the manuscript.

Funding: This research received no external funding.

References

1. Zrimec, J.; Kokina, M.; Jonasson, S.; Zorrilla, F.; Zelezniak, A. Plastic-Degrading Potential across the Global Microbiome Correlates with Recent Pollution Trends. *MBio* **2021**, *12*, e02155-21. [CrossRef] [PubMed]
2. Obalı, Z.; Sezgi, N.A.; Doğu, T. Catalytic Degradation of Polypropylene over Alumina Loaded Mesoporous Catalysts. *Chem. Eng. J.* **2012**, *207–208*, 421–425. [CrossRef]
3. Obali, Z.; Sezgi, N.A.; Doğu, T. Performance of acidic MCM-like aluminosilikate catalysts in pyrolysis of polypropylene. *Chem. Eng. Commun.* **2008**, *196*, 116–130. [CrossRef]
4. Furda, L.V.; Ryl'tsova, I.G.; Lebedeva, O.E. Catalytic Degradation of Polyethylene in the Presence of Synthetic Aluminosilicates. *Russ. J. Appl. Chem.* **2008**, *81*, 1630–1633. [CrossRef]
5. Furda, L.V.; Smalchenko, D.E.; Titov, E.N.; Lebedeva, O.E. Thermocatalytic Degradation of Polypropylene in Presence of Aluminum Silicates. *Izv. Vyss. Uchebnykh Zaved. Khimiya Khimicheskaya Tekhnologiya* **2020**, *63*, 85–89. [CrossRef]
6. Maity, A.; Chaudhari, S.; Titman, J.J.; Polshettiwar, V. Catalytic Nanosponges of Acidic Aluminosilicates for Plastic Degradation and CO_2 to Fuel Conversion. *Nat. Commun.* **2020**, *11*, 3828. [CrossRef] [PubMed]
7. Li, K.; Lei, J.; Yuan, G.; Weerachanchai, P.; Wang, J.-Y.; Zhao, J.; Yang, Y. Fe-, Ti-, Zr- and Al-Pillared Clays for Efficient Catalytic Pyrolysis of Mixed Plastics. *Chem. Eng. J.* **2017**, *317*, 800–809. [CrossRef]
8. Budsaereechai, S.; Hunt, A.J.; Ngernyen, Y. Catalytic Pyrolysis of Plastic Waste for the Production of Liquid Fuels for Engines. *RSC Adv.* **2019**, *9*, 5844–5857. [CrossRef]
9. Geng, J.; Sun, Q. Effects of High Temperature Treatment on Physical-Thermal Properties of Clay. *Thermochim. Acta* **2018**, *666*, 148–155. [CrossRef]
10. Peng, Y.; Wang, Y.; Ke, L.; Dai, L.; Wu, Q.; Cobb, K.; Zeng, Y.; Zou, R.; Liu, Y.; Ruan, R. A Review on Catalytic Pyrolysis of Plastic Wastes to High-Value Products. *Energy Convers. Manag.* **2022**, *254*, 115243. [CrossRef]
11. Fadillah, G.; Fatimah, I.; Sahroni, I.; Musawwa, M.M.; Mahlia, T.M.I.; Muraza, O. Recent Progress in Low-Cost Catalysts for Pyrolysis of Plastic Waste to Fuels. *Catalysts* **2021**, *11*, 837. [CrossRef]
12. Giese, R.F. Kaolin Group Minerals. In *Sedimentology*; Springer: Dordrecht, The Netherlands, 1978; pp. 651–655.
13. Liu, M.; Zhuo, J.K.; Xiong, S.J.; Yao, Q. Catalytic Degradation of High-Density Polyethylene over a Clay Catalyst Compared with Other Catalysts. *Energy Fuels* **2014**, *28*, 6038–6045. [CrossRef]
14. Kumar, S.; Singh, R.K. Optimization of Process Parameters by Response Surface Methodology (RSM) for Catalytic Pyrolysis of Waste High-Density Polyethylene to Liquid Fuel. *J. Environ. Chem. Eng.* **2014**, *2*, 115–122. [CrossRef]
15. Panda, A.K.; Singh, R.K. Thermo-Catalytic Degradation of Low Density Polyethylene to Liquid Fuel over Kaolin Catalyst. *Int. J. Environ. Waste Manag.* **2014**, *13*, 104. [CrossRef]
16. Luo, W.; Fan, Z.; Wan, J.; Hu, Q.; Dong, H.; Zhang, X.; Zhou, Z. Study on the Reusability of Kaolin as Catalysts for Catalytic Pyrolysis of Low-Density Polyethylene. *Fuel* **2021**, *302*, 121164. [CrossRef]
17. Soliman, A.; Farag, H.A.; Nassef, E.; Amer, A.; ElTaweel, Y. Pyrolysis of Low-Density Polyethylene Waste Plastics Using Mixtures of Catalysts. *J. Mater. Cycles Waste Manag.* **2020**, *22*, 1399–1406. [CrossRef]
18. Khan, K.; Hussain, Z. Comparison of the Catalytic Activity of the Commercially Available Clays for the Conversion of Waste Polyethylene into Fuel Products. *J. Chem. Soc. Pakistan* **2011**, *33*, 956–959.
19. Erawati, E.; Hamid; Martenda, D. Kinetic Study on the Pyrolysis of Low-Density Polyethylene (LDPE) Waste Using Kaolin as Catalyst. *IOP Conf. Ser. Mater. Sci. Eng.* **2020**, *778*, 012071. [CrossRef]
20. Attique, S.; Batool, M.; Yaqub, M.; Goerke, O.; Gregory, D.H.; Shah, A.T. Highly Efficient Catalytic Pyrolysis of Polyethylene Waste to Derive Fuel Products by Novel Polyoxometalate/Kaolin Composites. *Waste Manag. Res.* **2020**, *38*, 689–695. [CrossRef]
21. Attique, S.; Batool, M.; Jalees, M.I.; Shehzad, K.; Farooq, U.; Khan, Z.; Ashraf, F.; Shah, A.T. Highly Efficient Catalytic Degradation of Low-Density Polyethylene Using a Novel Tungstophosphoric Acid/Kaolin Clay Composite Catalyst. *Turkish J. Chem.* **2018**, *42*, 684–693. [CrossRef]
22. Hakeem, I.G.; Aberuagba, F.; Musa, U. Catalytic Pyrolysis of Waste Polypropylene Using Ahoko Kaolin from Nigeria. *Appl. Petrochem. Res.* **2018**, *8*, 203–210. [CrossRef]
23. Uzair, M.A.; Waqas, A.; Khoja, A.H.; Ahmed, N. Experimental Study of Catalytic Degradation of Polypropylene by Acid-Activated Clay and Performance of Ni as a Promoter. *Energy Sources Part A Recover. Util. Environ. Eff.* **2016**, *38*, 3618–3624. [CrossRef]
24. Panda, A.K.; Singh, R. Catalytic Performances of Kaoline and Silica Alumina in the Thermal Degradation of Polypropylene. *J. Fuel Chem. Technol.* **2011**, *39*, 198–202. [CrossRef]
25. Panda, A.K.; Singh, R. Conversion of Waste Polypropylene to Liquid Fuel Using Acid-Activated Kaolin. *Waste Manag. Res. J. Sustain. Circ. Econ.* **2014**, *32*, 997–1004. [CrossRef] [PubMed]
26. Panda, A.K.; Singh, R. Experimental Optimization of Process for the Thermo-Catalytic Degradation of Waste Polypropylene to Liquid Fuel. *Adv. Energy Eng.* **2013**, *1*, 74–84.
27. Uzair, M.A.; Waqas, A.; Afzal, A.; Ansari, S.H.; Anees ur Rehman, M. Application of Acid Treated Kaolin Clay for Conversion of Polymeric Waste Material into Pyrolysis Diesel Fuel. In Proceedings of the 2014 International Conference on Energy Systems and Policies (ICESP), Islamabad, Pakistan, 24–26 November 2014; pp. 1–4.
28. Ribeiro, A.M.; Machado Júnior, H.F.; Costa, D.A. Kaolin and Commercial fcc Catalysts in the Cracking of Loads of Polypropylene under Refinary Conditions. *Braz. J. Chem. Eng.* **2013**, *30*, 825–834. [CrossRef]

29. Hadi, B.; Sokoto, A.M.; Garba, M.M.; Muhammad, A.B. Effect of Neat Kaolin and Cuo/Kaolin on the Yield and Composition of Products from Pyrolysis of Polystyrene Waste. *Energy Sources Part A Recover. Util. Environ. Eff.* **2017**, *39*, 148–153. [CrossRef]
30. Eze, W.U.; Madufor, I.C.; Onyeagoro, G.N.; Obasi, H.C.; Ugbaja, M.I. Study on the Effect of Kankara Zeolite-Y-Based Catalyst on the Chemical Properties of Liquid Fuel from Mixed Waste Plastics (MWPs) Pyrolysis. *Polym. Bull.* **2021**, *78*, 377–398. [CrossRef]
31. Auxilio, A.R.; Choo, W.-L.; Kohli, I.; Chakravartula Srivatsa, S.; Bhattacharya, S. An Experimental Study on Thermo-Catalytic Pyrolysis of Plastic Waste Using a Continuous Pyrolyser. *Waste Manag.* **2017**, *67*, 143–154. [CrossRef]
32. Cho, K.-H.; Jang, B.-S.; Kim, K.-H.; Park, D.-W. Performance of Pyrophyllite and Halloysite Clays in the Catalytic Degradation of Polystyrene. *React. Kinet. Catal. Lett.* **2006**, *88*, 43–50. [CrossRef]
33. Altaner, S.P. Smectite Group. In *Sedimentology*; Springer: Dordrecht, The Netherlands, 1978; pp. 1120–1124.
34. Elordi, G.; Olazar, M.; Castaño, P.; Artetxe, M.; Bilbao, J. Polyethylene Cracking on a Spent FCC Catalyst in a Conical Spouted Bed. *Ind. Eng. Chem. Res.* **2012**, *51*, 14008–14017. [CrossRef]
35. Faillace, J.G.; de Melo, C.F.; de Souza, S.P.L.; da Costa Marques, M.R. Production of Light Hydrocarbons from Pyrolysis of Heavy Gas Oil and High Density Polyethylene Using Pillared Clays as Catalysts. *J. Anal. Appl. Pyrolysis* **2017**, *126*, 70–76. [CrossRef]
36. Hussain, Z.; Khan, K.; Jan, M.; Shah, J. Conversion of Low Density Polyethylene into Fuel Products Using Gachi Clay as Catalyst. *J. Chem. Soc. Pakistan* **2010**, *32*, 240–244.
37. Qureshi, M.; Nisar, S.; Shah, R.; Salman, H. Studies of Liquid Fuel Formation from Plastic Waste by Catalytic Cracking over Modified Natural Clay and Nickel Nanoparticles. *Pak. J. Sci. Ind. Res. Ser. A Phys. Sci.* **2020**, *63*, 79–88. [CrossRef]
38. hamouda, A.; Abdelrahman, A.; Zaki, A.; Mohamed, H. Studying and Evaluating Catalytic Pyrolysis of Polypropylene. *Egypt. J. Chem.* **2021**, *64*, 2593–2605. [CrossRef]
39. Dewangga, P.B.; Rochmadi; Purnomo, C.W. Pyrolysis of Polystyrene Plastic Waste Using Bentonite Catalyst. *IOP Conf. Ser. Earth Environ. Sci.* **2019**, *399*, 012110. [CrossRef]
40. Ahmad, I.; Khan, M.I.; Khan, H.; Ishaq, M.; Tariq, R.; Gul, K.; Ahmad, W. Influence of Metal-Oxide-Supported Bentonites on the Pyrolysis Behavior of Polypropylene and High-Density Polyethylene. *J. Appl. Polym. Sci.* **2015**, *132*, 41221. [CrossRef]
41. Narayanan, K.S.; Anand, R.B. Experimental Investigation on Optimisation of Parameters of Thermo-Catalytic Cracking Process for H.D.P.E. & P.P. Mixed Plastic Waste with Synthesized Alumina-Silica Catalysts. *Appl. Mech. Mater.* **2014**, *592–594*, 307–311. [CrossRef]
42. Sembiring, F.; Purnomo, C.W.; Purwono, S. Catalytic Pyrolysis of Waste Plastic Mixture. *IOP Conf. Ser. Mater. Sci. Eng.* **2018**, *316*, 012020. [CrossRef]
43. Panda, A.K. Thermo-Catalytic Degradation of Different Plastics to Drop in Liquid Fuel Using Calcium Bentonite Catalyst. *Int. J. Ind. Chem.* **2018**, *9*, 167–176. [CrossRef]
44. Borsella, E.; Aguado, R.; De Stefanis, A.; Olazar, M. Comparison of Catalytic Performance of an Iron-Alumina Pillared Montmorillonite and HZSM-5 Zeolite on a Spouted Bed Reactor. *J. Anal. Appl. Pyrolysis* **2018**, *130*, 320–331. [CrossRef]
45. De Stefanis, A.; Cafarelli, P.; Gallese, F.; Borsella, E.; Nana, A.; Perez, G. Catalytic Pyrolysis of Polyethylene: A Comparison between Pillared and Restructured Clays. *J. Anal. Appl. Pyrolysis* **2013**, *104*, 479–484. [CrossRef]
46. Olivera, M.; Musso, M.; De León, A.; Volonterio, E.; Amaya, A.; Tancredi, N.; Bussi, J. Catalytic Assessment of Solid Materials for the Pyrolytic Conversion of Low-Density Polyethylene into Fuels. *Heliyon* **2020**, *6*, e05080. [CrossRef] [PubMed]
47. Gobin, K.; Manos, G. Polymer Degradation to Fuels over Microporous Catalysts as a Novel Tertiary Plastic Recycling Method. *Polym. Degrad. Stab.* **2004**, *83*, 267–279. [CrossRef]
48. Manos, G.; Yusof, I.Y.; Gangas, N.H.; Papayannakos, N. Tertiary Recycling of Polyethylene to Hydrocarbon Fuel by Catalytic Cracking over Aluminum Pillared Clays. *Energy Fuels* **2002**, *16*, 485–489. [CrossRef]
49. Manos, G.; Yusof, I.Y.; Papayannakos, N.; Gangas, N.H. Catalytic Cracking of Polyethylene over Clay Catalysts. Comparison with an Ultrastable Y Zeolite. *Ind. Eng. Chem. Res.* **2001**, *40*, 2220–2225. [CrossRef]
50. Lal, S.; Anisia, K.S.; Jhansi, M.; Kishore, L.; Kumar, A. Development of Heterogeneous Catalyst by Ionically Bonding Macrocyclic Zr–Zr Complex to Montmorillonite Clay for Depolymerization of Polypropylene. *J. Mol. Catal. A Chem.* **2007**, *265*, 15–24. [CrossRef]
51. Cho, K.-H.; Cho, D.-R.; Kim, K.-H.; Park, D.-W. Catalytic Degradation of Polystyrene Using Albite and Montmorillonite. *Korean J. Chem. Eng.* **2007**, *24*, 223–225. [CrossRef]
52. Tomaszewska, K.; Kałużna-Czaplińska, J.; Jóźwiak, W. Thermal and Thermo-Catalytic Degradation of Polyolefins as a Simple and Efficient Method of Landfill Clearing. *PJCT* **2010**, *12*, 50–57. [CrossRef]
53. Gong, J.; Liu, J.; Jiang, Z.; Chen, X.; Wen, X.; Mijowska, E.; Tang, T. Converting Mixed Plastics into Mesoporous Hollow Carbon Spheres with Controllable Diameter. *Appl. Catal. B Environ.* **2014**, *152–153*, 289–299. [CrossRef]
54. Kebritchi, A.; Nekoomansh, M.; Mohammadi, F.; Khonakdar, H.A. Effect of Microstructure of High Density Polyethylene on Catalytic Degradation: A Comparison Between Nano Clay and FCC. *J. Polym. Environ.* **2018**, *26*, 1540–1549. [CrossRef]
55. Marcilla, A.; Gómez, A.; Menargues, S.; Ruiz, R. Pyrolysis of Polymers in the Presence of a Commercial Clay. *Polym. Degrad. Stab.* **2005**, *88*, 456–460. [CrossRef]
56. Chen, Z.; Wang, Y.; Sun, Z. Application of Co Ni Intercalated Vermiculite Catalyst in Pyrolysis of Plastics. *J. Phys. Conf. Ser.* **2021**, *1885*, 032030. [CrossRef]
57. Zhou, N.; Dai, L.; Lv, Y.; Li, H.; Deng, W.; Guo, F.; Chen, P.; Lei, H.; Ruan, R. Catalytic Pyrolysis of Plastic Wastes in a Continuous Microwave Assisted Pyrolysis System for Fuel Production. *Chem. Eng. J.* **2021**, *418*, 129412. [CrossRef]

58. Patil, L.; Varma, A.K.; Singh, G.; Mondal, P. Thermocatalytic Degradation of High Density Polyethylene into Liquid Product. *J. Polym. Environ.* **2018**, *26*, 1920–1929. [CrossRef]
59. Hussain, Z.; Khan, K.; Jan, M.; Shah, J. Conversion of Low Density Polyethylene into Fuel Products Using Indian Fuller's Earth as Catalyst. *J. Chem. Soc. Pak.* **2010**, *32*, 790–793.
60. Nguyen, L.T.T.; Poinern, G.E.J.; Le, H.T.; Nguyen, T.A.; Tran, C.M.; Jiang, Z. A LaFeO$_3$ Supported Natural-Clay-Mineral Catalyst for Efficient Pyrolysis of Polypropylene Plastic Material. *Asia-Pac. J. Chem. Eng.* **2021**, *16*, e2695. [CrossRef]
61. Ali, G.; Nisar, J.; Shah, A.; Farooqi, Z.H.; Iqbal, M.; Shah, M.R.; Ahmad, H.B. Production of Liquid Fuel from Polystyrene Waste: Process Optimization and Characterization of Pyrolyzates. *Combust. Sci. Technol.* **2021**, 1–14. [CrossRef]
62. Patil, V.; Adhikari, S.; Cross, P. Co-pyrolysis of Lignin and Plastics Using Red Clay as Catalyst in a Micro-Pyrolyzer. *Bioresour. Technol.* **2018**, *270*, 311–319. [CrossRef]
63. Kyaw, K.; Hmwe, C. Effect of Various Catalysts on Fuel Oil Pyrolysis Process of Mixed Plastic Wastes. *Int. J. Adv. Eng. Technol.* **2015**, *8*, 794.
64. Lei, J.; Yuan, G.; Weerachanchai, P.; Lee, S.W.; Li, K.; Wang, J.-Y.; Yang, Y. Investigation on Thermal Dechlorination and Catalytic Pyrolysis in a Continuous Process for Liquid Fuel Recovery from Mixed Plastic Wastes. *J. Mater. Cycles Waste Manag.* **2018**, *20*, 137–146. [CrossRef]
65. Filip, M.; Pop, A.; Perhaiṭa, I.; Trusca, R.; Rusu, T. The Effect of Natural Clays Catalysts on Thermal Degradation of a Plastic Waste Mixture. *Adv. Eng. Forum* **2013**, *8–9*, 103–114. [CrossRef]

Article

Simulation and Modelling of Hydrogen Production from Waste Plastics: Technoeconomic Analysis

Ali A. Al-Qadri [1], Usama Ahmed [1,2,*], Abdul Gani Abdul Jameel [1,3], Umer Zahid [1,4], Muhammad Usman [2] and Nabeel Ahmad [5]

1   Chemical Engineering Department, King Fahd University of Petroleum and Minerals,
    Dhahran 31261, Saudi Arabia; g201472160@kfupm.edu.sa (A.A.A.-Q.);
    a.abduljameel@kfupm.edu.sa (A.G.A.J.); uzahid@kfupm.edu.sa (U.Z.)
2   Interdisciplinary Research Center for Hydrogen and Energy Storage, King Fahd University of Petroleum &
    Minerals, Dhahran 31261, Saudi Arabia; muhammadu@kfupm.edu.sa
3   Center for Refining & Advanced Chemicals, King Fahd University of Petroleum and Minerals,
    Dhahran 31261, Saudi Arabia
4   Interdisciplinary Research Center for Membranes & Water Security, King Fahd University of Petroleum and
    Minerals, Dhahran 31261, Saudi Arabia
5   Department of Chemical Engineering, COMSATS University Islamabad, Lahore Campus,
    Islamabad 54000, Pakistan; nabeelahmad@cuilahore.edu.pk
*   Correspondence: usama.ahmed@kfupm.edu.sa

Citation: Al-Qadri, A.A.; Ahmed, U.;
Abdul Jameel, A.G.; Zahid, U.;
Usman, M.; Ahmad, N. Simulation
and Modelling of Hydrogen
Production from Waste Plastics:
Technoeconomic Analysis. *Polymers*
2022, *14*, 2056. https://doi.org/
10.3390/polym14102056

Academic Editor:
Sheila Devasahayam

Received: 7 April 2022
Accepted: 10 May 2022
Published: 18 May 2022

Publisher's Note: MDPI stays neutral
with regard to jurisdictional claims in
published maps and institutional affiliations.

Abstract: The global energy demand is expected to increase by 30% within the next two decades. Plastic thermochemical recycling is a potential alternative to meet this tremendous demand because of its availability and high heating value. Polypropylene (PP) and polyethylene (PE) are considered in this study because of their substantial worldwide availability in the category of plastic wastes. Two cases were modeled to produce hydrogen from the waste plastics using Aspen Plus®. Case 1 is the base design containing three main processes (plastic gasification, syngas conversion, and acid gas removal), where the results were validated with the literature. On the other hand, case 2 integrates the plastic gasification with steam methane reforming (SMR) to enhance the overall hydrogen production. The two cases were then analyzed in terms of syngas heating values, hydrogen production rates, energy efficiency, greenhouse gas emissions, and process economics. The results reveal that case 2 produces 5.6% more hydrogen than case 1. The overall process efficiency was enhanced by 4.13%. Case 2 reduces the CO_2 specific emissions by 4.0% and lowers the hydrogen production cost by 29%. This substantial reduction in the H_2 production cost confirms the dominance of the integrated model over the standalone plastic gasification model.

Keywords: gasification; reforming; plastic waste; H_2 production; CO_2 emissions

1. Introduction

Globally, 9% of plastics out of 6.3 billion tons have been recycled between 1950 and 2018. Additionally, 12% have been burnt [1]. However, the remaining 79% of plastics promote severely harmful pollutants. Those pollutants have different forms such as furans, dioxins, and mercury. The pollutants are highly hazardous, negatively affecting the environment and marine organisms [2]. Moreover, 4–12 million tons of plastics are annually thrown into the ocean [3]. Many countries are encouraging and legislating laws to minimize plastic usage, followed by recycling the plastics [4]. The efficient recycling of plastics to valuable products is essential to save the environment and utilize the energy from these huge amounts of waste. Several studies have confirmed the feasibility of plastic recycling [5,6].

The recycling process encompasses four main steps: collection, separation, manufacture, and marketing [7]. The most convenient technique is thermochemical recycling because it converts the plastics into synthesis gas, which could be used in synthesizing several valuable chemicals [8,9].

Gasification is a process that produces synthesis gas (CO_2, CO, H_2, CH_4, etc.) from carbon-based materials such as fossil fuels, and biomass [10–14]. The syngas can then be used to produce several fuels and chemicals [15]. The gasification process is usually promoted through a high-temperature reaction (>700 °C) using oxygen or steam as an auxiliary component (air gasification or steam gasification) [16,17]. Steam gasification, air gasification, co-gasification, pyrolysis, and plasma gasification are types of thermal recycling for plastics or any carbon-based feedstock [18]. Pyrolysis is a dry heating of the feed in the absence of air [19]. The pyrolysis produces syngas that is completely free of tar [20]; however, the hydrogen to carbon monoxide ratio is not high. Sometimes, it is considered as the first step in the gasification process because it maximizes the conversion of volatile materials (high carbon chain) to relatively low carbon hydrocarbons (<C25) [21]. Another process is co-gasification, which mixes two carbonaceous feedstocks such as plastic with coal or biomass to enhance the gas yield and suppress char formation [22]. However, this process increases the tar formation [23]. Air gasification produces less tar; nevertheless, it produces a lower hydrogen to carbon monoxide ratio [24]. Pure oxygen gasification is a very efficient process; however, the production of oxygen from air is highly expensive [25]. To effectively produce syngas with a high hydrogen to carbon (HCR) ratio in a quite simple process, the steam gasification of plastics is the optimal choice [26]. It is quite simple, and it produces a higher hydrogen to carbon monoxide ratio.

The production of syngas facilitates the production of essential chemicals and fuels, such as hydrogen, methanol, ethanol, DME, LPG, olefins, and gasoline [27–29]. Modeling the whole journey of plastics to clean fuels under several operational conditions is essential to support industrial applications, and to maximize the clean fuel production from a heterogeneous plastic mixture [30]. Antzela and Ioanna [31] conducted a pilot plant study on the techno economic evaluation of the conversion of plastics into heavy oil through pyrolysis using Aspen HYSYS. The production cost of the heavy oil was 0.87 EUR /kg, which is 58% higher than the market price. They suggested a more sophisticated study for large-scale data. Deng et al. [32] modeled the municipal solid plastic (MSW) to syngas using a combination of two technologies: pyrolysis (RYield + RGibbs), and gasification (RGibbs). The results show good agreement with the experimental data, where the temperature of 750 °C is considered the optimal gasification temperature, with a steam/plastic ratio of 0.4. Furthermore, they economically recommended the use of flue gas and steam as gasifying agents. Another study by Pravin et.al [33] accomplished the conversion of PE (polyethylene) to syngas through pyrolysis then gasification using Aspen plus. The results were not validated by the experiment due to the lack of resources; however, they claimed that the most convenient temperature, and equivalence ratio for the pyrolysis unit were 0.4–0.6, and 500–750 °C, respectively The catalytic approach has advantages over the thermal one in terms of reducing the sulfur content when special catalysts are used (i.e., CaS, and MgS) [34–36]. Several studies have been performed on the conversion of waste plastics to hydrogen along with other feedstocks [37,38]. The development of catalysts for plastic gasification in a cost-effective manner is still under research; therefore, the thermal gasification technique is considered, which is a well established process with fewer operational issues.

Fivga and Dimitriou [39] studied and analyzed the conversion of waste plastics to heavy fuel. They used a mixture of PE, PP, and PS as a feedstock at 530 °C, and 1 atm. They modeled their work using Aspen HYSYS based on the ultimate analysis of the plastics. The product of their pyrolysis reactor was basically n-C30, n-C25, n-C18, n-C14, n-octane, ethane, and a small proportion of gases. The remaining solids and gases were separated, then pyrolyzed liquid fuel was collected. They validated their results with plant data, and they performed cost analysis. Generally, the work is promising and has the idea of using plastic waste to generate liquid heavy fuel. Another study on plastic waste conversion to fuel was conducted by Emad and Vahid [40], which was basically on the production of hydrogen via the co-gasification of a mixture of asphaltene and plastics using Aspen Plus. They decomposed the feed on a pyrolysis reactor (RYield), and then they used an RGibbs reactor followed by CSTR to produce syngas. They studied some factors influencing the hydrogen production rate, namely, asphaltene to plastic ratio (A/P), equivalence ratio (ER), and steam to feed ratio (S/F). They found that A/P and steam to feed (S/F) have a positive impact on carbon conversion efficiency (CCE). The study provided the excellent idea of producing hydrogen from a co-gasification mixture. However, they did not produce pure hydrogen; it was a synthesis gas mixed with acid gases that should be removed. Additionally, they need to implement WGS to maximize hydrogen production and to suppress the carbon monoxide in the product.

There are limited studies on the production of hydrogen from plastic wastes. Therefore, investigating the hydrogen production from different feedstocks (i.e., coal or biomass) will assist hydrogen production from plastics. A study was performed by Noussa et al. [41] on the techno economic evaluation of producing H_2 from biomass. They investigated different gasifier agents and several types of feedstocks. They found that steam as gasification agent was better than other agents. Namioka et al. [42] studied the production of H_2-rich synthesis gas using pyrolysis, then low-temperature steam gasification. The study was focused on polystyrene (PS), and polyethylene (PE) as a feedstock. They performed the pyrolysis and steam reforming at 673 and 903 K, respectively. Ruthenium was used as a pyrolysis catalyst, and it enhanced the process performance. The study recommended combining the thermal and catalytic process. Similarly, Chaia et al. [43] studied the conversion of plastics to hydrogen using a combination of co-pyrolysis and gasification processes. Ni-CaO-C was tested as a novel catalyst to promote H_2 production. They claimed a hydrogen production efficiency of 87.7 mole %, controlling the greenhouse gas emissions. Consequently, the conversion of waste plastics into hydrogen is a practical process, proved theoretically and experimentally. Thus, the current study will focus on using a thermochemical approach based on steam gasification to convert plastics into hydrogen fuel [42].

2. System and Analysis Framework

2.1. Modelling and Simulation Approach

In this study, polyethylene (PE) and polypropylene (PP) were selected due to their availability and their higher heating value [44]. Aspen Plus® software V-12 was used as a simulation tool, selecting Peng Roberson (PR) as an appropriate property package. It is generally recommended for oil and gas systems [45,46]. There are several classifications of plastics in terms of composition. The approximate and ultimate analyses of the plastic feedstock are provided in Table 1. To specify the plastic heating value, the HCOALGEN model was selected. Prior to generating syngas, the RYield reactor was simulated to break down the solid feedstock, and then the outlet mixture was fed to the gasifier (i.e., RGibbs reactor). The products were mainly syngas containing CO, H_2, and CO_2. The RGibbs reactor operated at a high temperature (i.e., 900 °C). The outlet syngas was introduced to water gas shift (WGS) to convert CO to hydrogen via the WGS reaction in two REquil reactors. The reactions are given in Table A1.

Table 1. Plastics and natural gas composition.

	Plastic Composition Analysis	
	Proximate Analysis (Weight %)	
	PE	**PP**
Moisture	0.02	0
Ash	0.15	0.7
Volatile matter	99.83	99.30
Total	100	100
LHV (MJ/kg)	38.04	44.70
	Ultimate analysis (weight %)	
Carbon	85.81	86.23
Hydrogen	13.86	12.28
Nitrogen	0.12	0.62
Sulfur	0.06	0.17
Ash	0.15	0.7
Total	100	100
	Natural gas composition (mol %)	
CH_4	93.9	
C_2H_6	3.2	
C_3H_8	0.7	
C_4H_{10}	0.4	
CO_2	1.0	
N_2	0.8	
Total	100	
LHV(MJ/kg)	47.76	

The process operational conditions were set based on previous studies with several assumptions. Table 2 illustrates the major assumptions made in the whole process. The primer design of the model was based on a study conducted by Dang et al. (2019) [47].

Table 2. Design assumptions made for case 1 and case 2.

Equipment	Aspen Model	Assumption
Plastic Flow Rate	RYield/RGibbs	Plastics = 100 kg/h Entrained flow gasifier; steam:plastic = 1.25; Temperature = 900 °C; P = 1 atm
Pre-reformer	RStoic (reactor)	Heavier hydrocarbon hydrocracking
Reformer	RGibbs (reactor)	Temperature = 894.3 °C, pressure = 3 bar, Steam: NG = 1.6; nickel-based catalyst
Water Gas Shift (WGS)	REquil (reactor)	Two equilibrium reactors Steam:CO = 2:1 (molar basis)
Acid Gas Removal (AGR)	RadFrac and flash drums	Rectisol process; temperature = −30 °C, P = 1 bar CO_2 removal = 99%; H_2S removal = 10 ppm

Standalone models for polyethylene and polypropylene were developed and validated with the literature results based on the experiments [26,48]. For the purpose of validation, the same process conditions used in the simulation model were kept in the experimental setup. Table 3 represents the comparison between the experimental and the simulation results for plastic gasification. The simulation results are in good agreement with the experimental results and the simulation models can be used with confidence for hydrogen production.

Table 3. Polyethylene and polypropylene gasification validation.

Validation of Polypropylene Gasification			
Component	Reference Case	Base Case	Difference
H_2	68.3	66.4	1.9
CO	26.1	27.5	−1.4
CO_2	3.9	5.7	−1.8
CH_4	1.3	0.3	1.0
Others	0.3	0.1	0.2
Validation of Polyethylene Gasification			
Component	Reference Case	Base Case	Difference
H_2	68.6	67.4	1.2
CO	25.5	28.8	−3.3
CO_2	1.1	3.7	−2.6
CH_4	3.6	0.0	3.6
Others	1.2	0.0	1.2

2.2. Development and Validation of Case Studies

2.2.1. Case 1 (Base Case)

Figure 1 represents the general process flow diagram of case 1. The mixture of PE and PP in the equal weight ratio of 50:50 was crushed and fed to the steam gasification unit to generate the syngas. The solid plastics were first decomposed in the decomposer (RYield) and then fed to the gasification unit to produce the syngas at a temperature of 900 °C, where a hydrogen to CO ratio of 1.86 was achieved. Then, the syngas was quenched to sustain WGS reactions. The outlet stream from WGS reactors mainly included hydrogen, CO_2, and some traces of H_2S, where the ratio of H_2/CO_2 was obtained as 2.86. Methanol was selected as an absorbent in the AGR unit to remove hydrogen sulfide and carbon dioxide, where the methanol was recovered in the H_2S and CO_2 regenerator columns.

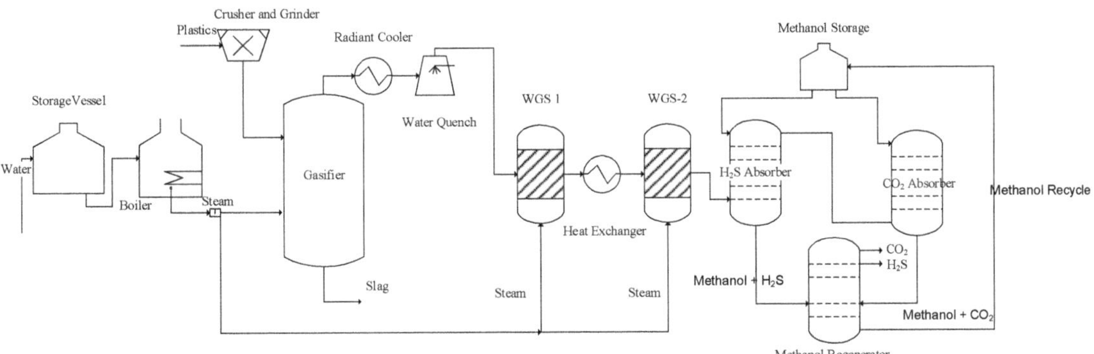

Figure 1. Hydrogen production from waste plastics: PE and PP (case 1).

2.2.2. Case 2 (Alternative Case)

Case 2 is similar to case 1 in terms of the gasification process; however, case 2 contains an additional process unit. The alternative case (case 2) represents the integration of the steam methane reforming (SMR) model with the plastic gasification model to utilize the gasifier's heat energy in the reforming unit, making it different from case 1. The process base flow diagram is provided in Figure 2. The steam to methane molar ratio was set as 1.50 and the inlet temperature was selected as ~900 °C. The process reactions are provided in Table A1. The SMR results were also validated with the literature in terms of hydrogen to carbon monoxide ratio, which was found to be around 3.0 [49]. The syngas mixture obtained from SMR and gasification was mixed and introduced to WGS with the same conditions applied in the base case design, and were also used in case 2. Finally, the acid gas removal unit was used to remove the CO_2 and H_2S from the gas streams to obtain pure hydrogen.

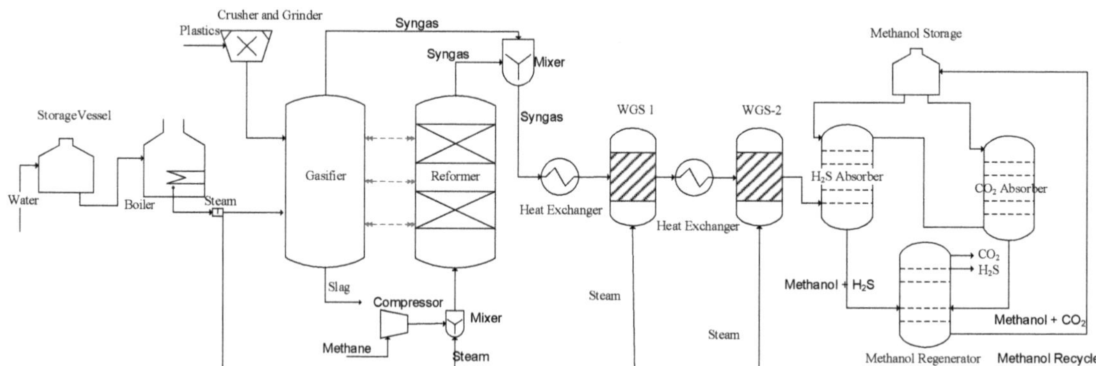

Figure 2. Hydrogen production from plastic waste gasification integrated with reforming (case 2).

2.3. Governing Equations for Technical Analysis

Table A2 represents some of the equations used in this study for technical and economical comparison between the two cases. The lower heating value (LHV) was calculated based on the mole fraction of hydrogen and carbon monoxide [47]. HPF is an indicator that represents the hydrogen per total feed in terms of mass basis. The hydrogen thermal energy was calculated from the lower hydrogen heating value considering the flow rate of hydrogen. The specific carbon dioxide emissions and process efficiency indicators were also used for the comparison between two cases [50]. The total investment cost (TIC) represents the capital cost with respect to the hydrogen production rate. The capital investment for each unit was calculated from previous similar studies considering the Chemical Engineering Plant Cost Index (CEPCI). To assess the operating expenditures, total manufacturing cost was computed as the sum of maintenance, administrative, labor, support, and overhead costs. The utility and labor costs were calculated based on Donald E. Garrett [51]. The levelized hydrogen production cost was estimated for 30 years considering the total hydrogen produced in a lifetime and the expense incurred.

3. Results and Discussion

Case 1 and case 2 are compared in terms of hydrogen production rates, syngas heating values, hydrogen purity, carbon emissions, production cost and the process feasibility. The equation given in Table A2 was used for the comparative analysis.

3.1. Technical Analysis

3.1.1. Syngas Production and Analysis

The feedstock mainly consists of polyethylene and polypropylene. The feedstock is fed with a mass ratio of 1:1. The total plastic flow rate is considered as 100 kg/h, where the steam to plastic ratio is maintained as 1.25:1. The natural gas flow rate in case 2 is taken as 42 kg/h, with a steam to natural gas ratio of 1.6:1. It can be seen from the results that the molar ratio of H_2/CO for case 1 and case 2 is 1.86 and 2.23, respectively. The hydrogen to carbon monoxide ratio was enhanced in the second case by 62% compared with case 1. The carbon dioxide emission after WGS reactors was lower in case 2 than case 1 by 11%. Overall, the results reveal that case 2 is more efficient than the base case in terms of syngas heating value and carbon dioxide emissions. Table 4 provides the operational conditions, and the stream flow rates at the outlet of all the essential units.

Table 4. Flow rates and stream compositions at the exit of each unit.

	Plastics	Steam (Gasifier)	Gasifier	Reformer	Cooling and Syngas Mixing		WGS Unit		AGR Unit (H$_2$ Storage)		CO$_2$ Storage	
	Case 1 and 2	Case 1 and 2	Case 1 and 2	Case 2	Case 1	Case 2	Case 1	Case 2	Case 1	Case 2	Case 1	Case 2
T (°C)	300	300	900	894.3	220	220	10	10	25	25	25	25
P (bar)	1.013	1.013	1.013	3	1	1	1	1	1	1	1	1
Mass Flow (kg/h)	100	125	225	109	224.58	333.58	469.38	578.38	49.58	75.22	226.40	337.53
Mole (%)												
H$_2$		-	0.636	0.683	0.636	0.653	0.579	0.654	0.976	0.978	0.0026	0.0026
CO		-	0.341	0.206	0.341	0.292	0.001	0.004	0.002	0.006	0.341	0.206
CO$_2$		-	0.002	0.020	0.002	0.008	0.202	0.206	0.003	0.003	0.993	0.994
H$_2$O		1	0.004	0.089	0.004	0.034	0.206	0.128	0	0	0	
CH$_4$		-	0.017	0.001	0.017	0.011	0.010	0.008	0.016	0.011	0.0018	0.0012
N$_2$		-	0.0008	0.0018	0.0008	0.0011	0.0004	0.0008	0.0007	0.0012	0.0008	0.0018
H$_2$S		-	0.0002	-	0.0002	0.0001	0.0001	0.0001		-	0.0002	-
CH$_3$OH		-	0.0000	-	0.0000		-		0.0018	0.0018	0.0000	-
Molar H$_2$/CO	-	-	-	3.32	1.86	2.23	-	-	-	-	-	-
Molar H$_2$/CO$_2$	-	-	-	33.34	389.50	77.43	2.86	3.18	-	-	-	-

To determine the efficiency of the process, syngas composition is a key parameter for such evaluation. The HCR at the inlet of WGS was evaluated for case 1 and case 2, as given in Figure 3. It was found that the H_2/CO was higher in case 2 than case 1, indicating a higher heating value for the integrated case.

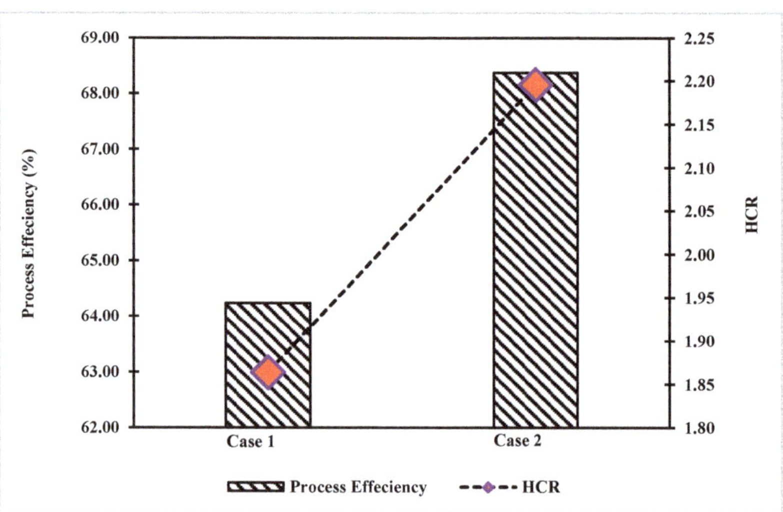

Figure 3. Comparison of process efficiency and hydrogen to carbon monoxide ratio for case 1 and case 2.

3.1.2. Overall Process Performance

The overall process efficiency was investigated for the two cases by considering several essential parameters, as given in Table 5. The first parameter is the hydrogen purity at the outlet of the acid gas removal unit, which is greater in the second case by 0.15% compared with the first case. The process's overall efficiency in feedstock conversion was calculated and is represented by HPF, which was found to be greater in case 2 by 5.6%. Additionally, the two cases were evaluated in terms of heating value (i.e., lower, and higher heating values). The results show that the second case had a lower heating value (LHV) and higher heating value (HHV) than case 1 by 5.7%, and 5.0%, respectively.

Table 5. Energy analysis.

Characteristic/Model Type	Case 1	Case 2
Hydrogen per feedstock HPF (mass %)	50	52.8
Hydrogen purity (mole %)	97.62	97.77
Syngas gross heating value GHV (MJ/kg)	26.18	27.67
Syngas net heating value LHV (MJ/kg)	23.55	24.73
Feed stock energy (kWth)	1198.61	1757.07
Thermal energy of produced H_2 (kWth)	1385.25	2060.59
Minimum hot utilities required (kW)	757.06	1069.75
Minimum cold utilities required (kW)	200.80	187.06
Total energy required after heat integration (kW)	957.86	1256.81
Process efficiency(ηnet) (%)	64.24	68.37

The integrated process produced more hydrogen than the classical one because the SMR unit had higher hydrogen production. The overall energy process efficiency was calculated and studied, and it was higher in case 2 than case 1 by 4.13%. Thus, case 2 is more efficient than case 1 in terms of syngas heating value. However, economic analysis will be performed to confirm the final preference for the alternative design.

3.1.3. CO_2 Specific Emissions

Another essential factor in the investigation and comparison of the two designed cases is the specific emission of carbon dioxide. Case 2 showed lower carbon emissions than case 1 by 1.2%. A study conducted by Usman et al. [37], about the conversion of coal to hydrogen, found that the specific CO_2 emissions were in the range of 0.70 on a mass basis. Figure 4 shows the specific carbon dioxide emissions for each case along with HPE (hydrogen per total feed; mass ratio). The HPF for case 2 is higher than case 1, with lower carbon emissions. The results show that the alternative case produces a higher amount of hydrogen with minimal carbon dioxide emissions.

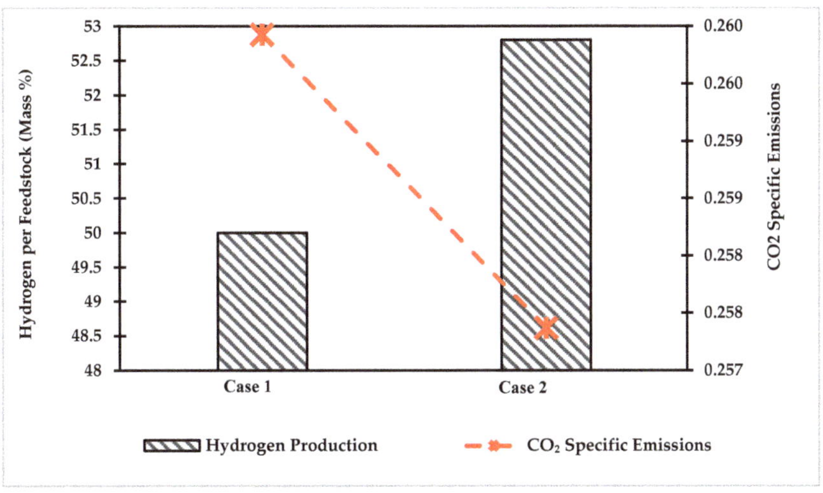

Figure 4. Comparison of hydrogen production and specific CO_2 emissions for case 1 and case 2.

3.1.4. Sensitivity Analysis on the Gasifier

To optimize the design parameters, sensitivity analysis was performed mainly on the gasification unit. The main factors affecting the process performance are the gasification temperature, pressure, steam to feed ratio, and PE and PP blending ratio.

3.1.4.1. Steam to Feed Ratio Effect on Syngas Composition

The steam to feed ratio has a strong effect on the gasification process because it significantly controls the outlet syngas composition. Figure 5 represents the sensitivity analysis of the steam gasification unit when investigating the impact of the steam to plastic ratio (S/P or S/F) on syngas composition. Increasing the steam to plastic ratio decreases the CO production rate; however, it dramatically enhances the hydrogen production. CH_4 is suppressed when S/P increases. The optimal steam to feed ratio at 900 °C, as deduced from the figure, is 1.25, because any further increase had a negligible impact on syngas LHV. The analysis was performed on the blend of PE and PP based on equal weight. The results show that increasing the steam to plastic ratio has a positive impact on enhancing the syngas heating value; however, going beyond a steam to plastic (S/P) ratio of 1.25 decreases the heating values of syngas by producing more CO_2.

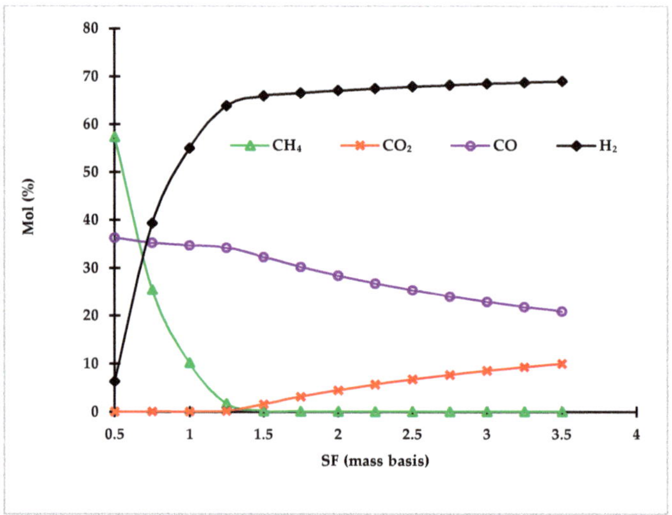

Figure 5. Effect of steam to plastic ratio on syngas composition.

3.1.4.2. Temperature Effect on Syngas Composition

A lower gasification temperature promotes higher carbon dioxide production due to the Boudouard reaction, which is the reaction of CO_2 with carbon to produce CO. It is an endothermic reaction; therefore, as the temperature increases, less carbon dioxide is produced [52]. Increasing the temperature has a positive effect on the heating value of syngas. Increasing the temperature up to 900 °C has a positive effect on the heating values and produces more hydrogen. Figure 6 shows the impact of gasifier temperature on the gasification process at an S/F of 1.0, and a PE/PP of 1:1. Therefore, the gasification temperature of 900 °C was considered for both cases to achieve maximum hydrogen production and a high heating value of syngas.

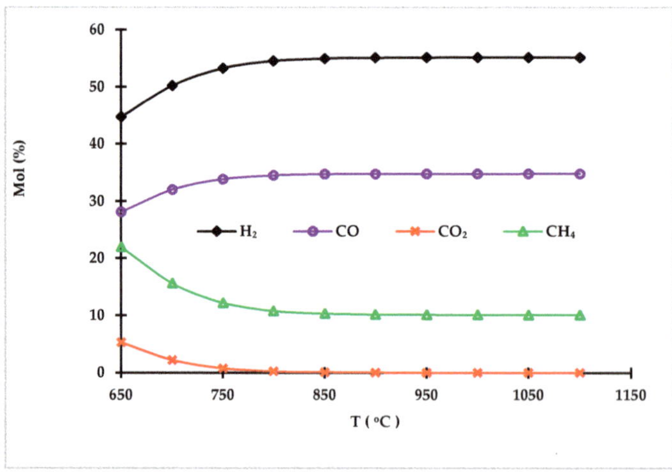

Figure 6. Effect of gasification temperature on syngas composition.

3.2. Economic Analysis

The economic analysis is essential to comprehend the process's feasibility and promotability and to precisely compare the two designs. The order-of-magnitude cost analysis has been used to determine the capital cost [53]. Several assumptions were considered to perform the economic analysis. Table A3 represents the assumptions to accomplish the economic analysis [50]. The waste plastic price was ignored in both cases, and the plant life was considered for 30 years using an exponent factor (x) of 0.6 for consistent analysis. Three shifts per day were considered with a stream factor of 0.95 to estimate the operational expenses. The working capital, land and salvage were each taken to be 10% from the fixed capital investment. The offsite unit and utilities were taken to be 25% of the equipment and installation cost. The contingency and permitting costs were chosen to be 15% and 5% of the equipment and installation cost, correspondingly. The discount rate and taxation rate were assumed to be 8% and 15%, respectively, in both cases.

3.2.1. Estimation of CAPEX and OPEX

The capital expenditure (CAPEX) and the operational expenditure (OPEX) are the two important parts of a project's economic evaluation. The total investment is impacted by various variables such as the capacity of the plant, raw materials, operational time, and the process efficiency. The fixed CAPEX predominantly comprises the equipment and plant facilities costs. This study used the power law to estimate the CAPEX with a capacity factor (x) of 0.6, as mentioned in Table A3. The power law uses the concept of Chemical Engineering Plant Cost Index (CEPCI). Table 6 represents the CAPEX cost summary, calculating some of the important parameters such as total investment cost (TIC). The total investment cost for the two cases has a huge difference due to the variation in the process configuration and the type and capacity of the plant. The total investment cost (FCI) in terms of MMEUR for case 1 and case 2 was calculated as 3.79 and 4.46, respectively. The FCI for the alternative case was higher than that of the base case because case 2 has an SMR process with an additional feedstock (i.e., natural gas). The TIC represents the total investment cost per hydrogen production rate in tons. The TIC for the alternative case was higher than the base case by 23%, indicating the cost-effectiveness of the alternative case in terms of capital expenditure while considering the production rate of hydrogen.

On the other hand, the operational cost of the project is represented by OPEX, which is classified into two different categories. The two categories are the fixed OPEX and variable OPEX. The fixed OPEX involves the maintenance, labor, and administrative costs, where the variable OPEX encompasses the fuel, catalysts, waste disposal, and boiler feed water costs. Table 6 shows the OPEX summary for the two designed cases. The total OPEX in MMEUR/year for case 1 and case 2 is calculated as 1.39 and 1.47, respectively. The total operational expenditures are higher in case 2 than case 1; however, when the production rate of the fuel is considered, case 2 shows a 30% reduction in the operational cost per ton of hydrogen production. The revenue calculated for both cases revealed that case 2 offers 51% higher revenue than case 1.

Table 6. CAPEX and OPEX for case 1 and case 2.

Capital Expenditure		
Equipment	Case 1 EUR (10^3)	Case 2 EUR (10^3)
Gasification price	110	110
Acid gas removal unit	1339	1624
Solid handling facility	522	522
Syngas processing unit	646	690
Reformer cost	0	128
Equipment and installation cost	2617	3074
Offsite unit and utilities	654	768
Contingency cost	393	461
Permitting	131	154
Total investment cost	3795	4457
TIC per ton of H_2 MMEUR /ton	76.53	59.25
Operational expenditure		
Cost sector/designed case	Case 1 EUR (10^3)/Year	Case 2 EUR (10^3)/Year
Maintenance cost (2% of equipment and installed cost)	52.3	61.5
Labor cost	459.4	472.9
Administrative, support and overhead cost	137.8	141.9
Total fixed manufacturing cost	649.6	676.2
Natural gas	0.0	16.5
WGS catalyst	16.6	18.0
Reforming catalyst	0.0	0.5
Solvent	39.0	57.8
Waste disposal	7.1	7.1
Utility costs	677.9	693.2
Total OPEX/year	1390.0	1469.3
Total OPEX/ton H_2	3.4	2.3
Revenue (MMEUR /year)	4.804	7.289
NPV	22.450	39.978
PVR	6.401	9.288

3.2.2. Cash Flow and Hydrogen Cost Analysis

The purpose of this section is to provide the cash flow diagram and to compare the levelized hydrogen production rate with the literature. The TIC per ton of hydrogen was calculated as 76.53 MMEUR /ton and 59.25 MMEUR /ton for case 1 and case 2, respectively. Additionally, the levelized lifetime hydrogen production cost was calculated as 3.78 EUR /kg and 2.56 EUR /kg for case 1 and case 2, respectively. This indicates that case 2 produces hydrogen with 1.22 EUR less compared with case 1 for every kilogram of hydrogen produced. Figure 7 shows the cash flow diagram over the lifetime of the project for both cases [54]. The cash flow return on investment was higher for case 2 than the base case by 52%. Additionally, the net present value (NPV) was higher in the alternative case when compared with the base case by 78%. The present value ratio (PVR) for case 2 was found to be higher than case 1 by 45%. Overall, case 2 offered better process economics compared with case 1.

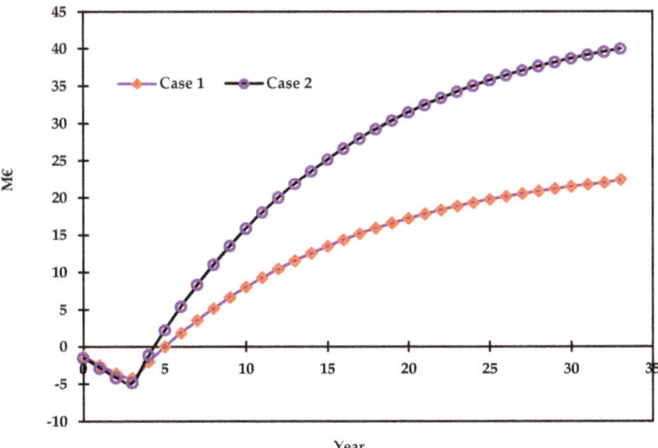

Figure 7. Cash flow diagram for case 1 and case 2.

3.2.3. Comparison of Hydrogen Cost with the Literature

Figure 8 compares the hydrogen cost obtained from our study with the literature considering different feedstocks [55,56]. The production of hydrogen via the solar and photovoltaic electrolysis of water has the highest hydrogen cost. It is considered as a green process; however, it consumes more energy. Biomass, coal, and heavy oil can produce hydrogen with a lower cost than the solar process. It was analyzed from the literature that the hydrogen production cost ranges from 5 to 8 EUR /kg [57] depending on the type of feedstock and the technology used for hydrogen production. From the comparative analysis, case 2 was found to be an attractive approach for hydrogen production with lower costs and shows potential to resolve the global plastic waste issue. The catalytic plastic gasification could also further reduce the hydrogen production cost because it is usually performed at a lower temperature [58]. Dan et al. [59] performed a study converting plastic wastes and biomass to hydrogen using $Ni/\gamma\text{-}Al_2O_3$ as a catalyst with a temperature of 800 °C. This might be a future direction in enhancing the conversion of waste plastic to hydrogen.

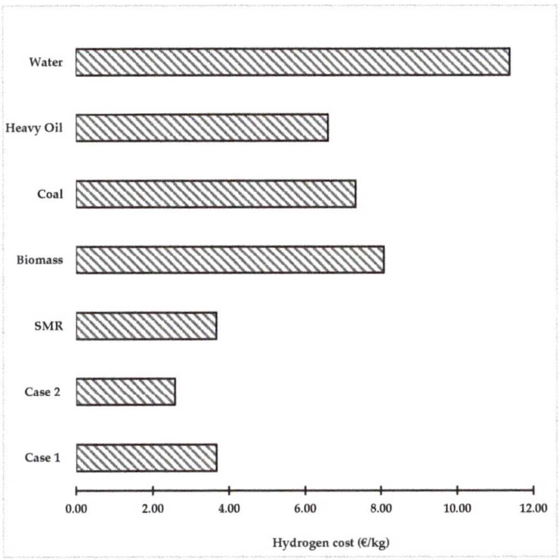

Figure 8. Comparison of hydrogen production costs from waste plastic with those of the literature.

4. Conclusions

The study represents the hydrogen production from plastics (polyethylene and polypropylene) through two developed cases in Aspen Plus®. Case 1 is the thermochemical steam gasification process for converting waste plastics to hydrogen. On the other hand, case 2 represents a modified version of case 1 by integrating the steam gasification model with the steam reforming model to enhance the overall hydrogen production. The technical and economic analyses were performed for both cases to evaluate the feasibility of the process and are summarized as follows:

1. The H_2/CO of the syngas for case 1 and case 2 is calculated as 1.86 and 2.23, respectively, whereas case 2 showed 19.78% higher values.
2. The hydrogen production rate per unit of feedstock for case 1 and case 2 is calculated as 50% and 52.2%, respectively.
3. The overall process efficiency for case 1 and case 2 is calculated as 64.24% and 68.37%, respectively, whereas case 2 shows 4.13% higher efficiency.
4. The TIC per ton of H_2 calculated for case 1 and case 2 is 76 and 59 EUR/ton, whereas case 2 has the potential to increase the revenue by 51.7%.
5. Case 2 showed the potential to lower CO_2 emissions by 1.0%.

Author Contributions: Conceptualization, A.A.A.-Q. and U.A.; software, A.A.A.-Q. and U.A.; writing—original draft preparation, A.A.A.-Q. and U.A.; formal analysis, A.G.A.J. and U.Z.; simulation, A.A.A.-Q. and U.A.; visualization, M.U. and N.A. All authors have read and agreed to the published version of the manuscript.

Funding: The authors would like to acknowledge the support provided by the Deanship of Research Oversight and Coordination (DROC) at King Fahd University of Petroleum & Minerals (KFUPM) for funding this work and supporting the APC through project no. DF201017.

Appendix A

Table A1. Chemical reactions comprehended in the process.

Gasification Reactor	
$C_{(s)} + H_2O \leftrightarrow CO + H_2$	$\Delta H = +131\,\frac{MJ}{kmol}$
$C_{(s)} + CO_2 \leftrightarrow 2CO$	$\Delta H = +172\,\frac{MJ}{kmol}$
$C_{(s)} + 2H_2 \leftrightarrow CH_4$	$\Delta H = -74.8\,\frac{MJ}{kmol}$
$CO + H_2O \leftrightarrow CO_2 + H_2$	$\Delta H = -41.2\,\frac{MJ}{kmol}$
$CH_4 + H_2O \leftrightarrow CO + 3H_2$	$\Delta H = +206\,\frac{MJ}{kmol}$
Steam Methane Reforming Reactor	
$3C_2H_6 + H_2O \rightarrow 5\,CH_4 + CO$	$\Delta H = +3.6460\,\frac{MJ}{kmol}$
$3C_3H_8 + 2H_2O \rightarrow 7\,CH_4 + 2\,CO$	$\Delta H = +16.607\,\frac{MJ}{kmol}$
$3C_4H_{10} + 3\,H_2O \rightarrow 9\,CH_4 + 3\,CO$	$\Delta H = +41.116\,\frac{MJ}{kmol}$
$CH_4 + 2\,O_2 \rightarrow CO_2 + 2\,H_2O$	$\Delta H = -802.54\,\frac{MJ}{kmol}$
$CH_4 + H_2O \rightarrow CO + 3\,H_2$	$\Delta H = +206.12\,\frac{MJ}{kmol}$
Water Gas Shift Reactor	
$CO + H_2O \leftrightarrow H_2 + CO_2$	$\Delta H = -41\,\frac{MJ}{kmol}$

Table A2. Equations used for technical and economic appraisal.

Equations	Eq. No.
$LHV_{Syngas} = 12.636\ y_{CO} + 10.798\ y_{H_2}$	(A1)
$HPF = \dfrac{\text{Produced } H_2\ (\frac{kg}{h})}{\text{Total Feed } (\frac{kg}{h})} \times 100\%$	(A2)
$H_2\ Thermal\ Energy = H_2 LHV\ (\frac{kJ}{kg}) \times Produed\ H_2\ (\frac{kg}{h});\ LHV = 100.539\frac{MJ}{kg}$	(A3)
$Total\ Consumed\ Energy\ (kW) = Hot\ Utility\ (kW) + Cold\ Utility\ (kW)$	(A4)
$CO_2\ specific\ Emissions = \dfrac{Uncaptured\ CO_2\ (\frac{kmol}{h})}{H_2\ Production\ (\frac{kmol}{h})}$	(A5)
$Process\ Efficiency\ (\eta_{net}) = \dfrac{H_2\ Thermal\ Energy\ (kW)}{Feed\ Thermal\ Energy\ (kW) + Energy\ Consumed\ (kW)} \times 100\%$	(A6)
$Cost_{new} = Cost_{old} \times \left(\dfrac{Capacity_{New}}{Capacity_{Old}}\right)^{x} \times \dfrac{CEPCI_{New}}{CEPCI_{Old}}$	(A7)
$Total\ Fixed\ Manufact\ Cost = Maintenance + Labor + Admin,\ support\ and\ overhead\ costs$	(A8)
$N_{OL} = (6.29 + 0.23\ N_{np})^{0.5}$ N_{OL} is the number of operators per shift and N_{np} is nonparticulate processing steps	(A9)
$TIC\ per\ ton\ of\ H_2 = \dfrac{Total\ investment\ cost}{Hydrogen\ generation}$	(A10)
$Hydrogen\ Cost\ [\frac{EUR}{kg}] = \dfrac{Hydrogen\ Life\ Cost\ (EUR)}{Hydrogen\ Life\ Production\ Flow\ Rate\ (kg)}$	(A11)

Table A3. The assumptions for economic analysis.

Economic Assumptions	
Waste plastics	Available free of charge
Natural gas (EUR /GJ)	5
Cooling water price EUR /ton	0.01
Waste disposal (EUR /t)	10
Plant construction time (year)	3
Plant life (years)	30
Maintenance	3.5% of OPEX
Discount rate	0.08
Administration	30% Labor Cost
Labor cost EUR /person	45,000
Offsite unit and utilities	25% from equipment cost
Stream factor	0.95
Daily number of shifts	3
Land and salvage (MMEUR)	10% of FCI
Working capital (MMEUR)	10% of FCI
Taxation rate (%)	15
Ratio of recycling methanol solvent	0.01
Price of methanol (EUR /ton)	400
Price of boiling water 2017 MEUR /ton	2.03
x	0.60
CEPCI (2021)	620

References

1. Okunola, A.A.; Kehinde, I.O.; Oluwaseun, A.; Olufiropo, E.A. Public and Environmental Health Effects of Plastic Wastes Disposal: A Review. *J. Toxicol. Risk Assess.* **2019**, *5*, 1–13. [CrossRef]
2. Thompson, R.C.; Moore, C.J.; Saal, F.S.V.; Swan, S.H. Plastics, the Environment and Human Health: Current Consensus and Future Trends. *Philos. Trans. R. Soc. B Biol. Sci.* **2009**, *364*, 2153–2166. [CrossRef] [PubMed]
3. Jambeck, J.R.; Geyer, R.; Wilcox, C.; Siegler, T.R.; Perryman, M.; Andrady, A.; Narayan, R.; Law, K.L. Plastic Waste Inputs from Land into the Ocean. *Science* **2015**, *347*, 768–771. [CrossRef] [PubMed]
4. d'Ambrières, W. Plastics Recycling Worldwide: Current Overview and Desirable Changes. *Field Actions Sci. Rep. J. Field Act.* **2019**, *19*, 12–21.
5. Francis, R. (Ed.) *Recycling of Polymers: Methods, Characterization and Applications*; John Wiley & Sons: Hoboken, NJ, USA, 2016.
6. Ahmad, N.; Ahmad, N.; Maafa, I.M.; Ahmed, U.; Akhter, P.; Shehzad, N.; Amjad, U.-E.; Hussain, M.; Javaid, M. Conversion of Poly-Isoprene Based Rubber to Value-Added Chemicals and Liquid Fuel via Ethanolysis: Effect of Operating Parameters on Product Quality and Quantity. *Energy* **2020**, *191*, 116543. [CrossRef]
7. Singh, N.; Hui, D.; Singh, R.; Ahuja, I.P.S.; Feo, L.; Fraternali, F. Recycling of Plastic Solid Waste: A State of Art Review and Future Applications. *Compos. Part B Eng.* **2017**, *115*, 409–422. [CrossRef]
8. Simões, C.L.; Pinto, L.M.C.; Bernardo, C.A. Environmental and Economic Analysis of End of Life Management Options for an HDPE Product Using a Life Cycle Thinking Approach. *Waste Manag. Res.* **2014**, *32*, 414–422. [CrossRef]
9. Ahmed, U.; Hussain, M.A.; Bilal, M.; Zeb, H.; Ahmad, N.; Ahmad, N.; Usman, M. Production of Hydrogen from Low Rank Coal Using Process Integration Framework between Syngas Production Processes: Techno-Economic Analysis. *Chem. Eng. Process. -Process Intensif.* **2021**, *169*, 108639. [CrossRef]
10. Díaz de León, J.N.; Loera-Serna, S.; Zepeda, T.A.; Domínguez, D.; Pawelec, B.; Venezia, A.M.; Fuentes-Moyado, S. Noble Metals Supported on Binary γ-Al$_2$O$_3$-α-Ga$_2$O$_3$ Oxide as Potential Low-Temperature Water-Gas Shift Catalysts. *Fuel* **2020**, *266*, 117031. [CrossRef]
11. Abdul Jameel, A.G.; Alkhateeb, A.; Telalović, S.; Elbaz, A.M.; Roberts, W.L.; Sarathy, S.M. Environmental Challenges and Opportunities in Marine Engine Heavy Fuel Oil Combustion. *Lect. Notes Civ. Eng.* **2019**, *22*, 1047–1055. [CrossRef]
12. Abdul Jameel, A.G.; Han, Y.; Brignoli, O.; Telalović, S.; Elbaz, A.M.; Im, H.G.; Roberts, W.L.; Sarathy, S.M. Heavy Fuel Oil Pyrolysis and Combustion: Kinetics and Evolved Gases Investigated by TGA-FTIR. *J. Anal. Appl. Pyrolysis* **2017**, *127*, 183–195. [CrossRef]
13. Ordonez-Loza, J.; Chejne, F.; Jameel, A.G.A.; Telalovic, S.; Arrieta, A.A.; Sarathy, S.M. An Investigation into the Pyrolysis and Oxidation of Bio-Oil from Sugarcane Bagasse: Kinetics and Evolved Gases Using TGA-FTIR. *J. Environ. Chem. Eng.* **2021**, *9*, 106144. [CrossRef]
14. Garba, M.D.; Usman, M.; Khan, S.; Shehzad, F.; Galadima, A.; Ehsan, M.F.; Ghanem, A.S.; Humayun, M. CO$_2$ towards Fuels: A Review of Catalytic Conversion of Carbon Dioxide to Hydrocarbons. *J. Environ. Chem. Eng.* **2021**, *9*, 104756. [CrossRef]
15. Al-Qadri, A.A.; Nasser, G.A.; Galadima, A.; Muraza, O. A Review on the Conversion of Synthetic Gas to LPG over Hybrid Nanostructure Zeolites Catalysts. *ChemistrySelect* **2022**, *7*, e202200042. [CrossRef]
16. Farzad, S.; Mandegari, M.A.; Görgens, J.F. A Critical Review on Biomass Gasification, Co-Gasification, and Their Environmental Assessments. *Biofuel Res. J.* **2016**, *3*, 483–495. [CrossRef]
17. Almohamadi, H.; Alamoudi, M.; Ahmed, U.; Shamsuddin, R.; Smith, K. Producing Hydrocarbon Fuel from the Plastic Waste: Techno-Economic Analysis. *Korean J. Chem. Eng.* **2021**, *38*, 2208–2216. [CrossRef]
18. Tian, H.; Li, J.; Yan, M.; Tong, Y.W.; Wang, C.H.; Wang, X. Organic Waste to Biohydrogen: A Critical Review from Technological Development and Environmental Impact Analysis Perspective. *Appl. Energy* **2019**, *256*, 113961. [CrossRef]
19. Lopez, G.; Artetxe, M.; Amutio, M.; Alvarez, J.; Bilbao, J.; Olazar, M. Recent Advances in the Gasification of Waste Plastics. A Critical Overview. *Renew. Sustain. Energy Rev.* **2018**, *82*, 576–596. [CrossRef]
20. Barbarias, I.; Lopez, G.; Artetxe, M.; Arregi, A.; Santamaria, L.; Bilbao, J.; Olazar, M. Pyrolysis and In-Line Catalytic Steam Reforming of Polystyrene through a Two-Step Reaction System. *J. Anal. Appl. Pyrolysis* **2016**, *122*, 502–510. [CrossRef]
21. Al-Haj Ibrahim, H. Introductory Chapter: Pyrolysis. In *Recent Advances in Pyrolysis*; IntechOpen: London, UK, 2020.
22. Kern, S.J.; Pfeifer, C.; Hofbauer, H. Cogasification of Polyethylene and Lignite in a Dual Fluidized Bed Gasifier. *Ind. Eng. Chem. Res.* **2013**, *52*, 4360–4371. [CrossRef]
23. Zaccariello, L.; Mastellone, M. Fluidized-Bed Gasification of Plastic Waste, Wood, and Their Blends with Coal. *Energies* **2015**, *8*, 8052–8068. [CrossRef]
24. Gil, J.; Corella, J.; Aznar, M.P.; Caballero, M.A. Biomass Gasification in Atmospheric and Bubbling Fluidized Bed: Effect of the Type of Gasifying Agent on the Product Distribution. *Biomass Bioenergy* **1999**, *17*, 389–403. [CrossRef]
25. Xiao, R.; Jin, B.; Zhou, H.; Zhong, Z.; Zhang, M. Air Gasification of Polypropylene Plastic Waste in Fluidized Bed Gasifier. *Energy Convers. Manag.* **2007**, *48*, 778–786. [CrossRef]
26. Erkiaga, A.; Lopez, G.; Amutio, M.; Bilbao, J.; Olazar, M. Syngas from Steam Gasification of Polyethylene in a Conical Spouted Bed Reactor. *Fuel* **2013**, *109*, 461–469. [CrossRef]
27. Dieterich, V.; Buttler, A.; Hanel, A.; Spliethoff, H.; Fendt, S. Power-to-Liquid via Synthesis of Methanol, DME or Fischer–Tropsch-Fuels: A Review. *Energy Environ. Sci.* **2020**, *13*, 3207–3252. [CrossRef]

28. Usman, M.; Zeb, Z.; Ullah, H.; Suliman, M.H.; Humayun, M.; Ullah, L.; Shah, S.N.A.; Ahmed, U.; Saeed, M. A Review of Metal-Organic Frameworks/Graphitic Carbon Nitride Composites for Solar-Driven Green H2 Production, CO_2 Reduction, and Water Purification. *J. Environ. Chem. Eng.* **2022**, *10*, 107548. [CrossRef]
29. Arslan, M.T.; Qureshi, B.A.; Gilani, S.Z.A.; Cai, D.; Ma, Y.; Usman, M.; Chen, X.; Wang, Y.; Wei, F. Single-Step Conversion of H_2-Deficient Syngas into High Yield of Tetramethylbenzene. *ACS Catal.* **2019**, *9*, 2203–2212. [CrossRef]
30. Yao, Z.; Yu, S.; Su, W.; Wu, W.; Tang, J.; Qi, W. Kinetic Studies on the Pyrolysis of Plastic Waste Using a Combination of Model-Fitting and Model-Free Methods. *Waste Manag. Res.* **2020**, *38*, 77–85. [CrossRef]
31. Vijayakumar, A.; Sebastian, J. Pyrolysis Process to Produce Fuel from Different Types of Plastic—A Review. In *IOP Conference Series: Materials Science and Engineering*; Institute of Physics Publishing: Bristol, UK, 2018; Volume 396.
32. Deng, N.; Li, D.; Zhang, Q.; Zhang, A.; Cai, R.; Zhang, B. Simulation Analysis of Municipal Solid Waste Pyrolysis and Gasification Based on Aspen Plus. *Front. Energy* **2019**, *13*, 64–70. [CrossRef]
33. Kannan, P.; Al, A.; Srinivasak, C. Optimization of Waste Plastics Gasification Process Using Aspen-Plus. In *Gasification for Practical Applications*; IntechOpen: London, UK, 2012.
34. Devasahayam, S.; Bhaskar Raju, G.; Mustansar Hussain, C. Utilization and Recycling of End of Life Plastics for Sustainable and Clean Industrial Processes Including the Iron and Steel Industry. *Mater. Sci. Energy Technol.* **2019**, *2*, 634–646. [CrossRef]
35. Devasahayam, S. Catalytic Actions of $MgCO_3$/MgO System for Efficient Carbon Reforming Processes. *Sustain. Mater. Technol.* **2019**, *22*, e00122. [CrossRef]
36. Devasahayam, S. Decarbonising the Portland and Other Cements—Via Simultaneous Feedstock Recycling and Carbon Conversions sans External Catalysts. *Polymers* **2021**, *13*, 2462. [CrossRef]
37. Devasahayam, S. Review: Opportunities for Simultaneous Energy/Materials Conversion of Carbon Dioxide and Plastics in Metallurgical Processes. *Sustain. Mater. Technol.* **2019**, *22*, e00119. [CrossRef]
38. Devasahayam, S.; Strezov, V. Thermal Decomposition of Magnesium Carbonate with Biomass and Plastic Wastes for Simultaneous Production of Hydrogen and Carbon Avoidance. *J. Clean. Prod.* **2018**, *174*, 1089–1095. [CrossRef]
39. Fivga, A.; Dimitriou, I. Pyrolysis of Plastic Waste for Production of Heavy Fuel Substitute: A Techno-Economic Assessment. *Energy* **2018**, *149*, 865–874. [CrossRef]
40. Emad, N.; Vahid, B. Hydrogen Production from Co-Gasification of Asphaltene and Plastic. *Pet. Sci. Technol.* **2019**, *37*, 1905–1909. [CrossRef]
41. AlNouss, A.; McKay, G.; Al-Ansari, T. Enhancing Waste to Hydrogen Production through Biomass Feedstock Blending: A Techno-Economic-Environmental Evaluation. *Appl. Energy* **2020**, *266*, 114885. [CrossRef]
42. Namioka, T.; Saito, A.; Inoue, Y.; Park, Y.; Min, T.J.; Roh, S.A.; Yoshikawa, K. Hydrogen-Rich Gas Production from Waste Plastics by Pyrolysis and Low-Temperature Steam Reforming over a Ruthenium Catalyst. *Appl. Energy* **2011**, *88*, 2019–2026. [CrossRef]
43. Chai, Y.; Gao, N.; Wang, M.; Wu, C. H2 Production from Co-Pyrolysis/Gasification of Waste Plastics and Biomass under Novel Catalyst Ni-CaO-C. *Chem. Eng. J.* **2020**, *382*, 122947. [CrossRef]
44. Hafiz Kamarudin, M.; Yaakob, M.Y.; Salit, M.S.; Haery, H.; Pieter, I.; Badarulzaman, N.A.; Sohaimi, R.M. A Review on Different Forms and Types of Waste Plastic Used in Concrete Structure for Improvement of Mechanical Properties. *J. Adv. Res. Appl. Mech. J. Homepage* **2016**, *28*, 9–30.
45. Jin, J.; Wei, X.; Liu, M.; Yu, Y.; Li, W.; Kong, H.; Hao, Y. A Solar Methane Reforming Reactor Design with Enhanced Efficiency. *Appl. Energy* **2018**, *226*, 797–807. [CrossRef]
46. Aspen Plus®. *Aspen Plus User Guide*; AspenTech: Bedford, MA, USA, 1981.
47. Saebea, D.; Ruengrit, P.; Arpornwichanop, A.; Patcharavorachot, Y. Gasification of Plastic Waste for Synthesis Gas Production. *Energy Rep.* **2020**, *6*, 202–207. [CrossRef]
48. Wu, C.; Williams, P.T. Hydrogen Production by Steam Gasification of Polypropylene with Various Nickel Catalysts. *Appl. Catal. B Environ.* **2009**, *87*, 152–161. [CrossRef]
49. Ghoneim, S.A.; El-Salamony, R.A.; El-Temtamy, S.A. Review on Innovative Catalytic Reforming of Natural Gas to Syngas. *World J. Eng. Technol.* **2016**, *4*, 116–139. [CrossRef]
50. Ahmed, U. Techno-Economic Analysis of Dual Methanol and Hydrogen Production Using Energy Mix Systems with CO_2 Capture. *Energy Convers. Manag.* **2021**, *228*, 113663. [CrossRef]
51. Garrett, D.E. *Chemical Engineering Economics*; Springer Science & Business Media: New York, NY, USA, 1989. [CrossRef]
52. Motta, I.L.; Miranda, N.T.; Maciel Filho, R.; Wolf Maciel, M.R. Biomass Gasification in Fluidized Beds: A Review of Biomass Moisture Content and Operating Pressure Effects. *Renew. Sustain. Energy Rev.* **2018**, *94*, 998–1023. [CrossRef]
53. Sorrels, J.L.; Walton, T.G. *Section 1 Introduction-2-Chapter 2 Cost Estimation: Concepts and Methodology*; EPA: Washington, DC, USA, 2017.
54. Panneerselvam, R. *Engineering Economics*; Prentice-Hall of India: Delhi, India, 2001; ISBN 9788120317437.
55. Hydrogen Basics-Production. Available online: http://www.fsec.ucf.edu/en/consumer/hydrogen/basics/production.htm (accessed on 11 October 2021).
56. Linde, Hydrogen Utopia to Deploy Plastic Waste-to-Hydrogen Technology in Poland | S&P Global Platts. Available online: https://www.spglobal.com/platts/en/market-insights/latest-news/electric-power/090621-linde-hydrogen-utopia-to-deploy-plastic-waste-to-hydrogen-technology-in-poland (accessed on 11 October 2021).

57. Technical Contact, N.; Ruth, M. Hydrogen Production Cost Estimate Using Biomass Gasification: Independent Review. 2010. Available online: https://www.osti.gov/biblio/1028523 (accessed on 15 March 2022).
58. Mourhly, A.; Kacimi, M.; Halim, M.; Arsalane, S. New Low Cost Mesoporous Silica (MSN) as a Promising Support of Ni-Catalysts for High-Hydrogen Generation via Dry Reforming of Methane (DRM). *Int. J. Hydrogen Energy* **2020**, *45*, 11449–11459. [CrossRef]
59. Xu, D.; Xiong, Y.; Ye, J.; Su, Y.; Dong, Q.; Zhang, S. Performances of Syngas Production and Deposited Coke Regulation during Co-Gasification of Biomass and Plastic Wastes over Ni/γ-Al$_2$O$_3$ Catalyst: Role of Biomass to Plastic Ratio in Feedstock. *Chem. Eng. J.* **2020**, *392*, 123728. [CrossRef]

Article

Fabrication of Recycled Polycarbonate Fibre for Thermal Signature Reduction in Camouflage Textiles

Asril Soekoco [1,2,*], Ateeq Ur Rehman [3], Ajisetia Fauzi [1], Hamdi Tasya [1], Purnama Diandra [1], Islami Tasa [1], Nugraha [2] and Brian Yuliarto [2,*]

1   Department of Textile Engineering, Politeknik STTT Bandung, Kota Bandung 40272, Indonesia; fauziaji98@gmail.com (A.F.); tasya.wirdati27@gmail.com (H.T.); diandra190896@gmail.com (P.D.); tasaislami7@gmail.com (I.T.)
2   Department of Engineering Physics, Faculty of Industrial Technology, Institut Teknologi Bandung, Kota Bandung 40132, Indonesia; nugraha@tf.itb.ac.id
3   School of Engineering, College of Science & Engineering, The University of Edinburgh, Edinburgh, EH8 9YL, UK; ateeq-ur-rehman@sms.ed.ac.uk
*   Correspondence: asril-s@kemenperin.go.id (A.S.); brian@tf.itb.ac.id (B.Y.)

Abstract: Thermal signature reduction in camouflage textiles is a vital requirement to protect soldiers from detection by thermal imaging equipment in low-light conditions. Thermal signature reduction can be achieved by decreasing the surface temperature of the subject by using a low thermally conductive material, such as polycarbonate, which contains bisphenol A. Polycarbonate is a hard type of plastic that generally ends up in dumps and landfills. Accordingly, there is a large amount of polycarbonate waste that needs to be managed to reduce its drawbacks to the environment. Polycarbonate waste has great potential to be used as a material for recycled fibre by the melt spinning method. In this research, polycarbonate roofing-sheet waste was extruded using a 2 mm diameter of spinnerette and a 14 mm barrel diameter in a 265 °C temperature process by using a lab-scale melt spinning machine at various plunger and take-up speeds. The fibres were then inserted into 1 × 1 rib-stitch knitted fabric made by Nm 15 polyacrylic commercial yarns, which were manufactured by a flat knitting machine. The results showed that applying recycled polycarbonate fibre as a fibre insertion in polyacrylic knitted fabric reduced the emitted infrared and thermal signature of the fabric.

Keywords: polycarbonate waste; recycled fibre; thermal signature; camouflage textiles

Citation: Soekoco, A.; Rehman, A.U.; Fauzi, A.; Tasya, H.; Diandra, P.; Tasa, I.; N.; Yuliarto, B. Fabrication of Recycled Polycarbonate Fibre for Thermal Signature Reduction in Camouflage Textiles. *Polymers* **2022**, *14*, 1972. https://doi.org/10.3390/polym14101972

Academic Editor: Sheila Devasahayam

Received: 8 April 2022
Accepted: 2 May 2022
Published: 12 May 2022

Publisher's Note: MDPI stays neutral with regard to jurisdictional claims in published maps and institutional affiliations.

1. Introduction

Heat migrates from hotter areas to cooler areas through thermal convection, thermal conduction, and thermal radiation [1]. This phenomenon is applied in every aspect of our daily lives, from the most simple parts of our routine to the most sophisticated military applications. The combination of heat migration from an object to the environment can create a thermal signature, measured by thermal imaging techniques [2]. Thermal signature is affected by the temperature of the object and background, and an effective way to reduce it is by reducing the emissivity [3,4]. The temperature difference (ΔT) between the object and background affects the thermal region in thermal imaging [5,6]. Thermal imaging is proven to produce an advanced spectral range that can be seen by a human and shows an obvious contrast between the environment and objects of high-temperature variance [7–11]. Currently, thermal imaging techniques are used incommercial activities, industrial processes and by the military. The military is one area that develops camouflage textiles to protect a soldier from detection by various equipment in many operations, including surveillance purposes at night or in low-light conditions [4]. Generally, there are several polymers used in camouflage textile materials because of their capability to minimize infrared visual detection by reducing the thermal signature of the soldier, one of them is polycarbonate.

Polycarbonate polymer was developed in 1953; it is tough, rigid, and has a relatively high melting temperature, from 225 °C to 250 °C, with a glass transition temperature of 145 °C. It also has flame retardant property, with an oxygen index of 26, in addition to thermal and electrical insulation [12]. Polycarbonate is made from a condensation reaction of bisphenol A with phosphorus (phosgene) in alkaline media [13]. The presence of bisphenol A makes polycarbonate capable of absorbing infrared, which can be utilized for military textile applications [13,14]. Due to the optical and mechanical properties of polycarbonate, Fujitsu developed and introduced polycarbonate optical fibre in 1986 [15]. Polycarbonate fibre was developing rapidly, particularly for high-temperature resistant polymer sensor applications, since it has a high glass transition temperature and high flexibility for bending. Since that time, polycarbonate fibre also started to develop rapidly in various applications such as high-temperature-resistant polycarbonate fibre, electrical conductive polycarbonate fibre (by adding multiwalled carbon nanotubes), and strain sensing polycarbonate fibre [15–17]. However, all of the existing polycarbonate fibres are developed from virgin polycarbonate material instead of waste polycarbonate material.

The global polycarbonate market significantly increased from 2000 to 2010, and it is forecasted to increase to 5.5 million tons by 2024, due to the high demand in the electrical, electronics, and automotive industries [13,18,19]. Polycarbonate is a hard type of plastic that is not easily recycled and polycarbonate waste generally ends up in dumps and landfills [20]. Accordingly, there is a huge amount of polycarbonate waste available which needs to be managed to reduce its drawbacks to the environment. Therefore, the fabrication of polycarbonate sheet waste for thermal signature reduction application is quite promising, particularly for incorporating microstructures into apparel for visible and infrared camouflage used in the military [21].

2. Materials and Methods

All of the polycarbonate fibres in this research were produced from multiwall polycarbonate roofing-sheet waste (grey colour, 5 mm thickness, from Twinlite, PT Impack Pratama, Jakarta, Indonesia). To identify and confirm the material, material characterization was conducted using an FTIR-8400 (Shimadzu, Kyoto, Japan). To produce recycled polycarbonate fibres, polycarbonate roofing-sheet waste was cleaned before the grinding process. The ground polycarbonate chips were then extruded using a 2 mm diameter of spinnerette and a 14 mm barrel diameter in a 265 °C temperature process, using a lab-scale melt spinning machine as can be seen in Figure 1. Variations in two processing parameters were applied to produce several polycarbonate fibre properties, plunger speed (0.10, 0.15 and 0.18 cm/s) and take-up speed (6.2, 6.9 and 9.2 m/min). The recycled polycarbonate fibres that were produced were then inserted into 1 × 1 rib-stitch knitted fabric made by Nm 15 polyacrylic commercial yarns, which were manufactured by a flat knitting machine with a fabric construction of 15 courses per inch and 18 wales per inch.

The fibre diameter measurement was performed using an Olympus CX-22 binocular microscope (Tokyo, Japan) assisted camera adapter, the fibre tensile strength and elongation were measured using an Instron tensile testing machine according to ASTM D 3822-07, the fabric area density was measured according to SNI-ISO 3801:2010, and the fabric bursting test was measured according to SNI 0561:2008. To demonstrate the thermal signature reduction in fabrics, we first covered the heat source (40 °C, 68 °C and 100 °C) with the various polycarbonate–polyacrylic fabrics, the fabric surface temperature and the thermal signature of each fabric was then captured using a Flir One thermal imaging camera (Teledyne FLIR LLC, Wilsonville, OR, USA), which can measure temperatures from −20 °C to 200 °C (2% instrumental error). To achieve a high accuracy result, the ambient temperature was maintained between 30 °C and 45 °C, and the measurement distance was 0.8 m [22,23]. The surface temperature reduction in each fabric was calculated to discover the emissivity decrement and confirmed the thermal signature reduction effect.

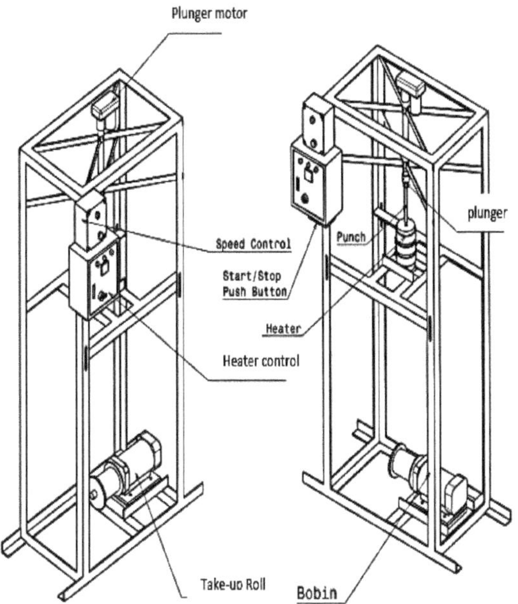

Figure 1. Schematic diagram of lab-scale melt spinning machine.

3. Results and Discussion

3.1. Fabrication of Polyacrylic–Polycarbonate Fabric

The FTIR spectrum of polycarbonate sheet waste is shown in Figure 2. Polycarbonate has principal peaks around 1015 cm^{-1} caused by symmetric O–C–O carbonate group deformations, CH3-vibrations around 1081 cm^{-1}, C=C-vibrations at 1506 cm^{-1}, C=O carbonate group deformations near 1775 cm^{-1} and peaks around 3.000 cm^{-1} caused by C–H aromatic ring deformations. The result of the characterization confirmed that the polymer of this material was polycarbonate.

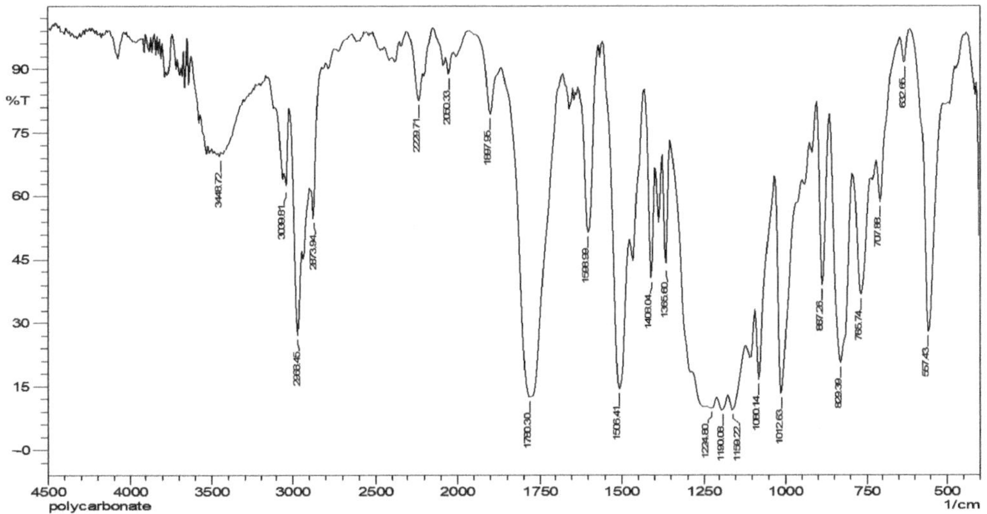

Figure 2. IR absorption spectrum of multiwall roofing-sheet waste.

Figure 3b–d show the several variations of fabric produced in this research, all of these variations were used in a 1 × 1 rib-stitch knit due to its structural dimension stability and extensive shrinkage in width [24]. To demonstrate the thermal signature reduction ability of the recycled polycarbonate fibre, we prepared three variations of polycarbonate composition in knitted fabric by inserting recycled polycarbonate fibres in polyacrylic knitted fabric. We prepared fabrics containing 0% polycarbonate–100% polyacrylic fabric (without polycarbonate fibre insertion); 50% polycarbonate–50% polyacrylic (with one strand of polycarbonate fibre insertion) and 66.7% polycarbonate–33.3% polyacrylic (with two strands of polycarbonate fibre insertion). This variation was the best option considering the structure of the 1 × 1 rib-stitch fabric construction and the fluency of the knitting process in knitted fabric fabrication, according to a preliminary research result. The amount of recycled polycarbonate fibre insertion influenced the fabrication process, excessive recycled polycarbonate fibre insertion caused a jam of the yarn feeding, whereas inadequate recycled polycarbonate fibre insertion may cause loose and stray fibre in the fabric due to improper yarn tension.

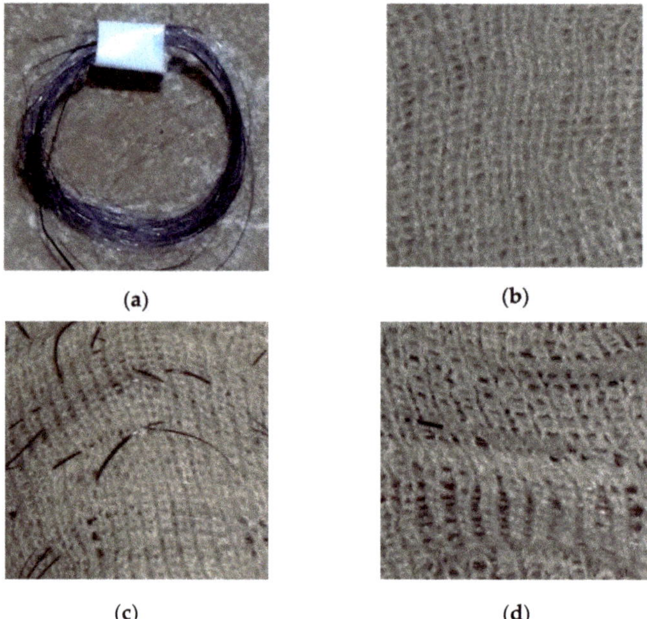

Figure 3. (**a**) Recyled polycarbonate fibre; (**b**) 0% Polycarbonate–100% Polyacrylic fabric; (**c**) 50% Polycarbonate–50% Polyacrylic; (**d**) 66.7% Polycarbonate–33.3% Polyacrylic.

3.2. Fibre Diameter

In the second part of the study, we examined the correlation of plunger speed and take-up speed with the recycled polycarbonate fibre diameter. Considering the available machine settings, we used five variations of processing parameters; the highest plunger speed was 0.18 m/min and the lowest plunger speed was 0.10 m/min, whereas the highest take-up speed was 9.2 cm/min and the lowest plunger speed was 6.9 cm/min. We determined that a high take-up speed resulted in a finer fibre diameter than a low take-up speed. Meanwhile, a low plunger speed produced a coarser fibre diameter than a fast plunger speed. This occurred because of the strain rate along the spin line [25]. Higher take-up speed leads to a higher strain rate, but otherwise higher plunger speed leads to a lower strain rate.

Figure 4 shows the finest fibre diameter in this study was produced by setting the plunger speed at 0.10 m/min and the take-up speed at 9.2 cm/min, whereas the coarsest fibre diameter was produced by setting the plunger speed at 0.18 m/min and the take-up speed at 6.9 cm/min. This occurred because a high take-up speed leads to a rapid decrease in the fibre diameter [26].

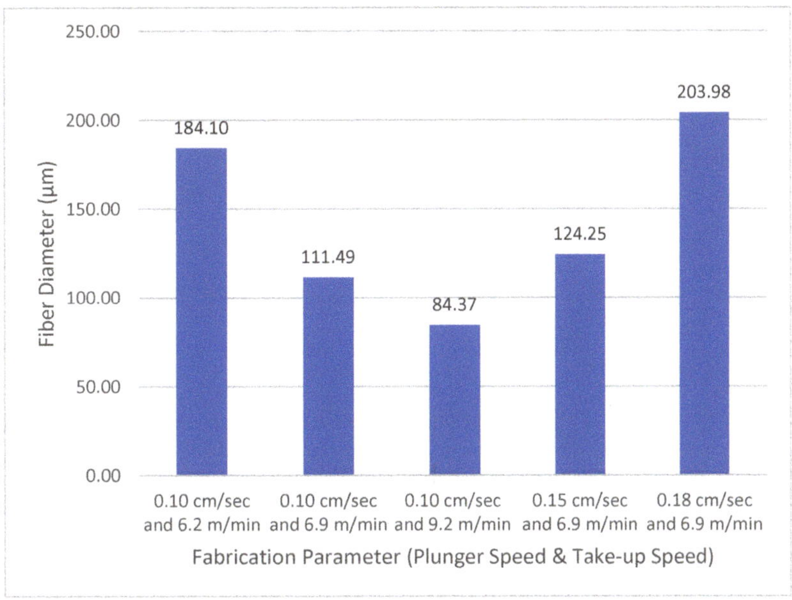

Figure 4. Correlation of plunger speed and take-up speed with polycarbonate fibre diameter.

3.3. Fibre Tenacity

Fibre tenacity has an inversely proportional value to fibre diameter, a fine diameter of fibre leads to higher fibre tenacity, since an increase in the strain rate creates a better molecular chain orientation and alignment. An increase in molecular chain alignment leads to an increase in tenacity because the molecular chains become tighter, as shown in Figure 5 [27,28]. Figure 6 shows that the tenacity of fibre produced from the combination of the highest take-up speed and lowest plunger speed was 1.69 g/denier. The combination of the highest take-up speed and lowest plunger speed creates the highest strain rate, which leads to improvement of polymer molecular orientation degree. The improvement of polymer molecular orientation degree results in the increase in fibre tenacity.

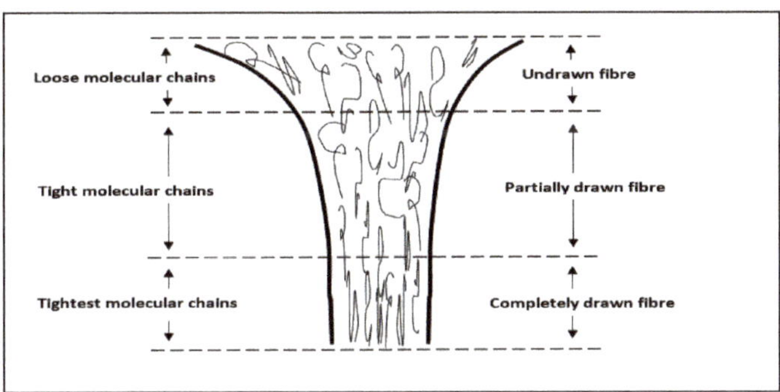

Figure 5. Change of molecular orientation in the take-up process.

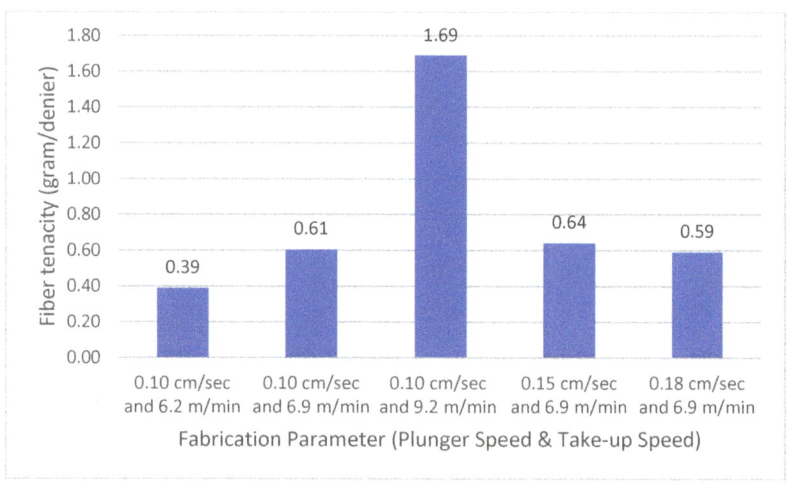

Figure 6. Correlation of plunger speed and take-up speed with polycarbonate fibre tenacity.

3.4. Fabric Areal Density

Fabric areal density is a measurement of mass per unit area of the fabric, it is affected by course per inch, wale per inch, yarn count, material insertion, and other knit structures that influence physical properties, such as shrinkage, water absorbency, and air permeability [29]. As can be seen in Figure 7, the 66.7% polycarbonate–33.3% polyacrylic fabric has the highest areal density, meanwhile, the 0% polycarbonate–100% polyacrylic fabric has the lowest areal density. This occurs because the 66.7% polycarbonate–33.3% polyacrylic fabric has two strands of recycled polycarbonate fibre insertion in every course, and this will improve fabric mass. Improvement of fabric mass affects the fabric's physical properties, which is very important, especially for protective garments [30].

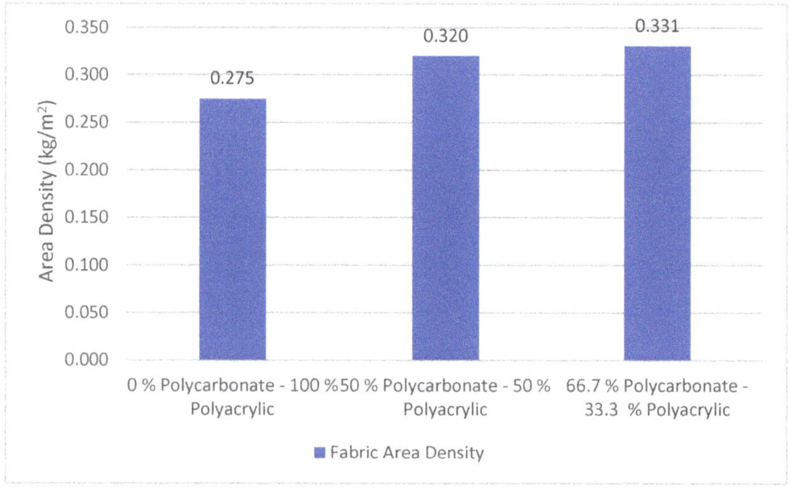

Figure 7. The areal density of polycarbonate–polyacrylic fabrics.

We determined that the 0% polycarbonate–100% polyacrylic fabric has the lowest areal density, more than 20% lighter than the areal density of the 66.7% polycarbonate–33.3% polyacrylic fabric. This may lead to poor cover opacity and bursting strength [31]. For some applications, lightweight knitted fabric such as the 0% polycarbonate–100% polyacrylic is preferred over other types of clothing [32].

3.5. Bursting Strength

There are several important mechanical properties in fabric, one of them is bursting strength, which is generally associated with knitted fabrics, or breaking strength with woven fabrics [33,34]. Figure 8 shows that the 66.7% polycarbonate–33.3% polyacrylic fabric has a higher bursting strength compared with the 0% polycarbonate–100% polyacrylic fabric and the 50% polycarbonate–50% polyacrylic fabric. The bursting strength of the 66.7% polycarbonate–33.3% polyacrylic fabric was 25% higher than the 0% polycarbonate–100% polyacrylic fabric. This occurred because the 66.7% polycarbonate–33.3% polyacrylic fabric has recycled polycarbonate fibre insertion that absorbs external force during the bursting strength test. A high tightness factor and high areal density of the fabric structure leads to a high bursting strength [35]. According to this study, we found that the insertion of recycled polycarbonate fibres can improve the mechanical properties of a knit-fabric and that this is suitable for a high-performance textile application.

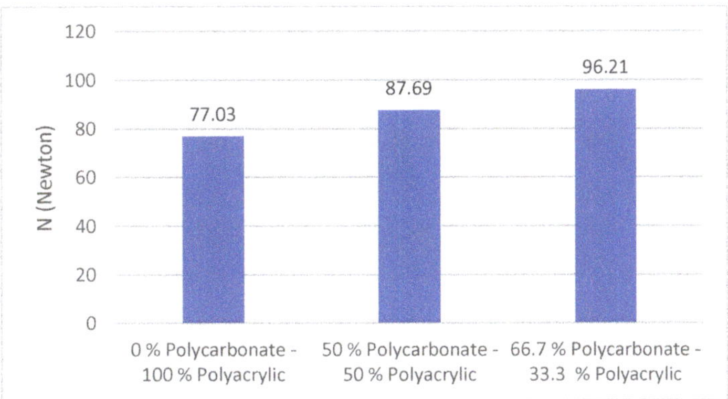

Figure 8. Bursting strength of polycarbonate–polyacrylic fabrics.

3.6. Thermal Signature Reduction

A thermal signature is created through an apparent temperature differential between an object and its background [6]. Thermal signature is one of the most critical aspects to prevent or delay detection in low-light military camouflage operations. Military camouflage has at least two requirements that need to be managed; the near-infrared and the visible regions of the spectrum [36]. Temperature regulation is required for reducing the emitted infrared waves in military camouflage, lower emitted infrared waves lead to higher thermal insulation properties [37]. Therefore, it is important to select a material that has a low infrared emittance and high thermal insulation to reduce the thermal signature of camouflage textile.

Figure 9a–c show the thermal signatures of knitted fabrics for three different recycled polycarbonate fibre contents using a 40 °C heat source. Figure 9c, showing the fabric with the highest recycled polycarbonate content (67.7%), has a darker thermal signature compared with Figure 9b,c, which have lower recycled polycarbonate contents. The result was similar when the temperature of the heat source was increased, Figure 9d–f are the thermal signatures of knitted fabrics with various recycled polycarbonate fibre contents using a 68 °C heat source. The thermal signature of the knitted fabric in Figure 9d, which had the lowest recycled polycarbonate fibre content (0%), is brighter than that of the other knitted fabrics, which had a higher recycled polycarbonate content. The thermal signatures of the knitted fabrics on a 100 °C heat source are shown in Figure 9g–i. The knitted fabrics which had a higher recycled polycarbonate content produced darker thermal signatures, as shown by Figure 9i,j, which had 67% and 50% recycled polycarbonate content. Figure 9a–g have no recycled polycarbonate fibre insertion and all show brighter thermal signatures compared with the other fabric variations, especially Figure 9c,f,i, which had the highest amount of polycarbonate fibre insertion. It was found that applying recycled polycarbonate fibres as an insertion in polyacrylic knitted fabrics reduced the emitted infrared waves captured by a thermal imaging camera.

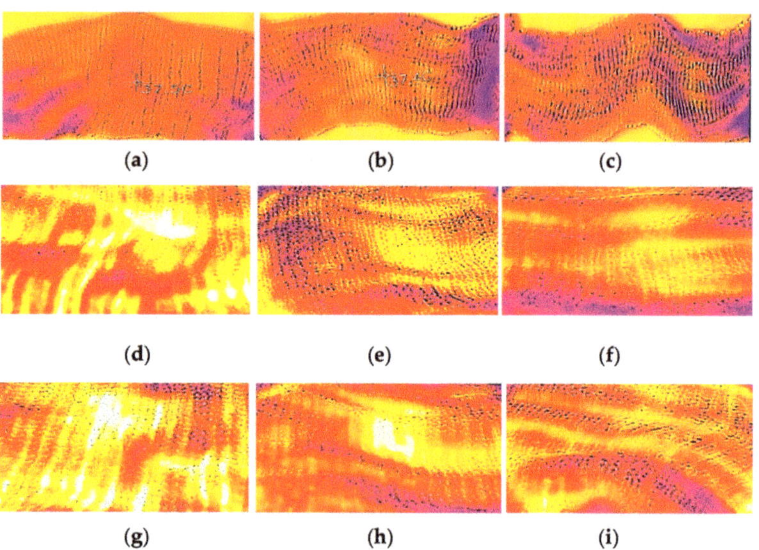

Figure 9. Thermography of knitted fabric: (**a**) 0% Polycarbonate–100% Polyacrylic on 40 °C; (**b**) 50% Polycarbonate–50% Polyacrylic on 40 °C; (**c**) 66.7% Polycarbonate–33.3% Polyacrylic on 40 °C; (**d**) 0% Polycarbonate–100% Polyacrylic on 68 °C; (**e**) 50% Polycarbonate–50% Polyacrylic on 68 °C; (**f**) 66.7% Polycarbonate–33.3% Polyacrylic on 68 °C; (**g**) 0% Polycarbonate–100% Polyacrylic on 100 °C; (**h**) 50% Polycarbonate–50% Polyacrylic on 100 °C; (**i**) 66.7% Polycarbonate–33.3% Polyacrylic on 100 °C.

Thermal signature reduction in infrared camouflage is reached by managing the surface temperature and the surface emittance of the object [38]. Decreasing the surface temperature of the object represents a reduction in the surface emittance and thermal signature. Figure 10 shows that the 66.7% polycarbonate–33.3% polyacrylic fabric has the highest surface temperature reduction in all heat source temperatures. At high operational temperature (100 °C), the surface temperature reduction in the 66.7% polycarbonate–33.3% polyacrylic fabric is almost 75% higher than the 0% polycarbonate–100% polyacrylic fabric. The effect is quite similar at other operational temperatures, for 40 °C and 68 °C heat source temperatures, the surface temperature reduction in the 66.7% polycarbonate–33.3% polyacrylic fabric is 27.6%, and 70.1% higher than the 0% polycarbonate–100% polyacrylic fabric. This is caused by the use of recycled polycarbonate fibre as an insertion in polyacrylic knitted fabric, since polycarbonate has a low thermal conductivity of 200 mW/m*K compared with polyacrylic yarn which has 310 mW/m*K of thermal conductivity [39,40]. The most substantial differences can be observed using a heat source temperature of 100 °C, where the knitted fabric thermal insulation is increased to 74.5% in the 66.7% polycarbonate–33.3% polyacrylic fabric. A higher operational temperature will lead to a higher thermal insulation difference. The thermal insulation of the fabric is proportional to operational temperature because thermal conductivity is dependent on temperature [39].

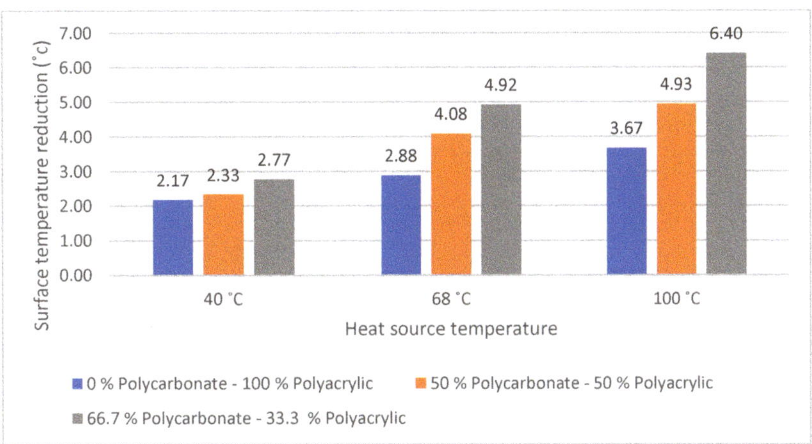

Figure 10. Surface temperature reduction in various polycarbonate-content fabrics.

The surface temperature reduction is directly proportional to the recycled polycarbonate fibre content, higher recycled polycarbonate fibre content leads to the increases in surface temperature reduction. This result confirms the advantage of applying recycled polycarbonate fibre to reduce the surface temperature and thermal signature of the fabric.

4. Conclusions

This research produced and evaluated recycled polycarbonate fibre from multiwall polycarbonate roofing-sheet waste. It was found that multiwall polycarbonate roofing-sheet waste was a potential material for synthetic fibres due to its spinnability. High take-up speeds produced fine fibres, whereas high plunger speeds lead to lower strain rates and created coarse fibrea. The coarsest fibre diameter in this study was produced by setting the plunger speed at 0.18 m/min and the take-up speed at 6.9 cm/min, whereas the finest fibre diameter was produced by setting the plunger speed at 0.10 m/min and the take-up speed in 9.2 cm/min; this setting also produced the highest fibre tenacity of 1.69 g/denier. The take-up speed value was proportionally linear to the resulting tenacity of fibre.

The insertion of recycled polycarbonate fibre can improve the mechanical properties of knitted fabrics and it is suitable for high-performance textile applications, the bursting strength of the 66.7% polycarbonate–33.3% polyacrylic fabric was 25% higher than the 0% polycarbonate–100% polyacrylic fabric. The fabric which had a higher recycled polycarbonate fibre content had a higher surface temperature and thermal signature reduction at all temperatures. The surface temperature reduction in the 66.7% polycarbonate–33.3% polyacrylic fabric was almost 75% higher than the 0% polycarbonate–100% polyacrylic fabric. We found that applying recycled polycarbonate fibres as a fibre insertion in polyacrylic knitted fabrics reduced the emitted infrared waves captured by a thermal imaging camera.

Author Contributions: Conceptualization, A.S.; methodology, A.S, A.F., H.T., P.D. and I.T.; formal analysis, A.S, A.F., H.T., P.D. and I.T.; investigation, A.S, A.F., H.T., P.D. and I.T.; data curation, A.F., H.T., P.D. and I.T.; writing—original draft preparation, A.S. and A.U.R.; writing—review and editing, A.S., A.U.R., N. and B.Y.; supervision, N. and B.Y.; project administration, A.F., H.T., P.D. and I.T. All authors have read and agreed to the published version of the manuscript.

Funding: This research was funded by BPSDMI—Kementerian Perindustrian Republik Indonesia.

Acknowledgments: The technical support provided by PT Superbtex Spinning Mill, Bandung, Indonesia.

Conflicts of Interest: The authors declare no conflict of interest.

References

1. Dai, G.; Shang, J.; Huang, J. Theory of Transformation Thermal Convection for Creeping Flow in Porous Media: Cloaking, Concentrating, and Camouflage. *Phys. Rev. E* **2018**, *97*, 022129. [CrossRef] [PubMed]
2. Negied, N.K.A.-W.; Hemayed, E.B.; Fayek, M. HSBS: A Human's Heat Signature and Background Subtraction Hybrid Approach for Crowd Counting and Analysis. *Int. J. Pattern Recognit. Artif. Intell.* **2016**, *30*, 1655025. [CrossRef]
3. Dev, O.; Dayal, S.; Dubey, A.; Abbas, S.M. Multi-Layered Textile Structure for Thermal Signature Suppression of Ground Based Targets. *Infrared Phys. Technol.* **2020**, *105*, 103175. [CrossRef]
4. Rubežienė, V.; Padleckienė, I.; Žuravliova, S.V.; Baltušnikaitė, J. Reduction of Thermal Signature Using Fabrics with Conductive Additives. *Mater. Sci.* **2013**, *19*, 409–414. [CrossRef]
5. Hixson, J.G.; Jacobs, E.L.; Vollmerhausen, R.H. *Target Detection Cycle Criteria When Using the Targeting Task Performance Metric*; Driggers, R.G., Huckridge, D.A., Eds.; International Society for Optics and Photonics: London, UK, 2004; p. 275.
6. Holst, G.C. *Common Sense Approach to Thermal Imaging*; SPIE: Bellingham, WA, USA, 2000; ISBN 978-0-8194-3722-8.
7. Liu, Q.; Zhuang, J.; Ma, J. Robust and Fast Pedestrian Detection Method for Far-Infrared Automotive Driving Assistance Systems. *Infrared Phys. Technol.* **2013**, *60*, 288–299. [CrossRef]
8. Bertozzi, M.; Broggi, A.; Caraffi, C.; Del Rose, M.; Felisa, M.; Vezzoni, G. Pedestrian Detection by Means of Far-Infrared Stereo Vision. *Comput. Vis. Image Underst.* **2007**, *106*, 194–204. [CrossRef]
9. Fang, Y.; Yamada, K.; Ninomiya, Y.; Horn, B.K.P.; Masaki, I. A Shape-Independent Method for Pedestrian Detection with Far-Infrared Images. *IEEE Trans. Veh. Technol.* **2004**, *53*, 1679–1697. [CrossRef]
10. Li, J.; Gong, W.; Li, W.; Liu, X. Robust Pedestrian Detection in Thermal Infrared Imagery Using the Wavelet Transform. *Infrared Phys. Technol.* **2010**, *53*, 267–273. [CrossRef]
11. Olmeda, D.; de la Escalera, A.; Armingol, J.M. Contrast Invariant Features for Human Detection in Far Infrared Images. In Proceedings of the 2012 IEEE Intelligent Vehicles Symposium, Alcal de Henares, Madrid, Spain, 3–7 June 2012; IEEE: Alcal de Henares, Madrid, Spain, 2012; pp. 117–122.
12. Kyriacos, D. Polycarbonates. In *Brydson's Plastics Materials*; Elsevier: Amsterdam, The Netherlands, 2017; pp. 457–485, ISBN 978-0-323-35824-8.
13. Muhamad, M.S.; Salim, M.R.; Lau, W.J.; Yusop, Z. A Review on Bisphenol a Occurrences, Health Effects and Treatment Process via Membrane Technology for Drinking Water. *Environ. Sci. Pollut. Res.* **2016**, *23*, 11549–11567. [CrossRef]
14. Kraus, R.G.; Emmons, E.D.; Thompson, J.S.; Covington, A.M. Infrared Absorption Spectroscopy of Polycarbonate at High Pressure. *J. Polym. Sci. Part B Polym. Phys.* **2008**, *46*, 734–742. [CrossRef]
15. Fasano, A.; Woyessa, G.; Stajanca, P.; Markos, C.; Stefani, A.; Nielsen, K.; Rasmussen, H.K.; Krebber, K.; Bang, O. Fabrication and Characterization of Polycarbonate Microstructured Polymer Optical Fibers for High-Temperature-Resistant Fiber Bragg Grating Strain Sensors. *Opt. Mater. Express* **2016**, *6*, 649. [CrossRef]
16. Pötschke, P.; Brünig, H.; Janke, A.; Fischer, D.; Jehnichen, D. Orientation of Multiwalled Carbon Nanotubes in Composites with Polycarbonate by Melt Spinning. *Polymer* **2005**, *46*, 10355–10363. [CrossRef]
17. Bautista-Quijano, J.R.; Pötschke, P.; Brünig, H.; Heinrich, G. Strain Sensing, Electrical and Mechanical Properties of Polycarbonate/Multiwall Carbon Nanotube Monofilament Fibers Fabricated by Melt Spinning. *Polymer* **2016**, *82*, 181–189. [CrossRef]
18. Polycarbonates Demand Worldwide from 2011 to 2024. Available online: https://www.statista.com/statistics/750965/polycarbonates-demand-worldwide/ (accessed on 22 November 2021).
19. Polycarbonate Market Research Report. Marketresearchfuture 2021. Available online: https://www.marketresearchfuture.com/reports/polycarbonate-market-1080 (accessed on 22 November 2021).
20. A New Way to Recycle Polycarbonates That Prevents BPA Leaching. Available online: https://phys.org/news/2016-06-recycle-polycarbonates-bpa-leaching.html (accessed on 22 November 2021).
21. Degenstein, L.M.; Sameoto, D.; Hogan, J.D.; Asad, A.; Dolez, P.I. Smart Textiles for Visible and IR Camouflage Application: State-of-the-Art and Microfabrication Path Forward. *Micromachines* **2021**, *12*, 773. [CrossRef] [PubMed]
22. Frequently Asked Questions: Thermal Imaging for Elevated Skin Temperature Screening. Available online: https://www.flir.com/discover/public-safety/faq-about-thermal-imaging-for-elevated-body-temperature-screening/ (accessed on 27 April 2022).
23. Ghassemi, P.; Pfefer, T.J.; Casamento, J.P.; Simpson, R.; Wang, Q. Best Practices for Standardized Performance Testing of Infrared Thermographs Intended for Fever Screening. *PLoS ONE* **2018**, *13*, e0203302. [CrossRef]
24. Singhal, K.; Mishra, S.; Kumar, B. A Study of Curling in Rib-Knit Constructions. *J. Text. Inst.* **2021**, *112*, 666–675. [CrossRef]
25. Nakajima, T.; Kajiwara, K. (Eds.) *Advanced Fiber Spinning Technology*; Woodhead Publishing: Cambridge, UK, 1994; ISBN 978-1-85573-182-0.
26. Teke, S.; Altun, S. CFD Modeling of the Melt Spinning of Poly (Ethylene Terepthalate) at Low Take-up Velocities. *Sci. Res. Essays* **2012**, *7*, 372–386.
27. Groover, M.P. *Fundamentals of Modern Manufacturing*, 4th ed.; Hohn Wilery & Sons, Inc.: Danver, CO, USA, 2010; pp. 40–49.
28. Eichhorn, S.J.; Hearle, J.W.S. *Handbook of Textile Fibre Structure*; Woodhead Publishing Limited: Cambridge, UK, 2009; pp. 217–219.
29. Kumar, V.; Sampath, V.R. Investigation on the Physical and Dimensional Properties of Single Jersey Fabrics made from Cotton Sheath—Elastomeric Core Spun. *Fibres Text. East. Eur.* **2003**, *21*, 73–75.
30. Arbataitis, E.; Mikucioniene, D.; Halavska, L. Flexible Theoretical Calculation of Loop Length and Area Density of Weft-Knitted Structures: Part I. *Materials* **2021**, *14*, 3059. [CrossRef]

31. Khalil, E. Effect of Stitch Length on Physical and Mechanical Properties of Single Jersey Cotton Knitted Fabric. *Int. J. Sci. Res. IJSR* **2012**, *3*, 4.
32. Rahman, S.; Smriti, S.A. Investigation on the Changes of Areal Density of Knit Fabric with Stitch Length Variation on the Increment of Tuck Loop Percentages. *J. Polym. Text. Eng.* **2015**, *2*, 1–14.
33. Chowdhary, U. Bursting Strength and Extension for Jersey, Interlock and Pique Knits. *Trends Text. Eng. Fash. Technol.* **2018**, *1*, 19–27. [CrossRef]
34. Degirmenci, Z.; Çelik, N. Relation between extension and bursting strength properties of the denim viewed knitted fabrics produced by cellulosic fibers. *Fibres Text. East. Eur.* **2016**, *24*, 101–106. [CrossRef]
35. Rashid, M.R.; Ahmed, F.; Azad, A.K. Bursting Strength and Pilling Properties of Weft Knitted Fabrics Made from Conventional Ring and Compact Spun Yarn. Available online: https://www.researchgate.net/publication/328419424_Bursting_Strength_and_Piling_Properties_of_Weft_Knitted_Fabrics_Made_from_Conventional_Ring_and_Compact_Spun_Yarn (accessed on 7 April 2022).
36. Adanur, S.; Tewari, A. *An Overview of Military Textiles*; NIScPR: New Delhi, India, 1997; pp. 348–352.
37. Steffens, F.; Gralha, S.E.; Ferreira, I.L.S.; Oliveira, F.R. Military Textiles—An Overview of New Developments. *Key Eng. Mater.* **2019**, *812*, 120–126. [CrossRef]
38. Zhu, H.; Li, Q.; Zheng, C.; Hong, Y.; Xu, Z.; Wang, H.; Shen, W.; Kaur, S.; Ghosh, P.; Qiu, M. High-Temperature Infrared Camouflage with Efficient Thermal Management. *Light Sci. Appl.* **2020**, *9*, 60. [CrossRef] [PubMed]
39. Weingart, N.; Raps, D.; Kuhnigk, J.; Klein, A.; Altstädt, V. Expanded Polycarbonate (EPC)—A New Generation of High-Temperature Engineering Bead Foams. *Polymers* **2020**, *12*, 2314. [CrossRef]
40. Tian, L.; Yuanyuan, Z.; Yingying, M.; Ran, H. Fabrication of Functional Silver Loaded Montmorillonite/Polycarbonate with Superhydrophobicity. *Appl. Clay Sci.* **2015**, *118*, 337–343. [CrossRef]

Review

Review of Soil Quality Improvement Using Biopolymers from Leather Waste

Daniela Simina Stefan [1], Magdalena Bosomoiu [1,*], Annette Madelene Dancila [1] and Mircea Stefan [2]

[1] Department of Analytical Chemistry and Environmental Engineering, Faculty of Chemical Engineering and Biotechnologies, University Politehnica of Bucharest, 1-7 Polizu Street, 011061 Bucharest, Romania; daniela.stefan@upb.ro (D.S.S.); madelene.dancila@upb.ro (A.M.D.)

[2] Pharmacy Faculty, University Titu Maiorescu, 22 Dâmbovnicului Street, 040441 Bucharest, Romania; stefan_apcpm@yahoo.com

* Correspondence: mbosomoiu@yahoo.com

Abstract: This paper reviews the advantages and disadvantages of the use of fertilizers obtained from leather waste, to ameliorate the agricultural soil quality. The use of leather waste (hides and skins) as raw materials to obtain biopolymer-based fertilizers is an excellent example of a circular economy. This allows the recovery of a large quantity of the tanning agent in the case of tanned wastes, as well as the valorization of significant quantities of waste that would be otherwise disposed of by landfilling. The composition of organic biopolymers obtained from leather waste is a rich source of macronutrients (nitrogen, calcium, magnesium, sodium, potassium), and micronutrients (boron, chloride, copper, iron, manganese, molybdenum, nickel and zinc), necessary to improve the composition of agricultural soils, and to remediate the degraded soils. This enhances plant growth ensuring better crops. The nutrient release tests have demonstrated that, by using the biofertilizers with collagen or with collagen cross-linked with synthetic polymers, the nutrient release can be controlled and slowed. In this case, the loss of nutrients by leaching into the inferior layers of the soil and ground water is minimized, avoiding groundwater contamination, especially with nitrate.

Keywords: biopolymers; leather waste; soil; fertilizers; industrial crops

Citation: Stefan, D.S.; Bosomoiu, M.; Dancila, A.M.; Stefan, M. Review of Soil Quality Improvement Using Biopolymers from Leather Waste. *Polymers* **2022**, *14*, 1928. https://doi.org/10.3390/polym14091928

Academic Editors: Swarup Roy and Sheila Devasahayam

Received: 14 February 2022
Accepted: 6 May 2022
Published: 9 May 2022

Publisher's Note: MDPI stays neutral with regard to jurisdictional claims in published maps and institutional affiliations.

1. Introduction

The leather industry is continuously increasing, due to the increased demand for finished leather products and meat for human consumption. This generates large quantities of waste (fleshings, hairs, shavings, dust, liquid waste which contains the tanning agent) from different steps of leather processing. These wastes have a rich content of proteins (collagen, gelatine and keratin) [1,2]. Recent strategies demand a transition towards zero landfill and waste in leather production by reusing the leather waste as secondary raw material [3].

The most fertile layer of the soil is the topsoil, which is the top layer of the soil and is rich in microorganisms, minerals and humus. This layer is considered the best location for crop development.

A soil is considered to be contaminated when a moderate increase in substances occurs, substances that are not harmful for a plant to grow at this stage. Soil degradation consists of the action (simultaneous or not) of several physical, chemical and biological factors on the soil. Intensive agriculture, climate change, acid rains, water shortage caused by drought, and accelerated growth of the population all consist of elements that determine the acceleration of soil erosion and nutrient depletion. The soil is continuously exposed to physical erosion by wind and water, a fact that causes the loss of fertile topsoil. The chemical factors consist of acid rains, accidental chemical pollution, and excessive use of chemical fertilizers and pesticides. Moreover, poor farming activities can lead to a decrease in microbial activity in the soil. Therefore, it is obvious that there is a necessity for soil quality improvement to obtain at least the original state of fertility, and productivity of agricultural soil.

Long-term prediction of soil fertility and biodiversity can be approximated using a series of indicators such as: soil structure, soil pH, soil erosion, soil organic matter, phosphorous and potassium content, humidity, etc. These parameters give both qualitative and quantitative information about the possible behavior of soil, over a more extended period [4]. Among these factors, soil organic matter and humidity play a key role in keeping the nutrients available and preventing soil erosion. In the countries with warm temperatures and dry winters, the loss of organic matter is accelerated by the enhanced decomposition of crop residues [5]. The pH controls the chemical and biochemical processes, by enhancing or not enhancing the availability of some nutrients to the plants, and by increasing micro-organism activity. The plant itself can change the soil pH value, e.g., vegetables such as soybeans lower the soil pH [4]. Soil structure is characterized by the porosity and pore size distribution in the soil layers. Soil compaction reduces crop productivity because the plant roots develop and grow in the pores of the soil [6]. An extensive study of Luvisoil type agricultural soil modification, when applying biochar, oyster shells, biopolymers (synthesized from lignin or starch), or synthetic polymer (polyacrylamide) was made by Awad et al. [7]. The authors have shown that the application of the polyacrylamide solely increased the percentage of macroaggregates (1 to 2 mm). The addition of oyster shell to biochar or biopolymer increased the percentage of microaggregates (size below 0.25 mm), having a positive effect on soil quality.

For a continuously increasing population estimated to achieve 11 billion people by the end of this century, it is mandatory to conserve the soil fertility, and to ensure food quality and people's safety [4,8–10]. The most common strategy to increase crop production is fertilization [11]. When looking at the statistics about fertilizers consumption worldwide, in the last 10 years, a trend of continuous increase in fertilizer consumption can be observed (Figure 1) [12]. The demand for healthier agricultural products is continuously increasing as a result of customer awareness increase. Kilic et al. compared the results obtained from GAP farms (Good Agricultural Practices program) with those obtained from farms not included in this program regarding the use of chemical fertilizers, pesticides, the yields and the gross profit [13]. The scope of this program is to reduce the use of pesticides and harmful substances with negative environmental impact which cause health problems, in view of making agriculture more sustainable.

Polymers are natural or synthesized large molecules made by linking repeating units, monomers; they are characterized by different physical and chemical properties than their constituent monomers. Biopolymers are polymers produced by living organisms (such as plants and animals), and they are a renewable resource of polymers. The repeating units can be nucleic acids, saccharides and amino acids having either linearly or branched structure molecules, and the formed biopolymers can be polysaccharides (carbohydrates, starch, cellulose), proteins (collagen, keratin, gelatine), polynucleotides (DNA, RNA) [1,2]. Natural polymers are used for medical and pharmaceutical applications [14–20], food additives [21], and, in recent years, they received attention for agricultural applications because of their particular properties such as: biodegradability, biocompatibility, non-toxic, bioactivity and hydrophilic character [22]. Both natural [23–28] and synthetic polymers [29–31] have reportedly been used to stabilize soils.

Over time, a number of studies have been dedicated to synthesizing and improving the characteristics of the fertilizers that are used to maintain the balance between human needs for consumption and natural available resources. One of these substances is superabsorbent hydrogels (SAHs), which have been proved to be beneficial for plant growth and soil health, and, consequently, have extensive applications in the agricultural field. The SAHs are polymeric materials that have the capacity to retain a large amount of water and nutrients and to slowly release the water along with the nutrients, to respond to the plant demand. The SAHs can be natural (collagen, gelatine), synthetic, or combined polymers (cross-linked). The natural polymers are easily biodegradable but have low functional properties needed for agricultural properties. Therefore, it is preferable to use natural polymers cross-linked with synthetic polymers, to combine the properties of the two categories of polymers [8,9,32]. Soil crusting impedes seedlings and accentuates water runoff. Polyelectrolytes (e.g., based on polyacrylamide), which are synthetic polymers, improve the soil quality by preventing the crust formation in a critical period between plant seeding and emergence, increasing the resistance to water and air erosion of soils, and improving the soil permeability by enhancing the formation of hydro-stable structural aggregates [8].

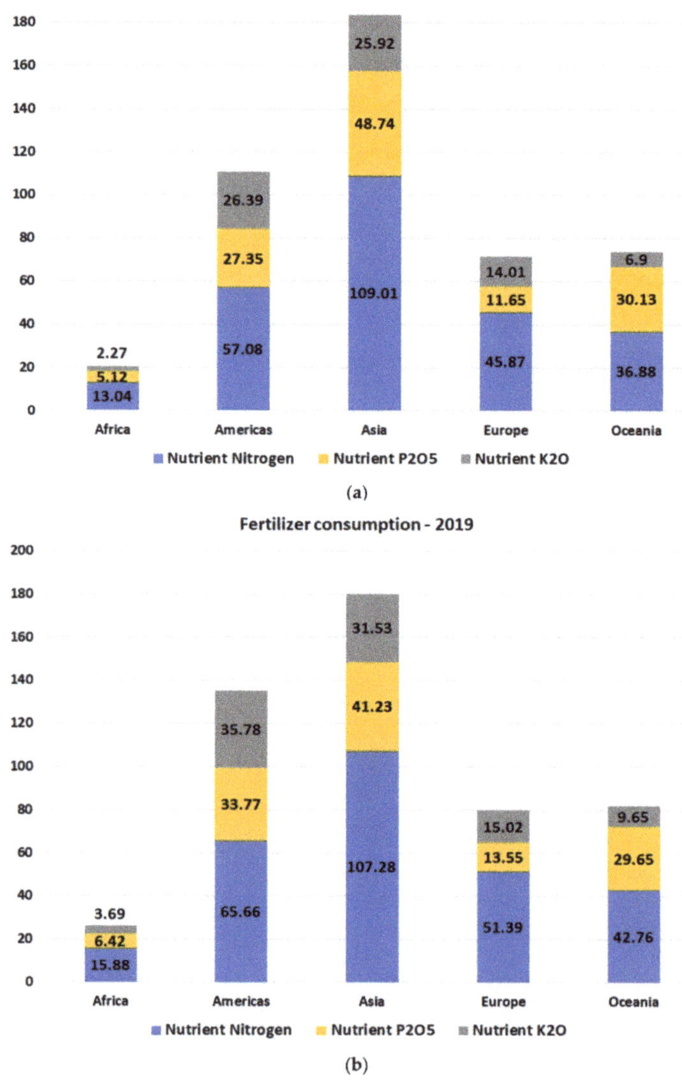

Figure 1. Fertilizer consumption expressed as nitrogen, phosphate, P_2O_5, potassium K_2O, in the last 10 years (**a**) 2010, (**b**) 2019 [12].

It has been reported that collagen SAHs obtained from leather waste have a swelling capacity of more than 2000% by weight [33,34], while gelatine SAHs synthesized from chicken waste have a swelling capacity of more than 800% by weight [35]. Another SAH category is the chitosan-based one. Unlike collagen and gelatine that are recovered from leather waste, chitosan can be extracted from squid bones, crustacean shells, and insects. Essawy et al. studied the cross-linking of chitosan with cellulose, to improve the resistance to acidic soil conditions, and increase water retention capacity [36].

The most used chemical fertilizer as a nitrogen and carbon source is urea, which has a relatively low cost [37]. However, only a small percentage of the applied urea will be effectively used for crop growth, because of urea's high volatility and solubility in water [38]. Urea hydrolysis produces ammonia, which, in turn, has a negative impact on seed germination, respectively, on seedling growth in soil. [39]. Another disadvantage is ammonia volatilization and nitrite accumulation in soils, which can be further leached and cause environmental problems. To overcome these disadvantages, new methods which consist of the controlled release of the fertilizer were developed. These methods consist of the deposition of organic/inorganic functional materials, coating with polymers, encapsulation in matrices, and copolymerization via immolable bonds [40–44].

Many industries, including the leather industry, produce large quantities of wastes that are rich in organic matter [11,45–47]. The skin consists of three main layers: the epidermis, dermis and hypodermis; the dermal layer, representing 85% of total skin thickness, is the main layer and consists of type I fibrillar collagen [48]. Leather waste is an important source of raw materials such as protein and gelatine, providing elements such as nitrogen and carbon essential for plant development. Biopolymer-based fertilizers can be applied either for the purpose of degraded soil rehabilitation or for crop quality and quantity improvement [1,49,50].

Obtaining biodegradable polymers with application in agriculture is multidisciplinary research that involves the steps of recovering the collagen or the gelatine from leather waste materials (obtaining the so-called protein hydrolysate, PH), enrichment of collagen/gelatine with P and K nutrients, functionalization with synthetic polymers and testing the obtained bio-fertilizer on different types of crops/soil type/application rate [51,52]. The implementation of this method also requires an economic analysis besides scientific acceptance.

Leather processing involves multiple preparatory steps: curing, soaking, painting, liming, fleshing, de-liming, degreasing, tanning, splitting, shaving, finishing, etc. [53,54] During these steps, about 35–40% of the raw material is found as waste in various stages of processing [55,56].

Depending on its provenance, leather waste can be classified into two main groups:

(a) untanned leather waste from the processing of raw and gray leather: wax (hypodermic layer) and gelatine skin (fringes and cuttings from the shaping of the leather contour), which represents the dermal layer without epidermis, hypodermis and hair;

(b) tanned leather waste from the processing of tanned and finished hides, from the leather footwear and clothing industry (tanned leather and finished leather).

Several proteins of high value can be recovered from leather waste: gelatine and collagen are the so-called protein hydrolysate part of the waste and the tanning agent (mostly chromium) that can be reused [51,52,57]. This is in line with the circular economy strategy allowing the recovery of valuable products that otherwise would be not only a loss of raw materials but also harmful to the environment. Moreover, the extraction of keratin from hairs with the purpose of using the keratin hydrolysate in agriculture has also been reported [58].

The protein hydrolysate is a mixture of peptides and amino acids that can be obtained either by chemical (basic or acid) or by enzymatic hydrolysis of leather waste [59,60].

Traditionally, chemical hydrolysis is achieved with strong acids (e.g., sulphuric acid, phosphoric acid) or alkaline bases and around 80 °C, allowing chromium removal without destroying the collagen tissue [61,62]. These methods are especially employed when it is necessary to remove the tanning agent (chromium). During the chemical hydrolysis part, the amino acids and peptides are lost; to overcome this disadvantage and for the non-tanned leather waste, enzymatic hydrolysis can be applied [63]. The enzymatic hydrolysis uses specific enzymes and lower temperatures (<60 °C) [64,65]. Due to the health and environmental problems generated by chromium use in the tanning step, low chromium methods [66] and eco-friendly alternatives have been developed consisting of using vegetable tannins [67,68], aluminum salts [69,70], titanium salts [71], combined vegetable-aluminum tanning agents [72], 4-(4,6-dimethoxy-1,3,5-triazin-2-yl)-4-methylmorpholinium chloride [73] or amino-acids [74–76]. Hide and leather waste characteristics and processing have been reviewed in a previous publication [60]. The methods of collagen recovery and chromium removal from tanned waste have been extensively discussed; therefore, this review is the second part and presents aspects related to the use of the extracted collagen as fertilizer. In Figure 2, the technology of obtaining a collagen-based fertilizer by using acid hydrolysis is exemplified [77]. For more details on this topic, the readers are advised to first lecture the article presenting the methods for leather-based fertilizers synthesis [60].

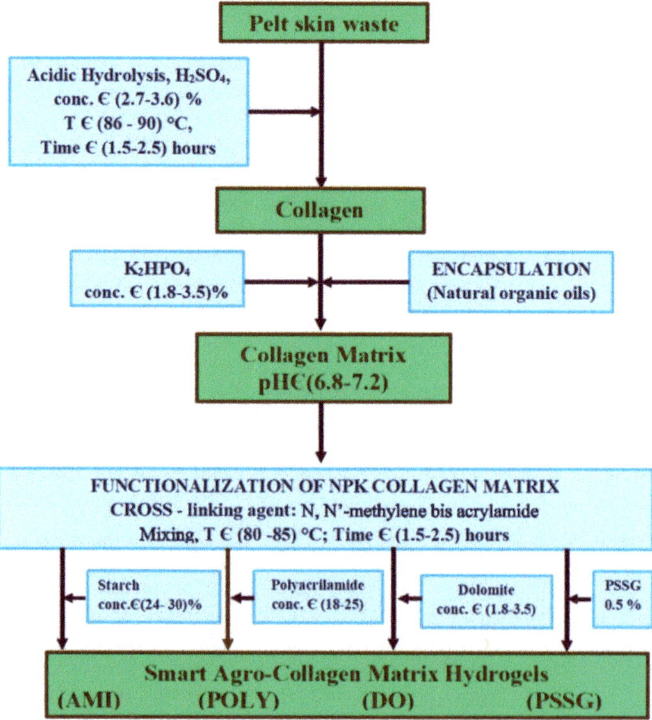

Figure 2. Technology scheme for obtaining smart-fertilizers by using acid hydrolysis, readapted from [77]; CH—collagen hydrolysate, Ref—CH—collagen hydrolysate with nutrients encapsulated as reference sample, PSSG—Ref—CH functionalized with P(SSNa—co—GMAx) copolymer, POLY—Ref—CH functionalized with poly-acrylamide, AMI—Ref-CH functionalized with starch, AMI—Ref-CH functionalized with dolomite).

2. Comparison among Biopolymer-Based Fertilizers Obtained from Leather Waste and Other Types of Fertilizers (Chemical Fertilizer, Compost, etc.) Applied for Crop Growth

Biopolymer-based fertilizers have the advantage of retaining large quantities of aqueous solutions, along with the slow release capacity of the water and of the nutrients, over extended periods of time [33–35]. On the contrary, chemical fertilizers do not have the capacity to retain water and therefore ensure a constant humidity level in conditions of drought. Moreover, chemical fertilizers are known to release their nutrients in the first few days after the fertilization, ensuring fast development of the plant in this period [78].

Collagen and gelatine are important resources of nitrogen and carbon. The main amino acids found in the composition of collagen powder, extracted from chromium leather scrap waste, are: aspartic acid, threonine, serine, glutamic acid, glycine, valine, isoleucine, tyrosine, phenylalanine, histidine, arginine, and proline [79].

Commercial hydrogels have been tested as a source of nutrients and moisture conservation, in the Semi-Arid Zone of Kongelai (Kenya), for the cultivation of *Cajanus cajan* [80]. The experiments were conducted both in a nursery and in a field, and the results showed that the use of hydrogels retards plant growth in nursery soils, but improves growth in the field by increasing the soil moisture.

Majee et al. synthesized a biopolymer-based fertilizer, recovering the collagen from tanned leather waste, and enriching this material with poultry bone meal as a source of phosphorous, and with water hyacinth ash as a potassium source [52]. This fertilizer was applied to the *Catharantus roseus* (Madagascar Periwinkle) plant, and a comparison was made with a commercial fertilizer, or with a plant without fertilizer. The results for the three cases are shown in Figure 3 together with the initial stage where the plants had no fertilizer applied. The authors compared the plant growth by considering plant length and diameter, leaf size, flower size, and total number of flowers. Both chemical and biopolymer-based fertilizers provide nutrients for good plant development, although in the initial stage, the growth of the plant with chemical fertilizer was accelerated, due to the fast nutrient release.

Figure 3. Plant growth comparison: (**1**) Control plant without fertilizer; (**2**) plant with collagen NPK fertilizer; (**3**) plant with commercial fertilizer; final stage is the 80th day after the soil fertilization. Reprinted with permission from ref. [52]. 2022, Elsevier.

These results are in agreement with a later study conducted by Majee et al. when a combined polymer-potato peel biochar fertilizer was synthesized and tested [81]. The source of phosphorous and potassium was the potato peel biochar, obtained at lower temperatures, to conserve several functional groups useful in agricultural applications.

The test experiments were conducted on *Abelmoschus esculentus* (okra plant), cultivated in pots containing soil without fertilizer, with biopolymer-based fertilizer, or with chemical fertilizer.

A comparison between green manure, mineral N-fertilizer, and biopolymer-based fertilizers obtained from leather waste, was made for the tomato crop [82]. Two types of biofertilizers were used: 5% N and 8% N, respectively. The results showed that the 5% N biofertilizer does not provide sufficient N for the plant to grow, while the 8% N gave similar results to the mineral fertilizer.

Table 1 presents different types of biofertilizers synthesized from leather waste that were tested on various plants, together with the applied amendment rates and the type of soil.

Table 1. Comparison between different biopolymer-based fertilizers obtained from leather waste.

Fertilizer Type	Soil Type	Crop	Rate	Comments
Poultry manure [82]	Central Italy—unspecified type	tomato	100 kg N/ha	Does not fulfill the crop demand in nutrients
Poultry manure and by-product from leather factory [82]	Central Italy—unspecified type	tomato	100 kg N/ha	The fertilizer gave the same efficacy as the mineral fertilizer.
Organic fertilizer by-product from leather factory [82]	Central Italy—unspecified type	tomato	100 kg N/ha	The fertilizer gave the same efficacy as the mineral fertilizer.
Mineral fertilization [82]	Central Italy—unspecified type	tomato	100 Kg N/ha and 200 Kg N/ha	The fertilizer gave the same efficacy as the fertilizers by-products from leather factory
Collagen-based biofertilizer [83]	stagnic albeluvisol, Romania; degraded soil classified as dusty clay soil	soybean	10 kg fertilizer/m^2 20 kg fertilizer/m^2	The second rate provided only a slightly higher production (about 0.2%), compared with the first-rate, and both gave about 20% more productivity, compared to unfertilized soil.
Collagen-based biofertilizer; collagen extracted from wet white leather waste [84]	Neutral or slightly alkaline soil	peas	0.25–0.50 kg fertilizer/m^2	Good results on soil quality improvement and crop quantity.
Collagen extracted from wet blue leather [85]	Yellow-Red Latosol, clayey texture, Oxisol, pH = 5.9	bean plants cultivated after the growth of elephant grass on the soil fertilized with collagen	4, 8, 16, or 32 t collagen/ha	Results similar to mineral fertilization.

Farneselli et al. (Table 1), conducted an extensive study, investigating for 2 years the efficiency of fertigation treatments on tomato crops using poultry manure, by-products from the leather factory, organic by-products from the leather factory, and mineral liquid fertilizer (7.5% NO_3-N + 7.5% NH_4-N + 15% urea, radicon N30).

Fertigation is applied at rates of 100 and 200 kg N/ha, in 10 splits (2 times/week for 5 weeks), according to the expected nitrogen uptake rate for tomato processing throughout the growing season. It was found that fertigation treatments, using mineral and organic fertilizer by-products from the leather factory in doses of 200 kg N/ha almost always caused luxury N consumption, for both situations, in the first year and a deficit in the second year for organic by-products from the leather factory. Reducing N rates, both for the mineral and for the organic one to 100 kg N/ha, ensured optimal N status for the main part of the crop cycle with a slight deficiency of growing at the end of the second year. The effect of mineral and organic by-products from the leather factory is similar.

The use of poultry manure and by-products from the leather factory is similar to other fertigations that used doses of 100 kg N/ha. In the first year of fertilization, the nitrogen uptake was 27% higher than the second year. N uptake with fertilization with mineral fertilizer at a dose of 200 kg N/ha was significantly higher than any other treatments but much larger than the optimum N status [82].

This study shows that, although mineral fertilizers are particularly effective, a well-known fact, they can be replaced with fertilizers derived from leather waste.

Table 2 gives the nutrient composition of several mineral fertilizers and leather-based biofertilizers.

Table 2. Composition of different biopolymer-based fertilizers obtained from leather waste.

Fertilizer Type	% N	P, (Expressed as % P_2O_5)	K, (Expressed as % K_2O)	Other Components	Comments	Reference
NPK, universal fertilizer	26	13	6	0.004% Cu, 0.037% Fe, 0.03% Mn, 0.0015% Mo, 0.015% Zn	it is used for any type of culture	Produced by Azomures S.A. [86]
Radicon N30	30	–	–	–	7.5% $NO_3 - N + 7.5\%$ NH_4-N + 15% urea	[82]
Urea	46	–	–	–	–	[40]
Ammonium sulfate	21	–	–	–	–	[40]
Ammonium nitrate	30.5	–	–	–	–	[40]
Floranid	32	–	–		Low solubility material containing (3% urea-N; 29% IBDU—isobutilidenediurea -N)	[40]
Fertilizer by-product from leather factory	5	–	–	C/N = 5.4	The fertilizer with higher N content gave better results for tomato crop.	[82]
Organic fertilizer by-product from leather factory	8	–		C/N = 2.8		[82]
Gelatine based fertilizer; gelatine extracted from leather waste	43.84 (weight)	–	Not specified	7.72% C; 40.26% O; 1.76% Na; 0.35% Al; 0.2% Si; 0.05% S; 5.28% Cl; 0.54% Ca		[87]
Collagen-based biofertilizer; collagen extracted from leather waste	11.14	2.43	3.77	0.127% Mg	pH of aqueous extract 7–7.5	[83]
Collagen based fertilizer cross-linked with different polymers: (a) collagen hydrolysate with nutrients encapsulated as reference sample (b) collagen hydrolysate functionalized with P(SSNa-co-GMAx) copolymer (c) collagen hydrolysate functionalized with poly-acrylamide (d) collagen hydrolysate functionalized with functionalized with starch	10.55 10.14 12.13 8.29	7.67 6.75 5.79 5.54	10.62 8.21 8.40 10.07	(expressed as % TOC) 45.2 37.56 48.1 64.32	pH = 7.2 pH = 6.87 pH = 6.76 pH = 6.20	[77]
Collagen extracted from wet blue leather	14.6	2.6	0.014		Collagen was applied on a soil having the pH 5.9, and only minor changes in the soil pH were observed, in the range of 5.9–6.1	[85]

It can be seen that the classic fertilizers urea, ammonium nitrate and NPK have a very high nitrogen content (over 21%) and can reach up to 46%. NPK fertilizers also contain macroelements such as phosphorus and potassium, which gives them extra efficiency in terms of plant growth and development. Conventional fertilizers release nutrients quickly, and consequently, losses are significant. Ammonium nitrate loses nitrogen the fastest, followed by urea, ammonium sulfate and Floranid (IBDU) [40].

Fertilizers obtained from leather factory by-products contain between 5 and 8% of N, while those obtained from hydrolysates of leather waste have a nitrogen content of up to 16%. These fertilizers have a complex composition containing, in addition to nitrogen, other macronutrients, such as phosphorus and potassium, that can be introduced by mixing, but also micronutrients absolutely necessary in plant growth and development. They also have a high carbon content in biodegradable organic compounds, that can grow soil fertility [82–85].

Due to the complexity of the composition, the fertilizers derived from leather waste have an action with a much wider spectrum, that aims at a positive action both on the growth of the plants, and on the improvement of the soil quality. Conventional mineral fertilizers are effective, but their action is one-sided and can be characterized by point-to-point hits.

Several alternatives to the use of the leather waste hydrolysate as fertilizers are the composted or vermicomposted leather waste. These methods have the advantage of lowering the carbon to nitrogen ratio, providing more nitrogen necessary for plant growth [88,89].

Silva et al. studied the use of tannery sludge for the cultivation of ornamental *Capsicum* plants [46]. The compost was prepared from tannery sludge, mixed with agricultural waste (sugarcane straw, and cattle manure, "carnauba" straw, and cattle manure, respectively) in different ratios. Results have shown that, when replacing the inorganic fertilizer with these composts, there was a significantly increased number of leaves and fruits, as well as a higher content of chlorophyll in the leaves. However, the concentrations of Cu, Cd, Cr, Zn, Pb, Ni, Mo, and Mn increased in soil, because of soil amendment with composted tannery sludge. This suggests that tannery waste is a good option when it is necessary for the amendment of soils to grow ornamental plants and not recommended for plants and crops intended for human consumption. A more recent study of leather biodegradability showed that the quality of the compost is influenced by the nature of the tanning agents (chromium or titanium salts) and that biodegradation is a complex process that could be achieved in the presence of food wastes [71]. Altogether, the titanium tanned hides were more biodegradable than the chrome tanned hides and vegetal tanned hides.

Vermicomposting supposes the use of several earthworm species such as: *Eisenia fetida, Eisenia andrea, Eudrilus eugeniae,* for the conversion of different waste types (including leather waste), into a product useful for soil amendment [90–97]. Ravindran et al. (2019) studied the amendment of soil with vermicompost hydrolyzed tannery animal fleshing on the growth and yield of commercial crop tomato plant (*Lycopersicon esculentum*) [88]. The quality of tomato fruit was assessed from the point of view of its size, weight, tomato juice pH, ascorbic acid, total sugar content, etc. These parameters indicated that plant growth, yield, fruit quantity, and nutrients in fruits were higher when the soil was treated with the leather waste vermicompost, compared to the control sample. Tannery waste vermicomposting, like composting, does not involve chromium recovery prior to its use in agriculture. Over time, this causes chromium to accumulate in the soil and is a major drawback. One method to reduce the chromium impact on the crop, and also on the earthworms (as chromium is toxic to earthworms), consists of the mixing of tannery sludge with other materials (e.g., manure) to reduce the concentration of chromium ions [89].

Vermicompost tannery sludge was compared with the conventional NPK fertilizer, and control sample (soil without fertilizer), in the cultivation of sweet pepper [98]. The addition of vermicompost stimulated the plant growth and enhanced the production of more fruits per plant (one fruit harvested per plant, for NPK fertilizer, vs. up to three fruits harvested per plant for vermicompost). The authors found similar chromium contents in all the fruits (control sample, NPK conventional, and tannery sludge vermicompost), indicating that there is not a major contamination of fruits with chromium, but did not study a possible chromium accumulation in time, in the soil amended with vermicompost tannery sludge.

A ten-year study on the early application of composted tannery sludge showed that soil properties change and that elements accumulate: organic matter, N and K content, increased over the 10 years of the study, showing a positive effect of this treatment [99]. However, the soil pH and chromium content also increased, which is not beneficial for soil used for agricultural purposes. During the experiment, it was found that chromium content increase took place mainly in the first 5 years and remained almost constant in the next 5 years. Another disadvantage of composted tannery sludge application was that the enzyme activity decreased. This affects a lot the complex process of transforming organic compounds into assimilable subunits (sugars, amino acids, NH_4^+, PO_4^{-3}).

A biofertilizer obtained from titanium tanned leather waste, by combined chemical-enzymatic hydrolysis (the so-called "wet white leather"), was tested for pea crop growth [82]. After the chemical hydrolysis, two phases are obtained: the liquid phase containing the collagen recovered, and an unhydrolyzed solid phase called "titanium-containing sludge". The two phases are separated. The resulted sludge is subjected to enzymatic hydrolysis, in the presence of lipase, cellulase, amylase and protease. The titanium salt was recovered for reuse, as a tanning agent. The protein hydrolysate is modified by chemical cross-linking with other polymers such as: polyacrylamide, acrylic polymer, maleic polymer, cellulose or starch. The cross-linking process gives the fertilizer resistance to water dissolution. Moreover, the addition of polyelectrolyte-type polymers improved the soil properties, by increasing the resistance to water and wind erosion for the soil located on slopes. It also prevents crust formation after sowing, which is essential for plants with small seeds [8,100]. The study of the amended soil indicated that the fertilizer was efficient not only for the peas' growth, but also for the remediation of the soil quality.

Keratin-based fertilizer has proved its efficiency in remediation of the soil contaminated by heavy metals, by fixing the chromium(III) contained in the soil [77]. The keratin was extracted from the waste bovine hair, and cross-linked with acrylic acid and N,N-methylene bis acrylamide, to form a keratin-based superabsorbent material. The core of the fertilizer spheres, made by a mixture of lignin powder and urea particles, is covered by an ethyl-cellulose layer, and finally, by a superabsorbent material (Figure 4). This fertilizer was tested in the wheat growth and gave better results than urea.

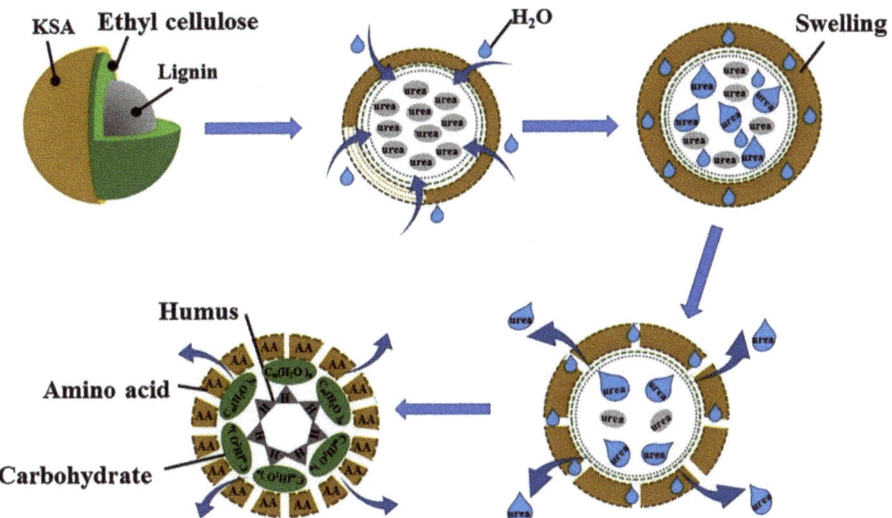

Figure 4. Schematic representation of keratin-based fertilizer nutrient release. Reprinted with permission from ref. [78]. 2022, Elsevier.

Constantinescu et al. tested a fertilizer obtained using the collagen recovered by alkaline hydrolysis of untanned leather waste using $K_2HPO_4 \cdot 3H_2O$ [83]. The fertilizer was applied at two rates: 10, 20 kg/m², respectively. The soil was amended with fertilizer before planting the seeds, to stimulate the processes of germination, seedling growth, deep rooting, and rigorous plant development. The growth of the plants was compared after 10, 25, and 40 days (Figure 5). It can be seen that there is almost no difference between the two rates of fertilizer (middle and right pots).

Figure 5. Soybean growth comparison (left—no fertilizer; middle—10 kg collagen NPK fertilizer/m^2; right—20 kg collagen NPKfertilizer/m^2) [83].

Hu et al., investigated how the structure of leather-based biofertilizer changes in soil after 60 days, by using the scanning electron microscopy technique [33]. The fertilizer porosity is an important parameter, as higher pore diameter values determine higher surface area per volume ratio, a fact which enhances the swelling rate and biodegradability [34]. The biodegradability tests were performed in the presence of *Ensifer* sp. Y1 bacterium, isolated from the soil. The selected samples were: control hydrogel in soil, fertilizer in soil, and fertilizer in *Ensifer* sp. Y1 medium (Figure 6). It can be seen that the fertilizer samples initially have a microporous structure that is lost after 60 days in soil, while the control hydrogel is slightly degraded. The presence of large colonies of *Ensifer* sp. Y1 enhanced the fertilizer biodegradation.

Nogueira et al. used tanned leather waste (wet blue leather) to synthesize a N$_{collagen}$PK biofertilizer tested for the growth of rice plants [101,102]. The chromium tanning agent was extracted according to the method presented in a previous study, which proved to be very efficient, recovering up to 99.6% of chromium contained in the waste [103,104]. The fertilizer was applied on a typical dystrophic Yellow-Red Latosol, clayey texture, Oxisol and its fertilization was compared with a commercial NPK fertilizer and urea, enriched with P and K, respectively. The biofertilizer showed activity similar to the urea enriched compound in the growth of rice plants, and slightly lower than the commercial NPK fertilizer.

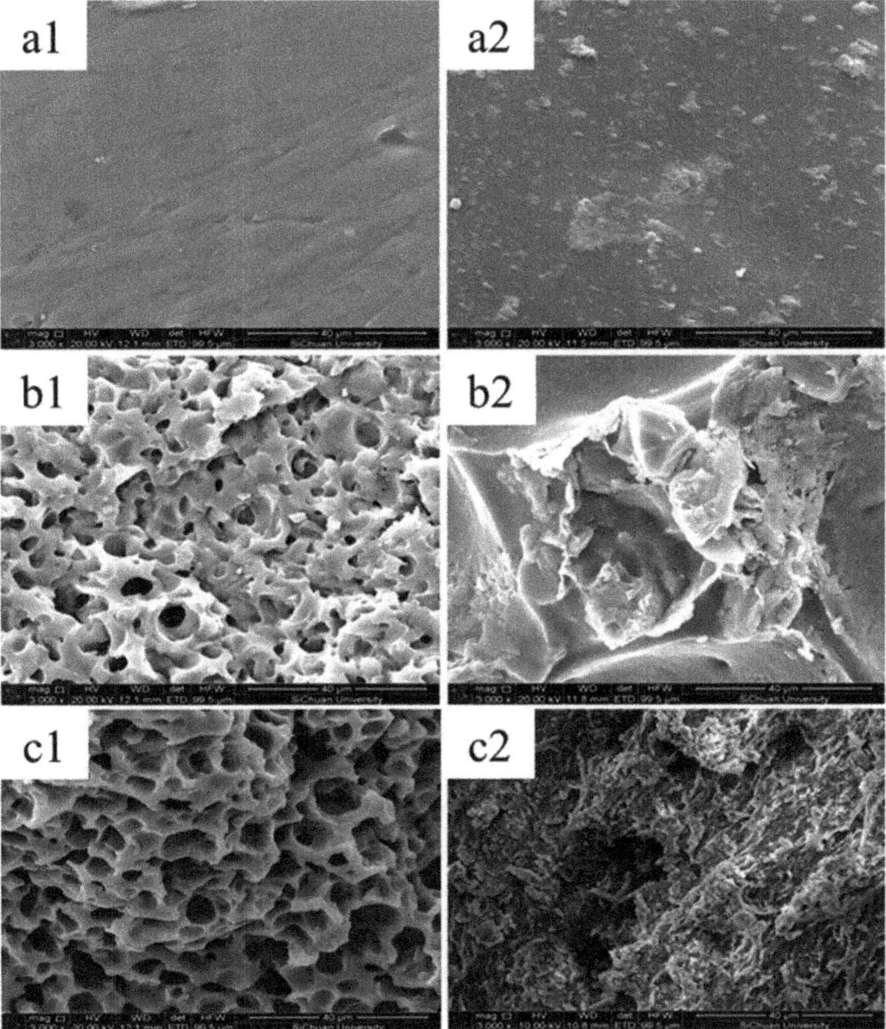

Figure 6. SEM images of control (**a**) and collagen-based fertilizer (**b,c**) samples before (**1**) and after (**2**) degradation in soils (**a,b**) and in *Ensifer* sp. Y1 medium (**c**). Reprinted with permission from ref. [33]. 2022, Elsevier.

3. Nutrient Releasing Processes

Fertile soils contain different inorganic mineral particles (sand, clay, silt); assimilable and non-assimilable organic matter; living organisms (earthworms, insects, bacteria, fungi), water, gases (O_2, CO_2, N_2, NO_x, CH_4). The inorganic materials are involved in the retaining processes of cations through ion exchange, and of anions and organic compounds through sorption (surface) reactions [105]. The temperature, soil pH, and humidity control the molecules/ions interchanges between soil phases (solid, liquid and gaseous) and soil biological activity. The living organisms contribute to topsoil regeneration by humus formation. They decompose the organic matter into assimilable forms, increase the soil porosity (and consequently, the aeration), and help the movement of organic matter and residues within the topsoil [106–108]. The microorganisms transform the nitrogen present in the fertilizers into nitrate, through the nitrification process. It has been reported that soil microorganisms are able to increase the carbon sequestration capacity, as CO_2 concentrations are continuously increasing [109]. Adding organic fertilizers helps beneficial microorganism development (*Fimicutes, Chloroflexi, Bacillus* and *Actinomadura*), increases enzyme activity (like sucrase enzyme), and increases *Fusarium* and *Phytophthora* pathogen mortality [107]. Organic matter contained in the biofertilizers is a resource of C, N, P, and S nutrients. The K and supplementary P are provided by enriching the biofertilizer with inorganic K and P.

Although there is great interest regarding the controlled release fertilizers [44,110–112], few studies have been dedicated to studying the nutrient release in leather-based biopolymers [77]. Most studies that have been published present the synthesis of the biopolymer fertilizer and its testing for different crops, being lesser dedicated to the nutrient release mechanisms. For a keratin-coated urea fertilizer, the schematic representation of nutrients' release steps is given in Figure 4 [78]. Once spread in the soil, the fertilizer particle starts to retain water from the soil until reaching the swelling equilibrium. The water passes through the ethyl cellulose and lignin layers, starting to dissolute the urea. The two layers act as a barrier to water passage, and they are meant to delay the penetration of water [113,114]. Dissolved urea diffuses to the exterior of fertilizer particles through the perforated layers of ethyl cellulose and lignin and is slowly released into the soil. The superabsorbent material and the ethyl cellulose and lignin layers are further degraded, to provide more nutrients (amino acids, carbohydrates and humus), under the action of soil microorganisms. The nutrient release experiments performed with uncoated urea and with keratin-based fertilizer showed that about 83% of uncoated urea was released in the first 24 h [78]. On the contrary, the biopolymer fertilizer released about 70% of encapsulated urea over 28 days, demonstrating its excellent performance in controlling the nutrient release.

The release of oxidable compounds (organic and inorganic) in water for different types of collagen-leather-based biopolymers has been presented by Stefan et al. [77]. Oxidable compounds released from tested fertilizers over a period of one month is shown in Figure 7. The release degree was evaluated in dynamic conditions, by the determination of chemical oxygen demand (CODMn) [115].

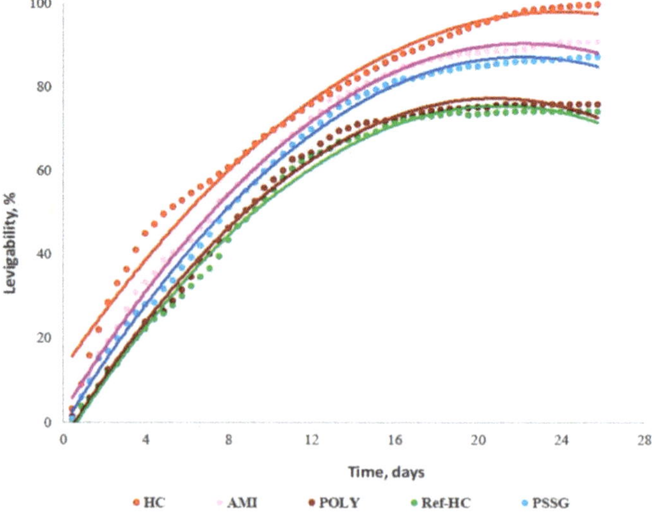

Figure 7. Oxidable nutrient release (organic and inorganic), in water for different collagen-leather-based biopolymers [77]; HC—collagen hydrolysate, Ref—HC—collagen hydrolysate with nutrients encapsulated as reference sample, PSSG—Ref—HC functionalized with P(SSNa—co—GMAx) copolymer, POLY—Ref—HC functionalized with polyacrylamide, AMI—Ref—HC functionalized with starch.

Collagen hydrolysate (HC) is practically completely degraded during chemical oxidation (>99%). Therefore, collagen cross-linking with a more stable polymer in aqueous conditions is needed. Among the functionalized biopolymers, collagen cross-linked with starch (AMI) has the highest nutrient release degree, over 25 days—namely about 90%. The lowest nutrient release is given by collagen functionalized with polyacrylamide (POLY). This suggests that synthetic polymers give a more stable structure to the fertilizer, slowing down the nutrients' release for a longer period.

Experiments of the nutrients' release in soil amended with NPK fertilizer without superabsorbent polymer showed a total release of nutrients (N, P, K) in the first 4 days [36].

Hu et al. tested the nutrients' release from a porous collagen-leather-based biofertilizer, over a more extended time period of 34 days [33]. About 45% of the total nitrogen and 30% of K is released in the first 2 days; the rate of K release is more accelerated, compared with N release—which is gradually released in time. This can be explained by the fact that the N from collagen is made accessible to the plants by biodegradation in the presence of microorganisms, and this is a slower process, compared to K dissolution in water (Figure 8).

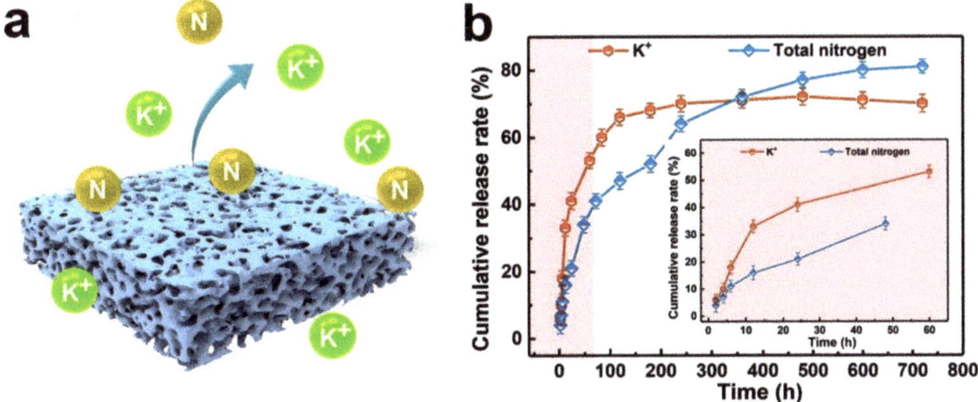

Figure 8. Biofertilizer synthesized from collagen-leather waste cross-linked with acrylic acid (AA) and maleic anhydride (MA), 15% leather waste hydrolysate: (**a**) porous structure; (**b**) N and K release in water. Reprinted with permission from ref. [33]. 2022, Elsevier.

4. Use of Biopolymers for Soil Remediation and Stabilization

Over recent years, there was an increasing interest in recovering biopolymers from different types of materials and wastes, to test them for soil remediation purposes. Among these biopolymers there are: agar gum, which is a polysaccharide extracted from of *Rhodophyta* (red algae), such as *Gelidium*, *Gracilaria*, and *Pterocladia* [116,117]; guar gum, which is a neutral polysaccharide extracted from the seeds of the leguminous shrub *Cyamopsis tetragonoloba* [118–120]; gellan gum, which is an anionic polysaccharide made by microbial fermentation of *Sphingomonas elodea* [121,122]; dextran, which is a group of glucose polymers made by lactic acid bacteria such as *Leuconostoc mesenteroides* and *Streptococcus mutans* from sucrose [123–126], xanthan, which is a polysaccharide biopolymer produced by *Xanthomonas campestris* [127–134]; chitosan, which is a polysaccharide extracted by alkaline hydrolysis of crustacean shells, insects, squid bones [135], starch -which is composed of monosaccharides found in seeds, grains, and roots of plants (maize, rice, wheat, corn, potatoes, cassava, etc.) [116,136,137]; casein, which is a phosphorous protein biopolymer contained in milk products, and is used for soil remediation, due to its hydrophobicity [138–140].

Dang et al. studied the use of graft copolymer extracted from leather solid waste for its application in chemical sand-fixation [141]. The by-product gelatine was extracted from the leather solid waste by alkaline hydrolysis, and, in a second step, the graft copolymer was synthesized by free-radical copolymerization with acrylamide and acrylic acid. The final product was tested and proved to have good water retention capacity, good biodegradability, and sand stabilization properties, due to the formation of adhesion forces among the copolymer and the sand particles.

Soil degradation (physical, chemical and biological) represents the loss of its productivity, following the action of natural and anthropogenic factors. The good quality of agricultural soils can be severely affected by phenomena such as drought, erosion, salinization, acidification, alkalinity or compaction [83]. Degradation of agricultural land takes place also through contamination/pollution processes with heavy metals such as: iron, manganese, copper, zinc, lead, cadmium, chromium, cobalt, nickel [46,142]. Soil restoration consists of the application of remedial methods, to obtain higher soil fertility and productivity, or at least a state closer to the initial one. These methods aim to improve the soil structure, microorganisms density, nutrient density, and overall carbon levels of soil. Therefore, soil quality can be improved by maintaining the humus layer, increasing microorganisms populations and biological diversity, as well as by reducing the use of chemicals (fertilizers, pesticides, herbicides). Soil degradation leads to a vicious cycle: low crop yields determine malnutrition, social disorders and unequitable distribution of wealth; this gives low agricultural resources needed to remediate the soil quality, thus inducing more severe soil degradation. Moreover, the soil quality is closely related to the environment quality, since degraded soil is an indicator of other environmental issues (such as water contamination, poor biodiversity, and dust in the air).

The aim of soil stabilization and remediation is to improve its mechanical properties, provide a proper quantity of nutrients, and regenerate microorganism populations. Huang et al. reviewed the use of biopolymers, geopolymers (inorganic polymers with different Si-Al backbone structures), and synthetic organic polymers for soil stabilization [31]. The main properties that are taken into account to evaluate the quality of polymer-stabilized soil are presented in Figure 9.

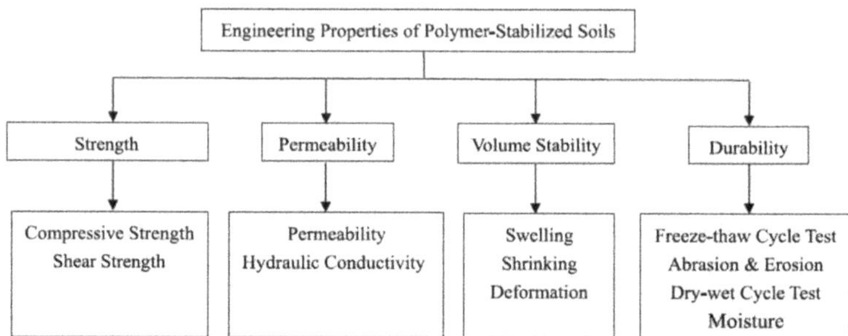

Figure 9. Methods to evaluate the effectiveness of soil stabilization using polymers. Reprinted with permission from ref. [31]. 2022, Elsevier.

Among the synthetic polymers employed for soil stabilization is polyacrylamide, which is also used in the copolymerization of collagen to produce biofertilizers. The PAM was applied to a wide range of soil types: silty gravel, clayey sand, clayey gravel [143–146].

Tingle et al. tested soils treated with different types of polymers and found a significant drop in strength under wet conditions [147,148]. The polymer addition improved the strength, compared to untreated soil.

Polyacrylamide alone applied to agricultural Luvisol soil type increased the proportion of large macroaggregates (1–2 mm), while the amendment with biochar, biopolymer and oyster shells increased the portion of microaggregates (<0.25 mm). The addition of biochar, biopolymer and oyster shells was found to contribute to the increase in the biological activity, by increasing the levels of leucine aminopeptidase and chitinase [7]. This was confirmed by higher NO_3^- concentrations, leucine aminopeptidase being involved in the N-cycle [149], whereas chitinase is a measure of increased fungal activity [150].

5. Conclusions

The collagen-based fertilizers have proved that they are a good candidate to replace conventional chemical fertilizers. They offer the advantage of recovering a valuable by-product (the collagen) from leather waste, and of transforming it into a valuable product.

When the collagen source is chromium-tanned leather, chromium content should be carefully checked, as its concentration in the soil must be in agreement with the imposed regulations.

The synthesized biopolymers have shown good fertilization, comparable, or even better, than the conventional chemical fertilizers, for a large range of crops (tomato, beans, peas, soybean, wheat, ornamental plants, etc.).

The nutrient release tests have demonstrated that, by adjusting the collagen content in the fertilizer, and by cross-linking the collagen with synthetic polymers, the N release can be controlled and slowed.

Another advantage given by the use of a controlled release fertilizer is that the loss of nutrients caused by leaching in the inferior layers of the soil or ground water is reduced, avoiding water contamination with nitrate.

The soil humidity is maintained for a longer time, because of the collagen capacity to retain large amounts of water, and in this way, the irrigation frequency is reduced.

The cross-linking of natural polymers with polyelectrolytes (polyacrylamide) has ameliorated the soil quality, by preventing crust formation, a fact which helps the plant seeding, especially for plants with small seeds.

Author Contributions: Conceptualization: M.B. and D.S.S.; validation: D.S.S. and M.B.; Formal analysis: M.S. and A.M.D.; Investigation: A.M.D. and M.S.; Resources: D.S.S. and M.B.; Data curation: A.M.D., M.S. and D.S.S.; Writing—original draft preparation: M.B. and D.S.S.; Writing—review and editing M.B., D.S.S. and M.S.; Vizualization: M.B., M.S. and A.M.D.; Supervision: D.S.S. and M.S.; Funding Acquisition: D.S.S. All authors have read and agreed to the published version of the manuscript.

Funding: This research received no external funding.

Institutional Review Board Statement: Not applicable.

Informed Consent Statement: Not applicable.

Data Availability Statement: Not applicable.

References

1. George, A.; Sanjay, M.R.; Srisuk, R.; Parameswaranpillai, J.; Siengchin, S. A comprehensive review on chemical properties and applications of biopolymers and their composites. *Int. J. Biol. Macromol.* **2020**, *154*, 329–338. [CrossRef]
2. Udayakumar, G.; Muthusamy, P.; Selvaganesh, S.B.; Sivarajasekar, N.; Rambabu, K.; Banat, F.; Sivamani, S.; Sivakumar, N.; Hosseini-Bandegharaei, A.; Show, L.P. Biopolymers and composites: Properties, characterization and their applications in food, medical and pharmaceutical industries. *J. Environ. Chem. Eng.* **2021**, *9*, 105322. [CrossRef]
3. Korhonen, J.; Snäkin, J.P. Quantifying the relationship of resilience and eco-efficiency in complex adaptive energy systems. *Ecol. Econ.* **2015**, *120*, 83–92. [CrossRef]
4. Mattsson, B.; Cederberg, C.; Blix, I. Agricultural land use in life cycle assessment (LCA): Case studies of three vegetable oil crops. *J. Clean. Prod.* **2000**, *8*, 283–292. [CrossRef]
5. Castro, C.; Logan, T.J. Liming effects on the stability and erodibility of some Brazilian oxisols. *Am. J. Soil Sci. Soc.* **1991**, *55*, 1407–1413. [CrossRef]
6. Håkansson, I.; Medvedev, V.W. Protection of soils from mechanical overloading by establishing limits for stresses caused by heavy vehicles. *Soil Tillage Res.* **1995**, *35*, 85–97. [CrossRef]
7. Awad, Y.M.; Lee, S.S.; Kim, K.-H.; Ok, Y.S.; Kuzyakov, Y. Carbon and nitrogen mineralization and enzyme activities in soil aggregate-size classes: Effects of biochar, oyster shells, and polymers. *Chemosphere* **2018**, *198*, 40–48. [CrossRef]
8. Zainescu, G.; Voicu, P.; Constantinescu, R.; Barna, E. Biopolymers from protein wastes used in industry and agriculture. *Rev. Ind. Text.* **2011**, *62*, 34–37.

9. Zainescu, G.A.; Stoian, C.; Constantinescu, R.R.; Voicu, P.; Arsene, M.; Mihalache, M. Innovative process for obtaining biopolymers from leather wastes for degraded soils remediation. In Proceedings of the 10th International Conference on Colloids and Surfaces Chemistry, Galati, Romania, 9–11 June 2011.
10. Tal, A. Making Conventional Agriculture Environmentally Friendly: Moving beyond the Glorification of Organic Agriculture and the Demonization of Conventional Agriculture. *Sustainability* **2018**, *10*, 1078. [CrossRef]
11. Hargreaves, J.C.; Adl, M.S.; Warman, P.R. A review of the use of composted municipal solid waste in agriculture. *Agric. Ecosyst. Environ.* **2008**, *123*, 1–14. [CrossRef]
12. FAO. Food and Agriculture Organization. Available online: http://www.fao.org/faostat/en (accessed on 25 January 2022).
13. Kılıc, O.; Boz, I.; Eryılmaz, G.A. Comparison of conventional and good agricultural practices farms: A socio-economic and technical perspective. *J. Clean. Prod.* **2020**, *258*, 120666. [CrossRef]
14. Horue, M.; Berti, I.R.; Cacicedo, M.L.; Castro, G.R. Microbial production and recovery of hybrid biopolymers from wastes for industrial applications—A review. *Bioresour. Technol.* **2021**, *340*, 125671. [CrossRef]
15. Rutz, A.L.; Shah, R.N. *Protein Based Hydrogels in Polymeric Hydrogels as Smart Biomaterials*, 1st ed.; Kalia, S., Ed.; Springer International Publishing: Cham, Switzerland, 2016; pp. 73–104.
16. Ucar, B. Natural biomaterials in brain repair: A focus on collagen. *Neurochem. Int.* **2021**, *146*, 105033. [CrossRef]
17. Ramakrishna, S.; Mayer, J.; Wintermantel, E.; Leong, K.W. Biomedical applications of polymer-composite materials: A review. *Compos. Sci. Technol.* **2001**, *61*, 1189–1224. [CrossRef]
18. Sionkowska, A. Collagen blended with natural polymers: Recent advances and trends. *Prog. Polym. Sci.* **2021**, *122*, 101452. [CrossRef]
19. Osmalek, T.; Froelich, A.; Tasarek, S. Application of gellan gum in pharmacy and medicine. *Int. J. Pharm.* **2014**, *466*, 328–340. [CrossRef]
20. Peppas, N.A.; Bures, P.; Leobandung, W.; Ichikawa, H. Hydrogels in pharmaceutical formulations. *Eur. J. Pharm. Biopharm.* **2000**, *50*, 27–46. [CrossRef]
21. Mudgil, D.; Barak, S.; Khatkar, B.S. Guar gum: Processing, properties and food applications—A Review. *J. Food Sci. Technol.* **2014**, *51*, 409–418. [CrossRef]
22. Van de Velde, K.; Kiekens, P. Biopolymers: Overview of several properties and consequences on their applications. *Polym. Test.* **2002**, *21*, 433–442. [CrossRef]
23. Gu, B.H.; Doner, H.E. The interaction of polysaccharides with silver hill illite. *Clays Clay Miner.* **1992**, *40*, 151–156. [CrossRef]
24. Chang, I.; Cho, G.C. Strengthening of Korean residual soil with β-1,3/1,6-glucan biopolymer. *Constr. Build. Mater.* **2012**, *30*, 30–35. [CrossRef]
25. Chang, I.; Lee, M.; Tran, A.T.P.; Lee, S.; Kwon, Y.M.; Im, J.; Cho, G.C. Review on biopolymer-based soil treatment (BPST) technology in geotechnical engineering practices. *Transp. Geotech.* **2020**, *24*, 100385. [CrossRef]
26. Liu, Y.; Chang, M.; Wang, Q.; Wang, Y.F.; Liu, J.; Cao, C.; Zheng, W.; Bao, Y.; Rocchi, I. Use of Sulfur-Free Lignin as a novel soil additive: A multi-scale experimental investigation. *Eng. Geol.* **2020**, *269*, 105551. [CrossRef]
27. Chang, I.; Im, J.; Cho, G.C. Introduction of microbial biopolymers in soil treatment for future environmentally-friendly and sustainable geotechnical engineering. *Sustainability* **2016**, *8*, 251. [CrossRef]
28. Latifi, N.; Horpibulsuk, S.; Meehan, C.L.; Abd Majid, M.Z.; Md Tahir, M.; Tonnizam Mohamad, E. Improvement of Problematic Soils with Biopolymer—An Environmentally Friendly Soil Stabilizer. *J. Mater. Civ. Eng.* **2017**, *29*, 04016204. [CrossRef]
29. Orts, W.J.; Roa-Espinosa, A.; Sojka, R.E.; Glenn, G.M.; Imam, S.H.; Erlacher, K.; Pedersen, J.S. Use of synthetic polymers and biopolymers for soil stabilization in agricultural, construction, and military applications. *J. Mater. Civ. Eng.* **2007**, *19*, 58–66. [CrossRef]
30. Georgees, R.N.; Hassan, R.A.; Evans, R.P. A potential use of a hydrophilic polymeric material to enhance durability properties of pavement materials. *Constr. Build. Mater.* **2017**, *148*, 686–695. [CrossRef]
31. Huang, J.; Kogbara, R.B.; Hariharan, N.; Masad, A.E.; Little, N.D. A state-of-the-art review of polymers used in soil stabilization. *Constr. Build. Mater.* **2021**, *305*, 124685. [CrossRef]
32. Rizwan, M.; Gilani, S.R.; Durani, A.J.; Naseem, S. Materials diversity of hydrogel: Synthesis, polymerization process and soil conditioning properties in agricultural field. *J. Adv. Res.* **2021**, *33*, 15–40. [CrossRef] [PubMed]
33. Hu, Z.Y.; Chen, G.; Yi, S.H.; Wang, Y.; Liu, Q.; Wang, R. Multifunctional porous hydrogel with nutrient controlled-release and excellent biodegradation. *J. Environ. Chem. Eng.* **2021**, *9*, 106146. [CrossRef]
34. Xu, S.; Li, X.; Wang, Y.; Hu, Z.; Wang, R. Characterization of slow-release collagen-g-poly(acrylic acid-co-2-acrylamido-2-methyl-1-propane sulfonic acid)–iron(III) superabsorbent polymer containing fertilizer. *J. Appl. Polym. Sci.* **2019**, *136*, 47178. [CrossRef]
35. Ibrahim, S.; Nawwar, G.A.M.; Sultan, M. Development of bio-based polymeric hydrogel: Green, sustainable and low cost plant fertilizer packaging material. *J. Environ. Chem. Eng.* **2016**, *4*, 203–210. [CrossRef]
36. Essawy, H.A.; Ghazy, M.B.M.; Abd El-Hai, F.; Mohamed, M.F. Superabsorbent hydrogels via graft polymerization of acrylic acid from chitosan-cellulose hybrid and their potential in controlled release of soil nutrients. *Int. J. Biol. Macromol.* **2016**, *89*, 144–151. [CrossRef]
37. Azeem, B.; KuShaari, K.Z.; Man, Z.B.; Basit, A.; Thanh, T.H. Review on materials & methods to produce controlled release coated urea fertilizer. *J. Control. Release* **2014**, *181*, 11–21.

38. Liu, X.; Zhang, Y.; Han, W.; Tang, A.; Shen, J.; Cui, Z.; Vitousek, P.; Erisman, J.W.; Goulding, K.; Christie, P.; et al. Enhanced nitrogen deposition over China. *Nature* **2013**, *494*, 459–463. [CrossRef]

39. Bremner, J.M. Problems in the use of urea as a nitrogen fertilizer. *Soil Use Manag.* **1990**, *6*, 70–71. [CrossRef]

40. Fernández-Escobar, M.; Benlloch, E.; Herrera, J. Effect of traditional and slow-release N fertilizers on growth of olive nursery plants and N losses by leaching. *Sci. Hortic.* **2004**, *101*, 39–49. [CrossRef]

41. Zhao, C.; Shen, Y.Z.; Du, C.W.; Zhou, J.M.; Wang, H.Y.; Chen, X.Y. Evaluation of waterborne coating for controlled-release fertilizer using wurster fluidized bed. *Ind. Eng. Chem. Res.* **2010**, *49*, 9644–9647.

42. Naz, M.Y.; Sulaiman, S.A. Slow release coating remedy for nitrogen loss from conventional urea: A review. *J. Control. Release* **2016**, *225*, 109–120. [CrossRef]

43. Yang, L.; An, D.; Wang, T.J.; Kan, C.Y.; Jin, Y. Photodegradation of polymer materials used for film coatings of controlled-release fertilizers. *Chem. Eng. Technol.* **2017**, *40*, 1611–1618. [CrossRef]

44. Vejan, P.; Khadiran, T.; Abdullah, R.; Ahmad, N. Controlled release fertilizer: A review on developments, applications and potential in agriculture. *J. Control. Release* **2021**, *339*, 321–334. [CrossRef]

45. Rajamani, S.; Chen, Z.G.; Zhang, S.H. Recent developments in cleaner production and environment protection in world leather sector. In Proceedings of the XXX Congress of the International Union of Leather Technologists and Chemists Societies, Beijing, China, 11–14 October 2009; pp. 69–73.

46. Silva, J.D.C.; Leal, T.T.B.; Araújo, A.S.F.; Araujo, R.M.; Gomes, R.L.F.; Melo, W.J.; Singh, R.P. Effect of different tannery sludge compost amendment rates on growth, biomass accumulation and yield responses of Capsicum plants. *Waste Manag.* **2010**, *30*, 1976–1980. [CrossRef]

47. Hu, J.; Xiao, Z.; Zhou, R.; Deng, W.; Wang, M.; Ma, S. Ecological utilization of leather tannery waste with circular economy model. *J. Clean. Prod.* **2011**, *19*, 221–228. [CrossRef]

48. Lai-Cheong, J.E.; McGrath, J.A. Structure and function of skin, hair and nails. *Medicine* **2021**, *49*, 337–342. [CrossRef]

49. Ni, M.; Liu, S.; Lu, L.; Xie, Y. Environmentally friendly slow-release nitrogen fertilizer. *J. Agric. Food Chem.* **2011**, *59*, 10169–10175. [CrossRef]

50. Hassan, M.; Bai, J.; Dou, D.-Q. Biopolymers; definition, classification and applications. *Egypt. J. Chem.* **2019**, *62*, 1725–1737. [CrossRef]

51. Sathish, M.; Madhan, B.; Raghava Rao, J. Leather solid waste: An eco-benign raw material for leather chemical preparation—A circular economy example. *Waste Manag.* **2019**, *87*, 357–367. [CrossRef]

52. Majee, S.; Halder, G.; Mandal, T. Formulating nitrogen-phosphorous-potassium enriched organic manure from solid waste: A novel approach of waste valorization. *Process Saf. Environ. Prot.* **2019**, *132*, 160–168. [CrossRef]

53. Al-Jabari, M.; Sawalha, H.; Pugazhendhi, A.; Rene, E.R. Cleaner production and resource recovery opportunities in leather tanneries: Technological applications and perspectives. *Bioresour. Technol. Rep.* **2021**, *16*, 100815. [CrossRef]

54. Hansen, E.; Monteiro de Aquim, P.; Gutterres, M. Environmental assessment of water, chemicals and effluents in leather post-tanning process: A review. *Environ. Impact Assess. Rev.* **2021**, *89*, 106597. [CrossRef]

55. Chojnacka, K.; Skrzypczak, D.; Mikula, K.; Witek-Krowiak, A.; Izydorczyk, G.; Kuligowski, K.; Bandrow, P.; Kułazynski, M. Progress in sustainable technologies of leather wastes valorization as solutions for the circular economy. *J. Clean. Prod.* **2021**, *313*, 127902. [CrossRef]

56. Mutlu, M.M.; Crudu, M.; Maier, S.S.; Deselnicu, D.; Albu, L.; Gulumser, G.; Bitlisli, B.O.; Basaran, B.; Tosun, C.C.; Adiguzel Zengin, A.C. Eco-leather: Properties of chromium-free leathers produced with titanium tanning materials obtained from the wastes of metal industry. *Ekoloji Int. J. Environ.* **2014**, *23*, 83–90. [CrossRef]

57. Dang, X.; Shan, Z.; Chen, H. The Preparation and Applications of One Biodegradable Liquid Film Mulching by Oxidized Corn Starch-Gelatin Composite. *Appl. Biochem. Biotechnol.* **2016**, *180*, 917–929. [CrossRef] [PubMed]

58. Niculescu, M.D.; Enascuta, C.E.; Stanca, M.; Gaidau, C.C.; Alexe, C.; Gidea, M.; Becheritu, M. Complexes based on collagen and keratin for applications in agriculture. In Proceedings of the 8th International Conference on Advanced Materials and Systems ICAMS2020, Bucharest, Romania, 1–3 October 2020; pp. 219–224.

59. Colla, G.; Nardi, S.; Cardarelli, M.; Ertani, A.; Lucini, L.; Canaguier, R.; Rouphael, Y. Protein hydrolysates as biostimulants in horticulture. *Sci. Hortic.* **2017**, *196*, 28–38. [CrossRef]

60. Stefan, D.S.; Bosomoiu, M.; Constantinescu, R.R.; Ignat, M. Composite Polymers from Leather Waste to Produce Smart Fertilizers. *Polymers* **2021**, *13*, 4351. [CrossRef]

61. Adeoye, D.T.; Adeyi, O.A.; Kathir, I.B.; Ejila, A. De-chroming of Chrome Tanned Leather Solid Waste using Modified Alkaline Hydrolysis Process. *Int. J. Eng. Res. Technol.* **2014**, *3*, 620–623.

62. Ferreira, M.J.; Almeida, M.F.; Pinho, S.C.; Gomes, J.R.; Rodrigues, J.L. Alkaline Hydrolysis of Chromium Tanned Leather Scrap Fibers and Anaerobic Biodegradation of the Products. *Waste Biomass Valoriz.* **2014**, *5*, 551–562. [CrossRef]

63. Tuomisto, H.L.; Teixeira De Mattos, M.J. Environmental impacts of cultured meat production. *Environ. Sci. Technol.* **2011**, *45*, 6117–6123. [CrossRef]

64. Nagaraju, R.K.; Khera, J.G.; Nielsen, P.H. The Combined Bioblasting Concept: Save Energy, Greenhouse Gases and Water. *Int. Dye. Tech. Brief.* **2013**, *4*, 36–38.

65. Nielsen, P.H.; Oxenbøll, K.M.; Wenzel, H. Cradle-to-Gate Environmental Assessment of Enzyme Products Produced Industrially in Denmark by Novozymes A/S. *Int. J. LCA* **2007**, *12*, 432–438. [CrossRef]

66. Rao, R.J.; Thanikaivelan, P.; Nair, B.U. An eco-friendly option for less-chrome and dye-free leather processing: In situ generation of natural colours in leathers tanned with Cr–Fe complex. *Clean Technol. Environ. Policy* **2002**, *4*, 115–121. [CrossRef]

67. China, C.R.; Hilonga, A.; Nyandoro, S.S.; Schroepfer, M.; Kanth, S.V.; Meyer, M.; Njau, K.N. Suitability of selected vegetable tannins traditionally used in leather making in Tanzania. *J. Clean. Prod.* **2020**, *251*, 119687. [CrossRef]

68. Carsote, C.; Sendrea, C.; Micu, M.-C.; Adams, A.; Badea, E. Micro-DSC, FTIR-ATR and NMR MOUSE study of the dose-dependent effects of gamma irradiation on vegetable-tanned leather: The influence of leather thermal stability. *Radiat. Phys. Chem.* **2021**, *189*, 109712. [CrossRef]

69. Shi, J.; Zhang, R.; Mi, Z.; Lyu, S.; Ma, J. Engineering a sustainable chrome-free leather processing based on novel lightfast wet-white tanning system towards eco-leather manufacture. *J. Clean. Prod.* **2021**, *282*, 124504. [CrossRef]

70. Gao, D.; Cheng, Y.; Wang, P.; Li, F.; Wu, Y.; Lyu, B.; Ma, J.; Qin, J. An eco-friendly approach for leather manufacture based on P(POSS-MAA)-aluminum tanning agent combination tannage. *J. Clean. Prod.* **2020**, *257*, 120546. [CrossRef]

71. Zuriaga-Agustí, E.; Galiana-Aleixandre, M.V.; Bes-Pia, A.; Mendoza-Roca, J.A.; Risueno-Puchades, V.; Segarra, V. Pollution reduction in an eco-friendly chrome-free tanning and evaluation of the biodegradation by composting of the tanned leather wastes. *J. Clean. Prod.* **2015**, *87*, 874–881. [CrossRef]

72. Pradeep, S.; Sundaramoorthy, S.; Sathish, M.; Jayakumar, G.C.; Rathinam, A.; Madhan, B.; Saravanan, P.; Rao, J.R. Chromium-free and waterless vegetable-aluminium tanning system for sustainable leather manufacture. *Chem. Eng. J. Adv.* **2021**, *7*, 100108. [CrossRef]

73. Beghetto, V.; Agostinis, L.; Gatto, V.; Samiolo, R.; Scrivanti, A. Sustainable use of 4-(4,6-dimethoxy-1,3,5-triazin-2-yl)-4-methylmorpholinium chloride as metal free tanning agent. *J. Clean. Prod.* **2019**, *220*, 864–872. [CrossRef]

74. Krishnamoorthy, G.; Sadulla, S.; Sehgal, P.K.; Mandal, A.B. Green chemistry approaches to leather tanning process for making chrome-free leather by unnatural amino acids. *J. Hazard. Mater.* **2012**, *215–216*, 173–182. [CrossRef]

75. Krishnamoorthy, G.; Sadulla, S.; Sehgal, P.K.; Mandal, A.B. Greener approach to leather tanning process: D-Lysine aldehyde as novel tanning agent for chrome-free tanning. *J. Clean. Prod.* **2013**, *42*, 277–286. [CrossRef]

76. Wu, X.; Qiang, X.; Liu, D.; Yu, L.; Wang, X. An eco-friendly tanning process to wet-white leather based on amino acids. *J. Clean. Prod.* **2020**, *270*, 122399. [CrossRef]

77. Stefan, D.S.; Zainescu, G.; Manea-Saghin, A.M.; Triantaphyllidou, I.E.; Tzoumani, I.; Tatoulis, T.I.; Syriopoulos, G.T.; Meghea, A. Collagen-Based Hydrogels Composites from Hide Waste to Produce Smart Fertilizers. *Materials* **2020**, *13*, 4396. [CrossRef]

78. Chen, Y.; Li, W.; Zhang, S. A multifunctional eco-friendly fertilizer used keratin-based superabsorbent as coatings for slow-release urea and remediation of contaminated soil. *Prog. Org. Coat.* **2021**, *154*, 106158. [CrossRef]

79. Dang, X.; Yang, M.; Zhang, B.; Chen, H.; Wang, Y. Recovery and utilization of collagen protein powder extracted from chromium leather scrap waste. *Environ. Sci. Pollut. Res.* **2019**, *26*, 7277–7283. [CrossRef]

80. Cheruiyot, G.; Sirmah, P.; Ng'etich, W.; Mengich, E.; Mburu, F.; Kimaiyo, S.; Bett, E. Effects of Hydrogels on Soil Moisture and Growth of Cajanus cajan in Semi Arid Zone of Kongelai, West Pokot County. *Open J. For.* **2014**, *4*, 34–37.

81. Majee, S.; Halder, G.; Mandal, D.D.; Tiwari, O.N.; Mandal, T. Transforming wet blue leather and potato peel into an eco-friendly bio-organic NPK fertilizer for intensifying crop productivity and retrieving value-added recyclable chromium salts. *J. Hazard. Mater.* **2021**, *411*, 125046. [CrossRef]

82. Farneselli, M.; Tosti, G.; Onofri, A.; Benincasa, P.; Guiducci, M.; Pannacci, E.; Tei, F. Effects of N sources and management strategies on crop growth, yield and potential N leaching in processing tomato. *Eur. J. Agron.* **2018**, *98*, 46–54. [CrossRef]

83. Constantinescu, R.R.; Zainescu, G.; Stefan, D.S.; Sirbu, C.; Voicu, P. Protein biofertilizer development and application on soybean cultivated degraded soil. *Leather Footwear J.* **2015**, *15*, 169–178. [CrossRef]

84. Zainescu, G.; Deselnicu, D.C.; Ioannidis, I.; Crudu, M.; Voicu, P. New versatile conversion technology for wet white waste transformation to biofertiliser. In Proceedings of the 4th International Conference on Advanced Materials and Systems ICAMS2012, Bucharest, Romania, 27–29 September 2012; Volume 1, p. 8, ISSN 2068-0783.

85. Lima, D.Q.; Oliveira, L.C.A.; Bastos, A.R.R.; Carvalho, G.S.; Marques, J.G.S.M.; Carvalho, J.G.; de Souza, G.A. Leather Industry Solid Waste as Nitrogen Source for Growth of Common Bean Plants. *Appl. Environ. Soil Sci.* **2010**, *2010*, 703842. [CrossRef]

86. Simple fertilizers. Available online: https://www.azomures.com/en/fertilizers/ (accessed on 29 April 2022).

87. Dang, X.; Shan, Z.; Chen, H. Biodegradable films based on gelatin extracted from chrome leather scrap. *Int. J. Biol. Macromol.* **2018**, *107*, 1023–1029. [CrossRef]

88. Ravindran, B.; Lee, S.R.; Chang, S.W.; Nguyen, D.D.; Chung, W.J.; Balasubramanian, B.; Mupambwa, H.A.; Arasu, M.V.; Al-Dhabi, N.A.; Sekaran, G. Positive effects of compost and vermicompost produced from tannery waste-animal fleshing on the growth and yield of commercial crop-tomato (*Lycopersicon esculentum* L.) plant. *J. Environ. Manag.* **2019**, *234*, 154–158. [CrossRef]

89. Vig, A.P.; Singh, J.; Wani, S.H.; Singh Dhaliwal, S. Vermicomposting of tannery sludge mixed with cattle dung into valuable manure using earthworm *Eisenia fetida* (Savigny). *Bioresour. Technol.* **2011**, *102*, 7941–7945. [CrossRef]

90. Ravindran, B.; Dinesh, S.L.; Kennedy, L.J.; Sekaran, G. Vermicomposting of solid waste generated from leather industries using epigeic earthworm *Eisenia foetida*. *Appl. Biochem. Biotechnol.* **2008**, *151*, 480–488. [CrossRef]

91. Singh, D.; Suthar, S. Vermicomposting of herbal pharmaceutical industry waste: Earthworm growth, plant-available nutrient and microbial quality of end materials. *Bioresour. Technol.* **2012**, *112*, 179–185. [CrossRef]

92. Molina, M.J.; Soriano, M.D.; Ingelmo, F.; Llinares, J. Stabilisation of sewage sludge and vinasse bio-wastes by vermicomposting with rabbit manure using *Eisenia fetida*. *Bioresour. Technol.* **2013**, *137*, 88–97. [CrossRef]

93. Singh, J.; Kaur, A. Vermidegradation for faster remediation of chemical sludge and spent carbon generated. *Soft Drink Ind.* **2013**, *1*, 13–20.

94. Singh, J.; Kaur, A.; Vig, A.P. Bioremediation of distillery sludge into soil-enriching material through vermicomposting with the help of *Eisenia fetida*. *Appl. Biochem. Biotechnol.* **2014**, *174*, 1403–1419. [CrossRef]

95. Ravindran, B.; Contreras-Ramos, S.M.; Wong, J.W.C.; Selvam, A.; Sekaran, G. Nutrient and enzymatic changes of hydrolysed tannery solid waste treated with epigeic earthworm *Eudrilus eugeniae* and phytotoxicity assessment on selected commercial crops. *Environ. Sci. Pollut. Res. Int.* **2014**, *21*, 641–651. [CrossRef]

96. Singh, S.; Bhat, S.A.; Singh, J.; Kaur, R.; Vig, A.P. Earthworms converting milk processing industry sludge into biomanure. *Open Waste Manag. J.* **2017**, *10*, 30–40. [CrossRef]

97. Malińska, K.; Golańska, M.; Caceres, R.; Rorat, A.; Weisser, P.; Ślęzak, E. Biochar amendment for integrated composting and vermicomposting of sewage sludge—The effect of biochar on the activity of *Eisenia fetida* and the obtained vermicompost. *Bioresour. Technol.* **2017**, *225*, 206–214. [CrossRef]

98. Nunes, R.R.; Pigatin, L.B.F.; Oliveira, T.S.; Bontempi, R.M.; Rezende, M.O.O. Vermicomposted tannery wastes in the organic cultivation of sweet pepper: Growth, nutritive value and production. *Int. J. Recycl. Org. Waste Agric.* **2018**, *7*, 313–324. [CrossRef]

99. Araujo, A.S.F.; de Melo, W.J.; Araujo, F.F.; Van den Brink, P.J. Long-term effect of composted tannery sludge on soil chemical and biological parameters. *Environ. Sci. Pollut. Res.* **2020**, *27*, 41885–41892. [CrossRef] [PubMed]

100. Zainescu, G.; Barna, V.; Constantinescu, R.; Barna, E.; Voicu, P.; Sandru, L. High yield biopolymer systems obtained from leather wastes. *Rev. Mater. Plast.* **2011**, *48*, 295–298.

101. Nogueira, F.G.E.; do Prado, N.T.; Oliveira, L.C.A.; Bastos, A.R.R.; Lopes, J.H.; de Carvalho, J.G. Incorporation of mineral phosphorus and potassium on leather waste (collagen): A new $N_{collagen}$PK-fertilizer with slow liberation. *J. Hazard. Mater.* **2010**, *176*, 374–380. [CrossRef]

102. Nogueira, F.G.E.; Castro, I.A.; Bastos, A.R.R.; Souza, G.A.; de Carvalho, J.G.; Oliveira, L.C.A. Recycling of solid waste rich in organic nitrogen from leather industry: Mineral nutrition of rice plants. *J. Hazard. Mater.* **2011**, *186*, 1064–1069. [CrossRef]

103. Oliveira, L.C.A.; Dallago, R.M.; Nascimento Filho, I. Processo de Reciclagem dos Resíduos Sólidos de Curtumes por Extração do Cromo e Recuperação do Couro Descontaminado. BR PI 001538, 2004.

104. Oliveira, D.Q.L.; Goncalves, M.; Oliveira, L.C.A.; Guilherme, L.R.G. Removal of As(V) and Cr(VI) from aqueous solutions using solid waste from leather industry. *J. Hazard. Mater.* **2008**, *151*, 280–284. [CrossRef]

105. Dunne, K.S.; Holden, N.M.; Daly, K. A management framework for phosphorus use on agricultural soils using sorption criteria and soil test P. *J. Environ. Manag.* **2021**, *299*, 113665. [CrossRef]

106. Bottinelli, N.; Maeght, J.L.; Pham, R.D.; Valentin, C.; Rumpel, C.; Pham, Q.V.; Nguyen, T.T.; Lam, D.H.; Nguyen, A.D.; Tran, T.M.; et al. Anecic earthworms generate more topsoil than they contribute to erosion—Evidence at catchment scale in northern Vietnam. *Catena* **2021**, *201*, 105186. [CrossRef]

107. Li, Q.; Zhang, D.; Song, Z.; Ren, L.; Jin, X.; Fang, W.; Yan, D.; Li, Y.; Wang, Q.; Cao, A. Organic fertilizer activates soil beneficial microorganisms to promote strawberry growth and soil health after fumigation. *Environ. Pollut.* **2022**, *295*, 118653. [CrossRef]

108. Treder, W.; Klamkowski, K.; Wójcik, K.; Tryngiel-Gać, A.; Sas-Paszt, L.; Mika, A.; Kowalczyk, W. Apple leaf macro- and micronutrient content as affected by soil treatments with fertilizers and microorganisms. *Sci. Hortic.* **2022**, *297*, 110975. [CrossRef]

109. Wang, C.; Sun, Y.; Chen, H.Y.H.; Ruan, H. Effects of elevated CO_2 on the C:N stoichiometry of plants, soils, and microorganisms in terrestrial ecosystems. *Catena* **2021**, *201*, 105219. [CrossRef]

110. Dong, L.; Changwen, D.; Fei, M.; Yazhen, S.; Ke, W.; Jianmin, Z. Interaction between polyacrylate coatings used in controlled-release fertilizers and soils in wheat-rice rotation fields. *Agric. Ecosyst. Environ.* **2019**, *286*, 106650.

111. Xie, L.; Liu, M.; Ni, B.; Wang, Y. New Environment-Friendly Use of Wheat Straw in Slow-Release Fertilizer Formulations with the Function of Superabsorbent. *Ind. Eng. Chem. Res.* **2012**, *51*, 3855–3862. [CrossRef]

112. Lubkowski, K.; Grzmil, B. Controlled release fertilizers. *Pol. J. Chem. Technol.* **2007**, *9*, 81–84. [CrossRef]

113. Ni, B.; Liu, M.; Lü, S. Multifunctional slow-release urea fertilizer from ethyl cellulose and superabsorbent coated formulations. *Chem. Eng. J.* **2009**, *155*, 892–898. [CrossRef]

114. Ge, J.J.; Wu, R.; Shi, X.H.; Yu, H.; Wang, M.; Li, W.J. Biodegradable polyurethane materials from bark and starch. II. Coating material for controlled-release fertilizer. *J. Appl. Polym. Sci.* **2002**, *86*, 2948–2952. [CrossRef]

115. Tzoumani, I.; Lainioti, G.C.; Aletras, A.J.; Zainescu, G.; Stefan, S.; Meghea, A.; Kallitsis, J.K. Modification of Collagen Derivatives with Water-Soluble Polymers for the Development of Cross-Linked Hydrogels for Controlled Release. *Materials* **2019**, *12*, 4067. [CrossRef]

116. Khatami, H.; O'Kelly, B. Improving mechanical properties of sand using biopolymers. *J. Geotech. Geoenviron. Eng.* **2012**, *139*, 1402–1406. [CrossRef]

117. Imeson, A. *Food Stabilisers, Thickeners and Gelling Agents*; John Wiley & Sons: Chichester, UK, 2010.

118. Chen, R.; Zhang, L.; Budhu, M. Biopolymer stabilization of mine tailings. *J. Geotech. Geoenviron. Eng.* **2013**, *139*, 1802–1807. [CrossRef]

119. Acharya, R.; Pedarla, A.; Bheemasetti, T.V.; Puppala, A.J. Assessment of guar gum biopolymer treatment toward mitigation of desiccation cracking on slopes built with expansive soils. *Transp. Res. Rec.* **2017**, *2657*, 78–88. [CrossRef]

120. Gupta, S.C.; Hooda, K.; Mathur, N.; Gupta, S. Tailoring of guar gum for desert sand stabilization. *Indian J. Chem. Technol.* **2009**, *16*, 507–512.

121. Chang, I.; Prasidhi, A.K.; Im, J.; Cho, G.-C. Soil strengthening using thermo-gelation biopolymers. *Constr. Build. Mater.* **2015**, *77*, 430–438. [CrossRef]

122. Chang, I.; Im, J.; Lee, S.-W.; Cho, G.-C. Strength durability of gellan gum biopolymer treated Korean sand with cyclic wetting and drying. *Constr. Build. Mater.* **2017**, *143*, 210–221. [CrossRef]

123. Naessens, M.; Cerdobbel, A.; Soetaert, W.; Vandamme, E.J. Leuconostoc dextransucrase and dextran: Production, properties and applications. *J. Chem. Technol. Biotechnol.* **2005**, *80*, 845–860. [CrossRef]

124. Pini, R.; Canarutto, S.; Guidi, G.V. Soil microaggregation as influenced by uncharged organic conditioners. *Commun. Soil Sci. Plant Anal.* **1994**, *25*, 2215–2229. [CrossRef]

125. Emerson, W.W.; Greenland, D.J. Soil Aggregates—Formation and Stability. In *Soil Colloids and Their Associations in Aggregates*; De Boodt, M.F., Hayes, M.H.B., Herbillon, A., De Strooper, E.B.A., Tuck, J.J., Eds.; Springer: Boston, MA, USA, 1990; pp. 485–511.

126. Bartoli, F.; Philippy, R.; Burtin, G. Aggregation in soils with small amounts of swelling clays. I. Aggregate stability. *Eur. J. Soil Sci.* **1988**, *39*, 593–616. [CrossRef]

127. Im, J.; Tran, A.T.P.; Chang, I.; Cho, G.-C. Dynamic properties of gel-type biopolymer treated sands evaluated by Resonant Column (RC) tests. *Geomech. Eng.* **2017**, *12*, 815–830. [CrossRef]

128. Lee, S.; Chang, I.; Chung, M.-K.; Kim, Y.; Kee, J. Geotechnical shear behavior of xanthan gum biopolymer treated sand from direct shear testing. *Geomech. Eng.* **2017**, *12*, 831–847. [CrossRef]

129. Becker, A.; Katzen, F.; Pühler, A.; Ielpi, L. Xanthan gum biosynthesis and application: A biochemical/genetic perspective. *Appl. Microbiol. Biotechnol.* **1998**, *50*, 145–152. [CrossRef]

130. Ayeldeen, M.K.; Negm, A.M.; El Sawwaf, M.A. Evaluating the physical characteristics of biopolymer/soil mixtures. *Arab. J. Geosci.* **2016**, *9*, 371. [CrossRef]

131. Qureshi, M.U.; Chang, I.; Al-Sadarani, K. Strength and durability characteristics of biopolymer-treated desert sand. *Geomech. Eng.* **2017**, *12*, 785–801. [CrossRef]

132. Lee, S.; Im, J.; Cho, G.-C.; Chang, I. Laboratory triaxial test behavior of xanthan gum biopolymer-treated sands. *Geomech. Eng.* **2019**, *17*, 445–452.

133. Bouazza, A.; Gates, W.; Ranjith, P. Hydraulic conductivity of biopolymer-treated silty sand. *Géotechnique* **2009**, *59*, 71–72. [CrossRef]

134. Cabalar, A.; Wiszniewski, M.; Skutnik, Z. Effects of Xanthan Gum Biopolymer on the Permeability, Odometer, Unconfined Compressive and Triaxial Shear Behavior of a Sand. *Soil Mech. Found. Eng.* **2017**, *54*, 356–361. [CrossRef]

135. Chatterjee, T.; Chatterjee, S.; Lee, D.S.; Lee, M.W.; Woo, S.H. Coagulation of soil suspensions containing nonionic or anionic surfactants using chitosan, polyacrylamide, and polyaluminium chloride. *Chemosphere* **2009**, *75*, 1307–1314. [CrossRef]

136. Reddy, I.; Seib, P.A. Paste properties of modified starches from partial waxy wheats. *Cereal Chem.* **1999**, *76*, 341–349. [CrossRef]

137. Lentz, R.D.; Sojka, R.E.; Carter, D.L.; Shainberg, I. Preventing irrigation furrow erosion with small applications of polymers. *Soil Sci. Soc. Am. J.* **1992**, *56*, 1926–1932. [CrossRef]

138. Fatehi, H.; Abtahi, S.M.; Hashemolhosseini, H.; Hejazi, S.M. A novel study on using protein based biopolymers in soil strengthening. *Constr. Build. Mater.* **2018**, *167*, 813–821. [CrossRef]

139. Chang, I.; Im, J.; Chung, M.-K.; Cho, G.-C. Bovine casein as a new soil strengthening binder from diary wastes. *Constr. Build. Mater.* **2018**, *160*, 1–9. [CrossRef]

140. McAuliffe, K.W.; Scotter, D.R.; MacGregor, A.N.; Earl, K.D. Casein Whey Wastewater Effects on Soil Permeability. *J. Environ. Qual.* **1982**, *11*, 31–34. [CrossRef]

141. Dang, X.; Yuan, H.; Shan, Z. An eco-friendly material based on graft copolymer of gelatin extracted from leather solid waste for potential application in chemical sand-fixation. *J. Clean. Prod.* **2018**, *188*, 416–424. [CrossRef]

142. Chiroma, T.M.; Ebewele, R.O.; Hymore, F.K. Comparative Assessment of Heavy Metal Levels in Soil, Vegetables and Urban Grey Waste Water Used for Irrigation in Yola and Kano. *Int. ref. J. Eng. Sci.* **2014**, *3*, 1–9.

143. Soltani, A.; Deng, A.; Taheri, A.; O'Kelly, B.C. Engineering reactive clay systems by ground rubber replacement and polyacrylamide treatment. *Polymers* **2019**, *11*, 1675. [CrossRef]

144. Georgees, R.N.; Hassan, R.A.; Evans, R.P.; Jegatheesan, P. An evaluation of performance-related properties for granular pavement materials using a polyacrylamide additive. *Int. J. Pavement Eng.* **2016**, *19*, 153–163. [CrossRef]

145. Georgees, R.N.; Hassan, R.A.; Evans, R.P.; Jegatheesan, P. Performance Improvement of Granular Pavement Materials Using a Polyacrylamide-Based Additive. In *Geo-China 2016: Advances in Pavement Engineering and Ground Improvement, Proceedings of the Fourth Geo-China International Conference, Shandong, China, 25–27 July 2016*; Khabbaz, H., Hossain, Z., Nam, B.H., Chen, X., Eds.; American Society of Civil Engineers: Reston, VA, USA, 2016; pp. 108–117.

146. Lentz, R.D. Polyacrylamide and biopolymer effects on flocculation, aggregate stability, and water seepage in a silt loam. *Geoderma* **2015**, *241–242*, 289–294. [CrossRef]

147. Tingle, J.S.; Newman, J.K.; Larson, S.L.; Weiss, C.A.; Rushing, J.F. Stabilization Mechanisms of Non-traditional Additives. *J. Transp. Res. Board* **2007**, *1989*, 59–67. [CrossRef]

148. Tingle, J.S.; Santoni, R.L. Stabilization of clay soils with non-traditional additives. *Transp. Res. Rec.* **2003**, *1819*, 72–84. [CrossRef]

149. Acosta-Martinez, V.; Tabatabai, M.A. Inhibition of arylamidase activity in soils by toluene. *Soil Biol. Biochem.* **2002**, *34*, 229–237. [CrossRef]
150. Nannipieri, P.; Kandeler, E.; Ruggiero, P. Enzyme activities and microbiological and biochemical processes in soil. In *Enzymes in the Environment*; Burns, R.G., Dick, R., Eds.; Marcel Dekker: New York, NY, USA, 2002; pp. 1–33.

Article

Multilayer Packaging in a Circular Economy

Jannick Schmidt [1,*], Laura Grau [1,*], Maximilian Auer [1], Roman Maletz [2] and Jörg Woidasky [1]

[1] Institut für Industrial Ecology, Hochschule Pforzheim, Tiefenbronner Straße 65, 75175 Pforzheim, Germany; maximilian.auer@hs-pforzheim.de (M.A.); joerg.woidasky@hs-pforzheim.de (J.W.)
[2] Institute of Waste Management and Circular Economy, Technische Universität Dresden, Pratzschwitzer Straße 15, 01796 Pirna, Germany; roman.maletz@tu-dresden.de
* Correspondence: jannick.schmidt@hs-pforzheim.de (J.S.); laura.grau@hs-pforzheim.de (L.G.)

Abstract: Sorting multilayer packaging is still a major challenge in the recycling of post-consumer plastic waste. In a 2019 Germany-wide field study with 248 participants, lightweight packaging (LWP) was randomly selected and analyzed by infrared spectrometry to identify multilayer packaging in the LWP stream. Further investigations of the multilayer packaging using infrared spectrometry and microscopy were able to determine specific multilayer characteristics such as typical layer numbers, average layer thicknesses, the polymers of the outer and inner layers, and typical multilayer structures for specific packaged goods. This dataset shows that multilayer packaging is mainly selected according to the task to be fulfilled, with practically no concern for its end-of-life recycling properties. The speed of innovation in recycling processes does not keep up with packaging material innovations.

Keywords: multilayer; lightweight packaging; circular economy; plastic; recycling

Citation: Schmidt, J.; Grau, L.; Auer, M.; Maletz, R.; Woidasky, J. Multilayer Packaging in a Circular Economy. *Polymers* **2022**, *14*, 1825. https://doi.org/10.3390/polym14091825

Academic Editor: Sheila Devasahayam

Received: 25 March 2022
Accepted: 26 April 2022
Published: 29 April 2022

Publisher's Note: MDPI stays neutral with regard to jurisdictional claims in published maps and institutional affiliations.

1. Introduction

For years, the quantities of plastic waste in Germany have been steadily increasing. In 1994, plastic waste generation was 2.8 million t [1], but by 2019 it had already risen to 6.28 million t [2]. The average annual growth of about 3.3% is almost exclusively due to the waste generated in the post-consumer sector, with the packaging industry being by far the largest consumer of plastics in Germany [2]. Consequently, more than 50% of the plastic waste generated today in Germany can be attributed to short-lived packaging [3]. Plastic packaging waste almost doubled from 1991 (1.64 million t) to 2017 (3.18 million t), even though individual plastic packaging items, in general, became on average 25% lighter during this period [4,5]. Accordingly, plastic packaging is perceived by the public as one of the biggest environmental problems [6], and has thus led to the adoption of stricter environmental laws to reduce plastic packaging and increase the recycling rate [7]. In Germany, the Packaging Act (VerpackG), which came into force in 2019, stipulates that mechanical recycling rates must be 63% by 2022 [8].

To strengthen a more sustainable approach to plastics, the EU presented "A European Strategy for Plastics in a Circular Economy (CE)" in 2018 [9]. This stipulates that all plastic packaging placed on the market in the EU should be either reusable or recyclable in a cost-effective manner by 2030 [9]. This approach is intended to break the current prevailing linear flow (open loop) of plastics along the value chain from production to use and disposal [10], as this is one of the main sources of CO_2 emissions and pollution [11]. It is estimated that 95% of the value of plastic packaging is lost after the first phase of use [12]. This is due to the use of mechanical recycling to reprocess mixed plastic waste streams, which leads to a decrease in molecular mass and thus limits the number of possible reprocessing processes [13]. Recycled plastics can therefore often only be downcycled [14] and are primarily used to produce products other than those originally made from the material [15]. Such intensive use of finite resources for a linear economic model of production, use, and disposal is proving to be unsustainable [16]. The implementation of a true CE (closed loop) in the field of plastics could help in reducing downcycling, incineration, and landfill, allowing plastic waste to be recycled back into the same or equivalent new products [11]. However, before a true CE based on a balance between economic, environmental, and social impacts can be achieved [17], the prevailing problems must be overcome. Well-functioning plastic recycling processes are needed, in order to move from a linear flow to a closed loop [18].

Along with the requirements for suitable CE packaging, the requirements for food packaging are also steadily increasing. The basic packaging requirements relate to strength and sealability, machinability (softening, slip, rigidity, pliability, and heat resistance), promotion, and convenience [19], as well as barrier properties against oxygen, water vapor, light, carbon dioxide, and flavoring substances, which enable a long shelf life and thus the current form of food trade and reduced food losses [20–22]. For this purpose, multilayer packaging is often used. Multilayer packaging combines different polymeric and non-polymeric materials such as paper or aluminum [23–25], which enables customized property profiles with low material consumption [26]. Multilayers can reduce the cost of existing film structures, e.g., by replacing expensive polymers with less costly ones, reducing film thickness, or using recycled materials [27]. Furthermore, the combination of different layers achieves a functionality that is not possible through the use of a single layer [28]. According to estimates, about 17% [23] to 20% [29] of plastic packaging consists of multilayer packaging, and it is increasingly used in the packaging of food, pharmaceuticals, medicines, cosmetics, and electronics [24,30].

While multilayer packaging does not differ from mono-material solutions in terms of use and collection, challenges become obvious in packaging sorting plants, as multilayer packaging is difficult to identify and hard to recycle [31]. Here, spectroscopic identification technology (NIR) can only identify surface properties, and thus, for physical reasons, does not identify multilayers properly. In subsequent recycling processes, recovery of multilayer material becomes possible if either the materials can be separated or they can be processed jointly. However, this is not easy to achieve because, on the one hand current recycling systems are aimed at recycling mono-materials, and on the other hand the different polymers or materials are often immiscible. Therefore, multilayer packaging is considered non-recyclable, and only thermal recovery or final disposal routes remain [14,19,32,33].

The wide variety of achievable properties through the combination of different numbers of layers, layer materials, and layer thicknesses cannot be properly managed with current waste management technologies and systems. Information regarding the individual material composition of packaging is required to improve current identification and recycling technologies. While this information is available from the producers, it is currently lost along the supply chain. Standardized recycling codes do not provide a sufficiently detailed level of information, and other indications provided by the manufacturer are rarely disclosed on the packaging media, as the composition often represents an important competitive advantage. Consequently, polymer multilayers are marked with the recycling code "Other" (07) according to DIN 6120 (German Institute for Standardization) (After the revision of DIN 6120, the additional designation 07 ("Other") was deleted.) or not marked at all [34]. This is insufficient for the future development of waste management systems and substantial material recycling quota increases, as technology and strategy development rely on large and precise amounts of data [35]. Currently, the lack of data leads to barriers in choosing the most appropriate strategy to close material and product loops [36]. Great potential is therefore seen in the area of big data applications, where comprehensive data-driven decision-making can also take into account the integration of sustainability approaches across supply chain networks, to pave the way toward a CE [37,38].

Waste characterization is key to reprocessing high-value end products [39]. This has become particularly clear since digitization has entered the waste management sector and knowledge in the form of data on the type and composition of individual LWP (lightweight packaging) wastes has come into focus. The combination with digital technologies enables companies to improve the circularity of their systems [40]. As early as 2017, a survey of

394 companies from the waste management and recycling technology sectors in Germany, Austria, and Switzerland revealed that 63% of those surveyed perceived digitization in the company as an opportunity for further development [41], and this was despite the fact that actual digital readiness, which describes the degree of digital transformation of companies in the waste management sector in German-speaking countries, is only at 30% [42]. This is reflected, for example, in the fact that many plant operators operate their sorting facilities with fixed parameter settings based on empirical values, without any physical or statistical proof of optimality, which is ultimately due to the complexity involved in the study of mixed solid waste [43]. There are already projects and companies working on developing new technological approaches to addressing the complexity of digitally assisted optimization of LWP waste sorting in an industrial setting. These approaches make an important contribution to the transition of the current linear system to a CE by capturing valuable resources that would normally be lost under the current linearity [44].

With this in mind, and motivated by the need for waste management to shift from a linear to a circular approach [45], the research question in this work concerns how multilayer LWP items, forming a substantial fraction of post-consumer packaging waste, can be characterized in detail. Consequently, the scope of this work is to collect, analyze, and publish key data on multilayer packaging applications as a fraction (class of items) of plastic LWP waste, as a basis for future waste management system developments and recycling rate increases. A more holistic view of the treatment of post-consumer LWP waste has been published in [46]. With these results, recommendations to the production industry can be made to increase the recyclability and circularity of this material stream.

2. Multilayer Packaging: State of the Art

2.1. Structure of Multilayer Packaging

Multilayer packaging can be flexible or semi-rigid and involve polymeric layers as well as inorganic layers such as Al or SiOx coatings. It usually consists of from 2 [47] up to 24 layers [48]. Each layer adds an important function to the overall architecture. The intended functions and the layers used to achieve them, as well as the materials used to fulfill these functions, are listed in Table 1.

The seal layer (1) is in direct contact with the packaged goods, therefore good migration limitation and an interaction barrier must be provided, for example by using a polyolefin-type inner layer [28]. For freshness conservation, the permeability of the packaging material against specific substances, or generally to prevent any gaseous and liquid exchange, often needs to be adjusted.

This can be enabled by increasing the layer thickness or just by using suitable *barrier layers* (2) in between to prevent oxidation, microbial spoilage, loss or gain of moisture, and both loss of flavor and aroma or gain of unwanted aromas from outside [26]. *The oxygen barrier* avoids the packaged goods becoming rancid, as well as the proliferation of aerobic microorganisms. A *light-barrier layer* such as an aluminum coating or a TiO_2-filled polymer layer can also help to conserve freshness [19,24,26]. *Migration barriers* may be of the utmost importance in multilayer films for sensitive foodstuffs, protecting the packaged goods from losing substances such as additives that are present in them or protecting them from the migration of unwanted residues in recycled intermediate layers [49].

If adjacent layers do not adhere to each other, a *tie layer* (3) can be applied to establish compatibility [50], often using PU (polyurethane) or acid/anhydride grafted polymers [19]. Alternatively, layers can be added by coating or lamination [50].

Better stability can be achieved by either increased thickness, using a (cheaper) filling layer, or by using a *structural layer* (4) with good mechanical properties, including tear and piercing strength [19,24].

Further processing requirements such as printability and print protection can be achieved by using a suitable outer *layer* (5) such as a cardboard layer or priming, or by modifying the surface of a polymer layer for good print adhesion, e.g., by using flame treatment, an electrochemical corona treatment, or fluorination [51,52]. The outer layer is often required to have a good mechanical performance [28]. Other requirements might be, for example, certain slip properties or shape stability [26], which would usually be achieved by including additives or by modifying the material preparation parameters. For the processing of the packaging materials, the inner layer and sometimes also the outer layer must often allow sealing of the contents.

Table 1. Layers, functions, and commonly used materials for multilayer food packaging, based on [19,20,24,26,28,50,53–61]. Abbreviations used: LLD (linear low-density), LD (low-density), HD (high-density), PE (polyethylene), EVA (ethylene-vinyl acetate), OPP (oriented polypropylene), OPA (oriented polyamide), OPET (oriented polyethylene terephthalate), PVDC (polyvinylidene chloride), EVOH (ethylene-vinyl alcohol).

Layer	Function	Material
(1) Seal Layer (innermost layer)	Heat sealability (low melting temperature), inert against filling goods	(LLD, LD) PE, EVA, ionomers, (O)PP, (O)PA, (O)PET
(2) Barrier Layer	Resistance against:	
	Moisture	(LD, LLD, HD) PE; (O)PP, EVA, ionomers, PVDC, PET
	Oil/grease	PET, HDPE, PA, Ionomers, EVOH, PVDC
	Water vapor	PP, HDPE, PELD, PVDC
	Aroma/flavor	PET, PA, EVOH, PVDC
	Oxygen	EVOH (standard), PA or PET (below standard), Aluminum (exceeding standard), PVDC, (biaxially oriented) PA, (oriented) PET, SiOx, or Al_2O_3 coatings
	Light	Aluminum, TiO_2-filled polymers
(3) Tie Layer	Combines two chemically incompatible materials	polyurethanes, acid/anhydride grafted polyolefins
(4) Structural layer	Provides shape: Toughness	PE, PET
	Puncture resistance	HDPE, PA
	Stiffness	PP, PET, HDPE, LDPE, PA, EVA, Ionomers, EVOH
	Stability	PP, PET, PA, EVA, ionomers, EVOH
(5) Outer layer	Provides printing surface and mechanical performance	PE or PET
(6) Coating (outermost layer facing environment)	Optional thin film to protect the printed material	Any specialized polymer

Furthermore, the print must be protected by a *coating* (6) such as a lacquer or other laminated layer. For print protection, reverse printing of the outer layer or the full composite is common [27].

Not all flexible packaging requires each of the above-mentioned layers. Aside from the purposes of the outer and inner layers, the intermediate layers can be randomly arranged, preferably in the way that is easiest to manufacture. While the amount, thickness, and arrangement of the packaging layers vary, some combinations of commonly used materials are frequently seen in cases where extreme requirements must be fulfilled, such as an extremely good moisture barrier for savory snacks, often provided by an aluminum layer on PP, or the usage of durable PA layers for packaging sharp-edged goods such as cheese blocks or T-bone steaks. Total packaging thicknesses usually range between 10 and

250 µm in multilayer food packaging [53]. Typical thicknesses of the individual layers are sometimes identified [26], but they are subject to frequent innovative changes.

2.2. Current End-of-Life Situation for Multilayer Packaging

After the packaging use phase, the packaging will be discarded by the consumer to the German LWP collection system. If a state-of-the-art waste management system is available, the packaging will be collected, ideally in a separate fraction, transported to a sorting facility that separates the different packaging materials and polymer types, and subsequently recycled via rinsing, shredding, and regranulation. Spectroscopic methods (VNIR (visible near-infrared), NIR (near-infrared), or MIR (mid-infrared) wavelength range) in particular are used for the automated sorting of plastic waste [62–64]. The reflection of infrared radiation by the packaging surface discloses information about the chemical composition of the plastics, based on which the sorting decisions can be made [65,66]. In the current LWP sorting of plastic waste, the FT–NIR technique is the most commonly used [67], applying up to 50 or more single NIR sorters to separate the plastic stream into individual types of plastic (HDPE, LDPE, PET, etc.) [68]. Nested, superimposed, black, dirty, wet, or printed packaging surfaces cannot be identified properly by NIR [14]. Differentiation by color would require additional technology (VIS cameras), and identification of filling goods is not possible; therefore, further downstream recycling steps become necessary [53,69]. NIR is a surface measurement technology that penetrates only 2 µm deep into the top layer of the material [70], resulting in the detection of only the polymer layer that is facing the sensor at the time [14,15], and not identifying the other materials used for the inner layers of composites [54]. Typically, however, NIR systems can achieve sorting purities of up to 96% [71]. In industrial processes, the sorting residues amount to up to 26% of the sorting plant input [72].

As well as these identification challenges, additional impurities due to mishandling by separation air blasts frequently occur [14,70]. All this leads to undesired impurities in the secondary raw materials, and even small quantities may result in poor adhesion, lower mechanical properties, or unwanted (dark) colors in the end products, which ultimately increase the recyclers' costs [53,73,74]. Therefore, "Der Grüne Punkt" (Cologne, Germany) usually allows impurities in the sorting fractions of about 2–8 wt% [75,76].

If a multilayer product has to be fully mechanically recycled, polymer miscibility is the key parameter. Here a distinction can be made between homogeneous (miscible), heterogeneous (immiscible), and limited miscibility systems (Table 2), with heterogeneous systems clearly predominating [77]. If the materials are compatible, i.e., fully miscible, direct regranulation is possible. The polymers PE, PP, PS, PA, and PET, which are often used in food packaging, are generally immiscible at the molecular level [78]. In the area of multilayer packaging, this heterogeneity is reinforced by the combination of other polymers and non-polymer materials [28]. Immiscible or limited miscibility multi-material structures must be separated into their layers, or at least into fractions of miscible components, which is a key waste management challenge [39,53].

Incompatibility in the reprocessing process is due to large differences in the specific melting temperatures or an overlap of melting and decomposition temperatures [79]. For example, the processing temperature of PET is between 270 and 300 °C, while the thermal degradation of ethylene-vinyl acetate (EVA) begins at 288 °C [79]. Immiscible polymers can be made miscible to a limited extent by compatibilizers. Such substances consist of copolymers, each with one end that is miscible and anchors itself in one of the two components of the blend. This creates strong bonds between different, incompatible polymers [77].

The better the upfront separation of the polymer waste into pure materials, the more efficient the recycling will be with conventional plastic recycling technologies [32,54]. With multilayers, this approach reaches its limits, because usually the layers cannot be separated with economically justifiable effort, and thus the low product mass along with the functionality of the packaging is achieved at the expense of recyclability [81]. Ultimately, the sum of the impurities and other contaminants in the post-consumer waste stream results in recyclates of degraded [82] or even undefinable quality [83]. Furthermore, there are currently hardly any possibilities, or only expensive test procedures, for proving the quality and composition of recyclates. This makes the use of recyclates an expensive and arduous endeavor for plastics manufacturers and recyclers, depending on who bears the costs. Furthermore, the lack of transparency and past uncertainties regarding the actual quality of plastic recyclates has led to a lack of trust between plastic recyclers and manufacturers and to insufficiently well-functioning markets for high-quality secondary raw materials [83,84]. Manufacturers are therefore increasingly turning to primary materials, as the costs are often lower and the quality fluctuates less than for secondary raw materials [83].

Table 2. Miscibility of different polymers, based on [80]. Abbreviations used: PS (polystyrene), PVC (polyvinyl chloride).

		Polymer-Matrix					
		PE	**PP**	**PET**	**PA**	**PS**	**PVC**
	PE	1	3–4	4	2–4	4	4
	PP	2–4	1	4	2–4	4	4
Additive	PET	4	4	1	3–4	4	4
material	PA	4	4	3	1	3–4	4
	PS	4	4	3	3–4	1	4
	PVC	4	4	4	4	2–4	1

Good compatibility (1); miscible up to approximately 20% (2); miscible up to approximately 5% (3); incompatible (4).

This highlights the need for a unified approach to recycling plastic packaging in a closed or open loop between recyclers and producers [54]. However, a corresponding system of closed material loops can only develop through the exchange of waste-related data and the cooperation of all participants along the supply chain [85]. Unlike for the recycling of mono-materials, to date there are no strategies for processing multilayer films in closed primary loops [32]. Multilayer packaging is thus emblematic of the problems that ultimately occur in waste management along the value chain of plastic packaging. The players involved focus on their interests and goals. For packaging manufacturers, for example, this leads to multilayer packaging being developed and marketed with a view to maximum functionality at minimum cost and not with a view to its recyclability. This is in contrast to the concept of CE, which aims to further develop the prevailing linear flow of plastics along the value chain by closing them into loops, so that plastic products and materials remain in the economic cycle for as long as possible [86,87]. To achieve a CE in the waste management of plastic packaging, the European Commission's "A European Strategy for Plastics in a Circular Economy" emphasizes, among other things, recycling-friendly design and the use of innovative sorting and recycling systems [9].

Although under discussion for decades, design for recycling has so far gained little influence on the numerous suppliers in the market, and the speed of innovation for new types of packaging such as multi-material packaging does not at all match that of innovations for methods and technologies for recycling [12]. In this context, Ceflex, an initiative representing the entire value chain of flexible packaging, is trying to make all flexible packaging recyclable by 2025 through its "Designing for a Circular Economy Guidelines" [88]. The focus of phase one is on polyolefin-based flexible packaging (mono-PE, mono-PP, and PE/PP blends), as this makes up the largest part of the flexible packaging waste stream, and sorting and mechanical recycling have already been demonstrated on an industrial scale. Phase two, which is currently underway, will then address the recycling of multilayer materials.

In addition, there are already projects working on new technological approaches that, among other things, make it possible to separate multilayer packaging from the post-consumer waste stream in a more targeted way than before. These include fluorescent marker particles, which are applied in low concentrations in or on packaging and emit a characteristic luminescence when excited in specific wavenumber ranges. The marker particles can thus contribute to improving sortability as a material-independent separation feature [89–91]. In the Holy Grail initiative, invisible digital water marks are applied to the surface of a package. Recycling machines can then read out the recycling information for the packaging in question [92].

Both approaches can be combined with the detection technologies typically used in waste management and have the potential to sort post-consumer waste streams into defined streams (e.g., food vs. non-food). More targeted sorting could enable better recycling of multilayer packaging, as it could then be treated with chemical or solvent-based delamination processes. This would pave the way to greater quantities of high-quality recyclates and enable the entire packaging value chain to take a step towards CE [93].

2.3. Recycling of Multilayer Materials

Each manufacturer has a choice of different material and layer combinations when developing an individual packaging solution, and this has led to an ever-increasing number of different multilayer packaging solutions (especially in the area of polymers). This prevents a clear separation into individual material groups, as is possible with packaging made of mono-materials. In the present work, the following multilayer packaging types could be identified, with the focus set on polymeric multilayers. By identifying the outer and inner layers of the packaging, a large number of sub-categories could be identified in the category *"Polymeric multilayer"* (e.g., PET–LDPE, PA–LDPE, PP–LDPE, PP–PP, PET–PET, PET–HDPE, PET–PP, LDPE–LDPE). Furthermore *"3-Composite paper/cardboard+metal+plastic"* and *"2-Composite paper/cardboard+plastic"* are mainly used for liquid cartons, *"2-Composite aluminum+paper/cardboard"* mainly consists of foils and pouches, *"2-Composite plastic+aluminum"* mainly consists of blisters and pouches, and *"Plastic+paper/cardboard unlaminated"* mainly consists of pouches, blisters, and cups.

For many of these multilayer products, there are already several promising approaches and processes for their recycling at various degrees of maturity. These are summarized in Table 3.

Table 3. Methods for the treatment of multilayers and their degrees of maturity.

Procedure/Company	Raw material/Recovery	Capacity	* TRL	Current Status
Solvent-Based Recycling Processes				
CreaSolv® (Fraunhofer IVV) [94–96]	PE from post-consumer multilayer pouches	1000 t/a	7	Pilot plant (2019) for recycling post-consumer multilayer pouches in Indonesia
	PE and PP from, e.g., multilayer (post-consumer) consisting of PE/PA, PP/PET, and aluminum content	Truckload per day (approx. 5 m^3 per day)	5	Construction of an industrial-scale pilot plant (2020) in Germany as part of the "Circular Packaging" project.
Newcycling® (APK AG) [33,97–99]	PE/PA and aluminum from multilayer films (post-industrial) Separation of PE from PP	8000 t/a	7	Operation of a pilot plant (2018) in Germany
Saperatec GmbH [98,100]	PET, PE, and aluminum from each other Paper, plastic, and aluminum (liquid cartons)	18,000 t/a	5–6	Pilot plant currently under construction (completion 2023)
Purecycle (Procter & Gamble) [96,101]	PP from, e.g., food and liquid packaging	48,000 t/a	6	Pilot plant currently under construction (completion end 2022)
Solvent-targeted recovery and precipitation [102]	PE, EVOH, and PET from each other	/	1	Release (solvent) of the target polymer from the composite system with subsequent precipitation and repetition for the next target polymer.
Recycling of post-consumer multilayer Tetra Pak® packaging with the selective dissolution–precipitation process [103]	LDPE from aluminum (Tetra Paks)	/	1	Separation through selective dissolution–precipitation process
Chemical recycling processes				
ChemCycling (BASF) [104]	Pyrolysis process enables recycling of post-consumer plastic waste (also multilayers)	/	3–4	/
ChemPET (Garbo) [105]	PET out of multilayer films (PET/PE/aluminum/PE) and multilayer trays (PET/PE/EVOH/PE)	1000 t/a	6	Operation of a pilot plant using glycolysis (3 t/day)
Other approaches				
Recycling of multilayer packaging using a reversible crosslinking adhesive [106]	PE/PET, PET/aluminum, and PE/aluminum from each other	/	1	Modification of the packaging adhesives. Separation by heated solvent from dimethylsulfoxide

* Technology Readiness Level [107]: TRL 1–3 represents proof of concept/research; TRL 4–6 represents development; and TRL 7–9 represents deployment.

With improved controllability and separability, multilayers would be of great importance for material recycling, as they represent not only a large but also a steadily growing (approximately 7% p.a. [108]) segment of the packaging market.

For a possible approximation of current multilayer shares, Equation (1) can be used. In 2010, multilayer films accounted for 17% (K0) of global film production [23]. If an annual growth rate of 7% (p) and a time horizon between the two studies of 9.5 years (n) is assumed (the LWP sorting took place in 2019), this yields a multilayer share of 32.33%. A second study assumed a multilayer share of 26 wt% [16] in the LWP stream in 2017. Taking into account a time horizon of 2.5 years, Equation (1) yields a multilayer share of 30.79%.

$$K_n = K_0 * (1 + \frac{p}{100})^n \tag{1}$$

3. Materials and Methods

3.1. Samples

The samples were taken from a Germany-wide collection of LWP covering a total of 350 participating households in 2019, as a part of the "MaReK" research project [46]. The participants were selected according to their household size and their place of residence. During a selectable two-week period from June to November 2019, they were asked not to dispose of their LWP in the usual manner but to collect it using an 80 L transparent HDPE collection bag they had been provided with, regardless of the local collection system. A total of 248 participants completed this field study and returned their collection bags via return mail to Pforzheim University. In total, 21,380 post-consumer LWP items with a total mass of 207 kg (188,869 g in this publication, as caps were not taken into account.) were analyzed, hereinafter referred to as the "original sample" [109] (see Tables 4 and 5). Further information about the collection method in the field study was published in [46].

Table 4. Packing material distribution of the original sample, based on a total of 188,869 g, based on [46].

Packaging Type	Mass Share [%]
Unmarked (no recycling code)	30.69
3-Composite paper/cardboard+metal+plastic	10.97
PP	10.46
Tinplate	8.53
PET	8.46
2-Composite paper/cardboard+plastic	6.77
Paper/cardboard (no compound)	5.71
HDPE	3.46
LDPE	3.14
PS	2.61
Aluminum	2.54
07-Other (recycling code)	2.41
2-Composite aluminum+paper/cardboard	1.57
Remaining small parts	1.27
2-Composite plastic+aluminum	0.80
Plastic+paper/cardboard unlaminated	0.53
PA	0.02
PLA	0.02
PVC	0.02
PMMA	0.01
Total	100.00

Table 5. Examined packaging of the IR sample, the ML sample, and the original sample.

Packaging Type	IR Sample		ML Sample		Original Sample	
	Count	Share	Count	Share	Count	Share [%]
Pouch	498	41.85	135	45.61	7351	34.38
Foil	156	13.11	85	28.72	2466	11.53
Tray	301	25.29	56	18.92	2197	10.28
Separate closure element	3	0.25			1865	8.72
Cup	42	3.53			1477	6.91
Bag	105	8.82			1466	6.86
Liquid packaging	2	0.17			1027	4.80
Bottle	7	0.59			610	2.85
Can					530	2.48
Blister	4	0.34			494	2.31
Non-packaging items	7	0.59			370	1.73
Skin packaging	36	3.03			355	1.66
Tube	3	0.25			250	1.17
Net	6	0.50			229	1.07
Remaining small parts					199	0.93
Folding box					189	0.88
Other packaging element	1	0.08			113	0.53
Rigid foil	17	1.43			98	0.46
Filling material	2	0.17			60	0.28
Wrap packaging					25	0.12
Screw-top jar					9	0.04
* Stand-up pouch			20	6.76		
Total	1190	100.00	296	100.00	21,380	100.00

* Stand-up pouches were treated separately in the ML sample to investigate possible differences.

This research on the original sample [46] showed that identification based on the recycling code leads to an unmarked material mass share of about 31 %, or 54 % based on a count share. To analyze the packaging materials used, Fourier transform infrared attenuated total reflectance (FTIR–ATR) measurements were carried out with 1190 randomly selected packages (3467 g, caps were not considered) and additionally evaluated based on the type of packaging and the packaged goods. This representative sample, taken from the original sample, which was analyzed via FTIR-ATR, is hereinafter referred as the "IR (infrared) sample". Table 5 shows the packaging types and their numbers within the IR sample.

Furthermore, a detailed analysis of 296 multilayer items (1828 g) was carried out. For sampling, a random selection was made from packages that were assumed to have multilayer content (see Table 6). The categories were identified through a literature review [22,27,28,110], expert interviews, and preliminary research on LWP. FTIR–ATR analysis and microscopy were used for the identification of outer and inner layers, layer thickness and number, and metallic content. This representative sample, taken from the original sample, is hereinafter referred as the "ML (multilayer) sample".

The IR sample and the ML sample originated from the population of the original sample but were, in addition, completely independent experimental series. The analytical setup was chosen based on its simplicity, speed, and availability, in order to generate the maximum information content from the samples.

Table 6. Examined packaging within the ML sample.

Packaging for	Pouch	Foil	Tray	Stand-Up Pouch	Total
Sliced cheese	14	18	21		53
Sausages and cold cuts	5	27	20		52
Baked goods	21	4	1		26
Meat substitutes		13	5		18
Non-food items	9			4	13
Salty biscuits	13			1	14
Feta in brine		12			12
Mozzarella in brine	12				12
Nuts	7			5	12
Dry food	11				11
Coffee, tea, spices	10				10
Sweets	8				8
Preserves	2	1		4	7
Ready meals		1	3	2	6
Soft cheese	2	3	1		6
Fresh meat + fish	1	2	2		5
Granulates	5				5
Animal feed	4			1	5
Dried fruits	3			2	5
Minced meat		2	2		4
Hard cheese	1	1	1	1	4
Grated cheese	4				4
Vegetables	2				2
Butter		1			1
Rice pudding	1				1
Total	135	85	56	20	296

3.2. FTIR–ATR Analysis

For each ML sample and IR sample item, without further sample modification, both sample sides ("inner" and "outer" layer, where the "inner layer" is the layer in direct contact with the filling good) were tested. To this end, each LWP was characterized at three different, preferably unprinted, locations each on the inner and the outer layer, using an FTIR Alpha Platinum ATR (attenuated total reflectance) spectroscope from Bruker (Billerica, MA, USA) with OPUS Version 7.5 software (Bruker) and the therein contained libraries BPAD.S01, Demolib.s01, and FILLER.S01. The measurements were made in the mid-infrared range (4000–400 cm^{-1}) with a resolution of 4 cm^{-1}. For each spectrum, 10 scans were performed, and the arithmetic mean value was calculated. The measurements, as well as the hit quality (on a scale from 100–1000) of the assigned database spectrum, were recorded to exclude false determinations in cases of low hit qualities (<500).

3.3. Microscopy Analysis

From every ML sample item, a 30 mm × 30 mm specimen was cut out with a scalpel. The specimen was set up using specimen embedding holders and examined under a microscope. A Leica DM RM light microscope (Type 301-371.010) from Leica (Wetzlar, Germany) with LasX software (Leica) was used to provide a compiled picture. The layers were measured with the aid of the image processing tool ImageJ. The number of layers, as well as the individual thickness of each layer, was recorded.

4. Research Results and Discussion

4.1. Analysis of the IR Sample

The material distributions of the IR sample (1190 items, 3467 g) given in mass shares based on the recycling code (printed on the packaging) characterization ("labeling") on the one hand and the IR characterization ("analysis") on the other hand are shown in Table 7. Using the recycling code for material determination, a share of 65.71% of unmarked packaging remains. Furthermore, 10.34% of the packaging marked with "07-Other" provided no precise information on the packaging material. Thus, in total, a share of 76.05% of the packaging did not provide clear identification of its material composition. In other words, three out of four plastic packaging items in the market do not provide the customer with any information about the packaging material.

Table 7. Packaging material proportions for the IR sample (1190 items), before and after their identification by IR analysis, based on [46].

Material/Category	Result of the Recycling Code Labeling	Result after IR Analysis Mass Share [%]	Percentage Change
PP	11.34	25.55	+125.30
LDPE	1.26	8.07	+540.50
PET	3.53	6.89	+95.20
PS	1.68	3.28	+95.20
PA	0.25	1.18	+372.00
PVC	0.34	2.35	+591.20
HDPE	0.76	4.96	+552.60
Multilayer	* 4.79	43.19	+801.70
Unmarked (no recycling code)	65.71	4.54 remain	−93.10
07-Other	10.34	not determinable	
Total	100.00	100.00	

* More recycling codes on the packaging or clear allocation to multilayer fraction (e.g., butcher film (plastic + paper)).

When FTIR analysis was applied, only 4.54% were not determinable, while a multi-layer packaging share of 43.19% was identified, which is by far the largest single fraction of all packaging (Table 7). As expected, the second-largest fraction was composed of PP packaging (25.55%), and all other packaging materials were only found in single-digit percentage shares. Nonetheless, if the material labeling is compared with the FTIR analysis, in the cases of LDPE, PVC, and HDPE, five to six times the mass could be identified by IR analysis compared with the amount identified by the recycling code labeling information. In contrast, within the PP, PET, and PS fractions, approximately the same mass that was identifiable via labeling by a recycling code could be identified by the IR analysis. This is because PET, PS, and PP are often used for cups and trays, which often carry a recycling code.

As an intermediate result, one might state that the recycling code labeling of LWP does not work well, as only about 25% of all packaging items bear recycling code information. Moreover, most of the multilayer materials remain unlabeled, and even in the best cases of mono-material packaging (PET, PS), only about 50% of the packaging items are clearly labeled for material identification. However, if a recycling code was present on the packaging, 93% of them (PP: 95%, LDPE: 93%, PET: 93%, PS: 100%, PA: 67%, PVC: 50%, HDPE: 78%) agreed with the IR analyses performed. While the labeling policy might be relevant for consumer decisions, in industrial sorting it does not play a role as the sorting relies on material properties (NIR reflectance) but not on labels. However, appropriate labeling could lead to an improvement in LWP input quality. The need to support consumers in sorting is shown by an online survey conducted by Kantar GmbH and commissioned by the dual system in Germany in 2020, in which almost 60% of respondents stated that they needed further information for the correct separation of all types of household waste [111]. This results in a 30% share of waste that mistakenly ends up in the yellow bin (or the yellow bag) [111]. This poses a great challenge to even the most modern sorting plant and can lead to a dramatic decrease in material quality.

4.2. Extrapolated Original Sample, Data Validation, and Recycling Approaches

The results of the IR sample (1190 items) allowed an extrapolation to the mass of the categories "07-Other" and "Unmarked (no recycling code)" (65.5 kg or 33% mass share) of the original sample (see Table 4), assuming that both had the same composition [46]. This reduced the proportion of non-identifiable items to 2.63% (see Figure 1). As can be seen, PP (17.4%), polymeric multilayers (12.5%), PET (12.4%), 3-Composites paper/cardboard+metal+plastic (11%), and tinplate (8.5%) were the most important fractions by mass. A comparison between sorting plant output results [72] and data regarding plastic processing in the packaging industry [112] showed good consistency, with only minor differences [46]. Mono-polyolefins, at 29.3% (PP: 17.4%; LDPE: 6.95%; HDPE: 4.91%) represented the largest fraction by mass. This is in line with statements from Ceflex and also clarifies the approach of placing polyolefins at the center of the recycling of flexible packaging [88].

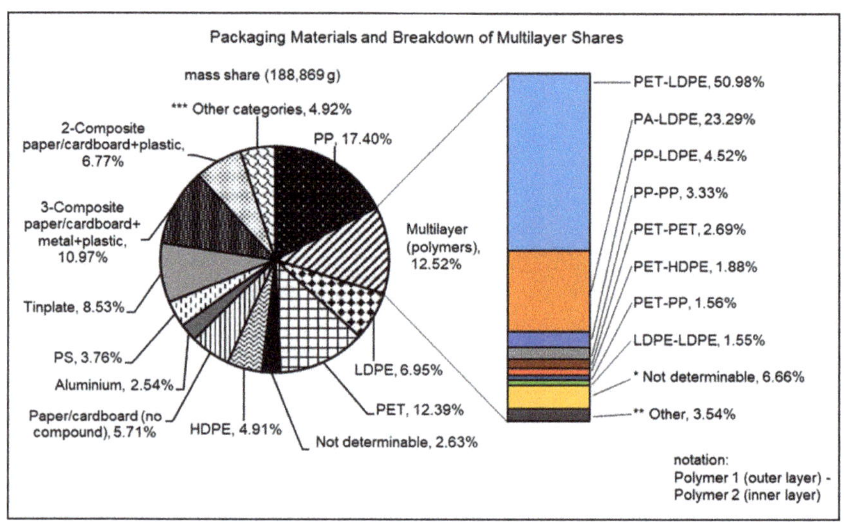

Figure 1. Original sample composition after IR analysis, based on [46]. * n.d.-n.d.; LDPE-n.d.; PET-n.d.; PA-n.d.; PP-n.d.; ** PP-PS; PA-PP; PA-HDPE; HDPE-HDPE; HDPE-PET; PA-PET; PA-PA; PA-PVC; PP-HDPE ***; 2-Composite aluminium+paper/cardboard, Remaining small parts, PA, Plastic+paper/cardboard unlaminated, PLA, PMMA, PVC, 2-Composite plastic+aluminium.

Furthermore, there was a total share of 33.16% of multilayers, consisting of polymeric multilayers at 12.52%, 3-Composite paper/cardboard+metal+plastic at 10.97%, 2-Composite paper/cardboard+plastic at 6.77%, and further combinations accounting for 2.90% (see "*** Other categories" within Figure 1). Measurements of the inner and outer layers of the 12.52% of polymer multilayer packaging further identified the polymers used, and these are shown as a bar chart in Figure 1. Here PET–LDPE (outer inner layer) makes up the largest share at 50.98%. This is followed by PA–LDPE at 23.29% and PP–LDPE at 4.52%. For the outer layer, PET (57.98%) and PA (24.12%) were mainly used, while the inner layer was mainly LDPE at 80.34%, followed by PP at 4.89%. For 6.66% of items, the outer, inner, or both layers could not be identified.

The 33.16% share of multilayers in the LWP stream determined in this study corresponds to the multilayer shares of 32.33% and 30.79% calculated in Equation (1). Deviations can be attributed to the authors' definition of the term multilayer. Validation of the presented multilayer proportions could not be performed, due to a lack of studies in the literature.

Taking into account the approaches and processes for multilayer recycling presented in Table 3, there are currently, or will be in the near future (all pilot plants will be in operation in 2023), possibilities for the recycling of all 20 multilayer combinations or the 33.16 wt% identified in the present work (Figure 1). However, processes such as Purecycle or ChemPet only recycle a target polymer from the multilayer, so further post-treatment steps or processes are necessary. In addition, the Newcycling process for separating PE and PA, for example, is currently only used in a post-industrial environment. Furthermore, some processes have not yet reached large-scale industrial maturity. For example, the pilot factory of Garbo GmbH recycles 1000 t/a of PET from multilayer films and trays using the ChemPET process [105], but faces 161,800 t/a of non-recyclable PET packaging waste in Germany [72]. A second example can be given for the recycling of liquid cartons, where the plant of Saperatec GmbH can handle 18,000 t/a [98], whereas the amount of liquid cartons in Germany is over 155,600 t/a [72]. This underlines the fact that there is no lack of innovative delamination processes for the recycling of multilayer materials, but there is a lack of market maturity and industrial-scale processes. Chemical recycling processes, for example, produce large quantities of CO_2 and are currently too expensive, due to their high energy requirements [113,114]. As a result, the accumulating quantities of multilayer packaging cannot yet be fully recycled, and the actual value for the recyclability of multilayers still has to be corrected downwards.

4.3. Depth Analysis of the ML Sample

The following section is intended to provide a descriptive insight into the wide variety of multilayer packaging structures. For this purpose, Figures 2–4 show the number of layers, total thickness, outer and inner polymer combinations, storage conditions for packaging types, and packaged goods for the ML sample (296 multilayer items, total mass 1828 g). These parameters were chosen as they can provide considerable information to assist the recyclability of multilayer packaging. Unless otherwise stated, the percentages in chapter 4.3 are to be understood as count shares.

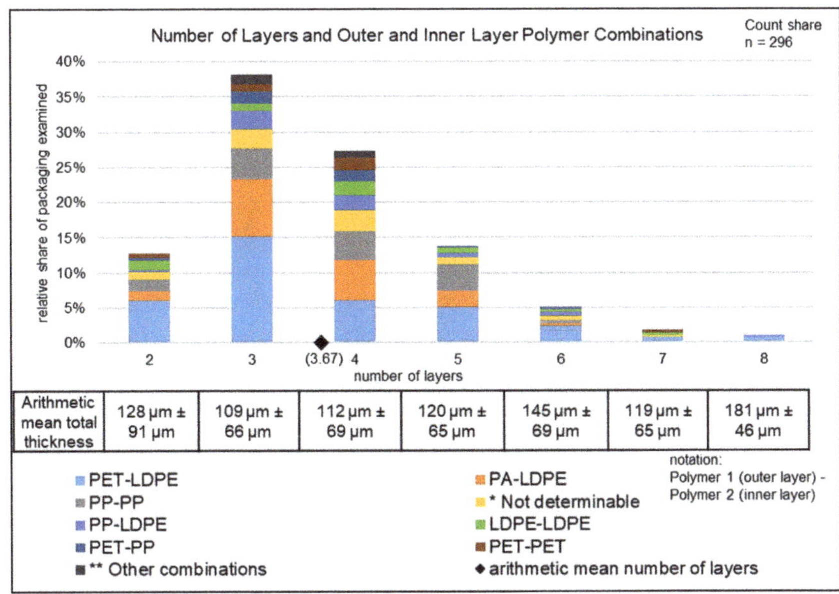

Arithmetic mean total thickness	128 µm ± 91 µm	109 µm ± 66 µm	112 µm ± 69 µm	120 µm ± 65 µm	145 µm ± 69 µm	119 µm ± 65 µm	181 µm ± 46 µm

Figure 2. Numbers of layers and outer and inner layer polymer combinations in the ML sample. * PA-n.d.; PET-n.d.; n.d.-n.d.; n.d.-PP; n.d.-LDPE; ** PP-PET; LDPE-PET; PET-HDPE; LDPE-PA; PA-PP; PA-PA.

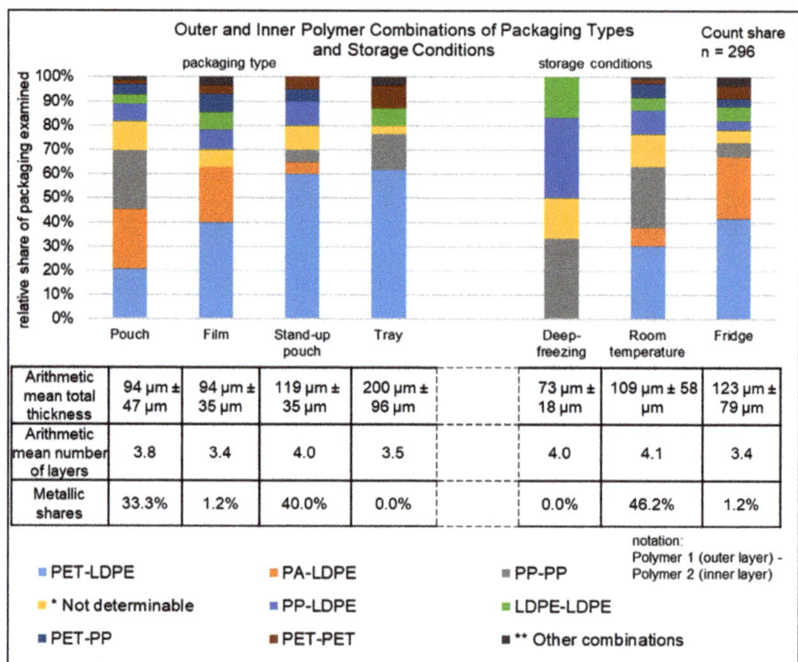

Figure 3. Analysis of the packaging with respect to packaging type and storage conditions. * PA-n.d.; PET-n.d.; n.d.-n.d.; n.d.-PP; n.d.-LDPE; ** PP-PET; LDPE-PET; PET-HDPE; PE-LD-PA; PA-PP; PA-PA.

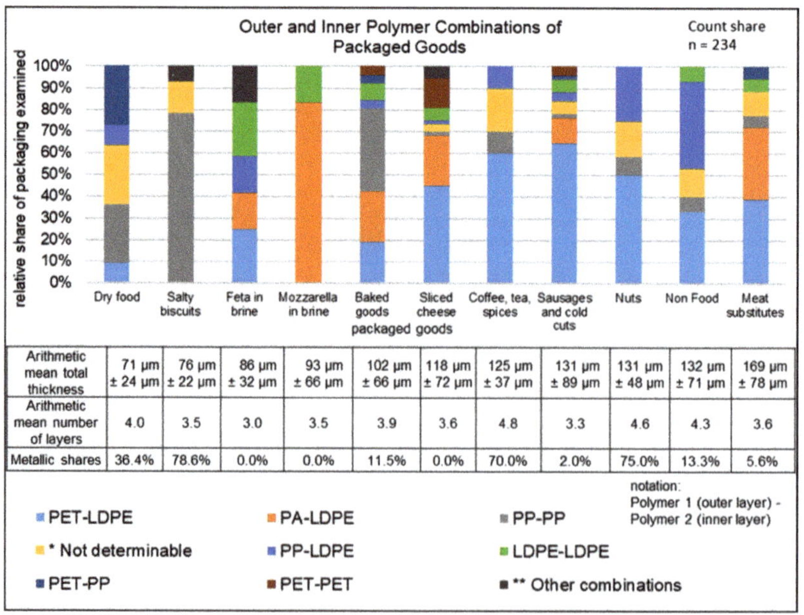

Figure 4. Outer and inner layer polymer combinations for different packaged goods. * PA-n.d.; PET-n.d.; n.d.-n.d.; n.d.-PP; n.d.-LDPE; ** PP-PET; LDPE-PET; PET-HDPE; HDPE-PA; PA-PA.

4.3.1. Number of Layers, Total Thickness, and Outer and Inner Layer Polymer Combinations

The number of layers, the total thickness, and the outer and inner layer polymer combinations used in the multilayer packaging examined can be seen in Figure 2. Overall, the packaging examined can be divided into systems with two to eight layers, with 92% consisting of two to five layers. With a share of 38%, three-layer systems are used most frequently. The arithmetical mean (AM) number of layers for all packaging was found to be 3.7, and the AM total thickness of all packaging was 116.7 μm, with the total layer thickness of the packages ranging from 71 to 200 μm. A clear linear correlation between the AM total thickness and the AM number of layers could be established (Pearson correlation coefficients: 0.65). Observation of the materials used for the outer and inner layers shows that the outer material of 45% (count of 134) of the packaging was PET, which was thus used twice as often as PP at 22% (64) or PA at 19% (57). For the inner layers, LPDE proved to be particularly dominant, being used in 71% (209) of the packaging. PP at 21% (62) and PET at 5% (14) were other notable polymers for inner layers. Overall, the most frequently used combination was PET–LDPE at 36% (107), followed by PA–LDPE at approximately 18% (53) and PP–PP at 14.2% (42). In 24% (71) of the packaging, the same polymer was used as the outer and inner layer (PP 14% (42); LDPE 6% (17); PET 4% (11); PA 0.3% (1)).

Furthermore, there was a 20% (58) share of non-polymeric composite materials. Metallic layers such as aluminum make up the largest share at 17% (51). In some cases, there was a paper layer (0.3% (1)) or both (2% (6)). Furthermore, only the two-layer systems, representing 13% (38) of all packaging, could be fully characterized. Polymer compatibility in terms of miscibility was found in 29% (11) of the two-layer systems, due to the same polymers being used in the outer and inner layers. Limited miscibility was found in 12% (PA–LDPE (4) and PP–LDPE (1)) of the two-layer systems with different outer and inner layers. This means that in total, only 5% of the packaging examined could be recycled using conventional recycling methods. The outer or inner layers of 9% (26) of the packaging could not be identified.

In the ML sample, 13 different polymer multilayer combinations could be identified, whereas 17 were found in the IR sample. Furthermore, there were similarities concerning the most common multilayer combinations identified (see Table 8). The similarities were due to the fact that the samples were taken from the same population (original sample) and the differences were due to the different mass proportions of the ML sample (1828 g) and the IR sample (3467 g).

Table 8. Polymer multilayer combinations of the ML sample and IR sample in wt%.

Material Combinations	PET–LDPE	PA–LDPE	PP–PP	PP–LDPE	LDPE–LDPE	PET–PET	PET–PP
ML sample (wt%)	38	15	14	8	6	5	3
IR sample (wt%)	51	23	3	5	2	3	2

4.3.2. Analysis of the Packaging Regarding Packaging Type and Storage Conditions

The outer and inner layer polymer combinations used, depending on the different packaging types and storage conditions, can be seen in Figure 3. The most common polymer combinations in the pouch fraction were PP–PP and PA–LDPE at 24% each (33) and PET–LDPE at 21% (28). In the film fraction, PET–LDPE at 40% (33), PA–LDPE at 23% (19), and PP–LDPE at 8% (7) were used frequently. In the stand-up pouch fraction, PET–LDPE was used most frequently at 60% (12), followed by PP–LDPE at 10% (2). In addition, within the tray fraction, PET–LDPE was used most frequently at 62% (34), followed by PP–PP at 15% (8). The highest AM total thickness of 200 μm was found within the analyzed trays, despite an AM number of layers of 3.5, which was below the AM number of layers for all packaging of 3.7. This was followed by stand-up pouches, with an AM total thickness of 119 μm, despite the highest AM number of layers of 4. The lowest AM total thickness of 94 μm was found in pouches and films. While films, with an AM number of layers of 3.4, were below the AM number of layers of all packaging, pouches had an AM number of layers of 3.8, which is above that for all packaging. Metallic components occurred in 40% (8) of the stand-up pouches and 33% (45) of the pouches. No metallic components were found in the tray fraction.

With a share of 58% (171), the majority of the ML sample packaging was stored in the refrigerator and 40% (119) at room temperature. Only 2% (6) of the packaging was stored in a deep freeze. The most frequent polymer combinations of the refrigerator fraction were PET–LDPE at 42% (71), PA–LDPE at 26% (44), and PP–PP and LDPE–LDPE at 6% each (10). In packaging stored at room temperature, PET–LDPE at 30% (36) and PP–PP at 25% (30) were mainly used. Frozen packaging consists mainly of PP–PP (33% (2)) and PP–LDPE (33% (2)). Packaging stored in the refrigerator had the highest AM total thickness of 123 μm, with the lowest AM number of layers of 3.4 of the storage options investigated. Packaging stored at room temperature had the highest AM number of layers of 4.1, but the AM total thickness of 109 μm was below that of the packaging stored in the refrigerator. Frozen packaging, despite an AM number of layers of 4, had the lowest AM total thickness of 73 μm. A total of 46% (55) of the products stored at room temperature had some metal content. In the refrigerator fraction, this was only 1% (2), while no metal content was found in frozen products.

4.3.3. Analysis of the Packaging with Respect to Packaged Goods

The proportions of polymer outer and inner layers, depending on the category of packaged goods, are illustrated in Figure 4. Only packaging items of which at least ten were available were selected (see Table 6). This resulted in a total of 234 packages divided into 11 of the 25 categories. Furthermore, the category "not determinable" could contain further combinations. In 10 of the 11 categories, more than two different combinations of outer and inner layers were used to package the same goods. For mozzarella in brine, only two different material combinations (PA–LDPE and LDPE–LDPE) were found. The greatest variety of packaging consisting of different material combinations was found in sliced cheese packages (nine), followed by baked goods and sausages and cold cuts (seven each).

A comparison of food and non-food packaging shows that non-food packaging mainly used PP–LDPE (40% (6)) and PET–LDPE (33% (5)), while food packaging mostly used PET–LDPE (40% (85)) and PA–LDPE (20% (42)). Furthermore, categories that at first glance appeared to be the same could also differ. For example, PA–LDPE was used in packaging for mozzarella in brine with a share of 83% (10), while it had only a 17% (2) share for feta in brine. LDPE–LDPE was also used in packaging for mozzarella, while PET–LDPE, PP–LDPE, PA–LDPE, PP–PET, and LDPE–PA, a much wider variety of material combinations, were all used for packaging of feta in brine. Meat substitutes had the thickest packaging with an AM total thickness of 169 μm and an AM layer number of 3.6, slightly below the AM layer number for all packaging of 3.7. Packaging for sausages and cold cuts had the second-lowest AM number of layers (3.3), despite an AM total thickness of 131 μm. Only feta in brine packaging had fewer layers (3.0). The lowest AM total thickness of 71 μm was found in dry food, despite a high AM number of layers of 4. The highest AM number of layers was found in coffee (4.8) and nuts (4.6). Metallic content was found in 8 of the 11 categories shown. However, this was particularly frequent in packaging for dry food (78%), nuts (75%), and coffee, tea, and spices (70%), while no metallic materials were used in cheese packaging.

4.3.4. Discussion of the ML Sample

In the investigated ML sample (296 items), systems with two to eight layers and an AM number of layers of 3.7 could be found. Furthermore, the AM total thickness of all packaging was 116.7 μm, with the total layer thickness of the packages ranging from 71 to 200 μm. The analysis of the packaging regarding the materials used showed that both PET as the outer layer and LDPE as the inner layer stood out as frequently used materials. The use of these materials was common to all the packaging types, storage conditions, and food categories investigated. The only exceptions were packaging for mozzarella and salty biscuits (crisps), as well as packaging that was stored in the freezer, for which no PET (outer layer) was used. Furthermore, LDPE (inner layer) was not used for salty biscuits. PET is used in packaging for mechanical stability, and as protection against moisture, oil/grease and aroma migration, and it also offers a good surface for printing. LDPE is suitable as an inner layer primarily because of its excellent heat sealability (low melting temperature) and inertness towards the contents. As a result, the PET–LPDE layer combination was the most commonly used across all packages at 36% (107). This combination was particularly dominant in packages for coffee, tea, and spice (60%), sausages and cold cuts (65%), nuts (50%), sliced cheese (45%), and meat substitutes (39%).

At 18% (53), the layer combination PA–LDPE was found to be the second most common. This combination is ideal for vacuum packaging of oxygen-sensitive foods such as ham, cheese, or sausages, and was therefore used in packaging for mozzarella (83%), meat substitutes (39%), sliced cheese (23%), and baked goods (23%).

The third most common material combination was PP–PP at 14% (42), with about 11% (5) of these being two-layer systems. Furthermore, PP was the only material worth mentioning besides LDPE that was used as an inner layer. The reason for the PP–PP combination is its high resistance to grease and moisture. This combination was therefore used particularly frequently in salty biscuits (79%), baked goods (39%), and dry food packaging (27%).

Aluminum was identified in 19% of the packaging within the ML sample. Aluminum protects against air, light, and moisture, and thus contributes to longer shelf life and aroma protection of the food. Therefore, aluminum was found in packaging for salty biscuits (79%), nuts (75%), and coffee, tea, and spices (70%). The same principle can be applied in reverse, as no aluminum was identified in packaging for feta in brine, mozzarella in brine, and sliced cheese, and only a 2% share in packaging was found for sausages and cold cuts. Considering the type of packaging, aluminum was detected in 40% of stand-up pouches, 30% of bags, and 1% of films, while trays contained none. In the analysis of storage conditions, it was found that 96% of the aluminum was used in food packaging stored at room temperature. While food packaging in the refrigerator contained 4% aluminum, no aluminum could be detected in packaging for frozen products. This is due to the fact that food in refrigerators and freezers is exposed to less moisture and light and requires less protection from flavor loss.

In the ML sample, only 5% (16) of the packaging examined could be recycled using conventional recycling methods. A package is defined as recyclable when all polymers can be correctly detected, and miscibility is present in the recycling process. This is only true for about 4% of the two-layer systems (13% of the ML sample) with the same outer and inner layer (PP–PP (5), LDPE–LDPE (4), and PET–PET (2)), and a proportion of about 2% (5) that despite different polymers in the outer and inner layers (PA–LDPE (4) and PP–LDPE (1)) can be recycled in a joint reprocessing process due to partial compatibility. For 9% (26) of the packaging, the outer or inner layer could not be identified, so an assessment of recyclability was not possible. This leaves a proportion of approx. 86% (254) that could not be recycled using conventional sorting methods. This is mainly due to a large number of layers with different materials that show no compatibility in the joint recycling process.

A comparison of the ML sample with the IR sample in terms of the identified polymeric multilayer combinations showed good consistency (see Table 8). Accordingly, the assessment of recyclability in Section 4.2 for the IR sample can be assumed for the proportion of conventionally unrecyclable fractions (86%) in the ML sample. This means that recycling processes or approaches for recycling the conventionally unrecyclable fractions already exist.

4.3.5. Limitations of the Study

The data collected in the present work are subject to limitations due to a variety of factors influencing the LWP generated. For example, seasonal variations were not considered, due to the collection period of the LWP (June–July). Furthermore, the mass fractions of packaging shown (Tables 4 and 7 and Figure 1) include residual content remaining in the packaging. In addition, a bias may have arisen due to the participants' knowledge of the subsequent characterization of their LWP waste or due to the participation of those who already had a strong awareness of the disposal of their LWP waste.

5. Outlook

LWP waste streams in the post-consumer sector are a complex mix of different, often contaminated, material types. In particular, the multilayer packaging contained in the waste stream poses major challenges to recycling companies, due to its limited detectability, sortability, and recyclability. At the same time, actors along the value chain have contrary interests. In the present work, a share of 33.16% of the multilayer packaging, divided into 17 different multilayer packaging solutions, could be identified (see Figure 1). This variety in multilayer packaging types is problematic because each must be fed into an appropriate reprocessing process (see Table 3).

Via the presented data, insights into the mass fraction, the recyclability, and suitable recycling processes for multilayer packages can be gained. Chemical recycling processes, which are criticized for generating particularly high CO_2 emissions, are often used. Here, the data provided can be used as input mass flows for the modeling of chemical recycling processes in the context of life cycle assessment analyses or cost calculations. In addition to the recyclability, the detection and sorting technologies represent a decisive aspect of the recyclability of packaging.

The problem of insufficient sortability in the waste management treatment of LWP still exists. Efficient sorting represents a key technology for the production of high-quality recycled polymers. The increasing digitalization of industrial processes also promises significant progress here, with further recorded information on the composition and type of waste available specifically for each item. For example, camera systems are increasingly being used in conjunction with machine learning algorithms to improve sortability. In some cases, the camera system is supplemented with other optical systems (NIR, VIS) [115]. The sorting task can include both a full sorting [116,117] of LWP streams and a sorting out of impurities [118] (silicone cartridges), the recognition of the brand, or the use of the stock-keeping unit on the packaging [117]. Insufficient sorting can, for example, lead to problems in the recycling process in the case of multilayer packaging containing aluminum, in line with the insufficient miscibility of polymers.

Flexible packaging containing aluminum is still considered by LWP waste sorters as a contaminant for recycled material and is the main cause of processing problems such as blockages of melt filters. In addition, aluminum in flexible packaging can lead to material losses during metal detection, as the metallic particles are sorted out of the line before the extruders and melt filters, in order to protect them [119]. Further, laminated and metalized aluminum leads to greying of the recyclate and is therefore not considered an optimal barrier material, but it is usually tolerated to some degree [119]. For example, AlOx coatings do not significantly affect the quality of the secondary materials, as they are typically only 1–10 nm thick and therefore often do not exceed the tolerable limit of a maximum of 5% of the total weight of the packaging structure [88]. Sorting solid metal objects, laminated films with solid aluminum layers, and laminated films with deposited aluminum layers into separate fractions would be desirable [119].

However, the creation of, e.g., object recognition systems would require a large and accurate amount of data. The additional packaging information collected in the present work, such as packaging type, filling material, or storage conditions, could contribute to a more optimal sorting of multilayer packaging, e.g., for items with aluminum content.

6. Conclusions

With the implementation of the European Strategy for Plastics in a Circular Economy in 2018 [9], the European Union set the course for the future achievement of a CE in the field of plastic packaging. This is urgently needed, as the recycling rate of plastic packaging in the EU is currently around 40% [87]. In the present work, the quantity and composition of multilayer packaging contained in the post-consumer waste stream was analyzed and identified as one of the problems limiting the achievement of future recycling rates, due to its limited recyclability.

Multilayer packaging enables tailor-made properties to protect a wide variety of packaged goods. Accordingly, multilayer packaging is selected based on the task to be fulfilled (protection against light, protective atmosphere, etc.), with practically no concern for its end-of-life recycling properties. The innovation speed in recycling processes does not keep up with the speed of packaging material innovations. Consequently, a lack of large-scale industrial sorting and recycling processes has led to the fact that multilayer packaging is not assigned to any specific sorting fraction but instead is dispersed into various recycling paths such as films, mixed plastics, or residual material. While residual materials, and in some cases mixed plastics, are recycled energetically, multilayer packaging represents a contaminant in the recycling of the film fraction, and in some cases also in mixed plastics, and thus it must be separated. The problem is based on the variety of polymers and non-polymeric materials used in food packaging and the differences in their specific melting temperatures or the overlap of melting and decomposition temperatures within the reprocessing process.

In overcoming these problems, the multilayer share of 33.16% in the LWP waste stream identified in this publication can not only be seen as showing the necessity for the further development of separation technologies but also as showing the potential to meet the increasing requirements of the Packaging Act (VerpackG: 63% by weight by 2022) for the recycling of plastic packaging and the potential for increasing quotas in the future.

Multilayer packaging in the area of post-consumer waste is a complex mixture of different types of materials, which are also contaminated. For the application of corresponding processes, the material composition of the multilayer packaging must be known, in order to feed it into the appropriate reprocessing process.

This problem was also recognized in this study, as IR spectrometry and microscopic examination of the outer and inner layers allowed only two-layer systems (13% of the packages in the ML sample) to undergo complete material determination.

In addition, there was also a share of 9% for which the outer or inner layer could not be identified. Furthermore, due to a multitude of different packaging solutions on the market and the sometimes too-small quantities of some multilayer packaging types, there are no economic processes for its recycling. Overall, the problem is not a lack of innovative approaches to recycling multilayer materials but rather their market maturity and industrial scale, which are not yet sufficient to recycle the current volumes of multilayers.

However, corresponding problems should not be addressed to the waste management sector alone, which, as the last link in the value chain, is often the focus of regulation and legislation. Insufficient consideration of other important factors within the areas of design, production, use, and disposal of LWP, accumulates along the value chain and makes it difficult to create a CE for plastic packaging, even with modern recycling facilities.

This underlines the fact that achieving future recycling rates, and especially a CE, will benefit from a diversity of different approaches and consideration of the entire value chain of LWP. Improvement could be provided by the introduction of innovative techniques and methods, the replacement of multilayer packaging with mono-material solutions, eco-design guidelines for distributors, the creation of a more transparent and harmonized system for all actors involved in the value chain of packaging, or legal requirements regarding standard packaging solutions per product category that would limit the possible material combinations to be managed.

To date, there are hardly any published detailed studies in the field of input analysis of LWP waste in Germany. This information gap hinders progress in waste management, as there are a large number of factors influencing the packaging waste generated, particularly in this area, which need to be investigated.

The next steps are to examine the factors influencing the packaging waste input in more detail. In particular, socio-demographic factors (e.g., household size, gender, level of education), the influence of rural or urban regions, the size of the municipality, the prevailing collection system, and the residual content remaining in the packaging could form the focus of investigations.

Author Contributions: Conceptualization, J.S., M.A. and J.W.; methodology, J.S., L.G., M.A., R.M. and J.W.; formal analysis, J.S., L.G. and J.W; investigation, J.S. and L.G.; writing—original draft preparation, J.S., writing—review and editing, J.S., M.A., R.M. and J.W.; visualization, J.S.; supervision, J.S. and J.W.; project administration, J.S. and J.W.; funding acquisition, J.W. All authors have read and agreed to the published version of the manuscript.

Funding: This research was funded by the German Federal Ministry for Education and Research (BMBF) as a part of the framework program "Research for Sustainable Development" (FONA3) on the topic "Plastics in the environment" with grants no. 033R195A-E under supervision of the project executing organization Jülich (PTJ).

Institutional Review Board Statement: Not applicable.

Informed Consent Statement: Not applicable.

Data Availability Statement: The data presented in this paper are used in the context of ongoing research projects. More in-depth data cannot yet be provided.

Conflicts of Interest: The authors declare no conflict of interest.

References

1. Lindner, C.; Hoffmann, O. Analyse/Beschreibung der Derzeitigen Situation der Stofflichen und Energetischen Verwertung von Kunststoffabfällen in Deutschland. 2015. Available online: https://docplayer.org/24614382-Endbericht-analyse-beschreibung-der-derzeitigen-situation-der-stofflichen-und-energetischen-verwertung-von-kunststoffabfaellen-in-deutschland.html (accessed on 3 February 2021).

2. Conversio GmbH. *Stoffstrombild der Kunststoffe in Deutschland 2019: Kurzfassung*; Conversio: Mainaschaff, Germany, 2020; Available online: https://www.vci.de/ergaenzende-downloads/kurzfassung-stoffstrombild-kunststoffe-2019.pdf (accessed on 2 May 2021).

3. Mellen, D.; Becker, T. Kunststoffe. In *Praxishandbuch der Kreislauf- und Rohstoffwirtschaft*; Kurth, P., Oexle, A., Faulstich, M., Eds.; Springer Fachmedien: Wiesbaden, Germany, 2018; pp. 327–345. ISBN 978-3-658-17044-8.

4. Schüler, K. *Aufkommen und Verwertung von Verpackungsabfällen in Deutschland im Jahr 2017 Abschlussbericht*; GVM: Mainz, Germany, 2019. Available online: https://www.umweltbundesamt.de/publikationen/aufkommen-verwertung-von-verpackungsabfaellen-in-12 (accessed on 3 February 2021).

5. Industrievereinigung Kunststoffverpackungen e.V. *Sustainability Report, 2018*; Industrievereinigung Kunststoffverpackungen e.V.: Bad Homburg vor der Höhe, Germany, 2018. Available online: http://kunststoffverpackungen.de/wp-content/uploads/2019/09/Sustainability-Report-2018 (accessed on 3 February 2021).

6. Dilkes-Hoffman, L.S.; Pratt, S.; Laycock, B.; Ashworth, P.; Lant, P.A. Public attitudes towards plastics. *Resour. Conserv. Recycl.* **2019**, *147*, 227–235. [CrossRef]

7. Pauer, E.; Tacker, M.; Gabriel, V.; Krauter, V. Sustainability of flexible multilayer packaging: Environmental impacts and recyclability of packaging for bacon in block. *Clean. Environ. Syst.* **2020**, *1*, 100001. [CrossRef]

8. Bundesministerium für Umwelt, Naturschutz, Nukleare Sicherheit und Verbraucherschutz. *Bundesgesetzblatt Teil I Nr. 45/Gesetz über das Inverkehrbringen, die Rücknahme und die hochwertige Verwertung von Verpackungen (Verpackungsgesetz–VerpackG)*; Bundesministerium für Umwelt, Naturschutz, Nukleare Sicherheit und Verbraucherschutz: Bonn, Germany, 2017.

9. European Commission. *A European Strategy for Plastics in a Circular Economy*; European Commission: Brussels, Belgium, 2018. Available online: https://eur-lex.europa.eu/resource.html?uri=cellar:2df5d1d2-fac7-11e7-b8f5-01aa75ed71a1.0001.02/DOC_1&format=PDF (accessed on 14 April 2022).

10. Blomsma, F.; Brennan, G. The Emergence of Circular Economy: A New Framing Around Prolonging Resource Productivity. *J. Ind. Ecol.* **2017**, *21*, 603–614. [CrossRef]

11. Johansen, M.R.; Christensen, T.B.; Ramos, T.M.; Syberg, K. A review of the plastic value chain from a circular economy perspective. *J. Environ. Manag.* **2022**, *302*, 113975. [CrossRef]

12. Ellen MacArthur Foundation. *The New Plastics Economy: Rethinking the Future of Plastics*; MacArthur Foundation: Cowes, UK, 2016.

13. Roux, M.; Varrone, C. Assessing the Economic Viability of the Plastic Biorefinery Concept and Its Contribution to a More Circular Plastic Sector. *Polymers* **2021**, *13*, 3883. [CrossRef]

14. Ragaert, K.; Delva, L.; van Geem, K. Mechanical and chemical recycling of solid plastic waste. *Waste Manag.* **2017**, *69*, 24–58. [CrossRef]
15. Faraca, G.; Astrup, T. Plastic waste from recycling centres: Characterisation and evaluation of plastic recyclability. *Waste Manag.* **2019**, *95*, 388–398. [CrossRef]
16. Ellen MacArthur Foundation. The New Plastics Economy: Catalysing Action. 2017. Available online: https://emf.thirdlight. com/link/u3k3oq221d37-h2ohow/@/preview/1?o (accessed on 10 April 2022).
17. Sassanelli, C.; Rosa, P.; Terzi, S. disassembly processes through simulation tools: A systematic literature review with a focus on printed circuit boards. *J. Manuf. Syst.* **2021**, *60*, 429–448. [CrossRef]
18. Picuno, C.; Alassali, A.; Chong, Z.K.; Kuchta, K. Flows of post-consumer plastic packaging in Germany: An MFA-aided case study. *Resour. Conserv. Recycl.* **2021**, *169*, 105515. [CrossRef]
19. Kaiser, K.; Schmid, M.; Schlummer, M. Recycling of Polymer-Based Multilayer Packaging: A Review. *Recycling* **2018**, *3*, 1. [CrossRef]
20. Wani, A.A.; Singh, P.; Langowski, H.-C. Introduction: Food Packaging Materials. In *Food Packing Materials: Testing & Quality Assurance*; Singh, P., Wani, A.A., Langowski, H.-C., Eds.; CRC Press: Boca-Rotan, FL, USA, 2017; pp. 1–9. ISBN 978-1-4665-5994-3.
21. Bishop, C.A.; Mount, E.M. Vacuum Metallizing for Flexible Packaging. In *Multilayer Flexible Packaging*; Elsevier: Amsterdam, The Netherlands, 2016; pp. 235–255. ISBN 9780323371001.
22. Gesellschaft für Verpackungsmarktforschung. *Entwicklung des Verpackungsverbrauchs Flexibler Kunststoffe nach Branchen: Auswertung des deutschen Marktes 2009, Prognose 2014*; Gesellschaft für Verpackungsmarktforschung: Mainz, Germany, 2010.
23. Tartakowski, Z. Recycling of packaging multilayer films: New materials for technical products. *Resour. Conserv. Recycl.* **2010**, *55*, 167–170. [CrossRef]
24. Anukiruthika, T.; Sethupathy, P.; Wilson, A.; Kashampur, K.; Moses, J.A.; Anandharamakrishnan, C. Multilayer packaging: Advances in preparation techniques and emerging food applications. *Compr. Rev. Food Sci. Food Saf.* **2020**, *19*, 1156–1186. [CrossRef] [PubMed]
25. Goulas, A. Overall migration from commercial coextruded food packaging multilayer films and plastics containers into official EU food simulants. *Eur. Food Res. Technol.* **2001**, *212*, 597–602. [CrossRef]
26. Dixon, J. *Packaging Materials: 9. Multilayer Packaging for Food and Beverages*; ILSI Europe: Brussels, Belgium, 2011; ISBN 9789078637264.
27. Butler, T.I.; Morris, B.A. PE-Based Multilayer Film Structures. In *Multilayer Flexible Packaging*; Elsevier: Amsterdam, The Netherlands, 2016; pp. 281–310. ISBN 9780323371001.
28. Morris, B.A. *The science and Technology of Flexible Packaging: Multilayer Films from Resin and Process to End Use*; William Andrew: Norwich, NY, USA, 2017; ISBN 978-0-323-24273-8.
29. Dahlbo, H.; Poliakova, V.; Mylläri, V.; Sahimaa, O.; Anderson, R. Recycling potential of post-consumer plastic packaging waste in Finland. *Waste Manag.* **2018**, *71*, 52–61. [CrossRef]
30. Tarantili, P.A.; Kiose, V. Effect of accelerated aging on the structure and properties of monolayer and multilayer packaging films. *J. Appl. Polym. Sci.* **2008**, *109*, 674–682. [CrossRef]
31. Bauer, A.-S.; Tacker, M.; Uysal-Unalan, I.; Cruz, R.M.S.; Varzakas, T.; Krauter, V. Recyclability and Redesign Challenges in Multilayer Flexible Food Packaging-A Review. *Foods* **2021**, *10*, 2702. [CrossRef]
32. Horodytska, O.; Valdés, F.J.; Fullana, A. Plastic flexible films waste management-A state of art review. *Waste Manag.* **2018**, *77*, 413–425. [CrossRef]
33. Riedl, F. Recyclingherausforderung Multi-Layer? Neuartiges Aufbereitungsverfahren bietet Lösungen. In *Vorträge-Konferenzband zur 14. Recy & DepoTech-Konferenz: Tracer Based Sorting–Innovative Sorting Options for Post Consumer Products*; Pomberger, R., Adam, J., Aldrian, A., Kranzinger, L., Lorber, K., Neuhold, S., Nigl, T., Pfandl, K., Sarc, R., Schwarz, T., et al., Eds.; Abfallverwertungstechnik & Abfallwirtschaft Eigenverlag: Leoben, Austria, 2018; pp. 269–274. ISBN 9783200058743.
34. Deutsches Institut für Normung. *Kennzeichnung von Packstoffen und Packmitteln-Packstoffe und Packmittel aus Kunststoff*; DIN 6120:2019-03; Beuth Verlag GmbH: Berlin, Germany, 2019.
35. Clauß, D. Abfallmenge und Abfallzusammensetzung. In *Einführung in die Kreislaufwirtschaft*; Kranert, M., Ed.; Springer Fachmedien: Wiesbaden, Germany, 2017; pp. 65–110. ISBN 978-3-8348-1837-9.
36. Acerbi, F.; Sassanelli, C.; Terzi, S.; Taisch, M. A Systematic Literature Review on Data and Information Required for Circular Manufacturing Strategies Adoption. *Sustainability* **2021**, *13*, 2047. [CrossRef]
37. Chiappetta Jabbour, C.J.; Fiorini, P.D.C.; Ndubisi, N.O.; Queiroz, M.M.; Piato, É.L. Digitally-enabled sustainable supply chains in the 21st century: A review and a research agenda. *Sci. Total Environ.* **2020**, *725*, 138177. [CrossRef]
38. Gupta, S.; Chen, H.; Hazen, B.T.; Kaur, S.; Santibañez Gonzalez, E.D.R. Circular economy and big data analytics: A stakeholder perspective. *Technol. Forecast. Soc. Chang.* **2019**, *144*, 466–474. [CrossRef]
39. Garcia, J.M.; Robertson, M.L. The future of plastics recycling. *Science* **2017**, *358*, 870–872. [CrossRef]
40. Rosa, P.; Sassanelli, C.; Urbinati, A.; Chiaroni, D.; Terzi, S. Assessing relations between Circular Economy and Industry 4.0: A systematic literature review. *Int. J. Prod. Res.* **2020**, *58*, 1662–1687. [CrossRef]
41. Sarc, R.; Hermann, R. Unternehmensbefragung zum Thema Abfallwirtschaft 4.0. In *Vorträge-Konferenzband zur 14. Recy & DepoTech-Konferenz: Tracer Based Sorting–Innovative Sorting Options for Post Consumer Products*; Pomberger, R., Adam, J., Aldrian,

A., Kranzinger, L., Lorber, K., Neuhold, S., Nigl, T., Pfandl, K., Sarc, R., Schwarz, T., et al., Eds.; Abfallverwertungstechnik & Abfallwirtschaft Eigenverlag: Leoben, Austria, 2018; pp. 805–812. ISBN 9783200058743.

42. Sarc, R.; Curtis, A.; Khodier, K.; Koinegg, J.; Ortner, M. Digitale Abfallwirtschaft. In *Vorträge-Konferenzband zur 14. Recy & DepoTech-Konferenz: Tracer Based Sorting–Innovative Sorting Options for Post Consumer Products*; Pomberger, R., Adam, J., Aldrian, A., Kranzinger, L., Lorber, K., Neuhold, S., Nigl, T., Pfandl, K., Sarc, R., Schwarz, T., et al., Eds.; Abfallverwertungstechnik & Abfallwirtschaft Eigenverlag: Leoben, Austria, 2018; pp. 793–798. ISBN 9783200058743.

43. Sarc, R.; Pomberger, R. „ReWaste4.0"–Abfallwirtschaftliches Kompetenzzentrum am AVAW der Montanuniversität Leoben. *Osterr. Wasser Abfallwirtsch.* **2022**, *74*, 39–50. [CrossRef]

44. Rocca, R.; Rosa, P.; Sassanelli, C.; Fumagalli, L.; Terzi, S. Integrating Virtual Reality and Digital Twin in Circular Economy Practices: A Laboratory Application Case. *Sustainability* **2020**, *12*, 2286. [CrossRef]

45. Kirchherr, J.; Reike, D.; Hekkert, M. Conceptualizing the circular economy: An analysis of 114 definitions. *Resour. Conserv. Recycl.* **2017**, *127*, 221–232. [CrossRef]

46. Schmidt, J.; Auer, M.; Moesslein, J.; Wendler, P.; Wiethoff, S.; Lang-Koetz, C.; Woidasky, J. Challenges and Solutions for Plastic Packaging in a Circular Economy. *Chem. Ing. Tech.* **2021**, *9*, 105. [CrossRef]

47. Langhe, D.; Ponting, M. Coextrusion Processing of Multilayered Films. In *Manufacturing and Novel Applications of Multilayer Polymer Films*; Ponting, M., Langhe, D., Eds.; William Andrew Publishing: Norwich, NY, USA, 2016; pp. 16–45. ISBN 9780323371254.

48. Häsänen, E. Composition Analysis and Compatibilization of Post-Consumer Recycled Multilayer Plastic Films. Master's Thesis, Tampere University of Technology, Tampere, Finland, 2016.

49. Franz, R. Migration of plastic constituents. In *Plastic Packaging Materials for Food: Barrier Function, Mass Transport, Quality Assurance, and Legislation*; Piringer, O.-G., Baner, A.L., Eds.; Wiley-VCH Verlag GmbH: Weinheim, Germany, 2000; pp. 287–357. ISBN 9783527613281.

50. Ajitha, A.R.; Aswathi, M.K.; Maria, H.J.; Izdebska, J.; Thomas, S. Multilayer Polymer Films. In *Multicomponent Polymeric Materials*; Kim, C.-k., Thomas, S., Saha, P., Eds.; Springer: Dordrecht, The Netherlands, 2016; pp. 229–258. ISBN 9789401773232.

51. Gutoff, E.B.; Cohen, E.D. Water- and Solvent-Based Coating Technology. In *Multilayer Flexible Packaging*; Elsevier: Amsterdam, The Netherlands, 2016; pp. 205–234. ISBN 9780323371001.

52. Mariam, M. Charakterisierung von Verbundfolien zur Evaluierunng von Recycling Potentialen. Ph.D. Thesis, Technische Universität Wien, Vienna, Austria, 2020.

53. Barlow, C.Y.; Morgan, D.C. Polymer film packaging for food: An environmental assessment. *Resour. Conserv. Recycl.* **2013**, *78*, 74–80. [CrossRef]

54. European Commission. *Plastics: Reuse, Recycling and Marine Litter: Final Report*; Publications Office of the European Union: Luxembourg, 2018; ISBN 978-92-79-93917-4.

55. Woidasky, J. Plastics Recycling. In *Ullmann's Encyclopedia of Industrial Chemistry*; Wiley: Hoboken, NJ, USA, 2000; pp. 1–29. ISBN 9783527303854.

56. Xiao, K.; Zatloukal, M. Multilayer Die Design and Film Structures. In *Film Processing Advances*; Kanai, T., Campbell, G.A., Eds.; Hanser Publications: Munich, Germany, 2014; pp. 68–102. ISBN 9781569905364.

57. Vera, P.; Canellas, E.; Nerín, C. Compounds responsible for off-odors in several samples composed by polypropylene, polyethylene, paper and cardboard used as food packaging materials. *Food Chem.* **2020**, *309*, 125792. [CrossRef]

58. Ashter, S.A. Matching Material Characteristics to Commercial Thermoforming. In *Thermoforming of Single and Multilayer Laminates*; Elsevier: Amsterdam, The Netherlands, 2014; pp. 193–209. ISBN 9781455731725.

59. Goetz, W. Polyamide for Flexible Packaging Film. Available online: https://www.tappi.org/content/enewsletters/eplace/2004/10-2goetz.pdf (accessed on 18 February 2021).

60. Marsh, K.; Bugusu, B. Food packaging–Roles, materials, and environmental issues. *J. Food Sci.* **2007**, *72*, R39–R55. [CrossRef]

61. Fereydoon, M.; Ebnesajjad, S. Development of High-Barrier Film for Food Packaging. In *Plastic Films in Food Packaging*; Elsevier: Amsterdam, The Netherlands, 2013; pp. 71–92. ISBN 9781455731121.

62. Becker, W.; Sachsenheimer, K.; Klemenz, M. Detection of Black Plastics in the Middle Infrared Spectrum (MIR) Using Photon Up-Conversion Technique for Polymer Recycling Purposes. *Polymers* **2017**, *9*, 435. [CrossRef]

63. Burns, D.A.; Ciurczak, E.W. *Handbook of Near-Infrared Analysis*; CRC Press: Boca Raton, FL, USA, 2007; ISBN 9780429123016.

64. Habich, U.; Beel, H. Modifizierung von Recyclingverfahren durch sensorbasierte Sortierung. In *Recycling und Rohstoffe*; Thomé-Kozmiensky, K.J., Goldmann, D., Eds.; TK Verlag: Neuruppin, Germany, 2014; pp. 471–482. ISBN 978-3-944310-09-1.

65. de Biasio, M.; Arnold, T.; McGunnigle, G.; Leitner, R.; Balthasar, D.; Rehrmann, V. Detecting and discriminating PE and PP polymers for plastics recycling using NIR imaging spectroscopy. In Proceedings of the SPIE Defense, Security, and Sensing, Thermosense XXXII, Orlando, FL, USA, 5 April 2010; Dinwiddie, R.B., Safai, M., Eds.; SPIE: Bellingham, WA, USA, 2010; Volume 7661.

66. Küppers, B.; Vollprecht, D.; Pomberger, R. Einfluss von Verschmutzungen auf die sensorgestützte Sortierung. In *Vorträge-Konferenzband zur 14. Recy & DepoTech-Konferenz: Tracer Based Sorting–Innovative Sorting Options for Post Consumer Products*; Pomberger, R., Adam, J., Aldrian, A., Kranzinger, L., Lorber, K., Neuhold, S., Nigl, T., Pfandl, K., Sarc, R., Schwarz, T., et al., Eds.; Abfallverwertungstechnik & Abfallwirtschaft Eigenverlag: Leoben, Austria, 2018; pp. 111–118. ISBN 9783200058743.

67. Hopewell, J.; Dvorak, R.; Kosior, E. Plastics recycling: Challenges and opportunities. *Philos. Trans. R. Soc. Lond. B Biol. Sci.* **2009**, *364*, 2115–2126. [CrossRef] [PubMed]

68. Kusch, A.; Gasde, J.; Deregowski, C.; Woidasky, J.; Lang-Koetz, C.; Viere, T. Sorting and Recycling of Lightweight Packaging in Germany—Climate Impacts and Options for Increasing Circularity Using Tracer-Based-Sorting. *Mater. Circ. Econ.* **2021**, *3*, 125. [CrossRef]
69. Brunner, S.; Fomin, P.; Zhelondz, D.; Kargel, C. Investigation of algorithms for the reliable classification of fluorescently labeled plastics. In Proceedings of the 2012 IEEE International Instrumentation and Measurement Technology Conference (I2MTC), Graz, Austria, 13–16 May 2012; IEEE: Piscataway, NJ, USA, 2012; pp. 1659–1664, ISBN 978-1-4577-1772-7.
70. Nonclercq, A. *Mapping Flexible Packaging in a Circular Economy [F.I.A.C.E]: Final Report*; Delft University of Technology: Delft, The Netherlands, 2016.
71. Briedis, R.; Syversen, F. *Plastic Packaging Recyclability in a Nordic Context*; Nordic Council of Ministers: Copenhagen, Denmark, 2019; ISBN 9789289362399.
72. Christiani, J.; Beckamp, S. Was können die mechanische Aufbereitung von Kunststoffen und das werkstoffliche Recycling leisten. In *Energie aus Abfall*; Thiel, S., Thomé-Kozmiensky, E., Quicker, P., Gosten, A., Eds.; Thomé-Kozmiensky Verlag GmbH: Neuruppin, Germany, 2020; pp. 139–152. ISBN 9783944310503.
73. Hahladakis, J.N.; Iacovidou, E. Closing the loop on plastic packaging materials: What is quality and how does it affect their circularity. *Sci. Total Environ.* **2018**, *630*, 1394–1400. [CrossRef] [PubMed]
74. Vilaplana, F.; Karlsson, S. Quality Concepts for the Improved Use of Recycled Polymeric Materials: A Review. *Macromol. Mater. Eng.* **2008**, *293*, 274–297. [CrossRef]
75. DerGrünePunkt. Produktspezifikation 03/2018 Fraktions-Nr. 325. Available online: https://www.gruener-punkt.de/fileadmin/Dateien/Downloads/PDFs/spezifikationen/325_PET-Flaschen_-_transparent.pdf (accessed on 10 March 2021).
76. DerGrünePunkt. Produktspezifikation 03/2018 Fraktions-Nr. 310-1. Available online: https://www.gruener-punkt.de/fileadmin/Dateien/Downloads/PDFs/spezifikationen/310-1_Kunststoff-Folien.pdf (accessed on 19 January 2022).
77. Ehrenstein, G.W. *Polymer Werkstoffe: Struktur Eigenschaften Anwendung, 3. Auflage*; Hanser Verlag: München, Germany, 2011; ISBN 978-3-446-42283-4.
78. Bonnet, M. *Kunststoffe in der Ingenieuranwendung: Verstehen und Zuverlässig Auswählen, 1. Aufl.*; Vieweg + Teubner Verlag/GWV Fachverlage: Wiesbaden, Germany, 2009; ISBN 9783834803498.
79. Jönkkäri, I.; Poliakova, V.; Mylläri, V.; Anderson, R.; Andersson, M.; Vuorinen, J. Compounding and characterization of recycled multilayer plastic films. *J. Appl. Polym. Sci.* **2020**, *137*, 49101. [CrossRef]
80. Nickel, W. *Recycling-Handbuch*; Springer: Berlin/Heidelberg, Germany, 1996; ISBN 978-3-642-95769-7.
81. Pilz, H.; Brandt, B.; Fehringer, R. *The Impact of Plastics on Life Cycle Energy Consumption and Greenhouse Gas Emissions in Europe*; PlasticsEurope: Vienna, Austria, 2010.
82. Knappe, F.; Reinhardt, J.; Kauertz, B.; Oetjen-Dehne, R.; Buschow, N.; Ritthoff, M.; Wilts, H.; Lehmann, M. *Technische Potenzialanalyse zur Steigerung des Kunststoffrecyclings und des Rezyklateinsatzes*; Wuppertal Institut für Klima, Umwelt, Energie: Wuppertal, Germany, 2021.
83. Elsner, P.; Müller-Kirschbaum, T.; Schweitzer, K.; Wolf, R.; Seiler, E.; Désilets, P.; Detsch, R.; Dornack, C.; Ferber, J.; Fleck, C.; et al. *Kunststoffverpackungen im Geschlossenen Kreislauf–Potenziale, Bedingungen, Herausforderungen*; Acatech: Munich, Germany, 2021.
84. Berg, H.; Kulinna, R.; Stöcker, C.; Guth-Orlowski, S.; Thiermann, R.; Porepp, N. *Overcoming Information Asymmetry in the Plastics Value Chain with Digital Product Passports*; Wuppertal Institut für Klima, Umwelt, Energie: Wuppertal, Germany, 2022. Available online: https://epub.wupperinst.org/frontdoor/index/index/docId/7940 (accessed on 14 April 2022).
85. Salmenperä, H.; Pitkänen, K.; Kautto, P.; Saikku, L. Critical factors for enhancing the circular economy in waste management. *J. Clean. Prod.* **2021**, *280*, 124339. [CrossRef]
86. Balwada, J.; Samaiya, S.; Mishra, R.P. Packaging Plastic Waste Management for a Circular Economy and Identifying a better Waste Collection System using Analytical Hierarchy Process (AHP). *Procedia CIRP* **2021**, *98*, 270–275. [CrossRef]
87. European Commission. *Closing the Loop–An EU Action Plan for the Circular Economy*; European Commission: Brussels, Belgium, 2015. Available online: https://eur-lex.europa.eu/resource.html?uri=cellar:8a8ef5e8-99a0-11e5-b3b7-01aa75ed71a1.0012.02/DOC_1&format=PDF (accessed on 14 April 2022).
88. CEFLEX. Designing for a Circular Economy: Recyclability of Polyolefin-Based Flexible Packaging. 2020. Available online: https://guidelines.ceflex.eu/resources/ (accessed on 27 January 2022).
89. Woidasky, J.; Schmidt, J.; Auer, M.; Sander, I.; Schau, A.; Moesslein, J.; Wendler, P.; Kirchenbauer, D.; Wacker, D.; Gao, G.; et al. Photoluminescent Tracer Effects on Thermoplastic Polymer Recycling. In *Advances in Polymer Processing 2020*; Hopmann, C., Dahlmann, R., Eds.; Springer: Berlin/Heidelberg, Germany, 2020; pp. 1–13. ISBN 978-3-662-60808-1.
90. Brunner, S.; Fomin, P.; Kargel, C. Automated sorting of polymer flakes: Fluorescence labeling and development of a measurement system prototype. *Waste Manag.* **2015**, *38*, 49–60. [CrossRef]
91. Ahmad, S.R. A new technology for automatic identification and sorting of plastics for recycling. *Environ. Technol.* **2004**, *25*, 1143–1149. [CrossRef]
92. AIM-European Brands Association. Pioneering Digital Watermarks for Accurate Sorting and High Quality Recycling–HolyGrail 2.0. Available online: https://www.aim.be/priorities/digital-watermarks/ (accessed on 16 April 2021).
93. Meys, R.; Frick, F.; Westhues, S.; Sternberg, A.; Klankermayer, J.; Bardow, A. Towards a circular economy for plastic packaging wastes–The environmental potential of chemical recycling. *Resour. Conserv. Recycl.* **2020**, *162*, 105010. [CrossRef]

94. Agulla, K. Circular Packaging–Bau einer Industriellen Demonstrationsanlage für das Recycling von Kunststoffverpackungen; Fraunhofer IVV, 14 January 2019. Available online: https://www.ivv.fraunhofer.de/de/presseinformationen/circular-packaging.html (accessed on 26 January 2022).

95. CreaCycle GmbH. CreaSolv®Demonstrationsanlage für Kunststoff Verpackungsabfälle–Lober. 2018. Available online: https://www.creacycle.de/de/creasolv-werke/circular-packaging-2018.html (accessed on 26 January 2022).

96. Schlummer, M.; Fell, T.; Mäurer, A.; Altnau, G. Die Rolle der Chemie beim Recycling: Physikalisches und chemisches Kunststoffrecycling im Vergleich. *Kunststoffe* **2020**, *6*, 51–54.

97. Vollmer, I.; Jenks, M.J.F.; Roelands, M.C.P.; White, R.J.; Harmelen, T.; Wild, P.; Laan, G.P.; Meirer, F.; Keurentjes, J.T.F.; Weckhuysen, B.M. Die nächste Generation des Recyclings–neues Leben für Kunststoffmüll. *Angew. Chem.* **2020**, *132*, 15524–15548. [CrossRef]

98. Lovis, F.; Seibt, H.; Kernbaum, S. Method and Apparatus for Recycling Packaging Material. U.S. Patent No. 10,682,788, 16 June 2020.

99. Wohnig, K.; Kaina, M.; Fleig, M.; Hanel, H. Solvent and Method for Dissolving at Least Two Plastics from a Solid within a Suspension. Patent No. DE.102.016.015.199.A1, 21 June 2018.

100. Saperatec GmbH. Homepage: Applications. Available online: https://www.saperatec.de/en/technology.html (accessed on 19 January 2022).

101. Purecycle. News-Seite. Available online: https://purecycle.com/2021/11/purecycle-technologies-provides-third-quarter-2021-update/ (accessed on 26 January 2022).

102. Walker, T.W.; Frelka, N.; Shen, Z.; Chew, A.K.; Banick, J.; Grey, S.; Kim, M.S.; Dumesic, J.A.; van Lehn, R.C.; Huber, G.W. Recycling of multilayer plastic packaging materials by solvent-targeted recovery and precipitation. *Sci. Adv.* **2020**, *6*, eaba7599. [CrossRef] [PubMed]

103. Georgiopoulou, I.; Pappa, G.D.; Vouyiouka, S.N.; Magoulas, K. Recycling of post-consumer multilayer Tetra Pak® packaging with the Selective Dissolution-Precipitation process. *Resour. Conserv. Recycl.* **2021**, *165*, 105268. [CrossRef]

104. BASF. Chemical Recycling of Plastic Waste. Available online: https://www.basf.com/global/de/who-we-are/sustainability/we-drive-sustainable-solutions/circular-economy/mass-balance-approach/chemcycling.html (accessed on 21 January 2022).

105. Garbo. Homepage. Available online: https://garbo.it/en/chempet/ (accessed on 21 January 2022).

106. Kaiser, K.M.A. Recycling of multilayer packaging using a reversible cross-linking adhesive. *J. Appl. Polym. Sci.* **2020**, *137*, 49230. [CrossRef]

107. Engel, D.W.; Dalton, A.C.; Anderson, K.K.; Sivaramakrishnan, C.; Lansing, C. Development of Technology Readiness Level (TRL) Metrics and Risk Measures. 2012. Available online: https://www.osti.gov/biblio/1067968 (accessed on 26 January 2022).

108. European Commission. A New Method for Separation Full Recovery of Multilayer Packaging Waste to Create High Value Materials—LAMPACK. Available online: https://cordis.europa.eu/project/id/736010 (accessed on 22 February 2021).

109. Schmidt, J.; Auer, M. Analyse von Leichtverpackungsabfällen aus deutschen Haushalten. In *Markerbasiertes Sortier und Recyclingsystem für Kunststoffverpackungen: Schlussbericht des BMBF-Forschungsvorhabens "MaReK". Förderkennzeichen: 033R195A bis E*; Technische Informationsbibliothek: Pforzheim, Germany, 2021; in press.

110. Berndt, D.; Sellschopf, L. Packstoffe, Packmittel und Packhilfsmittel. In *Grundlagen der Verpackung: Leitfaden für die Fächerübergreifende Verpackungsausbildung, 1. Aufl.*; Kaßmann, M., Ed.; Beuth: Berlin, Germany, 2011; pp. 19–96. ISBN 978-3410204923.

111. Duale Systeme Deutschland. *Neue Mülltrennungsstudie: Die Deutschen Brauchen Nachhilfe im Mülltrennen*; Köln, Germany, 2020. Available online: https://www.muelltrennung-wirkt.de/neue-muelltrennungsstudie-die-deutschen-brauchen-nachhilfe-im-muelltrennen/ (accessed on 9 April 2021).

112. Conversio GmbH. Material Flow Analysis Plastics in Germany 2019; Conversio: Mainaschaff, Germany. 2020. Available online: https://www.bkv-gmbh.de/files/bkv-neu/studien/Summary_Material_Flow_Analysis_Plastics_Germany_2019_EN.pdf (accessed on 8 June 2021).

113. Matthews, C.; Moran, F.; Jaiswal, A.K. A review on European Union's strategy for plastics in a circular economy and its impact on food safety. *J. Clean. Prod.* **2021**, *283*, 125263. [CrossRef]

114. Partridge, C.; Medda, F. Opportunities for chemical recycling to benefit from waste policy changes in the United Kingdom. *Resour. Conserv. Recycl. X* **2019**, *3*, 100011. [CrossRef]

115. Baldt, T. Robotersortierlösung von ZenRobotics. In *Vorträge-Konferenzband zur 14. Recy & DepoTech-Konferenz: Tracer Based Sorting–Innovative Sorting Options for Post Consumer Products*; Pomberger, R., Adam, J., Aldrian, A., Kranzinger, L., Lorber, K., Neuhold, S., Nigl, T., Pfandl, K., Sarc, R., Schwarz, T., et al., Eds.; Abfallverwertungstechnik & Abfallwirtschaft Eigenverlag: Leoben, Austria, 2018; pp. 695–700. ISBN 9783200058743.

116. ZenRobotics. Fast Picker: High-Speed Robot for Maximizing Material Recovery. Available online: https://zenrobotics.com/solutions/fast-picker/ (accessed on 5 May 2021).

117. Machinex. SamurAI. Available online: https://www.machinexrecycling.com/samurai/ (accessed on 5 May 2021).

118. Steinert. UniSort PR EVO 5.0: NIR-Sortieraggregat mit Hyper Spectral Imaging-Kameratechnik. State of the Art–Kamerabasierte NIR-Technologie für Noch bessere Sortierergebnisse. Available online: https://steinertglobal.com/de/magnete-sensorsortierer/sensorsortierung/nir-sortiersysteme/unisort-pr/ (accessed on 5 May 2021).

119. van Velzen, U.T.; de Weert, L.; Molenveld, K. *Flexible Laminates within the Circular Economy*; Wageningen University & Research: Wageningen, The Netherlands, 2020.

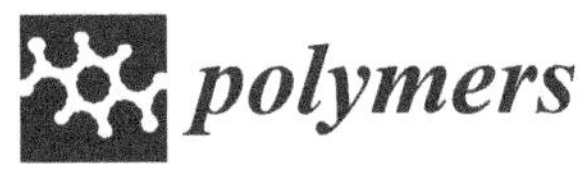

Article

Pulverization of Waste Polyvinyl Chloride (PVC) Film by Low Temperature Heat Treatment and Properties of Pulverized Product for Blast Furnace Injection as Alternative Fuel

Guang Wang *, Sixian Liu, Hongqiang Zhang *, Jingsong Wang and Qingguo Xue

State Key Laboratory of Advanced Metallurgy, University of Science and Technology Beijing, Beijing 100083, China; liusixian2022@126.com (S.L.); wangjingsong@ustb.edu.cn (J.W.); xueqingguo@ustb.edu.cn (Q.X.)
* Correspondence: wangguang@ustb.edu.cn (G.W.); zhanghongqiangustb@163.com (H.Z.)

Abstract: Recycling of waste plastics is of great significance for human society. The pulverization of waste film plastics is a key technical link in the development of collaborative utilization of waste plastics in the steel industry. In this study, waste polyvinyl chloride film plastics were first heated at different temperatures; then the de-chlorination ratio pulverization and the properties of the pulverized products closely related to blast furnace injection, such as powdery properties, combustion and explosiveness, were further analyzed. The weight loss ratio increased significantly with an increase in temperature and was not obvious between 370 °C and 400 °C. The highest de-chlorination ratio was approximately 84% at 370 °C, and the relative chlorine content in the product was 9%. The crushing performance of heat-treated polyvinyl chloride film increased with increasing temperature. Before 370 °C, there were more pores in the samples, and the surface of the sample seemed to be damaged with the temperature was further increased. The pulverized polyvinyl chloride had better fluidity and strong jet flow compared to industrial injection coals. At the same time, compared with other carbonaceous materials, it also exhibited better combustion performances. The pulverized polyvinyl chloride belonged to non explosiveness substance despite its high volatile content. The obtained results demonstrated that the pulverized polyvinyl chloride obtained under the present conditions could be used for blast furnace injection to some extent.

Keywords: waste polyvinyl chloride (PVC); pulverization; heat treatment; de-chlorination; blast furnace injection

Citation: Wang, G.; Liu, S.; Zhang, H.; Wang, J.; Xue, Q. Pulverization of Waste Polyvinyl Chloride (PVC) Film by Low Temperature Heat Treatment and Properties of Pulverized Product for Blast Furnace Injection as Alternative Fuel. *Polymers* **2022**, *14*, 1689. https://doi.org/10.3390/polym14091689

Academic Editors: Sheila Devasahayam, Raman Singh and Vladimir Strezov

Received: 21 March 2022
Accepted: 18 April 2022
Published: 21 April 2022

Publisher's Note: MDPI stays neutral with regard to jurisdictional claims in published maps and institutional affiliations.

1. Introduction

Polyvinyl chloride (PVC) is the second-most produced thermoplastic by volume, after polyethylene. It has the characteristics of easy processing, wear resistance, acid and alkali resistance, flame retardancy, and excellent electrical insulation. Therefore, it is widely used in pipes, window framing, floor coverings, roofing sheets, and cables [1,2]. Due to the strong demand for PVC, China's PVC production capacity has maintained an annual growth ratio of 20% since 2000. Notably, it is expected that by the end of 2050, the cumulative PVC waste in the environment will exceed 600 million tons in China [3,4].

In recent years, the question of the disposal of PVC waste has gained increasing importance in the public discussion. At present, there are four commonly used PVC waste processing technologies: mechanical recycling, landfilling, incineration, and chemical recycling [5]. The mechanical recycling method involves directly using PVC waste plastics after simple pretreatment (such as collection, sorting, washing, and grinding of the material) or mixing them with other polymers to produce blends. This method is simple and feasible, but requires high quality waste plastics [6]. Landfill treatment is common, but landfilled waste plastics can cause serious problems, such as land occupation, soil structure damage, environmental pollution, and the loss of chemical calorific value of waste plastics [7]. Incineration of waste PVC produces a large amount of HCl, which can damage the incineration equipment, and increase the investment and operation costs of the incineration and disposal process. In addition, PVC waste incineration will inevitably produce dioxins and other toxic gases, which will result in environmental pollution [8]. Chemical recycling is the conversion of PVC back into shorter chains for reuse in petrochemical or polymerization processes following cracking, gasification, hydrogenation, or pyrolysis [2]. Compared with other waste plastic processing technologies, this type of recycling has high potential for heterogeneous and contaminated plastic waste material, where separation is either not economically viable or not completely technically feasible [6].

The fossil fuel energy-intensive blast furnace ironmaking process of steel industry operates at high temperature and under high reduction potential with the function of energy conversion, which can provide an easier path for the collaborative utilization of waste plastics in large quantity and low cost. Up to the present, the application of waste plastics in the iron and steel industries has been studied extensively. In 1996, the steel plant in Germany achieved blast furnace injection of waste plastics for the first time in the world [9]. After sorting and removing harmful impurities, the waste plastics were pulverized into plastic particles (smaller than 10 mm) and injected into the blast furnace having an injection capacity of 70,000 tons per year. Japan first implemented blast furnace injection of waste plastics in the Keihin plant. In this method, the chlorine-containing plastics were removed in advance and the rest was crushed and granulated (the maximum particle size was approximately 6 mm) for being injected into the blast furnace together with hot air. The experimental injection amount was as high as 200 kg/tHM [10].

Whether waste plastics are used for blast furnace injection or gasification reactions to produce gas fuel, the particle size of waste plastics is very important. Particle size affects the reaction rate and conversion efficiency by affecting the mass and heat transfer between the particles [11]. Different processes have different requirements of the particle size of the raw material. The blast furnace requires that 80% of the total mass of injection coal should a particle size smaller than 0.074 mm. However, a smaller particle size increases the production cost and technical difficulty. Asanuma et al. [12] pulverized a mixture of various plastics into 0.2–0.4 mm through heat treatment, however, they didn't pay attention to the behavior of PVC plastics. Wang et al. [13] obtained low chlorine hydrochar from PVC by hydrothermal carbonization; however, the hydrothermal treatment equipment of PVC is significantly eroded by the formed HCl, as it is in the solution state.

To the best of our knowledge, few studies have reported the dry pulverization of waste PVC film and the application characteristics of the pulverized product in the blast furnace ironmaking process. In the present research, waste PVC film plastics were pulverized by heat treatment at different temperatures without using water, and the de-chlorination ratio and properties of the pulverized products closely related to blast furnace injection as solid fuel were further analyzed. The process not only help efficiently use the PVC waste plastics resources in the municipal solid waste, but also avoids bringing a large number of harmful substances into the next process.

2. Materials and Methods

2.1. Raw Materials

The PVC film plastic used in this experiment was a decoration material, also known as PVC foam board, widely used in model making and advertising. The thickness was 0.1 cm, and the chemical composition is shown in Table 1. Before the experiment, the PVC film was cut into pieces of about 2.0 cm × 2.0 cm for subsequent use. The coke, graphite, and anthracite were crushed into the size smaller than 180 μm and dried at 100 °C for 10 h for subsequent use.

Table 1. Proximate and elemental analysis of raw materials.

Material	Proximate Analysis (wt %)			Elemental Analysis (wt %)					
	V_d	A_d	FC_d	C	H	O	N	S	Cl
PVC film	70.10	16.57	13.33	33.67	2.68	17.19	<0.3	0.61	33.09
Coke	1.64	12.16	86.20	83.81	1.66	3.85	0.62	0.50	-
Graphite	-	-	100	-	-	-	-	-	-
Anthracite	6.4	11.1	81.4	77.71	1.21	8.19	0.55	1.12	-

Note: V_d was the volatile in dry basis, A_d, was the ash in dry basis, FC_d was fixed carbon in dry basis.

2.2. Experimental Methods

First, the PVC film pieces were subjected to low temperature heat treatment in N_2 atmosphere; then, the cooled heat-treated products were ground and sieved. The macroscopic morphology, microstructure, de-chlorination ratio, and properties for blast furnace injection of the pulverized heat treatment products were studied.

2.2.1. Low Temperature heat Treatment Experiment

A total of 10 g of PVC film were placed in stainless steel crucibles. The crucible was a cylinder having a diameter of 5 cm and height of 9 cm. The heating equipment was a tubular shaft furnace, and its heating element were U-shaped $MoSi_2$ rods, which were arranged around the furnace. During the test, high purity N_2 flow at 3 L/min was introduced from the bottom of the furnace tube to prevent oxidation in the furnace. The gas outlet from the top of the furnace tube was connected to the tail gas filtering device to completely absorb the HCl from the tail gas. The schematic of the setup is shown in Figure 1. Before the experiments, the furnace temperature was set to a fixed temperature required for each run (280, 310, 340, 370, 400, 430, and 460 °C). When the furnace reached the desired temperature, the stainless steel crucible containing the PVC film pieces was placed into the furnace and heated for 30 min.

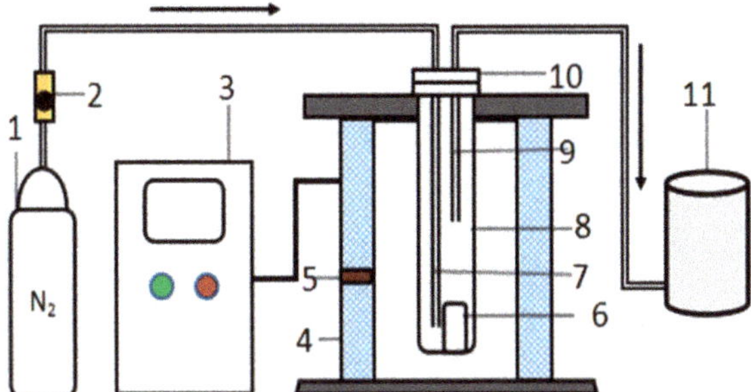

Figure 1. Schematic diagram of the experimental setup. 1: gas cylinder, 2: flowmeter, 3: control cabinet, 4: furnace body, 5: heating element; 6: crucible; 7: gas inlet, 8: furnace tube, 9: gas outlet, 10: flange, 11: tail gas washing tank.

2.2.2. Pulverization and Screening

After heat treatment, the PVC film samples were cooled to room temperature in an N_2 atmosphere. The weight of the PVC film products was recorded before pulverization to calculate the weight loss ratio. The crusher crushed the sample continuously with a rotating speed of 20,000 rpm. The heat treatment products obtained at each temperature were crushed for 20 s, and the crushed samples were screened using sieves of different sizes (12, 16, 28, 45, 60, 80, and 150 mesh).

2.2.3. Microstructure Characterization

The microstructure of the PVC film heated at different temperatures was compared using scanning electron microscope (SEM) images, and the distribution and content of selected elements in the products were analyzed using an electron probe microanalyzer (EPMA). The samples for microstructure characterization were cut and drowned in the resin. The cross section was further polished and sprayed a carbon film before the characterization test.

2.2.4. Thermogravimetric Analysis

The combustion experiment of the carbonaceous raw materials was performed by thermogravimetric analysis. The equipment used was an thermal analyzer (SDT Q600, TA Instruments, New Castle, DE, USA). Approximately 10 mg samples were taken and heated from room temperature to 1000–1250 °C at 10 °C/min in air, with a gas flow rate of 100 mL/min. The combustion performance of pulverized PVC, coke, graphite, and anthracite were analyzed.

2.2.5. Powdery Property Analysis

The fluidity and jet flow properties of carbonaceous raw materials were tested by a multifunctional powder physical property tester (MT-02, SEISHIN, Japan). The particle sizes of raw materials were smaller than 180 μm. The fluidity and jet flow of each sample were measured three times, and their average values were recorded.

2.3. Evaluation Indicator

2.3.1. De-Chlorination Ratio of Polyvinyl Chloride (PVC) Film (α)

To evaluate the de-chlorination of PVC films after heat treatment at different temperatures, the chlorine content of PVC raw materials and products after heat treatment was determined by ion chromatography (IC) designated ion test. The de-chlorination ratio at different heat treatment temperatures was calculated, using the expression given below:

$$\alpha = (M_0 - M_T)/M_0 \times 100\% \tag{1}$$

where M_0 was the chlorine amount in the PVC film (%), and M_T was the chlorine amount in the heated PVC film at a certain temperature.

2.3.2. Crushing Performance Index (φ)

To compare the crushing performance of PVC heat treatment products under different conditions. and considering that the product having particle size less than 0.18 mm is easy to be used for thermochemical reaction, the proportion of crushing products having a particle size less than 0.18 mm in the total material mass is defined as the crushing performance index, and its expression is given below:

$$\varphi = W_{0.18}/W \tag{2}$$

where $W_{0.18}$ was the mass having particle size smaller than 0.18 mm in the heat treatment product (g), and W was the total mass of the heat treatment product (g).

2.3.3. Combustion Performance Index (S_N)

To compare the combustion performance of pulverized PVC heat treatment product and various carbonaceous materials, the combustion performance index was introduced to characterize the combustion performance of each carbonaceous material, and its value was calculated using the following formula [14]:

$$S_N = (DTG_{max} \times DTG_{mean})/(T_i^2 \times T_f) \tag{3}$$

where S_N was the combustion performance index, $min^{-2}.°C^{-3}$; DTG_{max} was the maximum weight loss rate (%/min); DTG_{mean} was the mean weight loss rate (%/min); T_i was the ignition temperature (°C); and T_f was the burnout temperature (°C). Notably, the greater the S_N value, the better the combustion performance of the fuel, and vice versa.

3. Results and Discussion

3.1. Low Temperature Heat Treatment Experiment of Polyvinyl Chloride Film

The PVC film was subjected to constant temperature heat treatment at 280, 310, 340, 370, 400, 430, and 460 °C for 30 min. The morphology of the PVC obtained at different temperatures is shown in Figure 2. It can be seen that the PVC film shrank from film pieces to strips and the formed strips closely gathered together. At 280 °C, the PVC film in the center was light yellow, indicating that the temperature was not high enough. The PVC film was not carbonized. With the increase in temperature, the light-yellow part of the heat treatment products gradually disappeared, and the volume reduction was more.

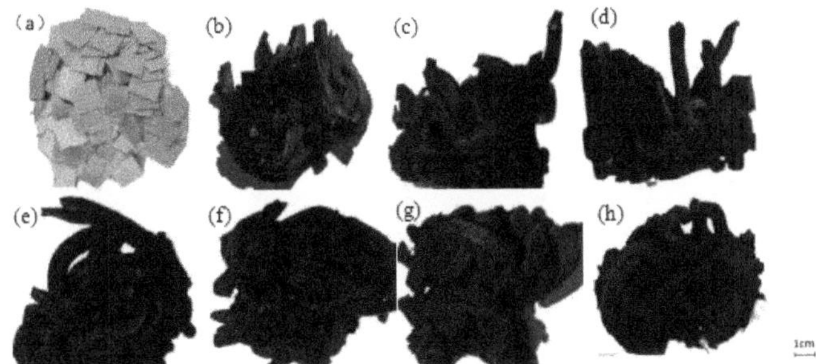

Figure 2. Morphology of PVC film heated at different temperatures: (**a**) PVC film pieces, (**b**) 280 °C, (**c**) 310 °C, (**d**) 340 °C, (**e**) 370 °C, (**f**) 400 °C, (**g**) 430 °C, (**h**) 460 °C.

The relationship between the heat treatment temperature and the weight loss ratio of the PVC samples is shown in Figure 3. It can be seen that the weight loss ratio of PVC film was very low at 280 °C. With the increase of temperature, the weight loss ratio continued to increase, and a flat region began to appear at 370 °C. After 400 °C, the weight loss ratio increased immediately with increasing temperature. The two weight loss stages (<370 and >400 °C) of the PVC films were also accompanied by different reactions. In the first stage, the main reaction was the removal of HCl from the PVC film. With increasing temperature, the internal energy of PVC film increased, the fracture of the C-Cl bond intensified, and the amount of removed HCl increased. It was also accompanied by the volatilization of small molecular substances during thermal degradation; thus, the weight loss ratio increased significantly [15]. After 400 °C, the weight of the PVC film entered the second loss stage, the cracking of the main chain (crosslinked polyene) in the PVC film intensified, many small molecular compounds were produced, and the weight loss ratio increased sharply. At 370–400 °C, the change in the weight loss ratio was very small, which meant no decomposition reaction occurred.

Figure 4 shows the de-chlorination ratio of the PVC film and the chlorine content in the product at different heat treatment temperatures. It can be seen from the figure that at a temperature lower than 370 °C, the de-chlorination ratio increased significantly with the increasing temperature. When the temperature was 370 °C, the de-chlorination ratio reached a maximum value of 84%. After 370 °C, the de-chlorination ratio decreased, portraying a stable trend at approximately 80%. The C-Cl bonds in the structure of PVC have a relatively lower binding energy than the C-C and C-H bonds. Therefore, the C-Cl bond in the PVC broke first with the increasing temperature. Moreover, the HCl released from PVC can catalyze the de-chlorination reaction; therefore, the ratios of de-chlorination portrayed an increasing trend [16]. After 370 °C, hydrocarbons were mainly produced, and the de-chlorination ratios decreased slightly and then showed a stable trend [17]. Notably, the chlorine content in the product was negatively correlated with the de-chlorination ratio. The chlorine content obviously decreased before 370 °C and gradually increased after 370 °C. At 370 °C, the chlorine content of the product was approximately 9%. Some chlorine should have been trapped in the residue as a result of interaction with the additives in the PVC film sample [18]. It can be concluded that the most suitable heat treatment temperature for the de-chlorination of PVC film in the present experimental condition was approximately 370 °C.

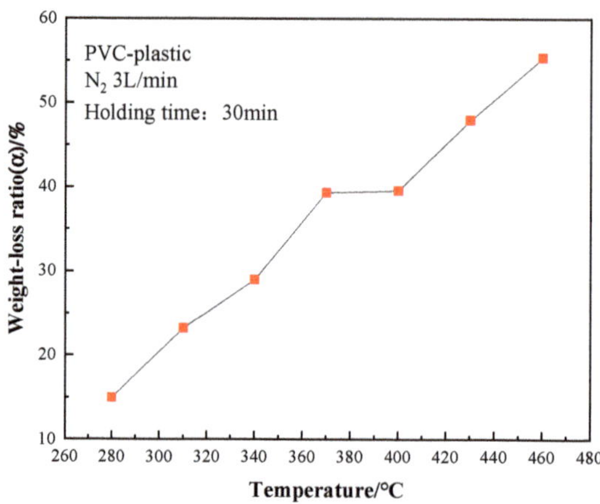

Figure 3. Relationship between heat treatment temperature and weight loss ratio of PVC film.

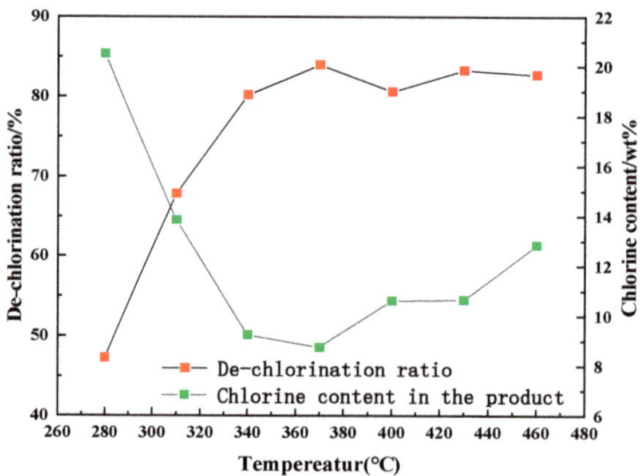

Figure 4. Chlorine content and de-chlorination ratio of PVC film at different temperatures.

3.2. Pulverization of Polyvinyl Chloride after Heat Treatment

The pulverization experiment was performed to crush and screen the heat treatment products of the PVC film. The influence of temperature on the crushing performance index of the products was determined. The optimal heat treatment conditions of PVC film could be obtained by considering the aforementioned de-chlorination results at the same time.

After being heated at various temperatures, the PVC film was crushed and sieved into different particle sizes. As shown in Figure 5, with an increase in temperature, the proportion of powder having a small particle size, such as 0.10 mm, gradually increased. At the temperature higher than 340 °C, the proportion of powder with the size larger than 0.25 mm became very little, which was around 2.50 to 3.00%. In the present work, the mass proportion of powder having particle size less than 0.18 mm was defined as the crushing performance index to roughly illustrate the particle size of the crushed products in total. The variation of the crushing performance index with temperature is shown in Figure 6. Notably, the crushing performance index was small at lower temperatures, such as 0.3 at 280 °C. Before 340 °C, the temperature had a significant influence on the crushing performance index. With increasing temperature, the crushing performance index increased rapidly due to obviously improved embrittlement by pyrolysis compared to the initial polymer state. The crushing performance index (>0.9) tended to be stable after 340 °C. Because the de-chlorination ratio of the PVC film reached its peak at 370 °C, this temperature of 370 °C was considered to be ideal for obtaining the optimal crushing performance of the PVC film. For the convenience of the following description, PVC heat treatment products at 370 °C were denoted as PVC370. Furthermore, the proximate analysis of newly obtained PVC370 was performed in dry basis and the content of fixed carbon, volatile matter, and ash of PVC370 were 19.56%, 54.75%, and 25.69%, respectively.

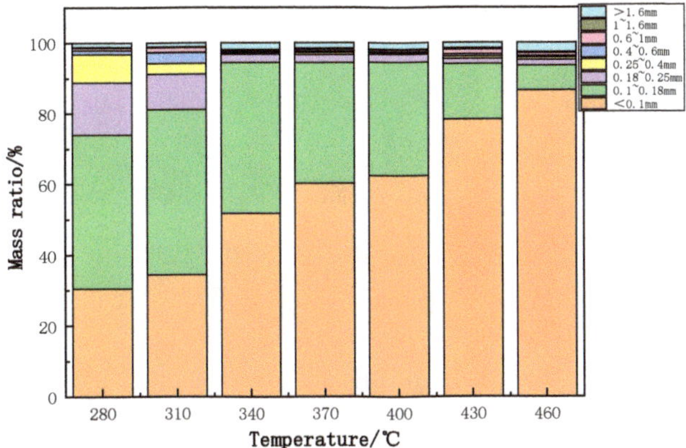

Figure 5. Particle size distribution of crushed PVC film heated at different temperatures.

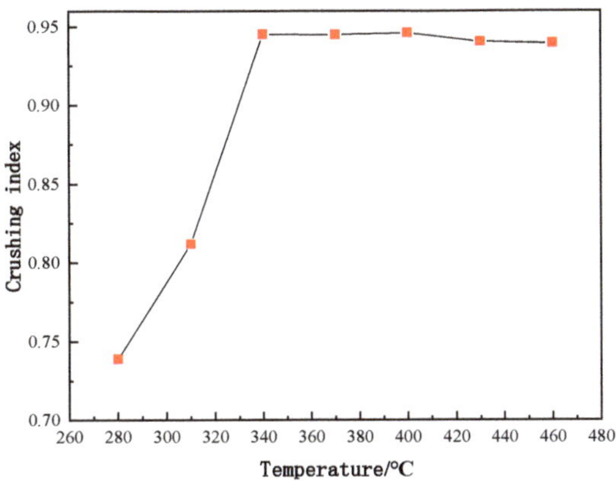

Figure 6. Crushing performance index of PVC film heated at different temperatures.

3.3. Microstructure Analysis of Polyvinyl Chloride after Heat Treatment

SEM images of the PVC film heated at various temperatures are shown in Figure 7. Many unevenly sized pores could be observed at 280 °C. The width of the large pores was approximately 200 μm, and the width of the small pores was around 30 μm. This indicated that only a part of the volatiles in the PVC film was released at 280 °C and it was mainly the lateral chain in the PVC macromolecules being broken [19]. As the temperature increased, the pore size of the samples became larger, which indicated that the amount of volatiles that diffusing outward increased [20]. Before 370 °C, there were more pores in the samples, and they were deeper. At 370 °C, the samples had much larger pores, with the size of approximately 400 µm. When the temperature was further increased, the surface of the sample seemed to be damaged. Many small pores were connected together and became shallower, and some disappeared at 430 °C. Notably, at 460 °C, the cracks began to appear, and the carbon matrix structure was much denser. Only a small part of the main chain in the PVC film broke before 370 °C. After 400 °C, a large number of carbon frameworks were thermally cracked into smaller molecular substances and volatilized, which resulted in the denser residual carbon matrix [21].

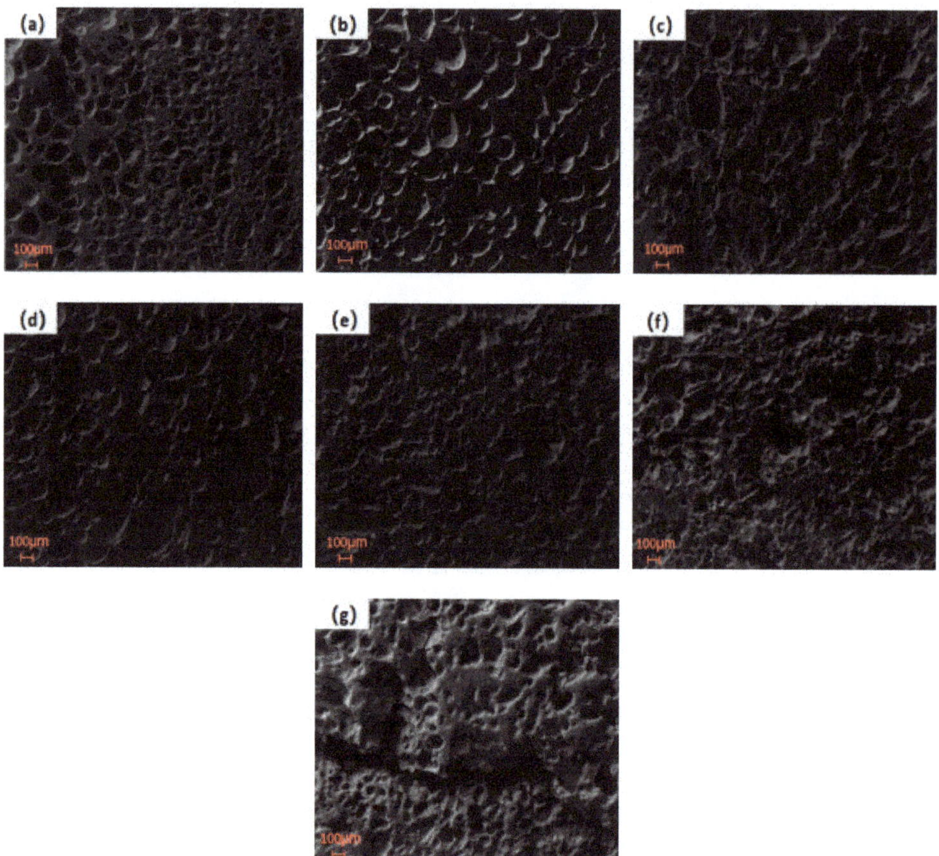

Figure 7. Scanning electron microscope (SEM) images of heated of PVC film at different temperatures: (**a**) 280 °C; (**b**) 310 °C; (**c**) 340 °C; (**d**) 370 °C; (**e**) 400 °C; (**f**) 430 °C; (**g**) 460 °C.

The PVC film sample heated at 370 °C was analyzed using EPMA to investigate the existing state and content of chlorine in the product. Figure 8 shows the distribution of Ca, Cl, and O. Cl was evenly distributed in the matrix around the holes on the sample surface (as shown in Figure 8b). Notably, Cl was widely distributed in the carbon matrix, indicating that residual chlorine was trapped in the residual matrix, which was difficult to remove. The Ca content in the sample was distributed in small pieces and concentrated at the edge of the hole (as shown in Figure 8c). Generally, some additives are added into the pure raw polyvinyl chloride material during the production of commercial polyvinyl chloride.

However, the PVC manufacturers are reluctant to tell the exact chemical composition of the additives. In the present work, the used PVC film sample contained a little CaCO$_3$ additive. Cl also appeared where the Ca was distributed, but not in the place where Ca was most concentrated.

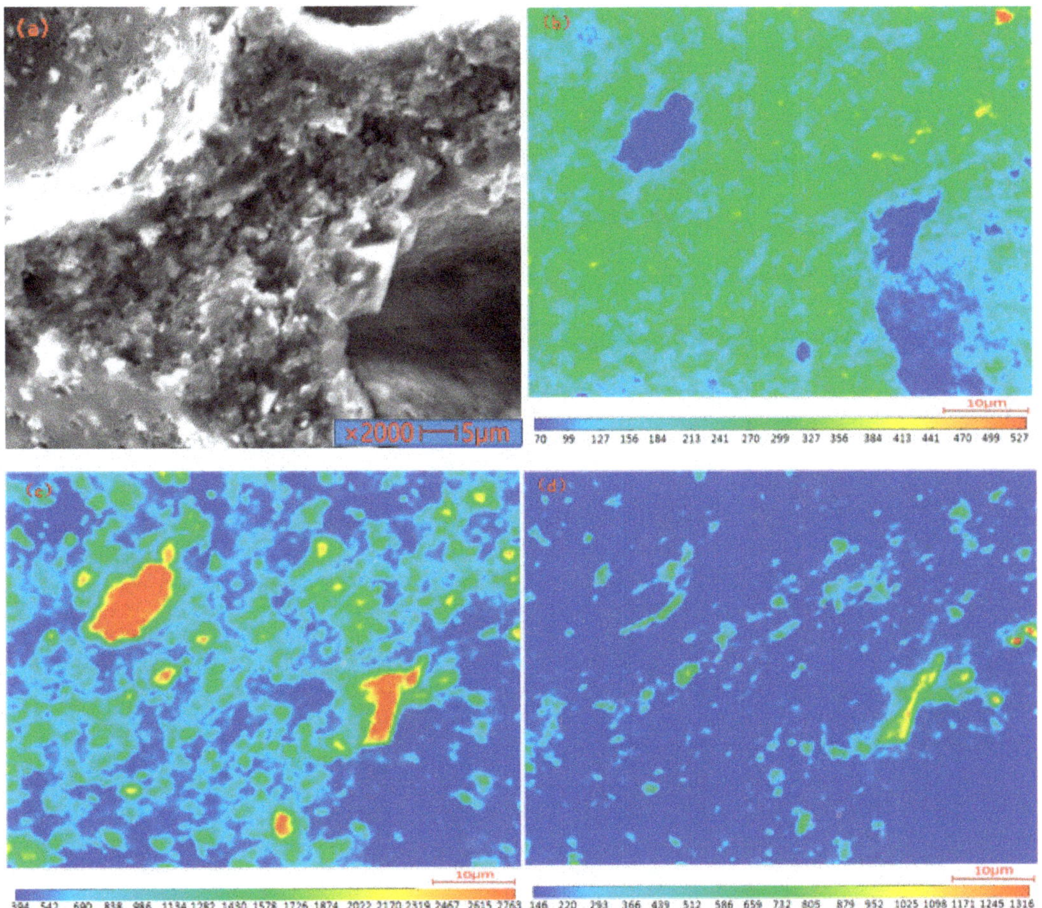

Figure 8. Mapping analysis of heated PVC film (370) by EPMA: (**a**) original image; (**b**) Cl element distribution; (**c**) Ca element distribution; (**d**) O element distribution.

Micro zone analysis was further conducted on the sample, as shown in Figure 9. The corresponding element content was obtained at the selected position on the sample, as listed in Table 2. It can be seen that the carbon matrix of the sample contained a large amount of C element and less O, Cl, and Ca elements. There were many Ca and O elements in the white zone, such as point 2, indicating that the white zone was mainly a compound of Ca. By comparing the amounts of Cl and Ca elements of point 2, it could be concluded that part of the Ca existed in the form of compounds combined with Cl, and the remaining Ca existed in the form of CaCO$_3$ or CaO.

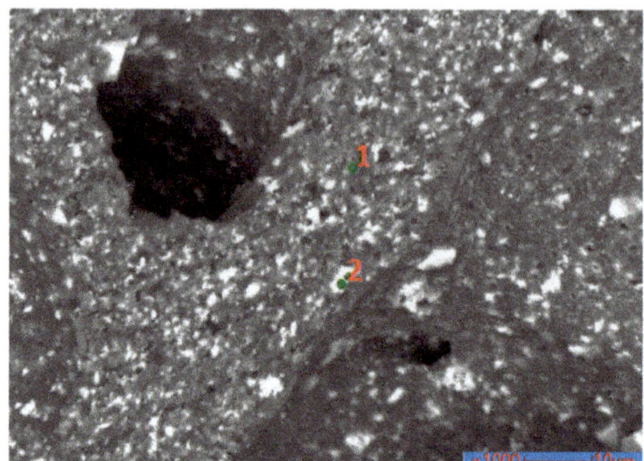

Figure 9. Micro zone analysis of heated PVC film (370 °C) by EPMA: (**1**) point 1; (**2**) point 2.

Table 2. Chemical composition of selected points (at%).

Point No.	C	O	Cl	Ca
1	87.68	6.69	2.76	2.87
2	55.52	23.37	2.93	17.88

3.4. Analysis of Pulverized Polyvinyl Chloride for Blast Furnace Injection

The injection of pulverized coal into a blast furnace can replace coke, reduce the cost of pig iron, and enrich reducing gas. It has become the prevailing technology in the blast furnace ironmaking process [22]. In the present study, the pulverized polyvinyl chloride after heat treatment was proposed to replace the coal or coke. With the help of the new technology, the consumption of fossil fuel can be reduced and the waste plastics will also be recovered. Therefore, it is meaningful to systematically study the powdery, combustion, and explosive characteristics of the pulverized polyvinyl chloride.

(1) Analysis of powdery properties of pulverized polyvinyl chloride

In a system containing powder, the particles of the powder are interrelated, and some special flow characteristics appear when the particles rub against each other. Therefore, the study of fluidity and jet flow is of great significance for the processing, transportation, storage, and packaging of powder. Studying these parameters can also provide practical reference for the metallurgical, chemical, and other industries, which consume powdery raw materials. The fluidity and jet flow of pulverized polyvinyl chloride and two injection coals were measured. The fluidity performance of the powder included four factors: natural slope angle, compression ratio, scoop angle, and uniformity. The compression ratio was obtained by the loose density and tap density, and the other factors were directly measured by the test. Additionally, the jet flow characteristics, including the crash angle, angle of difference, dispersity, and fluidity, were also studied. The crash angle, angle of difference, and dispersity were obtained directly from the test. Based on the principle of the Carr index [23], the properties of pulverized polyvinyl chloride (i.e., PVC370) and two injection coals in transportation, storage, and production were evaluated, and the results were listed in Table 3. As seen in the table, PVC370 showed better fluidity performance than the two kinds of injection coals. The natural slope angle of PVC370 was the smallest, indicating that the relative friction between the materials of this sample was small, which was conducive to improve the flow performance. The jet flow performance of PVC370 was weaker than those of the injection coals, but it still reached the strong jet flow degree. This indicated that the powdery properties of pulverized PVC product after heat treatment could meet the injection requirements of blast furnace ironmaking operation to partially replace coal or coke.

Table 3. Powdery properties of pulverized PVC370 and injection coal.

Physical Property	Pulverized PVC Heat Treatment Product	Injection Coal 1	Injection Coal 2
Tap-density (g/mL)	0.611	0.89	0.801
Compression ratio (%)	24	21	33
Natural slope angle (°)	22.3	37	26.2
Crash angle (°)	17.6	19	11
Angle of difference (°)	4.7	18	15.2
Scoop angle (°)	22.8	43	15.5
Dispersity (%)	26.9	25	33.9
Uniformity (D_{60}/D_{10})	6.55	5.94	7.86
Fluidity index	87	75	76
Degree of fluidity	Good	Good	Good
Jet flow index	68	82	81
Degree of jet flow	Strong	Very strong	Very strong

(2) Combustion properties of pulverized polyvinyl chloride

The weight loss curves of PVC370 and other carbonaceous materials during combustion are shown in Figure 10. The corresponding reaction performance parameters were obtained according to the weight loss curve, as shown in Table 4. It can be seen from the figure that coke, graphite, and anthracite lost the weight in only one stage. Graphite almost lost all the weight at the temperature of 997 °C. The weight loss ratio of coke was approximately 83% at the temperature of 880 °C. The weight loss ratio of anthracite was approximately 87% at the temperature of 749 °C. Compared with other carbonaceous materials, PVC370 was more prone to react. The weight loss curve of PVC370 can be divided into three stages: (1) 264–508 °C (weight loss ratio 42.58%); (2) 508–586 °C (weight loss ratio 3.36%); and (3) 586–703 °C (weight loss ratio 7.20%). The weight loss of the first stage consisted of the devolatilization and combustion of small molecules, and the removal of HCl, and the second stage mainly consisted of the combustion of the residual carbon backbone. The combustion of PVC370 was almost completed in the third stage; therefore, the rate of weight loss decreased [24]. The order of initial combustion temperatures increased as follows: PVC370 (264 °C) < anthracite (446 °C) < coke (533 °C) < graphite (660 °C). The maximum reaction rate of PVC370 was similar to that of anthracite and higher than those of graphite and coke. The combustion performance index (S_n) of each carbon material was calculated using Equation (3). The combustion performance index of PVC370 was significantly higher than those of the other carbon materials. Coke and graphite had almost the same combustion indexes, and both were much smaller than those of the other two carbonaceous materials. The combustion characteristic indexes decreased in the sequence of PVC370 > anthracite > coke > graphite.

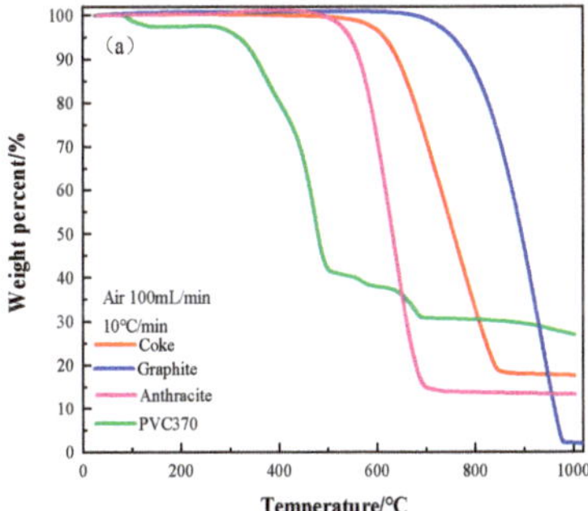

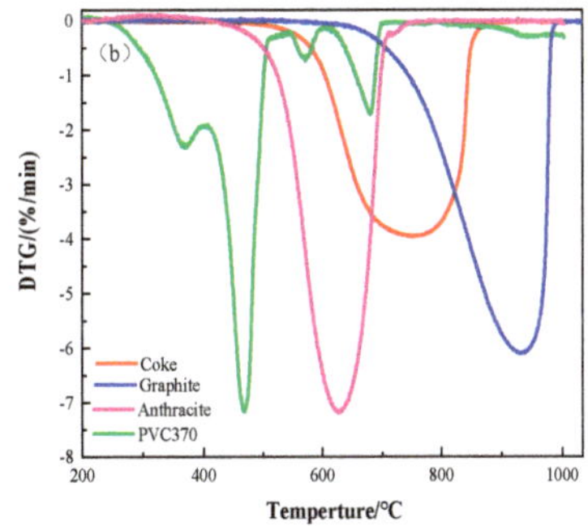

Figure 10. Combustion curves of each carbonaceous material: (**a**) weight loss curves, (**b**) weight loss rate curves.

Table 4. Combustion characteristic parameters of different samples.

Materials	Peak	T_i (°C)	T_f (°C)	T_m (°C)	DTG_{max} (%/min)	$S_N \times 10^9$	t_g (min)
	1	264	513	467	7.16	456.58	24.78
PVC370	2	542	606	572	0.71	1.56	6.38
	3	606	703	680	1.60	4.46	9.78
Anthracite	1	446	749	628	7.14	136.10	30.43
Graphite	1	660	997	934	6.12	40.87	33.83
Coke	1	523	880	760	3.95	37.25	35.84

Note: T_m was the temperature corresponding to the peak of the weight loss rate, t_g was the reaction time.

(3)   Explosive property of pulverized polyvinyl chloride

In the blast furnace ironmaking process, the injection coal is ground into the grain size 80% less than 0.074 mm to improve the combustion rate. However, the pulverized coal after fine grinding has a large specific surface area, which is prone to explosion during or after the grinding. Therefore, blast furnace injection is very concerned with the explosiveness of coal, especially for highly volatile bituminous coal [25]. The explosiveness of coal is in direct proportion to the level of volatile matter. The higher the volatile content, the stronger the explosiveness of pulverized coal [26]. The volatile content of PVC heat treatment products is higher than that of bituminous coal, so it is necessary to study its explosiveness.

Generally, the explosiveness of pulverized coal is reflected by the length of flame. The flame length of pulverized coal is measured by long tube explosion performance tester. In practical engineering applications, it is generally believed that if the length of the return flame formed by the detonation of the measured pulverized coal is greater than 600 mm, the coal can be classified as highly explosive. If it is between 400~600 mm, the coal has medium-intensity explosiveness, and if it is less than 400 mm, the coal has weak explosiveness. If there are only rare sparks or no sparks at the fire source, it exhibits non-explosiveness [27,28]. The return flame length of coal injected into the blast furnace is between 20~50 mm, which can realize safe injection [26]. The test results of explosiveness of PVC370, anthracite, mixed coal, and bituminous coal are shown in Table 5. It can be seen that return flame length of PVC370 and anthracite was almost zero and was much smaller than traditional bituminous coal although the volatile content of PVC370 was much higher. This was a very interesting experimental result. It might be related to the flame retardants that added into PVC during its production to prevent fire for safety [29]. Furthermore, the ignition point of PVC370 was 326 °C, which was lower than 398 °C for anthracite and 334 °C for mixed coal. The combustion characteristics of PVC370 were better than anthracite, and PVC370 was non-explosive. Therefore, PVC370 was more suitable for blast furnace injection than anthracite if the chlorine amount injected into blast furnace together with pulverized PVC was controlled to meet the requirements of total chlorine load of blast furnace ironnaking operation.

Table 5. Results of explosiveness of carbonaceous materials.

Materials	T/°C	L/mm	V_d/%
PVC370	326	0	54.75
Injection anthracite	398	0	7.4
Injection mixed coal	334	30	18.18
Bituminous coal [30]	293	>700	33.25

Note: T was the ignition point temperature, L was the explosion flame length.

4. Conclusions

(1)   The PVC film samples slowly shrank in volume with increasing temperature from 280 °C to 460 °C. The de-chlorination ratio increased significantly with increasing temperature before 370 °C. The de-chlorination rate was up to 84% at 370 °C, and the chlorine content in the product was 9%. The de-chlorination ratio decreased slightly

as the temperature continued to rise, then showed no change. The pulverization performance increased with the temperature. Overall, the optimum heat treatment temperature of the PVC film was 370 °C.

(2)   Pores would form during the heat treatment of PVC film samples due to the emission of volatile. Before 370 °C, there were more pores in the samples, and they were deeper. At 370 °C, the samples had much larger pores, with the size of approximately 400 μm. When the temperature was further increased, the surface of the sample seemed to be damaged. The microstructure of the PVC heat treatment product obtained above 400 °C was denser. Some of the Cl remained in the residual matrix or combined with Ca, and it was difficult to remove this part of Cl.

(3)   After heat treatment at 370 °C (i.e., PVC370), the fluidity of pulverized polyvinyl chloride was better than the two kinds of injection coals. The jet flow was weaker than that of the injection coals, but it still reached a strong degree. PVC370 had a lower initial combustion temperature and higher combustion rate than other carbonaceous materials. PVC370 was classified as a non-explosive substance despite its high volatility.

(4)   The low temperature heat treatment of PVC can help remove chlorine and improve the pulverization performance of thermoplastic PVC film. The pulverized PVC heat treatment product can meet the relevant requirements of blast furnace injection to replace coal or coke from the powdery, combustion, and explosive points of view based on present study.

Author Contributions: G.W. designed the experiments and results analysis, acquired the funding and prepared the draft. S.L. and H.Z. performed the experiments and prepared the original draft. J.W. and Q.X. administrated the project, supervised the experiment and reviewed the manuscript. All authors have read and agreed to the published version of the manuscript.

Funding: The work was supported by the National Natural Science Foundation of China, grant numbers 51804024, U1960205 and the Fundamental Research Funds for the Central Universities, grant number FRF-IC-20-09 and the State Key Laboratory of Advanced Metallurgy of University of Science and Technology Beijing, grant number 41621002.

Institutional Review Board Statement: Not applicable.

Informed Consent Statement: Not applicable.

Data Availability Statement: The data presented in this study are available on request.

Acknowledgments: The authors wish to acknowledge the contributions of associates and colleagues at University of Science and Technology Beijing of China and the financial support of National Natural Science Foundation of China and State Key Laboratory of Advanced Metallurgy of University of Science and Technology Beijing.

Conflicts of Interest: The authors declare no conflict of interest.

References

1.   Xu, S.L.; Li, D.G.; Yu, X.J.; Zhang, Y.L.; Yu, Y.Z.; Zhou, M.; Tang, S.Y. Study on pentaerythritol-zinc as a novel thermal stabilizer for rigid poly(vinyl chloride). *J. Appl. Polym. Sci.* **2012**, *126*, 569–574. [CrossRef]
2.   Janajreh, I.; Alshrah, M.; Zamzam, S. Mechanical recycling of PVC plastic waste streams from cable industry—A case study. *Sustain. Cities Soc.* **2015**, *18*, 13–20. [CrossRef]
3.   Li, Y.F.; Wu, X.M. Development of production technology of PVC and its market analysis. *China Chlor-Alkali* **2008**, *10*, 1–3.
4.   Zhou, Y.C.; Yang, N.; Hu, S.Y. Industrial metabolism of PVC in China: A dynamic material flow analysis. *Resour. Conserv. Recy.* **2013**, *73*, 33–40. [CrossRef]
5.   Sadat-Shojai, M.; Bakhshandeh, G.R. Recycling of PVC wastes. *Polym. Degrad. Stabil.* **2011**, *96*, 404–415. [CrossRef]
6.   Kim, R.; Laurens, D.; Kevin, V.G. Mechanical and chemical recycling of solid plastic waste. *Waste Manag.* **2017**, *69*, 24–58.
7.   Zhang, J.P.; Zhang, C.C.; Zhang, F.S. A novel process for waste polyvinyl chloride recycling: Plant growth substrate development. *J. Environ. Chem. Eng.* **2021**, *9*, 105475. [CrossRef]
8.   Buekens, A.; Huang, H. Comparative evaluation of techniques for controlling the formation and emission of chlorinated dioxins/furans in municipal waste incineration. *J. Hazard. Mater.* **1998**, *62*, 1–33. [CrossRef]

9. Babich, A.; Senk, D.; Knepper, M.; Benkert, S. Conversion of injected waste plastics in blast furnace. *Ironmak. Steelmak.* **2016**, *43*, 11–21. [CrossRef]

10. Asanuma, M.; Ariyama, T.; Sato, M.; Mural, R.; Nonaka, T.; Okochi, L.; Tsukiji, H.; Nemoto, K. Development of waste plastics injection progress in blast furnace. *ISIJ Int.* **2000**, *40*, 224–251. [CrossRef]

11. Herndndez, J.J.; Aranda-Almansa, G.; Bula, A. Gasification of biomass wastes in an entrained flow gasifier: Effect of the particle size and the residence time. *Fuel Process. Technol.* **2010**, *91*, 681–692. [CrossRef]

12. Asanuma, M.; Terada, K.; Inoguchi, T.; Takashima, N. Development of waste plastics pulverization for blast furnace injection. *JFE Tech. Rep.* **2014**, *19*, 110–117.

13. Wang, Q.; Wang, E.L. Kinetic analysis of the pyrolysis of hydrochar derived from PVC and its thermochemical behaviors in a blast furnace. *Ind. Eng. Chem. Res.* **2021**, *60*, 5102–5113. [CrossRef]

14. Li, X.G.; Ma, B.G.; Li, X.; Hu, Z.W.; Wang, X.G. Thermogravimetric analysis of the co-combustion of the blends with high ash coal and waste tyres. *Theramochim. Acta.* **2006**, *441*, 79–83. [CrossRef]

15. Pasek, R.J.; Chang, D.P.Y.; Jones, A.D. Investigation of thermal decomposition of chlorinated polymers. *Hazard. Waste Hazard. Mater.* **1996**, *13*, 23–38. [CrossRef]

16. Yu, J.; Sun, L.X.; Ma, C.; Qiao, Y.; Yao, H. Thermal degradation of PVC: A review. *Waste Manag.* **2016**, *48*, 300–314. [CrossRef]

17. Cao, B.; Sun, Y.K.; Gun, J.J.; Wang, S.; Yuan, J.P.; Esakkimuthu, S.; Uzoejinwa, B.B.; Yuan, C.; Abomohra, A.E.F.; Qian, L.L.; et al. Synergistic effects of co-pyrolysis of macroalgae and polyvinyl chloride on bio-oil/bio-char properties and transferring regularity of chlorine. *Fuel* **2019**, *246*, 319–329. [CrossRef]

18. Kim, S. Pyrolysis kinetics of waste PVC pipe. *Waste Manag.* **2001**, *21*, 609–616. [CrossRef]

19. Mcneill, I.C.; Cole, W.J.; Memetea, L. A study of the products of PVC thermal degradation. *Polym. Degrad. Stabil.* **1995**, *49*, 181–191. [CrossRef]

20. Sugano, M.; Hara, M.; Ichikawa, R.; Shitara, N.; Saitoh, Y.; Kakuta, Y.; Hirano, K. Inhibition of chlorinated organic compounds production by co-pyrolysis of poly (vinyl chloride) with cation exchanged coal. *Fuel* **2015**, *151*, 164–171. [CrossRef]

21. Li, B.Q.; Zhang, B.J.; Tian, F.J.; Liao, H.Q. Increasing oil and decreasing water copyrolysis of coal with coke-oven gas by adding waste plastics. *J. Fuel Chem. Technol.* **1999**, *5*, 385–388.

22. Naito, M.; Takeda, K.; Matsui, Y. Ironmaking technology for the last 100 years: Deployment to advanced technologies from introduction of technological know-how, and evolution to next-generation process. *ISIJ Int.* **2015**, *55*, 7–35. [CrossRef]

23. Zhang, P. Application of Carr index method in comprehensive evaluating properties of coal powder. *China Powder Sci. Technol.* **2000**, *6*, 33–36.

24. Zhang, Y.P.; Salovey, R. Pyrolysis of poly(vinyl chloride). *J. Polym. Sci.* **1974**, *12*, 2927–2941.

25. Pu, Y.K.; Hu, J.; Jia, F. Experimental studies of explosion characteristics of bituminous coal dust-air mixtures injected in blast furnace. *Explos. Shock. Waves* **2000**, *20*, 303–312.

26. Zhang, C.L.; Yang, J.L.; Xiang, D.W.; Yang, L.B. Analysis of influence factors of pulverized coal use in blast furnace injection. *Coal Qual. Technol.* **2016**, 64–67.

27. Jin, L.Z.; Jin, Y.H.; Zhang, J.Y.; Wang, L.Y. Experimental study on lower explosion limit and length of return flame when blasting Luan barren coal into blast furnace. *Chin. Saf. Sci. J.* **2005**, 61–64. [CrossRef]

28. Du, G.; Ying, Z.W. Effect of preheating on explosiveness of mixed pulverized coal. *China Metall.* **2007**, 9–12. [CrossRef]

29. Basfar, A.A. Flame retardancy of radiation cross-linked poly(vinyl chloride) (PVC) used as an insulating material for wire and cable. *Polym. Degrad. Stably* **2002**, *77*, 221–226. [CrossRef]

30. Yu, C.Y.; Sheng, J.W.; Li, S.B. Research and application of injection high volatile matter bituminous coal in a large proportion way in 2000 m^3 BF of Nanjing Steel. *Metall. Collect.* **2014**, 41–44. [CrossRef]

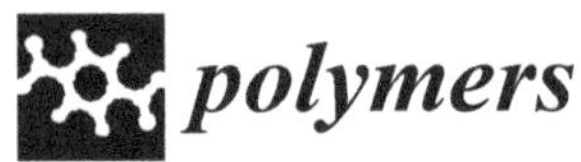

Communication

A Magic Filter Filled with Waste Plastic Shavings, Loofah, and Iron Shavings for Wastewater Treatment

Zengrui Pan [1], Jianlong Sheng [1], Chong Qiu [2], Hongtang Wei [2], Qianjin Yang [2], Jinbo Pan [2] and Jun Li [1,*]

[1] Key Laboratory of Microbial Technology for Industrial Pollution Control of Zhejiang Province, College of Environment, Zhejiang University of Technology, Hangzhou 310014, China; pzr19970701@163.com (Z.P.); sjl19970316@163.com (J.S.)

[2] Zhejiang Shuanglin Environment Co., Ltd., Hangzhou 311100, China; qiuchong0409@hotmail.com (C.Q.); weihongtang@slhj5.wecom.work (H.W.); 15868118495@163.com (Q.Y.); panjinbo@slhj5.wecom.work (J.P.)

* Correspondence: tanweilijun@zjut.edu.cn

Abstract: Integrated sewage treatment equipment has been widely used, but the commonly used fillers for wastewater treatment are not suitable in rural areas due to their price and performance issues. In this study, an integrated magic filter filled with waste fillers was proposed and established for wastewater treatment. The filter was composed of functional modules and an equipment room, and the fillers in each module can be taken out separately and changed arbitrarily according to the needs of specific treatment conditions. The fillers used include waste plastic shavings, loofah, and waste iron shavings, generated during the processing of plastic, crop, and steel. At the same time, a 91 d experiment was performed for real wastewater treatment, and a satisfactory removal performance was obtained, with average removal rates of COD, TP, NH_4^+-N, TN, and SS being 83.3%, 89.6%, 93.8%, 74.7%, and 94.0%, respectively. Through microscope observation, a large number of microorganisms were attached to the surface of the fillers, which was conducive to the simultaneous removal of nitrogen and phosphorus. The micro-electrolysis of waste iron shavings can produce Fe^{2+} and Fe^{3+}, which would combine with PO_4^{3-} to form $Fe_3(PO_4)_2$ and $FePO_4$ precipitates, enhancing the removal of phosphorus. In addition, the filled fillers have an excellent physical filtering effect, which can reduce the effluent SS. The magic filter achieves both the recycling of wastes and the treatment of wastewater.

Keywords: filter; waste fillers; pollutants removal; wastewater treatment

Citation: Pan, Z.; Sheng, J.; Qiu, C.; Wei, H.; Yang, Q.; Pan, J.; Li, J. A Magic Filter Filled with Waste Plastic Shavings, Loofah, and Iron Shavings for Wastewater Treatment. *Polymers* **2022**, *14*, 1410. https://doi.org/ 10.3390/polym14071410

Academic Editors: Sheila Devasahayam, Raman Singh and Vladimir Strezov

Received: 28 February 2022
Accepted: 29 March 2022
Published: 30 March 2022

Publisher's Note: MDPI stays neutral with regard to jurisdictional claims in published maps and institutional affiliations.

1. Introduction

Due to the advantages of a small footprint and short construction period, integrated sewage treatment equipment has become one of the main options for rural sewage treatment [1–4]. In practice, biological fillers are often added to the integrated equipment for wastewater treatment [5–7]. Commonly used fillers include ceramsite, quartz sand, activated carbon, polyvinyl chloride, etc., but high costs, being easy to plug, and poor film hanging performance are the main problems for practice application [8–12]. In this regard, the filler with low cost, a wide source, high efficiency, and simple operation is still limited for rural sewage treatment.

Waste plastic shavings are generated during the processing of plastic, and its recycling has always been a problem. Plastic recycling plants process about 30% of the material received, while the remaining 70% is disposed of in landfills [13,14]. However, it should be noted that waste plastic shavings have many advantages, such as high strength, stable chemical properties, and easy availability [15,16], which basically meet the conditions for being used as biological fillers. In addition, waste iron shavings are produced in the process of steel processing and utilization, which is an easily available scrap metal in the form of rolled flakes, with Fe^0 as the main component. Previous studies have confirmed that waste iron shavings have an excellent effect on phosphorus removal [17,18]. Loofah, an agricultural waste, is featured by its multi-layered fibrous network structure, making it an ideal microbial carrier. Meanwhile, loofah will slowly release the carbon source during the reaction process, thereby saving costs [19–21].

To this end, an integrated magic filter filled with waste fillers was proposed and established for real wastewater treatment. The fillers used included waste plastic shavings, loofah, and waste iron shavings. The structure of fillers and sludge morphology were observed and analyzed, and the pollutants removal performance of the magic filter was investigated. This work presents a valuable effort in expanding the practical application of waste fillers in wastewater treatment.

2. Materials and Methods

2.1. Design Thought

The thought of a magic filter is inspired by the Rubik's cube, which can be combined freely and installed modularly. The magic filter includes an equipment room and functional modules. The equipment room is mainly responsible for controlling the operation of the magic filter, such as influent flow distribution, aeration, sedimentation, water flow direction, reflux, and disinfection. The functional modules are filled with different fillers according to the process and needs. A box-type aquatic plant can be placed on the top layer of the filter, which has both landscape and ecological pollutant removal effects. The functional modules are connected by connecting pipes, and the connection methods include welding, flange connection, flexible connection with rubber parts, etc.

In a practical application, the number of functional modules can be set flexibly according to the site conditions. Filler installation and replacement steps are as follows: filler is added to the filler frame according to the process requirements firstly, and then the filler frame is placed into the functional module in sequence, and finally, the landscape plant is set on the top of the filter. When the replacement of filler is required, the filler can be taken out simultaneously by taking out the filler frame, which is convenient for use.

2.2. Reactor Setup and Operation

According to the idea described above, a magic filter filled with various types of waste fillers was established and operated for real wastewater treatment, and a 91 d experiment was performed (Figure 1). The effective volume was 12 m^3 (2.4 m × 2.4 m × 2.4 m), and the hydraulic retention time was 2 d. The reactor was composed of eight modules and an equipment room of the same size (0.8 m × 0.8 m × 2.4 m). The equipment room was located in the center of the reactor, module 8 was set for disinfection and sedimentation, and the remaining seven modules were filled with corresponding fillers according to processing requirements. A fine grille system (screen pore size of 1 cm), with the same size as the filler frame, was set in module 1.1 for eliminating the large-size particles in raw wastewater. Each module was loaded with three layers of filler frames (0.7 m × 0.7 m × 0.7 m), and each filler frame was equipped with brushes to prevent short flow. To isolate the aerator and support the filler frame, a support structure was provided at the bottom of the functional module. At the top of each module, boxed aquatic plants for landscape and biological deodorization were placed.

The following operational mode was determined via continuous optimization and debugging in the early operation (Figure 1c). The influent water was distributed to modules 1–3 using a three-stage non-uniform distribution, according to the 3:1:1 flow rate. There were two independent reflux pipelines connected by pipeline pumps in the equipment room, and the return paths were as follows: module 5 to module 1 and module 7 to module 3. In addition, air distribution pipes were installed in the equipment room and connected to aerators at the bottom of each module through an air pump and regulating valve. During operation, module 1 was set for consuming dissolved oxygen in raw wastewater through waste iron shavings, leaving module 2 in an anaerobic state. The dissolved oxygen of modules 3, 5, and 7 was controlled at 2–3 mg/L with an aerobic state, and modules 4 and 6 were controlled at 0.2–0.5 mg/L with an anoxic state.

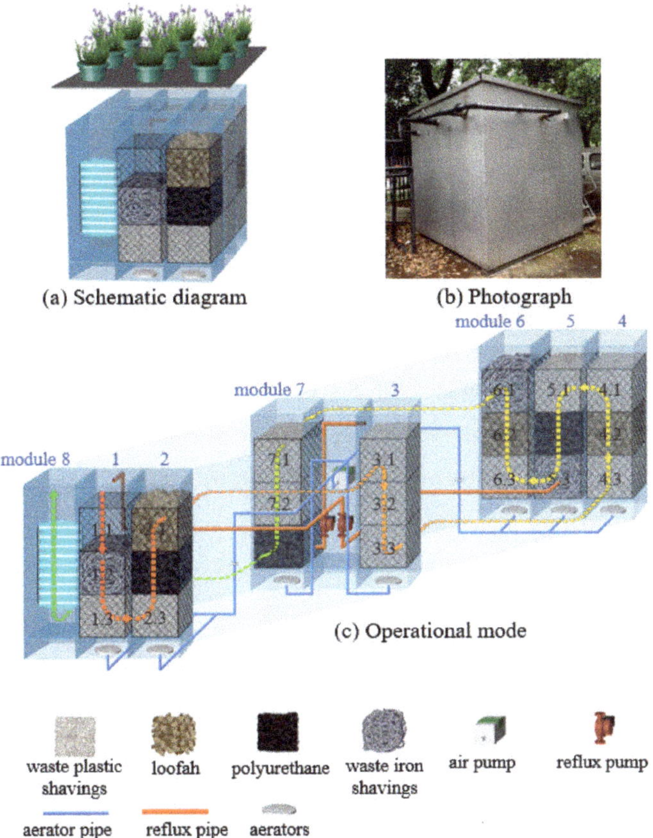

Figure 1. Schematic diagram, photograph, and operation mode of the magic filter.

2.3. Filler Selection and Characteristic

The selection of filler should consider its physical filtration and biochemical effect, and from this, the following four fillers were identified: waste plastic shavings, waste iron shavings, loofah, and polyurethane (Figure 2). Among them, waste iron shavings are mainly used for phosphorus removal, and loofah is also a solid carbon source. Waste plastic shavings and polyurethane can be used as microbial attachment carriers. In addition, the above fillers also have a physical filtering effect.

Thereinto, the waste plastic shavings are obtained from a plastic product factory, in the shape of pleated waves and long strips, with a rough surface, and can produce large gaps between individuals after stacking. Furthermore, it has the advantages of high strength, excellent chemical stability, and easy to obtain, etc. The waste iron shavings are taken from a processing steel plant, with a spiral-shaped and 98.2% Fe content. The loofah comes from agricultural waste, which is an interwoven mesh of multi-layer filamentous fibers, with a long shuttle shape, lightweight, hard texture, slightly curved, thin at both ends, yellowish-white, etc. The polyurethane is purchased from an environmental protection enterprise, with a 99% open-pore rate, and has the advantages of a large specific surface area and easy biological adhesion.

The specific filler configuration is shown in Figure 2. Each filler basically fills the entire filler frame, and the mass of waste plastic shavings, waste iron shavings, loofah, and polyurethane for a single filler frame is about 20 kg, 200 kg, 5 kg, and 10 kg, respectively, and the filling rates are about 58.3 g/L, 583.1 g/L, 14.6 g/L, and 29.2 g/L, respectively.

Module 1	Module 2	Module 3	Module 4	Module 5	Module 6	Module 7
1.1 Fine grille	2.1 c	3.1 a	4.1 a	5.1 a	6.1 b	7.1 a
1.2 b	2.2 d	3.2 a	4.2 c	5.2 d	6.2 c	7.2 a
1.3 a	2.3 a	3.3 a	4.3 a	5.3 b	6.3 a	7.3 d
a: Waste plastic shavings		b. Waste iron shavings		c. Loofah		d. Polyurethane

Figure 2. Filler composition of the magic filter.

2.4. Wastewater Characteristics and Seed Sludge

The wastewater comes from the domestic sewage of a company's dormitory, which is collected in the septic tank and regulating tank before entering the magic filter. The raw wastewater quality varied greatly, and the main parameters of the real wastewater are as follows: COD (274.4 ± 100.0 mg/L), TN (121.9 ± 26.6 mg/L), NH_4^+-N (102.8 ± 16.1 mg/L), TP (7.9 ± 1.8 mg/L), SS (84.5 ± 38.6 mg/L). The reactor was inoculated with dewatered sludge obtained from a municipal WWTP in Hangzhou, China.

2.5. Analytical Methods

Wastewater samples were filtered through 0.45 µm filter paper before analysis. Parameters, such as COD, NH_4^+-N, TP, TN, and SS were measured using the standard methods [22]. Photographs of the fillers were taken using a stereo microscope (Olympus SZ61).

2.6. Statistical Analysis

An analysis method of the cumulative frequency with reference to the German ATV- DVWK-A 131E standard [23] was used to evaluate the reactor performance in pollutants removal.

3. Results and Discussion

3.1. Variation and Replacement of Fillers

The filler structure and sludge morphology were observed, as shown in Figure 3. The surface of the waste plastic shavings was wrinkled and wavy before use (Figure 3a), and a thick biofilm can be observed on the surface after use (Figure 3e), indicating that waste plastic shavings have good bioadhesive properties. In addition, the surface of waste iron shavings was shiny and spiral before use (Figure 3b), but the surface was continuously corroded due to micro-electrolysis during operation. As a result, the surface of the waste iron shavings was rough, and iron deposition and sludge could be clearly observed after use (Figure 3f). Polyurethane was a commonly used filler, and it has a porous mesh structure that makes it easy to adhere to the organism. As shown in Figure 3g, a large number of microorganisms were attached to the polyurethane from inside to outside after use. Loofah had a rough surface and porous structure, which was easily attached to by microorganisms (Figure 3d,h). In addition, carbon sources would be released slowly during operation to save carbon source input costs.

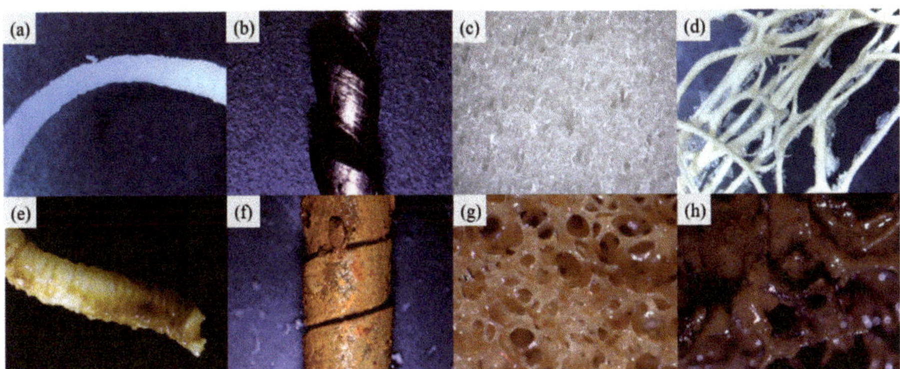

Figure 3. Images of the fillers before and after use (**a–h**).

With the progress of the reaction, the cellulose, and other components in the loofah were continuously decomposed, so the carbon sources released gradually decreased and the structure gradually collapsed. Similarly, the surface of waste iron shavings would be gradually corroded under the effect of micro-electrolysis. It is time to replace waste iron shavings and loofah when the phosphorus and nitrogen removal performance is poor [18]. The replacement cycle of loofah and waste iron shavings is about 80 d in this study, while waste plastic shavings and polyurethane generally do not need to be replaced due to high wear resistance and strength [15,16,24,25].

3.2. Pollutants Removal

Figure 4 depicts COD, TP, NH_4^+-N, and SS removal performance profiles of the reactor. Since this study treated real wastewater, the influent COD fluctuated significantly between 145 and 516 mg/L, resulting in a certain fluctuation of the effluent. During the first 16 d, the COD removal rate was 59.8–68.3%. With the continuous and stable operation of the reactor, the effluent COD concentrations were basically below 50 mg/L, and the removal rate of COD was 82.0–96.3%.

During the initial stage of operation, the removal rate of phosphorus fluctuates between 18.5% and 64.3% due to the release of a large amount of phosphorus from the seed sludge. With the reactor reaching the stable operation stage, the TP concentrations in the final effluent showed a significant decrease from 8.43 to 0.09 mg/L, with an average removal rate of 98.5%. It was speculated that the excellent removal rate of TP was related to the addition of waste iron shavings, which was confirmed by previous studies [17,18].

Moreover, the effluent NH_4^+-N concentrations (6.2 ± 0.8 mg/L) were relatively stable during the entire experimental period, with a wonderful removal rate of 93.8%, which showed that the filling fillers have an excellent removal performance on real wastewater. Despite the high TN concentrations (121.9 ± 26.6 mg/L) in the influent, the removal rate still reached 74.7%, which is related to the operation conditions, such as influent flow distribution, the addition of loofah, and the reflux of nitrifying liquid. In addition, the fillers used can effectively avoid the excessive wash-out of biomass at the initial stage of operation because of the good physical filtration performance. Thus, a stable and high removal efficiency of NH_4^+-N and TN can be maintained during the experimental period.

The influent SS fluctuated significantly during operation, with a maximum of 120 mg/L and a minimum of 36 mg/L, and an average of 84.5 mg/L. During the entire operation, the effluent SS remained stable below 15 mg/L, and the average removal rate was 94.0%. The low effluent SS was related to the fillers used in the reactor, particularly waste plastic shavings and polyurethane, which could effectively trap impurities and macromolecules in wastewater due to the microporous structure of the filler and the biological effect of surface attachment.

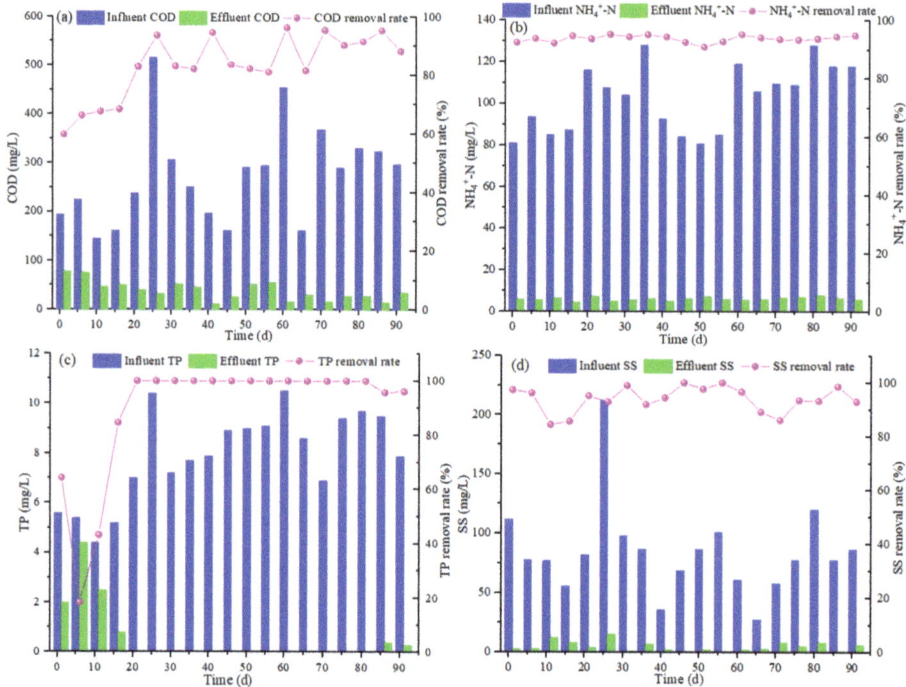

Figure 4. Variation of (**a**) COD; (**b**) NH_4^+-N; (**c**) TP; (**d**) SS.

Due to the large variation of pollutant concentration in the influent water, the water quality data was statistically analyzed using the cumulative frequency method with reference to the German ATV- DVWK-A 131E standard [23], and the removal rate of COD, TN, NH_4^+-N, TP, and SS could be stabilized at 84.6%, 65%, 91.5%, 89.5%, and 92.0%, respectively. The above analysis shows that the reactor has a reliable and robust performance in pollutants removal.

3.3. Proposed Hypothesis of Pollutants Removal Path

As mentioned above, excellent removal performance was obtained, which is related to the operation mode and the fillers configuration of the reactor. The three-stage non-uniform distribution, and simultaneously two independent reflux pipelines (module 5–1 and module 7–3) can save the carbon sources of raw water, enhancing the removal of NO_3^-. Our previous study [18] has confirmed that the effect of iron shavings micro-electrolysis can produce Fe^{2+} and Fe^{3+}, which would combine with PO_4^{3-} to form $Fe_3(PO_4)_2$ and $FePO_4$ precipitates, enhancing the removal of phosphorus. In addition, the morphology of fillers after use showed that a large number of microorganisms are attached to the surface of the fillers, which was conducive to the formation of anaerobic–anoxic and aerobic environments and the simultaneous removal of nitrogen and phosphorus. At the same time, the filled fillers also have a physical filtering effect, which can retain iron precipitation and microorganisms, reducing the effluent SS.

4. Conclusions

In this study, an integrated magic filter filled with waste fillers was proposed and established. The filter consists of an equipment room and functional modules. The equipment room is mainly for controlling the operation of each module, and the functional modules can be filled with different fillers according to the process and needs. In addition, the reliable, and robust removal performance was obtained for wastewater treatment, with average removal rates of COD, TP, NH_4^+-N, TN, and SS being 83.3%, 89.6%, 93.8%, 74.7%, and 94.0%, respectively. The magic filter can be freely combined and installed according to the site conditions and obtain better sewage treatment performance by filling different types of waste fillers. Thus, the magic filter achieves both the recycling of wastes and the treatment of wastewater.

Author Contributions: Z.P.: conceptualization, investigation, writing—original draft, writing—review and editing; J.S.: investigation, writing—original draft; C.Q.: investigation, resources; H.W.: investigation; Q.Y.: investigation; J.P.: investigation; J.L.: fund acquisition, supervision, writing—review and editing. All authors have read and agreed to the published version of the manuscript.

Funding: This work was supported by the Zhejiang Key Research and Development Program (No. 2021C03171), the Zhejiang Key Research and Development Program (No. 2018C03006), and the National Natural Sciences Foundation of China (No. 51478433).

Institutional Review Board Statement: Not applicable.

Informed Consent Statement: Not applicable.

Data Availability Statement: The data presented in this study are available on request from the corresponding author.

Conflicts of Interest: The authors declare no conflict of interest.

References

1. Chen, J.; Liu, Y.-S.; Deng, W.-J.; Ying, G.-G. Removal of steroid hormones and biocides from rural wastewater by an integrated constructed wetland. *Sci. Total Environ.* **2019**, *660*, 358–365. [CrossRef]
2. Gu, B.; Fan, L.; Ying, Z.; Xu, Q.; Luo, W.; Ge, Y.; Scott, S.; Chang, J. Socioeconomic constraints on the technological choices in rural sewage treatment. *Environ. Sci. Pollut. Res.* **2016**, *23*, 20360–20367. [CrossRef]
3. Han, Y.; Ma, J.; Xiao, B.; Huo, X.; Guo, X. New Integrated Self-Refluxing Rotating Biological Contactor for rural sewage treatment. *J. Clean. Prod.* **2019**, *217*, 324–334. [CrossRef]
4. Li, Z.; Hu, X.; Zhang, X.; Gong, L.; Jiang, Z.; Xing, Y.; Ding, J.; Tian, J.; Huang, J. Distributed treatment of rural environmental wastewater by artificial ecological geographic information system. *J. King Saud Univ.-Sci.* **2022**, *34*, 101806. [CrossRef]
5. Li, H.; Liu, F.; Luo, P.; Xie, G.; Xiao, R.; Hu, W.; Peng, J.; Wu, J. Performance of integrated ecological treatment system for decentralized rural wastewater and significance of plant harvest management. *Ecol. Eng.* **2018**, *124*, 69–76. [CrossRef]
6. Xu, D.; Liu, J.; Ma, T.; Zhao, X.; Ma, H. Coupling of sponge fillers and two-zone clarifiers for granular sludge in an integrated oxidation ditch. *Environ. Technol. Innov.* **2022**, *26*, 102264. [CrossRef]
7. Yang, F.; Zhang, H.; Zhang, X.; Zhang, Y.; Li, J.; Jin, F.; Zhou, B. Performance analysis and evaluation of the 146 rural decentralized wastewater treatment facilities surrounding the Erhai Lake. *J. Clean. Prod.* **2021**, *315*, 128159. [CrossRef]
8. Bassin, J.P.; Rachid, C.T.; Vilela, C.; Cao, S.M.; Peixoto, R.S.; Dezotti, M. Revealing the bacterial profile of an anoxic-aerobic moving-bed biofilm reactor system treating a chemical industry wastewater. *Int. Biodeterior. Biodegrad.* **2017**, *120*, 152–160. [CrossRef]
9. Dong, J.; Wang, Y.; Wang, L.; Wang, S.; Li, S.; Ding, Y. The performance of porous ceramsites in a biological aerated filter for organic wastewater treatment and simulation analysis. *J. Water Process Eng.* **2020**, *34*, 101134. [CrossRef]
10. Gidstedt, S.; Betsholtz, A.; Falås, P.; Cimbritz, M.; Davidsson, Å.; Micolucci, F.; Svahn, O. A comparison of adsorption of organic micropollutants onto activated carbon following chemically enhanced primary treatment with microsieving, direct membrane filtration and tertiary treatment of municipal wastewater. *Sci. Total Environ.* **2021**, *811*, 152225. [CrossRef] [PubMed]
11. Zhao, J.; Deng, Y.; Dai, M.; Wu, Y.; Ali, I.; Peng, C. Preparation of super-hydrophobic/super-oleophilic quartz sand filter for the application in oil-water separation. *J. Water Process Eng.* **2022**, *46*, 102561. [CrossRef]
12. Vítězová, M.; Jančiková, S.; Dordević, D.; Vítěz, T.; Elbl, J.; Hanišáková, N.; Jampílek, J.; Kushkevych, I. The Possibility of Using Spent Coffee Grounds to Improve Wastewater Treatment Due to Respiration Activity of Microorganisms. *Appl. Sci.* **2019**, *9*, 3155. [CrossRef]
13. Dahlbo, H.; Poliakova, V.; Mylläri, V.; Sahimaa, O.; Anderson, R. Recycling potential of post-consumer plastic packaging waste in Finland. *Waste Manag.* **2018**, *71*, 52–61. [CrossRef] [PubMed]
14. Ragaert, K.; Delva, L.; Van Geem, K. Mechanical and chemical recycling of solid plastic waste. *Waste Manag.* **2017**, *69*, 24–58. [CrossRef] [PubMed]
15. Demets, R.; Van Kets, K.; Huysveld, S.; Dewulf, J.; De Meester, S.; Ragaert, K. Addressing the complex challenge of understanding and quantifying substitutability for recycled plastics. *Resour. Conserv. Recycl.* **2021**, *174*, 105826. [CrossRef]
16. Shamsuyeva, M.; Endres, H.-J. Plastics in the context of the circular economy and sustainable plastics recycling: Comprehensive review on research development, standardization and market. *Compos. Part C Open Access* **2021**, *6*, 100168. [CrossRef]

17. Ma, Y.; Dai, W.; Zheng, P.; Zheng, X.; He, S.; Zhao, M. Iron scraps enhance simultaneous nitrogen and phosphorus removal in subsurface flow constructed wetlands. *J. Hazard. Mater.* **2020**, *395*, 122612. [CrossRef]
18. Pan, Z.; Guo, T.; Sheng, J.; Feng, H.; Yan, A.; Li, J. Adding waste iron shavings in reactor to develop aerobic granular sludge and enhance removal of nitrogen and phosphorus. *J. Environ. Chem. Eng.* **2021**, *9*, 106620. [CrossRef]
19. Akhtar, N.; Saeed, A.; Iqbal, M. Chlorella sorokiniana immobilized on the biomatrix of vegetable sponge of Luffa cylindrica: A new system to remove cadmium from contaminated aqueous medium. *Bioresour. Technol.* **2003**, *88*, 163–165. [CrossRef]
20. Bai, Y.; Su, J.; Ali, A.; Wen, Q.; Chang, Q.; Gao, Z.; Wang, Y. Efficient removal of nitrate, manganese, and tetracycline in a novel loofah immobilized bioreactor: Performance, microbial diversity, and functional genes. *Bioresour. Technol.* **2021**, *344*, 126228. [CrossRef]
21. Zhang, C.; Yuan, B.; Liang, Y.; Yang, L.; Bai, L.; Yang, H.; Wei, D.; Wang, F.; Wang, Q.; Wang, W.; et al. Carbon nanofibers enhanced solar steam generation device based on loofah biomass for water purification. *Mater. Chem. Phys.* **2021**, *258*, 123998. [CrossRef]
22. Apha, A. *Standard Methods for the Examination of Water and Wastewater*, 9th ed.; American Public Health Association: Washington, DC, USA, 2005.
23. ATV-DVWK-A 131. *Dimensioning of Single-Stage Activated Sludge Plants*; GFA Publishing Company of ATV-DVWK Water: Hennef, Germany, 2000.
24. Hu, S.; He, S.; Wang, Y.; Wu, Y.; Shou, T.; Yin, D.; Mu, G.; Zhao, X.; Gao, Y.; Liu, J.; et al. Self-Repairable, Recyclable and Heat-Resistant Polyurethane for High-Performance Automobile Tires. *Nano Energy* **2022**, *95*, 107012. [CrossRef]
25. Pedrazzoli, D.; Manas-Zloczower, I. Understanding phase separation and morphology in thermoplastic polyurethanes nanocomposites. *Polymer* **2016**, *90*, 256–263. [CrossRef]

polymers

Article

Towards Circular Economy by the Valorization of Different Waste Subproducts through Their Incorporation in Composite Materials: Ground Tire Rubber and Chicken Feathers

Xavier Colom [1], Javier Cañavate [1,*] and Fernando Carrillo-Navarrete [1,2]

1 Department of Chemical Engineering, Universitat Politècnica de Cataluna—BarcelonaTECH, Colom 1, 08222 Terrassa, Barcelona, Spain; xavier.colom@upc.edu (X.C.); fernando.carrillo@upc.edu (F.C.-N.)
2 Institut d'Investigació Tèxtil i Cooperació Industrial de Terrassa (INTEXTER), Universitat Politècnica de Catalunya (UPC)—Barcelona TECH, Colom 15, 08222 Terrassa, Barcelona, Spain
* Correspondence: francisco.javier.canavate@upc.edu

Abstract: Incorporation of residua into polymeric composites can be a successful approach to creating materials suitable for specific applications promoting a circular economy approach. Elastomeric (Ground Tire Rubber or GTR) and biogenic (chicken feathers or CFs) wastes were used to prepare polymeric composites in order to evaluate the tensile, acoustic and structural differences between both reinforcements. High-density polyethylene (HDPE), polypropylene (PP) and ethylene vinyl acetate (EVA) polymeric matrices were used. EVA matrix defines better compatibility with both reinforcement materials (GTR and CFs) than polyolefin matrices (HDPE and PP) as it has been corroborated by Fourier transform infrared spectroscopy (FTIR), termogravimetric analysis (TGA) and scanning electron microscopy (SEM). In addition, composites reinforced with GTR showed better acoustic properties than composites reinforced with CFs, due to the morphology of the reinforcing particles.

Keywords: waste; composites; tire rubber; chicken feathers; acoustic properties

Citation: Colom, X.; Cañavate, J.; Carrillo-Navarrete, F. Towards Circular Economy by the Valorization of Different Waste Subproducts through Their Incorporation in Composite Materials: Ground Tire Rubber and Chicken Feathers. *Polymers* 2022, 14, 1090. https://doi.org/10.3390/polym14061090

Academic Editor: Sheila Devasahayam

Received: 14 February 2022
Accepted: 4 March 2022
Published: 9 March 2022

Publisher's Note: MDPI stays neutral with regard to jurisdictional claims in published maps and institutional affiliations.

1. Introduction

Nowadays, one of the greatest concerns of humanity is the huge amounts of waste that are produced year after year around the world. Therefore, strategies for the valorization of wastes or by-products based on recovering and recycling have been developed during the last years in order to reduce the negative environmental impact caused by the use and transformation of resources [1,2]. There are many kinds of waste, but special attention is paid to materials that are not biodegradable, such as plastics [3,4], or to residua that, even being biodegradable, are produced in great quantities and can cause massive environmental impacts, including climate-harming emissions by illegal dumping or burning, such as crop wastes, eggshells, chicken feathers, crustacean shells and others [5].

The transition from waste disposal to a circular economy is in evolution. Consequently, many alternatives of waste valorization are currently under research and development. Hence, research all over the world focuses on the development of strategies for recycling [6] and upcycling [7] plastic wastes and, in the case of biowastes, on the advance of integrated biorefinery concepts or the production of basic chemicals and specialized fibers among others [5].

In parallel to the aforementioned approaches, several research groups are working with waste materials to obtain newly added value materials [8,9]. In this regard, a commonly used approach is to recycle wastes by mixing them with polymeric matrices to obtain new composite materials. On one hand, the use of natural fibers, from plants (jute, sisal, flax, etc.) or animals (wool, hair, feathers, etc.), to prepare polymer composites have gained attention during the last decades [10,11] and it has expanded considerably in some sectors, such as in automotive industry [12]. However, natural fibers are mainly hydrophilic, hindering the interfacial adhesion between the fiber and the hydrophobic polymeric matrices resulting in composite materials with weak mechanical and physical properties unless some treatments are performed on the fibers [13]. On the other hand, non-biodegradable wastes have also been proposed as reinforcements or fillers for the preparation of polymeric composites [14,15].

In this regard, tire rubber has been proposed as filler or reinforcement in order to manufacture composites with thermoplastic, thermosets and rubber matrixes with interesting tensile, electrical, or acoustical properties [16–18]. This approach allows the recovery and reuse of part of a tire rubber avoiding their disposal in landfills. Following the same idea, there are several interesting biogenic abundant and biodegradable wastes that would end in a landfill unless they were reused somehow, for example, by mixing the biogenic waste with polymeric matrices. Among the several biogenic wastes generated by industrial processes, chicken feathers (CFs) are generated in large amounts and they do not have any practical application, so they are a potential candidate for developing composite materials [19].

The main challenge for composite preparation by using both non-biodegradable and biodegradable wastes is to investigate how properties and development of these materials are influenced by the compatibility between the composite components since the fiber-matrix interaction can significantly affect the final macroscopic properties of the composite product. In the case of tire rubber and CFs, when they are blended with polymeric matrices (HDPE, PP, EVA, etc.), the fiber-matrix compatibility is expected to be low [19,20]. One way to increase the compatibility between both components is to carry out a pretreatment of the waste. For example, acid pretreatment with nitric and/or sulfuric acids due to chemical reactions produces a microporous surface that improves mechanical adhesion. The chemical attack leads to cavities that develop porosity on the waste particles, allowing a good interlocking with polymeric matrices [20]. In addition, grafting of polymers and compatibilizing agents have been used on tire powder to obtain useful materials, improving the interfacial adhesion [21]. In the same way, pretreatments of CFs have been proposed to prepare composite materials with enhanced mechanical properties [19]. In any case, the nature of each particular waste will have an impact on the final fiber-matrix compatibility and it is worth studying. Consequently, a comparative study is proposed in this manuscript in order to evaluate the inherent differences in compatibility and processing when using either GTR or CFs without the help of chemical compatibilization pretreatments.

Taking into account these premises, the main purpose of this study is to compare elastomeric (Ground Tire Rubber or GTR) and biogenic waste (chicken feathers or CFs) composite materials in order to understand the tensile and structural differences between both and which is the best for specific industrial applications. The aim is also to provide examples of two very different residua that can be useful in order to undertake related cases promoting a circular economy approach. High-density polyethylene (HDPE), polypropylene (PP) and ethylene vinyl acetate (EVA) polymeric matrices were used. Once the composites were prepared, mechanical and structural characterization were carried out by using tensile test, scanning electron microscopy (SEM), Fourier-transform infrared spectroscopy (FTIR) and thermogravimetric analysis (TGA) techniques to find a relationship between the macroscopic properties, i.e., tensile strength, and microstructure. In addition, the acoustical properties of the materials were determined to evaluate the viability of using them for sound absorber applications. Finally, the properties of the composites were compared in order to discern which type of waste provided better performance.

2. Materials and Methods

2.1. Materials

High-density polyethylene (HDPE, ALCUDIA® 4810-B, Repsol, Tarragona, Spain) with a melt flow index of 1.35 g/min and density of 960 kg/m^3, Polypropylene (PP Isplen® 099 K2M, Repsol, Tarragona, Spain) with a melt flow index of 1.15 g/min and density of 913 kg/m^3 and Ethylene vinyl acetate (EVA ALCUDIA® PA 539, Repsol, Tarragona, Spain) with a melt flow index of 1.18 g/min and density of 937 kg/m^3, were used in this study.

Ground Tire Rubber (GTR) with a size lower than 400 μm was supplied by GMN Company from Maials (Lleida, Spain) and were used without treatment.

Chicken feathers (CFs) were kindly supplied from a slaughterhouse located in Catalonia (Spain). CFs from slaughterhouses are unstable, unsafe and biodegradable, so pretreatment is mandatory to stabilize and sanitize the waste. Therefore, the CFs were first frozen at −20 °C and subsequently washed in a washing machine at 35 °C with a 3300 ppm H_2O_2 solution (hydrogen peroxide 35% *w/v*, Chem-Lab NV, Zedelgem, Belgium), in a 5/1 (*vol/wt*) liquor ratio for 50 min. After that, CFs were dried in an air oven at 60 °C for 24 h. Cleaned CFs were later chopped with a shredder (Retsch SM100, Haan, Germany) until each particle size was 2 mm or less. Finally, the CFs were air-dried at 105 °C for 4 h and kept under a dry atmosphere (desiccator) just before the composites were compounded.

2.2. Composite Preparation

Composite specimens were obtained by mixing the ground and dried particles of CFs and GTR previously prepared, with HDPE, PP and EVA matrices. Different compositions were studied: 5, 10, 20, and 40% *w/w* and controls of pure HDPE, PP and EVA were used as references.

The components were mixed using a Brabender mixer type W 50 EHT PL (Brabender® GmbH & Co. KG, Duisburg, Germany) heated at different temperatures for each matrix: 170 °C for PP; 160 °C for HDPE and 120 °C for EVA, respectively, and at a mixed speed of 50 rpm. The HDPE, PP and EVA matrices were melted for a minute and then, the particles (GTR or CFs) were added and mixed for a period of 5 min.

These blends were then consolidated in a hot plates press machine type Collin Mod. P 200E (Dr. Collin GmbH, Maitenbeth, Germany) forming square sheets, measuring $160 \times 160 \times 2.2$ mm^3. Consolidation was carried out at a pressure of 100 kN for 5 min using temperatures of 180 °C, 150 °C and 100 °C for PP, HDPE and EVA composites, respectively. Finally, the square sheets were cooled under pressure using cool water.

Test samples were properly dumbbell shaped according to the ASTM 412 specifications to carry out tensile test measurements [22].

2.3. Tensile Tests

The tensile test measurements were carried out in an Instron 3366 universal machine (Instron, High Wycombe, UK). The fabricated composites can be considered isotropic since the GTR particles are meanly spherical and the chicken feathers, although they are fibers, are randomly oriented, so loading direction was not taken into account. The testing speed was 20 mm/min at room temperature. The samples cross-sections were 6.1×2.2 mm^2. Young's modulus, tensile strength, elongation at break and toughness were calculated using Bluehill version 2 software. Five replicate samples were analyzed for each test and average and standard deviation percentages were calculated.

2.4. Structural Characterization by Fourier Transform Infrared Spectroscopy (FTIR)

The structural properties have been obtained using FTIR by means of a Nicolet Avatar spectrometer with CsI optics. Samples of the powdered rubber (400–600 μm average particle size) and powdered chicken feathers were ground and dispersed in a matrix of KBr (9 mg in 300 mg KBr), followed by compression at 167 MPa to consolidate the formation of the pellet.

2.5. Characterization by Scanning Electron Microscopy (SEM)

SEM was used to qualitatively examine the fracture surface of the samples broken by the mechanical tests to study the compatibility at the different samples reinforced by CFs or the GTR interface. Several images of the samples were taken in a JEOL 5610 microscope at an accelerating voltage of 30 kV and a working distance of 6 mm. Previously to the observations, the samples were covered with a fine layer of gold-palladium in order to increase their conductivity.

2.6. Characterization by Thermogravimetricy Analysis (TGA)

Thermal treatments were performed by means of TGA. TGA was performed in a TG/SDTA851 Mettler Toledo equipment at 10 °C/min heating rate in N_2 atmosphere, to obtain the TGA degradation process curve. The temperature range was from 30 to 600 °C, the typical degradation range of the thermoplastic matrix composite. A mass in a range of 10–12 mg of sample was analyzed in order to guarantee sample homogeneity that is of particular importance for analytical techniques, such as TGA.

2.7. Characterization of Acoustic Properties by Impedance Tube

The acoustic properties were measured using a two microphone impedance tube Brüel & Kjaer type 4206 ((Brüel & Kjaer, Virum, Denmark) in the frequency range 100–6500 Hz, according to the specification ASTM E 1050, which describes the standard test method for impedance and absorption of acoustical materials using a tube, two microphones and a digital frequency analysis system. Cylindrical samples (2.2 mm of thickness) were prepared by cutting the material and then submitted to a plane sound wave. The sound pressures were measured at the same time in two microphone positions and the relationship between the acoustic energy that is absorbed by the material and the total incident energy resulted in the normal incidence sound absorption coefficient (α).

3. Results and Discussion

3.1. Mechanical Properties

Figures 1–3 show the evolution of tensile strength, Young's Modulus and elongation at break of two different kinds of composite made by polyolephynic matrix and two waste reinforcements (GTR and CFs).

The values of tensile strength showed similar behavior for all composites (Figure 1). All of them decreased as a function of reinforced content, although the evolution is quite different for each one. The composites reinforced with GTR had a continuous decrease; meanwhile, the composites made by CFs had a decrease in two steps (contents of 5% and contents from 5% to 40%). Only with 5% of CFs, the value of tensile strength decreased 14.2% for the HDPE/CFs; 30% for the PP/CFs and 56% for EVA/CFs and after that the values decreased more slightly for the different contents of CFs. The values for the composites reinforced with GTR had a continuous evolution from pure matrix to compositions of 40% showing a total decrease of 49.5% for HDPE/GTR; 62.5% for PP/GTR and 66.3% for EVA/GTR.

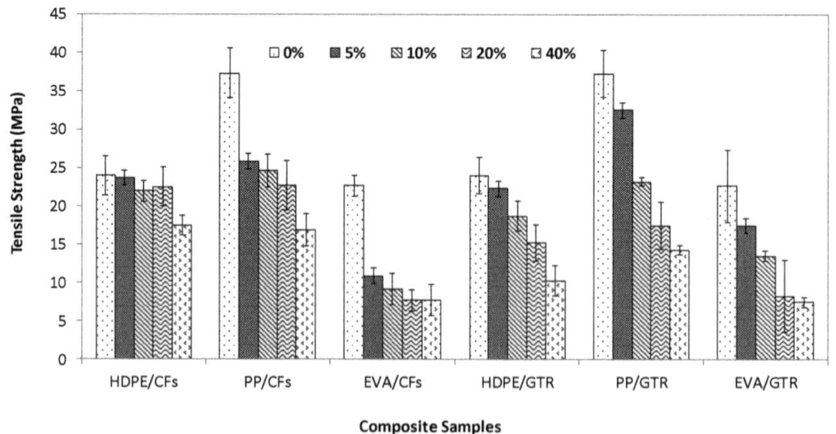

Figure 1. Tensile strength of the polymeric composites at compositions between 0 and 40% *w/w*.

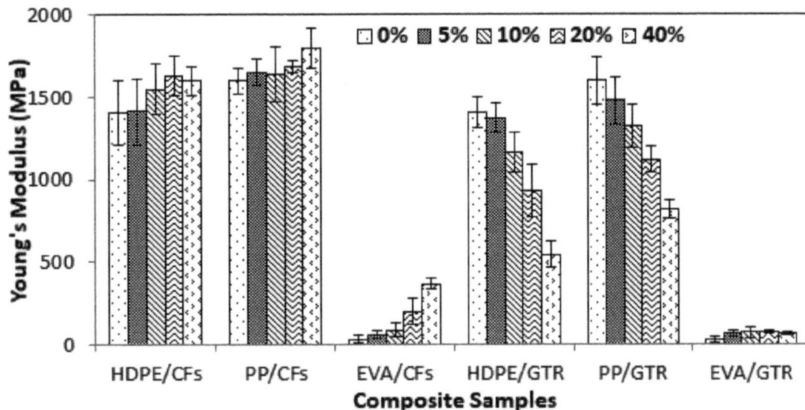

Figure 2. Young's Modulus of the polymeric composites at compositions between 0 and 40% w/w.

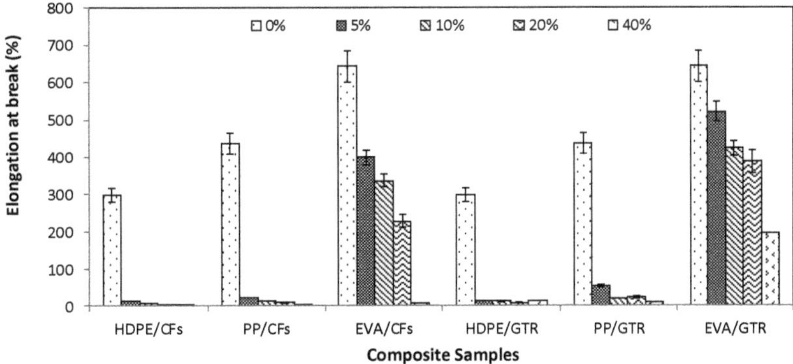

Figure 3. Elongation at break of the polymeric composites at compositions between 0 and 40% w/w.

According to the results shown in Figure 1, it was confirmed that the tensile strength of PP matrix composites was higher than other thermoplastic matrices used. This fact was very evident in CFs reinforcement, although it also took place in GTR reinforced composites.

The differences in behavior between both kinds of reinforcements are due to their dissimilar chemical and morphological nature and their compatibility with the matrices, which defines more or less lack of interfacial adhesion. CFs are hydrophilic compared to the highly hydrophobic nature of the GTR and the morphology of the chopped chicken feathers is long and sharp meanwhile the GTR has essentially irregular polyhedron geometry.

Young's modulus of CFs reinforced composites increased slightly as a function of the percentage of CFs (Figure 2). The most important increase took place in composites of EVA/CFs with an increase of 1200% whereas for HDPE/CFs and for PP/CFs the increase was only 13.7 and 12.4%, respectively. Instead, Young's modulus in composites reinforced by GTR decreases significantly. This means that the compatibility between thermoplastic matrices was better in CFs than in GTR. The Young's Modulus of CFs composites increased slightly at low contents of CFs, except in EVA/CFs, where the increase was significant. This is due to its ability to interact with CFs. The presence of GTR with a maximum size of 200 μm increased only the rigidity of EVA/GTR composites since the particle-matrix compatibility was not good enough, as has been demonstrated in previous works [23,24]. Only, in EVA/GTR Young's modulus increased 157% due to the different chemical composition of ethylene vinyl acetate with respect to HDPE and PP, with ester and carbonyl groups that are more reactive than methyl and methylene groups of HDPE and PP.

In all cases (CFs and GTR based composites), mechanical properties depend also on the dispersion of the reinforcement in the HDPE, PP and EVA matrices, since the CFs and GTR particles are responsible for the decrease in the deformation capacity in the elastic zone. Thermoplastic matrices provide ductility whereas the reinforced particles exhibit brittle behavior with a subsequent loss of toughness to the composite material. The elongation at the breaking point mainly depends on the reinforced particle content, obtaining better behavior for composites with EVA matrix (Figure 3). This mechanical property decreased from 297% to 14% (HDPE/CFs), 436% to 23% (PP/CFs), 642% to 401% (EVA/CFs), 297% to 13.4% (HDPE/GTR), 436% to 53%(HDPE/GTR) and 642% to 521% (HDPE/GTR) when incorporating only 5% of reinforced particles. When reinforced particles increase until contents were higher than 20% w/w, there was not a significant difference of elongation for a polyolephynic matrix. For all reinforcement contents, EVA matrix composites showed a better performance. As shown in Figure 3, the elongation at break using EVA as a matrix in composites with 20% w/w of reinforcement had important differences using GTR or CFs.

3.2. SEM Characterization of Interfacial Adhesion

As shown in Figure 4, SEM micrographs of fracture surfaces of different composites (all containing 40% of CFs and GTR particles), clearly indicate that the differences in microstructure of the various composites are significant. First of all, the samples containing GTR particles (Figure 4a,c,e) show some cracks and pores big enough to be observed at this level of magnification. The GTR particle is unlinked to the matrix, as can be observed by the deep voids around its contour. The rubber seems to be resting on the thermoplastic matrices, without being properly attached to them. On the other hand, the matrices have been strained and deformed plastically, independently of the GTR, which remains unchanged. The samples containing CFs (Figure 4b,d,f) show fibers cleanly extracted from the matrices. No residues or portions of matrix material have adhered to the fiber surface. There are also voids subsequent to the pull-out of CFs that can be observed on the surface. In EVA matrices, although there is fiber pullout, some of them are coated with the EVA matrix. It can also be observed that the break of the composites takes place mainly by shear yield and tearing. The differences between the failure surface of different matrix composites are attributed to the different chemical natures of the matrices and different adhesion mechanisms.

The SEM images corroborate that the interaction is considerably less intense than the cohesion forces of the matrices.

3.3. FTIR Characterization

Results from FTIR analysis indicated that only a weak adhesion is expected for GTR and CFs based composites, corroborating the results of the SEM characterization. Figure 5 compares the spectral evolution of HDPE/CFs and HDPE/GTR composites, respectively. The main differences between both composites are: (i) the doublet 1464/1474 cm^{-1}, (ii) the peak of 1370 cm^{-1} and (iii) the band at 1635 cm^{-1}, assigned to –CH$_2$–, –CH$_3$ and water absorption, respectively. Analyzing the band at 1635 cm^{-1}, it can be seen that the highest absorbance value corresponds to the CFs reinforced composites and the lowest to the GTR reinforced composite. The different patterns of this band, in relation to the types of reinforcement used, are due to the hydrophilic nature of the keratin fiber. Another difference between both kinds of HDPE matrix composite materials is the band's relation 1464/1474 cm^{-1} associated with the crystalline phase of HDPE. The spectra show that the evolution of the doublet in the HDPE matrix is larger in CFs than in GTR, and this means that the CFs create vibrational perturbations originated by the peptidic bonding in the HDPE backbone, affecting the degree of crystallinity in the matrix of CFs composite. However, these structural changes do not significantly improve the compatibility between the matrix and CFs to obtain composites with improved tensile properties. The band at 1370 cm^{-1}, assigned to tensile stretch in methyl groups is higher in GTR composites than in CFs composites and overall HDPE matrix. This is due to the composition of GTR with several elastomers with methyl groups (i.e., natural rubber, polybutadiene, styrene-butadiene rubber).

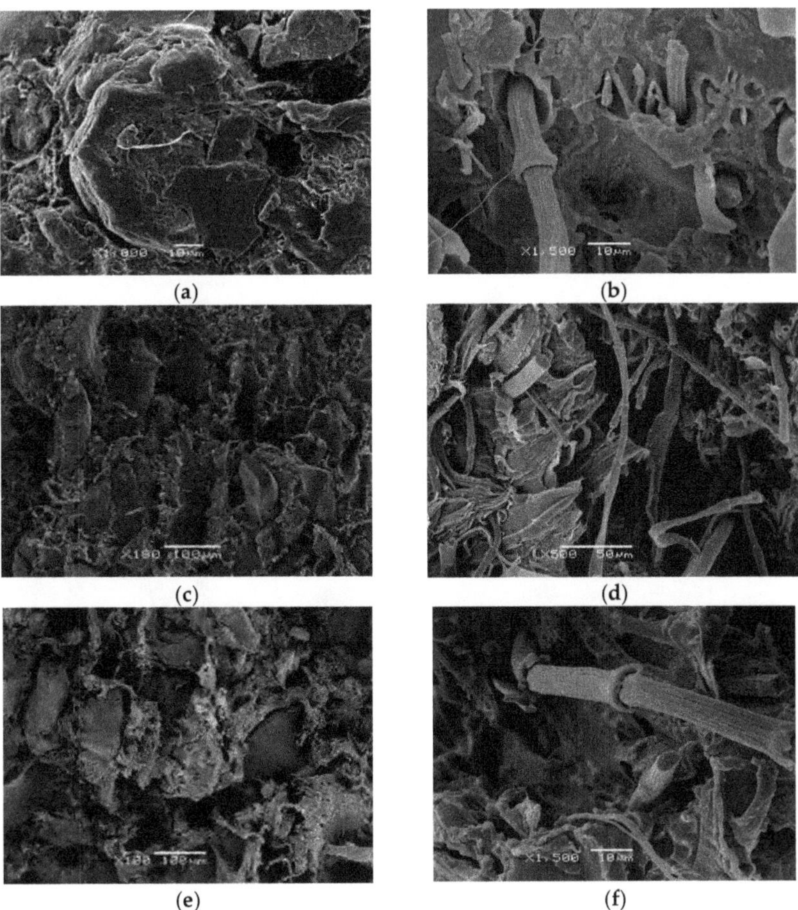

Figure 4. Scanning electron microscopy of the fracture surfaces of CFs and GTR based composites at composition of 40%: (**a**) HDPE/GTR, (**b**) HDPE/CFs, (**c**) PP/GTR, (**d**) PP/CFs, (**e**) EVA/GTR and (**f**) EVA/CFs.

Figure 6 shows the spectra of the EVA matrix and composite EVA/CFs and EVA/GTR. The comparative analysis of different spectra shows that in composite EVA/CFs, there is a difference in the maximum absorption of the band assigned to the carbonyl group of acetate component (1755 cm^{-1}), which interacts with the amine group (1537 cm^{-1}) moving the carbonyl group to a higher frequency. The band at 1150 cm^{-1} assigned to the COC group is another difference between both spectra composites. This band appears in the spectra of EVA/CFs and it is due to the interaction that take place between the acetate group of EVA and the queratinic groups of CFs.

This observation allows us to state that these two components present the best compatibility. As we can observe in the next section, these results are according to obtained mechanical properties of EVA/CFs, where all tensile properties were higher in value in EVA/CFs composites than in the materials obtained by the two other polyolephynic matrices.

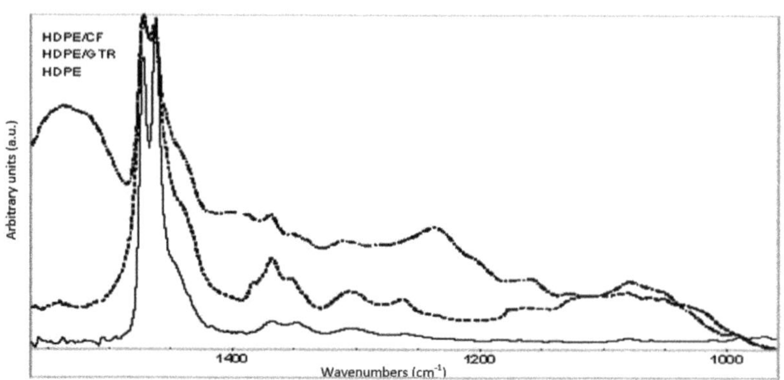

Figure 5. FTIR spectra of HDPE matrix and composites HDPE/GTR40 and HDPE/CFs40.

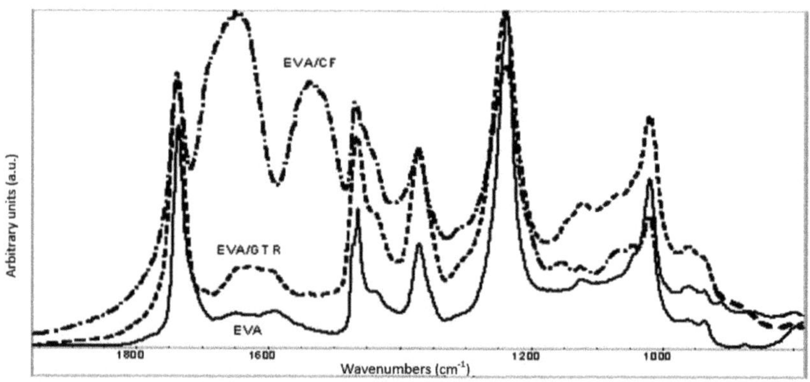

Figure 6. FTIR spectra of EVA matrix and composites EVA/GTR40 and EVA/CFs40.

Figure 7 shows the spectra of the PP matrix and composite PP/CFs and PP/GTR. Analyzing the spectra comparatively, neither changes in the absorption bands nor band shifts were observed, verifying the lack of interaction between the particles and the matrix.

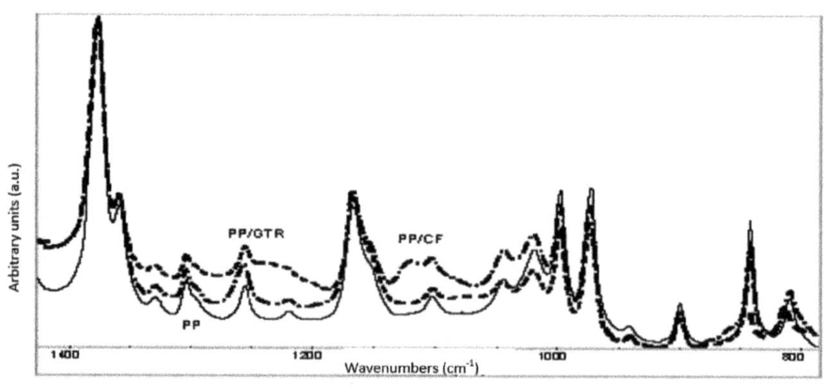

Figure 7. FTIR spectra of PP matrix and composites PP/GTR20 and PP/CFs20.

3.4. Thermogravimetric Analysis (TGA)

Figure 8 shows the thermograms of EVA/GTR and EVA/CFs composites and their corresponding components separately (EVA, GTR and CFs). The thermal behavior of CFs can be described in three main steps which are consistent with what has already been published by Tesfaye et al. [25]. In the first step, a loss of moisture was observed in the 30–200 °C temperature range. A second step was observed in the 200–360 °C temperature range corresponding to the partial decomposition of feather fractions that particularly consists in the denaturation of peptide bridges and protein chain linkages. In the third step, the feather fractions were decomposed from 360 °C to 550 °C. The residua obtained for the CFs (~20%) has been attributed to the inorganic components of the feathers [26]. For the GTR reinforcement, the thermogram showed the first loss of weight which starts at 280 °C corresponding to the release of volatile hydrocarbons and then continues until 350 °C. The second stage describes the release of the rest of the hydrocarbons with higher degradation temperatures (natural rubber, butadiene rubber, styrene butadiene rubber) until reaching a final mass that is 30–40% of the initial one [27]. The residua obtained for the GTR in the N_2 atmosphere are composed mainly of carbon and SiO_2 [28]. On the other hand, the EVA matrix showed a two-step thermal degradation process. The first stage, completed at around 370 °C, describes the deacetylation process in the vinyl acetate fraction. The second stage has been identified as complete chain scission of the residual main chain (within the interval of 380–480 °C) [29]. The comparative study of the thermal behavior of both composites allowed us to say that the first step of the thermograms (assigned to the decomposition of vinyl acetate of EVA) is completely different. The first step extension of EVA/CFs was higher than for EVA/GTR and this means that CFs interact with the EVA matrix. Analyzing both composites it can see that EVA/CFs had more compatibility than EVA/GTR.

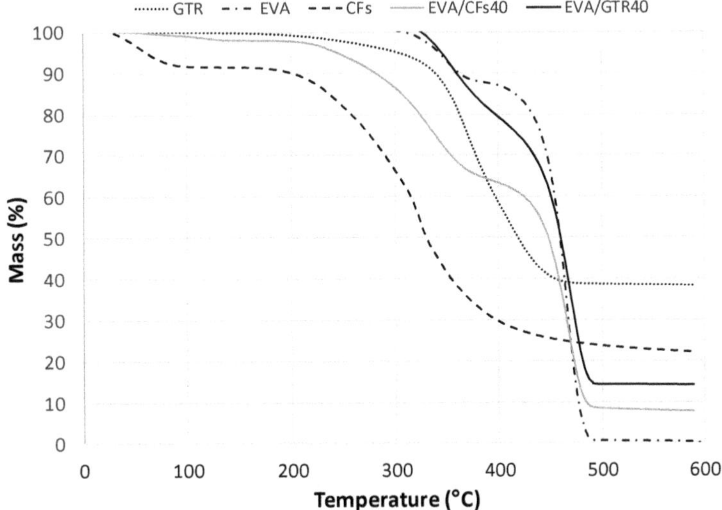

Figure 8. Thermograms of EVA/GTR and EVA/CFs composites at 40% composition and their corresponding components separately (EVA, GTR and CFs).

Figures 9 and 10 show the thermograms of both kinds of composite using HDPE and PP as a matrix, respectively. The loss of HDPE and PP mass occurred in a single-stage degradation process that occurred over the temperature range of 400–500 °C [30,31]. Comparing the thermograms of both kinds of composites it can be concluded that the interaction between polyolephynic matrix (HDPE and PP) is higher with CFs reinforcement than GTR. In both cases, the first step that relates to reinforcement degradation is larger in CFs than GTR, which means that the degree of interaction is lower in GTR than CFs.

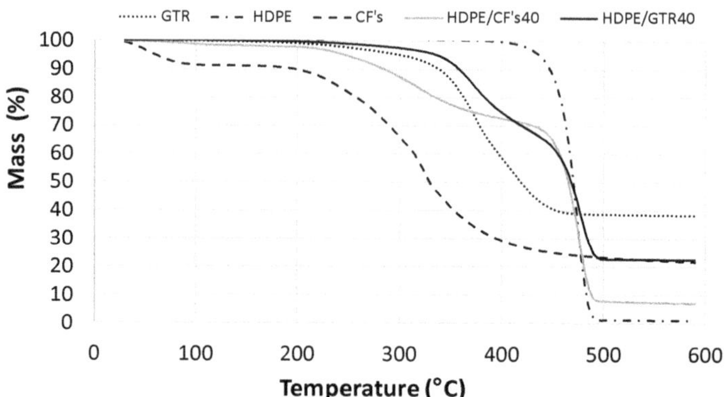

Figure 9. Thermograms of GTR/HDPE and CFs/HDPE composites at 40% composition and their corresponding components separately (EVA, GTR and CFs).

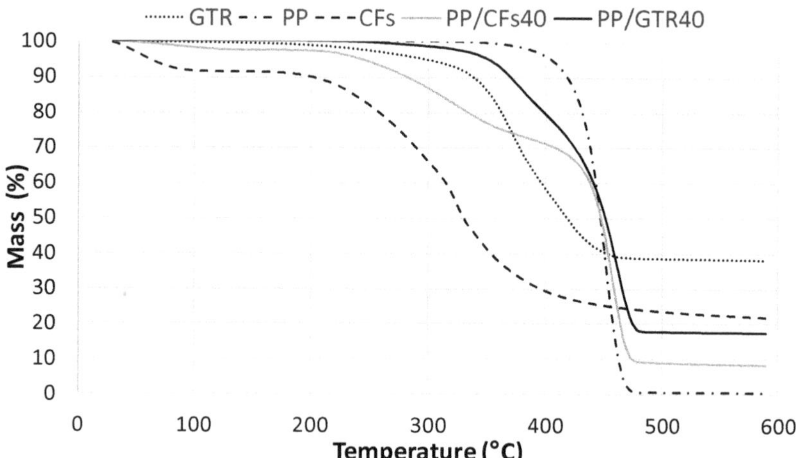

Figure 10. Thermograms of GTR/PP and CFs/PP composites at 40% composition and their corresponding components separately (EVA, GTR and CFs).

3.5. Acoustical Characterization

The measured sound absorption coefficients, in 1/3 octave bands, of the composites samples reinforced with 40% of GTR and 40% of CFs are presented in Figures 11–13. Only the HDPE/GTR40 show good results at low frequencies and mainly all the samples have acceptable results above 2500–3000 Hz. Figure 11 shows the absorption coefficient of the HDPE matrix and HDPE/CFs and HDPE/GTR. The results show that composite samples have a better noise absorption coefficient than the HDPE matrix and the best results take place using HDPE/GTR with two maximum values in 2000 and 5000 Hz. However, for EVA and PP composites the better noise reduction happens also for GTR reinforcement but at a frequency of 5500 Hz (Figures 12 and 13).

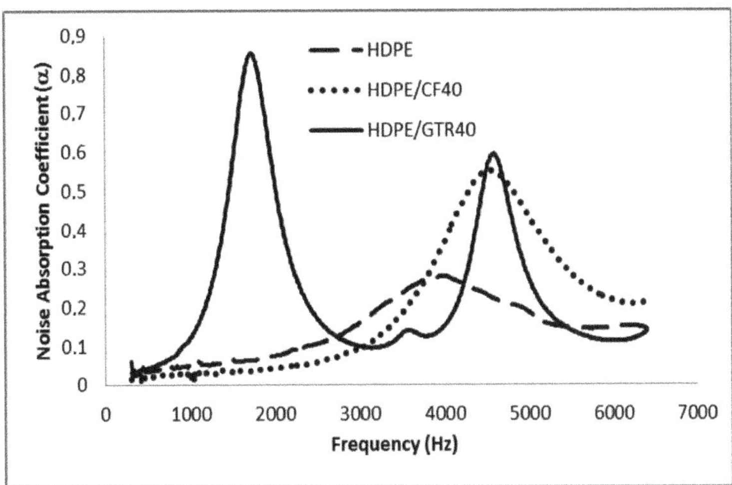

Figure 11. α values for HDPE and GTR/HDPE and CFs/HDPE composites at 40% composition.

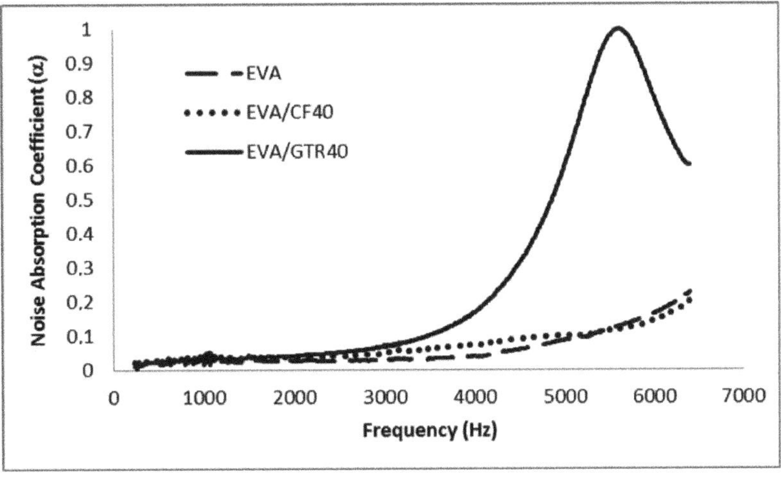

Figure 12. α values for EVA and GTR/EVA and CFs/EVA composites at 40% composition.

To know why the composite reinforced with GTR has better acoustical properties than the composite reinforced with CFs, we can observe that the composite reinforced with GTR has more crazes, holes and is more porous than composites reinforced CFs. This is due mainly to the morphology of the reinforcement since GTR has an irregular polyhedron geometry with flat faces, straight edges and sharp corners, meanwhile CFs have a fibrous morphology with a smooth surface. This kind of morphology allowed the formation of holes, porous and crazes between them and between GTR and matrix. To explain this behavior, a model was used defined as "rigid-framed porous materials" [32]. This model considers that the cavity walls are non-deforming, and the increase in acoustic absorption takes place due to the viscous losses and thermo-elastic damping where the sound propagates through a large number of air cavities in the composite. The sound propagation is, therefore, governed by the effective density and effective bulk modulus of the air in the air spaces cavities. Very often it can be assumed that these composites are microporous materials, due mainly at the interphase area between thermoplastic matrix and GTR, where several microcavities appear around the surface of the GTR. When the volume of these cavities increases, the structure factor increases, resulting in a greater effective porosity, thereby increasing the maximum acoustic absorption.

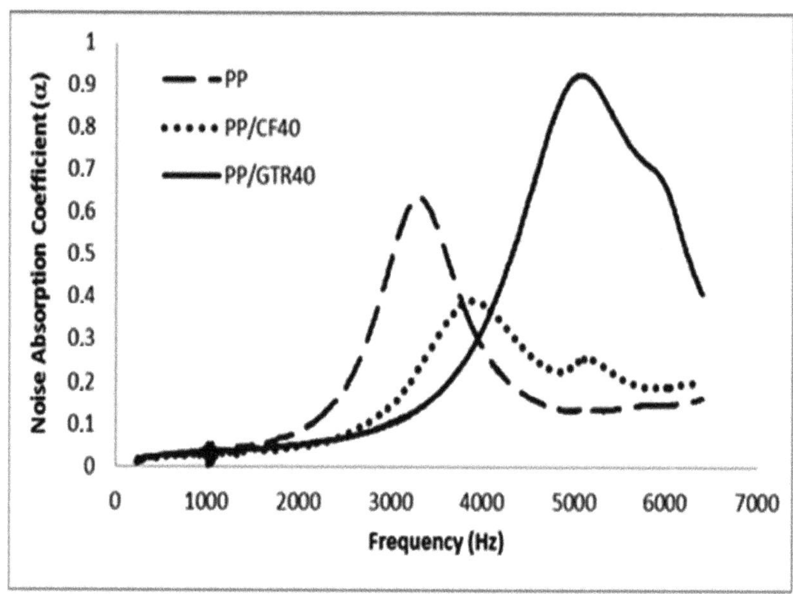

Figure 13. α values for PP and GTR/PP and CFs/PP composites at 40% composition.

4. Conclusions

Comparative analysis of different thermoplastic matrix composites reinforced with elastomeric and biogenic waste shows that: (i) the type of matrix is very important in determining the compatibility between the two components, seeing that EVA matrix defines better compatibility with both reinforced materials than polyolefin matrices (HDPE and PP); (ii) biogenic reinforcement (CFs) has better structural behavior than elastomeric reinforcement (GTR) due to higher compatibility of keratin (largely component of biogenic reinforced) with polyolephynic matrices, but mainly with EVA, as has been corroborated by FTIR and TGA; (iii) composites reinforced with GTR has better functional behavior (acoustic properties) than composites reinforced with CFs, due to the morphology of the reinforcing particles, where GTR defines an apparently irregular polyhedron geometry with a large number of microroughness in the surface that improves noise absorption when compared with the geometry of CFs particles (cylindrical, striated and very smooth).

The study of the behavior of two very different types of waste, with non-related chemical structures, biogenic and synthetic origins, dissimilar geometries and completely diverse properties shows that the incorporation of such residua in composites can be a successful approach to create materials suitable for specific applications, for construction, transport, entertainment and others. Nowadays, several proposals related to the use of fabricated composites, such as urban furniture, acoustic isolation panels, or non-structural panels for the automotive industry are under study. This path, leading to converting the subproducts of industry into useful products in other fields can be widened considering other matrices and the use of pretreatments, providing a way to fulfill the requirements of a circular economy concept.

Author Contributions: Conceptualization, X.C. and F.C.-N.; methodology, X.C. and J.C.; investigation, X.C.; resources, F.C.-N.; writing—original draft preparation, X.C. and F.C.-N.; writing—review and editing, J.C.; project administration, F.C.-N.; funding acquisition, F.C.-N. All authors have read and agreed to the published version of the manuscript.

Funding: This research was funded by MINECO/FEDER, grant number MAT2015-65392-C2-1-R.

Institutional Review Board Statement: Not applicable.

Informed Consent Statement: Not applicable.

Data Availability Statement: The data presented in this study are available on request from the corresponding author.

Conflicts of Interest: The authors declare no conflict of interest. The funders had no role in the design of the study; in the collection, analyses, or interpretation of data; in the writing of the manuscript, or in the decision to publish the results.

References

1. Blanco, I.; Siracusa, V. The Use of Thermal Techniques in the Characterization of Bio-Sourced Polymers. *Materials* **2021**, *14*, 1686. [CrossRef] [PubMed]
2. Blanco, I.; Ingrao, C.; Siracusa, V. Life-Cycle Assessment in the Polymeric Sector: A Comprehensive Review of Application Experiences on the Italian Scale. *Polymers* **2020**, *12*, 1212. [CrossRef] [PubMed]
3. Roy, P.S.; Garnier, G.; Allais, F.; Saito, K. Strategic Approach Towards Plastic Waste Valorization: Challenges and Promising Chemical Upcycling Possibilities. *ChemSusChem* **2021**, *14*, 4007–4027. [CrossRef] [PubMed]
4. Zhou, X.-L.; He, P.-J.; Peng, W.; Yi, S.-X.; Lü, F.; Shao, L.-M.; Zhang, H. Upcycling waste polyvinyl chloride: One-pot synthesis of valuable carbon materials and pipeline-quality syngas via pyrolysis in a closed reactor. *J. Hazard. Mater.* **2022**, *427*, 128210. [CrossRef]
5. Schüch, A.; Morscheck, G.; Nelles, M. Technological Options for Biogenic Waste and Residues-Overview of Current Solutions and Developments. In *Waste Valorisation and Recycling*; Ghosh, S., Ed.; Springer: Berlin/Heidelberg, Germany, 2019. [CrossRef]
6. Shen, L.; Worrell, E. Chapter 13—Plastic Recycling. In *Handbook of Recycling*; Worrell, E., Reuter, M.A., Eds.; Elsevier: Amsterdam, The Netherlands, 2014; pp. 179–190. [CrossRef]
7. Zhao, X.; Boruah, B.; Chin, K.F.; Đokić, M.; Modak, J.M.; Soo, H.S. Upcycling to Sustainably Reuse Plastics. *Adv. Mater.* **2021**, 2100843. [CrossRef]
8. Ashori, A. Hybrid Composites from Waste Materials. *J. Polym. Environ.* **2009**, *18*, 65–70. [CrossRef]
9. Seghiri, M.; Boutoutaou, D.; Kriker, A.; Hachani, M.I. The Possibility of Making a Composite Material from Waste Plastic. *Energy Procedia* **2017**, *119*, 163–169. [CrossRef]
10. Mohammed, L.; Ansari, M.N.M.; Pua, G.; Jawaid, M.; Islam, M.S. A Review on Natural Fiber Reinforced Polymer Composite and Its Applications. *Int. J. Polym. Sci.* **2015**, *2015*, 1–15. [CrossRef]
11. Girijappa, Y.G.T.; Rangappa, S.M.; Parameswaranpillai, J.; Siengchin, S. Natural Fibers as Sustainable and Renewable Resource for Development of Eco-Friendly Composites: A Comprehensive Review. *Front. Mater.* **2019**, *6*, 226. [CrossRef]
12. Davoodi, M.; Sapuan, M.S.; Ahmad, D.; Aidy, A.; Ali, A.; Jonoobi, M. Concept selection of car bumper beam with developed hybrid bio-composite material. *Mater. Des.* **2011**, *32*, 4857–4865. [CrossRef]
13. Senthilkumar, K.; Saba, N.; Rajini, N.; Chandrasekar, M.; Jawaid, M.; Siengchin, S.; Alotman, O.Y. Mechanical properties evaluation of sisal fibre reinforced polymer composites: A review. *Constr. Build. Mater.* **2018**, *174*, 713–729. [CrossRef]
14. Girge, A.; Goel, V.; Gupta, G.; Fuloria, D.; Pati, P.R.; Sharma, A.; Mishra, V.K. Industrial waste filled polymer composites—A review. *Mater. Today Proc.* **2021**, *47*, 2852–2863. [CrossRef]
15. Tasnim, S.; Shaikh, F.U.A. Effect of chemical exposure on mechanical properties and microstructure of lightweight polymer composites containing solid waste fillers. *Constr. Build. Mater.* **2021**, *309*, 125192. [CrossRef]
16. Mujal-Rosas, R.; Orrit-Prat, J.; Ramis-Juan, X.; Marin-Genesca, M.; Rahhali, A. Study on Dielectric, Mechanical and Thermal Properties of Polypropylene (PP) Composites with Ground Tyre Rubber (GTR). *Polym. Polym. Compos.* **2012**, *20*, 797–808. [CrossRef]
17. Colom, X.; Cañavate, J.; Carrillo, F.; Lis, M. Acoustic and mechanical properties of recycled polyvinyl chloride/ground tyre rubber composites. *J. Compos. Mater.* **2013**, *48*, 1061–1069. [CrossRef]
18. Sienkiewicz, M.; Janik, H.; Borzędowska-Labuda, K.; Kucinska-Lipka, J. Environmentally friendly polymer-rubber composites obtained from waste tyres: A review. *J. Clean. Prod.* **2017**, *147*, 560–571. [CrossRef]
19. Casadesús, M.; Macanás, J.; Colom, X.; Cañavate, J.; Álvarez, M.D.; Garrido, N.; Molins, G.; Carrillo, F. Effect of chemical treatments and additives on properties of chicken feathers thermoplastic biocomposites. *J. Compos. Mater.* **2018**, *52*, 3637–3653. [CrossRef]
20. Colom, X.; Cañavate, J.; Carrillo, F.; Suñol, J.J. Effect of the particle size and acid pretreatments on compatibility and properties of recycled HDPE plastic bottles filled with ground tyre powder. *J. Appl. Polym. Sci.* **2009**, *112*, 1882–1890. [CrossRef]

21. Yagneswaran, S.; Storer, W.J.; Tomar, N.; Chaur, M.N.; Echegoyen, L.; Smith, D.W., Jr. Surface-grafting of ground rubber tire by poly acrylic acid via self-initiated free radical polymerization and composites with epoxy thereof. *Polym. Compos.* **2013**, *34*, 769–777. [CrossRef]
22. *ASTM D412-16*; Standard Test Methods for Vulcanized Rubber and Thermoplastic Elastomers-Tension. ASTM International: West Conshohocken, PA, USA, 2021.
23. Archibong, F.N.; Sanusi, O.M.; Médéric, P.; Hocine, N.A. An overview on the recycling of waste ground tyre rubbers in thermoplastic matrices: Effect of added fillers. *Resour. Conserv. Recycl.* **2021**, *175*, 105894. [CrossRef]
24. Sonnier, R.; Leroy, E.; Clerc, L.; Bergeret, A.; Lopez-Cuesta, J.-M.; Bretelle, A.-S.; Lenny, P. Compatibilizing thermoplastic/ground tyre rubber powder blends: Efficiency and limits. *Polym. Test.* **2008**, *27*, 901–907. [CrossRef]
25. Tesfaye, T.; Sithole, B.; Ramjugernath, D.; Mokhothu, T. Valorisation of chicken feathers: Characterisation of thermal, mechanical and electrical properties. *Sustain. Chem. Pharm.* **2018**, *9*, 27–34. [CrossRef]
26. Aranberri, I.; Montes, S.; Azcune, I.; Rekondo, A.; Grande, H.-J. Fully Biodegradable Biocomposites with High Chicken Feather Content. *Polymers* **2017**, *9*, 593. [CrossRef]
27. Januszewicz, K.; Klein, M.; Klugmann-Radziemska, E.; Kardas, D. Thermogravimetric analysis/pyrolysis of used tyres and waste rubber. *Physicochem. Probl. Miner. Pro.* **2017**, *53*, 802–811. [CrossRef]
28. Colom, X.; Faliq, A.; Formela, K.; Cañavate, J. FTIR spectroscopic and thermogravimetric characterization of ground tyre rubber devulcanized by microwave treatment. *Polym. Test.* **2016**, *52*, 200–208. [CrossRef]
29. Badiee, A.; Ashcroft, I.; Wildman, R. The thermo-mechanical degradation of ethylene vinyl acetate used as a solar panel adhesive and encapsulant. *Int. J. Adhes. Adhes.* **2016**, *68*, 212–218. [CrossRef]
30. Koffi, A.; Mijiyawa, F.; Koffi, D.; Erchiqui, F.; Toubal, L. Mechanical Properties, Wettability and Thermal Degradation of HDPE/Birch Fiber Composite. *Polymers* **2021**, *13*, 1459. [CrossRef]
31. Gao, Z.; Kaneko, T.; Amasaki, I.; Nakada, M. A kinetic study of thermal degradation of polypropylene. *Polym. Degrad. Stab.* **2003**, *80*, 269–274. [CrossRef]
32. Brennan, M.; To, W. Acoustic properties of rigid-frame porous materials—An engineering perspective. *Appl. Acoust.* **2001**, *62*, 793–811. [CrossRef]

Article

Sustainable Cellulose-Aluminum-Plastic Composites from Beverage Cartons Scraps and Recycled Polyethylene

Irene Bonadies [1], **Roberta Capuano** [1,2], **Roberto Avolio** [1,*], **Rachele Castaldo** [1], **Mariacristina Cocca** [1], **Gennaro Gentile** [1] and **Maria Emanuela Errico** [1]

[1]   National Research Council of Italy, Institute for Polymers Composites and Biomaterials (IPCB-CNR), Via Campi Flegrei 34, 80078 Pozzuoli, Italy; irene.bonadies@cnr.it (I.B.); roberta.capuano@ipcb.cnr.it (R.C.); rachele.castaldo@ipcb.cnr.it (R.C.); cocca@ipcb.cnr.it (M.C.); gennaro.gentile@cnr.it (G.G.); mariaemanuela.errico@cnr.it (M.E.E.)

[2]   Department of Mechanical and Industrial Engineering—DIMI, University of Brescia, Via Branze 38, 25121 Brescia, Italy

*   Correspondence: roberto.avolio@cnr.it

Abstract: The sustainable management of multilayer paper/plastic waste is a technological challenge due to its composite nature. In this paper, a mechanical recycling approach for multilayer cartons (MC) is reported, illustrating the realization of thermoplastic composites based on recycled polyethylene and an amount of milled MC ranging from 20 to 90 wt%. The effect of composition of the composites on the morphology and on thermal, mechanical, and water absorption behavior was investigated and rationalized, demonstrating that above 80 wt% of MC, the fibrous nature of the filler dominates the overall properties of the materials. A maleated polyethylene was also used as a coupling agent and its effectiveness in improving mechanical parameters of composites up to 60 wt% of MC was highlighted.

Keywords: recycling; beverage cartons; composites; polymer processing; cellulose

Citation: Bonadies, I.; Capuano, R.; Avolio, R.; Castaldo, R.; Cocca, M.; Gentile, G.; Errico, M.E. Sustainable Cellulose-Aluminum-Plastic Composites from Beverage Cartons Scraps and Recycled Polyethylene. *Polymers* **2022**, *14*, 807. https://doi.org/10.3390/polym14040807

Academic Editor: Sheila Devasahayam

Received: 24 November 2021
Accepted: 14 February 2022
Published: 19 February 2022

Publisher's Note: MDPI stays neutral with regard to jurisdictional claims in published maps and institutional affiliations.

1. Introduction

Plastic-paper multilayer materials find wide application in the packaging sector. Among the most common examples of such systems are the so-called beverage cartons or multilayer cartons (MC). They are widely used for the storage of dairy products, juices and many other liquid foods, such as pre-cooked vegetables and soups.

MCs are constituted by a structural paperboard core sandwiched between plastic and aluminum layers with barrier and sealing functions (Figure 1) [1]. Non aseptic MCs for dairies are constituted only by paper (79 wt%) and polyethylene (PE, 21 wt%) layers, while an aluminum foil (5 wt%, partially substituting paper) is used to provide high protection against light and oxygen is used in aseptic cartons for long shelf-life products (UHT milk, juices).

With the increasing awareness on the environmental and management issues caused by single use disposable materials (i.e., by definition, most packaging materials), a large effort has been devoted to increase the collection and recycling rate of packages, including MCs. The constituents of multilayer cartons (high quality paper, virgin PE, and aluminum foil) are fully recyclable materials. From a theoretical point of view, MCs can therefore be recycled. However, their composite nature implies a non-straightforward separation of the different layers in order to have an effective recycling. Paper mills are currently the usual destination of waste MCs, as paper accounts for 75–80 wt%, but the pulping conditions used to recycle regular wastepaper are not suited to efficiently separate the cellulose from PE/Al layers in MCs. In fact, dedicated pulping stations must be implemented for the effective recovery of cellulose fibers from MCs [2,3]. Consequently, a separated collection and dedicated processing lines for such materials is required, strongly limiting the recycling rates of MCs [2]. The byproducts-including polyethylene, aluminum, and a variable amount of residual cellulose-are usually dried, ground, and processed to obtain composites.

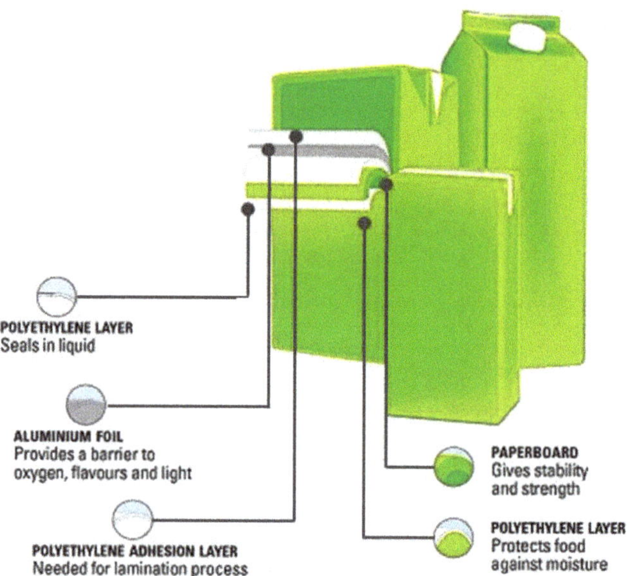

Figure 1. Structure of multilayer cartons, adapted from [1].

The problems related to the recycling of multilayer paper/plastic materials are expected to increase in the next few years due to the increased use of paper-based packaging materials [4]. This is in part driven by the ban on some single use plastic items recently enforced in the EU [5] which is boosting the use of alternative solutions, such as plastic-lined paper dishes and cups.

As an alternative to dedicated collection and pulping, the grinding of MCs to obtain a (mainly) cellulosic filler employed for the fabrication of polymer-based composites has been investigated by several authors (see review [3] and references therein). The realization of MC-based composites can represent a convenient recycling route as it does not require layers' separation. The resulting materials will show reasonable properties and low cost, with foreseen applications similar to well-known wood-plastic composites [6,7]. Polyolefins are the most investigated matrices for such composites. However, different polymers have been proposed, including thermosets. Due to the highly polar and hydrophilic nature of cellulosic materials, the ability of the selected polymer to form strong, well adhered interfaces with such fillers is a key factor to define processing and additivation strategies [8,9]. Polar matrices such as polyvinyl alcohol showed good adhesion and improved mechanical response when reinforced with MC [10], while studies on HDPE composites filled with up to 60 wt% of milled MCs [6] pointed out the need for an intense mixing to help separate paper fibers and the important role of coupling agents in improving polymer/fibers adhesion.

In this paper, these concepts are further developed. MC based materials have been investigated over a wide compositional range, demonstrating the effective production of panels with up to 90 wt% MC and extending the investigation to cartons containing aluminum foil (Al). A recycled PE was employed as the polymeric phase, leading to the realization of fully recycled, sustainable composites. Thermal, mechanical, and water sorption properties were investigated as a function of composition, and the effectiveness of a maleated polyethylene coupling agent was pointed out.

2. Materials and Methods

2.1. Materials

Non aseptic (C) and aseptic (C-Al) pre-consumer carton scraps were kindly provided by Italpack Cartons S.R.L., Lacedonia, Italy. Post-consumer high density polyethylene (PE) flakes were kindly supplied by a local selection platform. Maleated linear low-density polyethylene (MAPE, density 0.92 g/cm^3 grafted maleic anhydride 1 wt%) was kindly supplied by Agricola Imballaggi S.R.L., Pagani, Italy.

2.2. Composites Preparation

Both C and C-Al cartons were finely ground by means of a Retsch SM100 rotary knife mill (Retsch GmbH, Haan, Germany) with a 1 mm bottom sieve. The obtained powders were then mixed with PE at 175 °C in a Brabender Plastograph internal mixer (Brabender GmbH and co KG, Duisburg, Germany), equipped with a 55 cm^3 mixing chamber and two counter-rotating blades. The following procedure was adopted. First, the appropriate amount of polymeric phase (PE + MAPE) was introduced in the chamber and allowed to melt at a reduced speed for 1 min. Then, the MC powder was slowly added. The chamber was then sealed and the speed was raised to 40 rpm, mixing the materials for a further 8 min.

After melt mixing, the resulting materials were allowed to cool at room temperature and pelletized using hand cutters. They were then compression molded by means of a Collin P200 hot press equipped with a water circulated cooling system (COLLIN Lab and Pilot Solutions GmbH, Maitenbeth, Germany), using a temperature of 180 °C, a pressure of 50 bar and a permanence time of 5 min, followed by cooling to room temperature that was obtained in approximately 5 min. Sheets with a thickness of either 1 or 3 mm were obtained.

2.3. Characterization

The particle size distribution of the milled MC powders was determined by sieving through a stack of metal sieves (FILTRA Vibracion, Badalona, Spain) with nominal mesh sizes of 1000, 500, 200, 100, and 50 μm, and using a Retsch AS 200 vibratory sieve shaker (Retsch GmbH, Haan, Germany).

Morphological analysis was carried out on impact-fractured surfaces by means of a Phenom compact scanning electron microscope (Thermo Fisher Scientific Inc., Waltham, MA, USA). Samples were sputtered with a thin Au/Pd layer before analysis, by means of a Emitech K575X sputtering device (Quorum Technologies Ltd., Laughton, UK).

Differential Scanning Calorimetry (DSC) analyses were carried out by means of a Mettler Toledo 822 DSC (Mettler-Toledo, LLC, Columbus, OH, USA). Samples were sealed in aluminum pans and analyzed under a nitrogen flux, using the following the temperature program: heating from 25 °C to 180 °C, cooling from 180 °C to 0 °C, and heating from 0 °C to 180 °C at a heating/cooling rate of 10 °C/min. The percent crystallinity content (X_c) was calculated according to the following equation:

$$X_c = \Delta Hm / \Delta Hm^\circ \times 100 \tag{1}$$

where ΔHm is the heat of melting recorded on the sample, normalized on the content of polyethylene, and ΔHm° is the melting enthalpy of fully crystalline polyethylene, equal to 293 J/g [11,12].

Due to the hygroscopic nature of cellulose, the physical and mechanical properties of composites containing cellulosic fillers can be influenced by the humidity absorbed onto cellulose in ambient conditions [13]. Therefore, all samples were conditioned at 25 °C and 50% relative humidity (RH) for at least 24 h before mechanical testing.

Tensile tests were carried out on dumb-bell specimens with a cross section of 4 mm^2 using a gauge length of 26 mm and a deformation speed of 5 mm/min, by means of an Instron 4505 testing machine (ITW Inc., Glenview, IL, USA). Young's modulus (E), peak stress (σ_{max}), and elongation at break (ε_R) were calculated from stress/strain curves as average values over at least 10 tested specimens.

Charpy impact tests were carried out on notched specimens (notch depth to width ratio of 0.3, span length 48 mm) by means of a CEAST Resil Impactor pendulum (ITW Inc., Glenview, IL, USA), equipped with a DAS 4000 Acquisition System, using an impact energy of 3.6 J and an impact speed of 0.99 m/s. Impact toughness values were calculated as average over a least five tested specimens.

Water absorption tests were carried out by immersion of samples (30 mm × 10 mm × 3 mm) in distilled water, recording the mass difference at regular intervals. All specimens were padded with dry filter paper before weighing to remove excess water from the external surfaces.

3. Results

3.1. Filler Size Distribution, Composites Preparation and Morphology

In this paper a mechanical recycling strategy of MCs has been proposed, developing materials containing up to 90 wt% of MCs in combination with recycled polyethylene. Exploring such a wide range of compositions, fully recycled materials with properties ranging from lightly-filled polymer composites to lignocellulose-based products (e.g., fiberboards) have been realized.

The size distribution of C and C-Al materials after milling was measured by sieving the powders through multiple metal sieves and recording the weight of the different fractions. The results obtained are reported in Figure S1 in the Supplementary Materials and reveal a qualitatively similar distribution for both samples, with the maximum amount of particles retained by the 500 μm sieve. C cartons showed a larger fraction of particles with size < 200 μm (44 wt%), with respect to C-Al (31 wt%).

It is worth noting that in these systems, the size distribution of the cellulosic component is expected to change significantly during the melt processing step. In fact, the shear forces exerted by the molten polymer will induce a progressive fragmentation of paper particles, ideally approaching a separation into single cellulose fibers. A precise determination of the final particle size distribution is beyond the scope of this work, as it would require the separation of fibers from the polymeric fraction after processing. Optical micrographs of thin (100 μm) films were recorded for materials at moderate MC content, as reported in Figure S2 in the Supplementary Materials, revealing the coexistence of both single cellulosic fibers and residual compact, paper-like cellulose clusters and, where present, Al-foil fragments.

MC was successfully mixed with PE using a conventional polymer processing apparatus and then easily molded by compression molding to obtain composites sheets. Given the polar nature of cellulosic fibers and aluminum fragments contained in MC, a very low adhesion with the PE matrix is expected. Therefore, maleated polyethylene (MAPE) was selected and added as a coupling agent. The modification of polymers with maleic anhydride polar groups is well known as an effective strategy to improve dispersion and interfacial adhesion in composites containing fillers with polar surfaces (i.e., cellulose [14,15], inorganics such as glass fibers [16], and mineral micro/nano particles [17]). Maleated polyolefins are available commercially and can be conveniently used, in appropriate contexts, as processing additives in polyolefin and recycled polyolefin-based composites [18]. The amount of MAPE was varied among 2.5 and 10 wt% (calculated with respect to the PE matrix weight) to study and optimize its effect on composite properties. At a high MC content (80 and 90 wt%), due to the high content of polar fillers and, as a consequence, high polymer/filler contact surface, MAPE was only added at the maximum percentage. Materials without MAPE were also prepared for comparison. Table 1 resumes codes and composition of all of the materials realized.

Table 1. Compositions and codes of all materials realized. Sample codes are in the format MxCy and MxCAly for cartons without and with Al-foil, respectively. The number x indicates MAPE content while y indicates carton content.

PE + MAPE	MAPE (% vs. PE)	MC (%)	Sample Codes	
			Cartons without Al (C)	Cartons without Al (C-Al)
100	-	-	PE	-
80	-	20	M0 C20	M0 CAl20
60	-	40	M0 C40	M0 CAl40
40	-	60	M0 C60	M0 CAl60
20	-	80	M0 C80	M0 CAl80
10	-	90	M0 C90	M0 CAl90
100	2.5	-	M2.5	-
80	2.5	20	M2.5 C20	M2.5 CAl20
60	2.5	40	M2.5 C40	M2.5 CAl40
40	2.5	60	M2.5 C60	M2.5 CAl60
100	5	-	M5	-
80	5	20	M5 C20	M5 CAl20
60	5	40	M5 C40	M5 CAl40
40	5	60	M5 C60	M5 CAl60
100	10	-	M10	-
80	10	20	M10 C20	M10 CAl20
60	10	40	M10 C40	M10 CAl40
40	10	60	M10 C60	M10 CAl60
20	10	80	M10 C80	M10 CAl80
10	10	90	M10 C90	M10 CAl90

The effectiveness of the mixing procedure and the effect of the coupling agent were investigated analyzing the morphology of impact fracture surfaces by means of SEM analyses (details on impact testing are reported in Section 3.3). Micrographs of samples at the lowest and highest MC content-with and without MAPE-are reported in Figure 2 as representative examples of the composites realized.

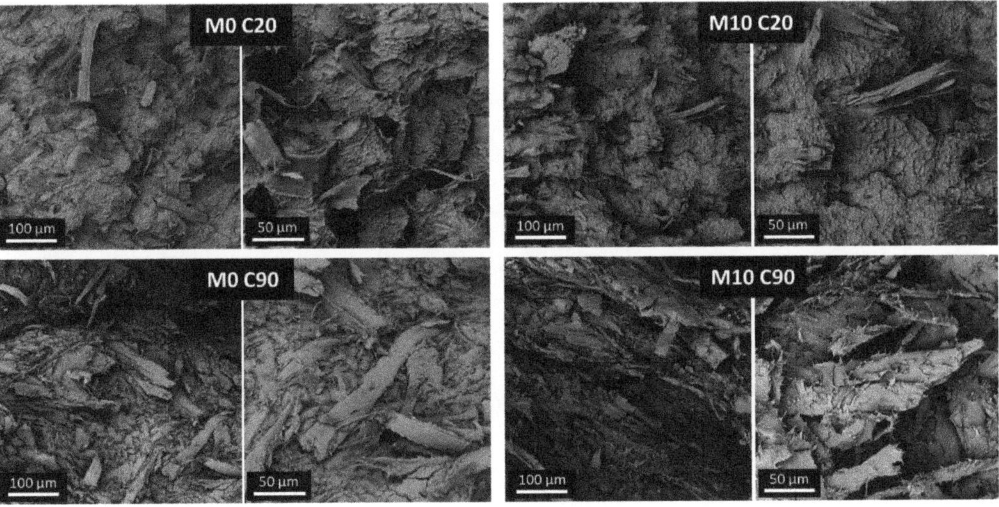

Figure 2. SEM micrographs of impact fracture surfaces of samples without (M0 C20, M0 C90) and with (M10 C20, M10 C90) coupling agent, at different magnification levels.

Micrographs of samples without MAPE, at the lowest and highest MC content, show a clear separation between the polymeric fraction, deformed during the impact test, and the cellulose fibers that appear almost completely "clean" (that is, not covered by the polymer). During deformation, the matrix/fiber interfaces were largely broken, resulting in extensive debonding. In both M0 C20 and M0 C90 samples the fracture surfaces appear irregular; moreover, large fiber bundles are observed in the C90 sample, along with some single fiber, indicating a partly ineffective mixing at high MC content. In the samples containing MAPE, a different scenario is observed. The fracture surface of the M10 C20 shows fewer fibers exposed with respect to M0 C20, showing that fracture does not propagate preferentially through the matrix/fiber interfaces. This finding can be attributed to a stronger polymer/filler adhesion induced by the coupling agent. In fact, visible fibers appear well bonded and covered by a polymer layer. These considerations partly hold also for the M10 C90 sample; a polymeric layer is observed onto fibers. However, a large number of exposed fibers are also observed, indicating an improved but insufficient adhesion. In the composites based on Al-containing cartons, a similar morphology was found (see Figure S3 in the Supplementary Materials). Morphological analyses confirm that at high MC content, the fibrous nature of the filler governs the structure of the materials.

3.2. Thermal Analysis

Differential scanning calorimetry was carried out on the composites to analyze the effect of fillers (cellulose, aluminum) and MAPE on the thermal behavior of the polymeric phase. The main thermal parameters obtained are reported in Table 2, while the thermograms of representative samples are illustrated in Figure 3 and in Figure S4 in the Supplementary Materials.

The presence of MC, with or without Al, does not show a strong influence on the melting/crystallization behavior of the recycled PE matrix up to 60 wt% of the load, as inferred by the minor changes observed in the melting and crystallization temperatures. Interestingly, at 80 and 90 wt% MC, the appearance of a low temperature crystallization/melting peak was observed (Figure 3). This signal can be attributed to the phase transition of LDPE, contained in MCs (at high carton content, LDPE represents more than 50% of the polymeric phase). The presence of separated phase transitions indicates a probable phase separation of the different polyethylene species, due to their limited compatibility [19,20]. The low-temperature peak is slightly visible at 60 wt% MC (see high magnification inserts in Figure 3), but is much more evident in the 80 and 90% materials. This phenomenon is expected to reduce the homogeneity of the polymeric phase in samples at high MC load, with consequences on the mechanical behavior (as discussed in next section). The crystallinity index decreases in all samples as a function of the MC content. This is due to the geometrical constraint of the fibrous fraction that hinders polymer crystallization.

Table 2. Results of DSC analysis of all materials realized: crystallization temperature (T_c), melting temperature (T_m), and crystallinity (X_c). The crystallinity content is calculated on the basis of the polymeric content (recycled PE + the PE fraction of MC + MAPE).

Code	T_c (°C)	T_m (°C)	X_c (%)	Code	T_c (°C)	T_m (°C)	X_c (%)
PE	116	138	71	-			
M0 C20	113	140	67	M0 CAl20	113	140	66
M0 C40	115	137	59	M0 CAl40	115	137	59
M0 C60	116	136	57	M0 CAl60	116	136	56
M0 C80	116	108–133	44	M0 CAl80	115	108–132	54
M0 C90	115	106–130	36	M0 CAl90	114	106–128	48
M2.5	116	137	71	-			
M2.5 C20	113	140	66	M2.5 CAl20	116	137	71
M2.5 C40	116	136	65	M2.5 CAl40	113	140	62
M2.5 C60	115	136	54	M2.5 CAl60	116	135	59

Table 2. *Cont.*

Code	T_c (°C)	T_m (°C)	X_c (%)	Code	T_c (°C)	T_m (°C)	X_c (%)
M5	116	137	68	-			
M5 C20	114	140	67	M5 CAl20	115	138	69
M5 C40	115	137	60	M5 CAl40	114	138	54
M5 C60	116	135	56	M5 CAl60	115	136	63
M10	115	137	71	-			
M10 C20	114	139	67	M10 CAl20	115	138	64
M10 C40	114	137	62	M10 CAl40	114	138	63
M10 C60	116	135	53	M10 CAl60	115	136	59
M10 C80	115	132	42	M10 CAl80	116	109–132	54
M10 C90	115	131	31	M10 CAl90	116	106–128	45

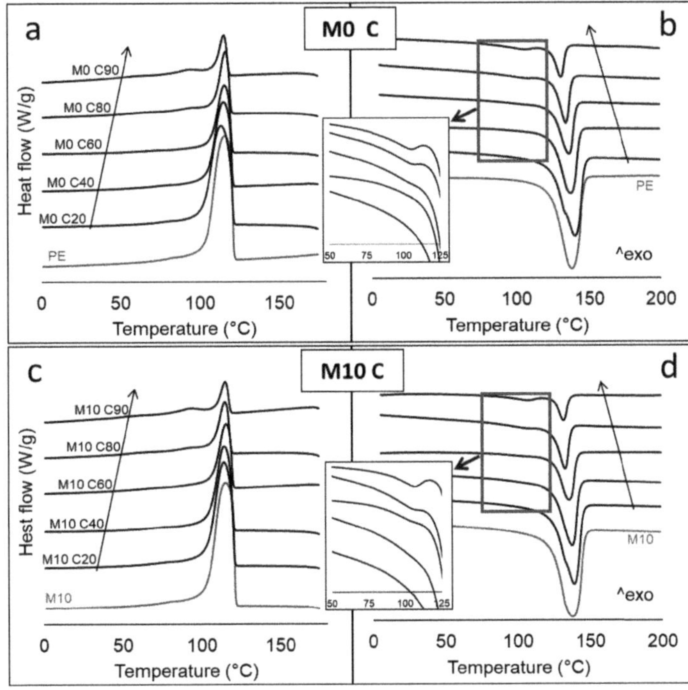

Figure 3. DSC thermograms showing the crystallization and melting peaks of M0 C (**a**,**b**) and M10 C (**c**,**d**) systems, respectively. Arrows indicate increasing MC content.

3.3. Mechanical Analysis

Mechanical parameters of the prepared composites were analyzed by means of tensile and Charpy impact tests. The results of mechanical testing are reported in Figure 4.

From the mechanical results, different observations can be pointed out. The recycled PE matrix has relatively high stiffness and low elongation at break, corresponding to rigid, high crystallinity HDPE grades. By increasing the MC content, the elongation (Figure 4e,f) further decreases coupled with a strong increase in elastic modulus (Figure 4a,b), up to 220% with respect to neat PE. These findings are expected as both cellulose and aluminum act as rigid fillers increasing the elastic modulus of the composites and, as a consequence, reducing ultimate elongation. The addition of MAPE slightly decreases stiffness and increases ultimate elongation in most materials. This is due to the nature of the additive backbone (LLDPE has generally lower stiffness than HDPE). At high MC content, the modulus stopped increasing in composites containing aluminum: this can be a symptom of ineffective mixing, due to the presence of large (hundreds of μm) Al-foil particles not fragmented during melt processing, as shown in Figures S1 and S7 in the Supplementary Materials.

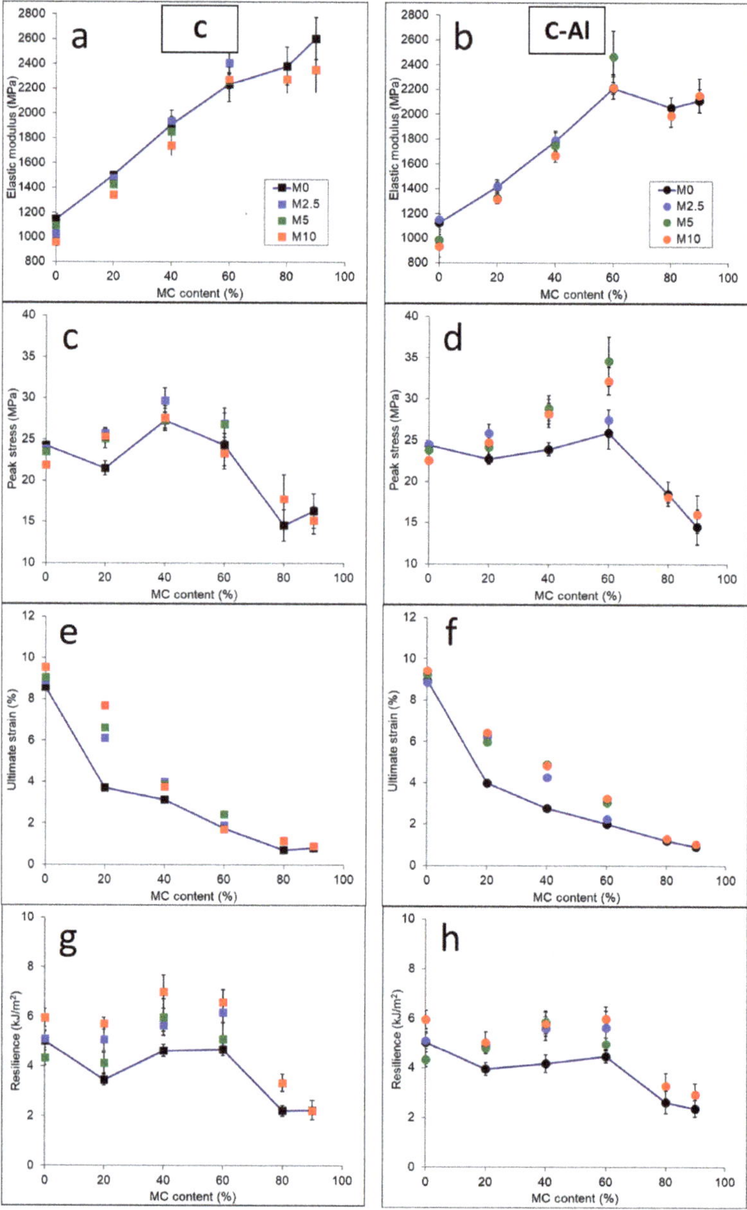

Figure 4. Tensile and impact parameters of the prepared composites as a function of composition: elastic modulus (**a**,**b**), peak stress (**c**,**d**), ultimate strain (**e**,**f**) and impact resilience (**g**,**h**). Lines connecting points of the M0Cx-CAlx systems are reported as a guide for the eye.

The peak stress of composites (Figure 4c,d) is strongly influenced by MAPE, as this parameter is more sensitive than modulus to adhesion at the polymer/filler interface [6,21]. MAPE generally led to an increase in peak stress for composites up to 60 wt% MC, in particular this effect was more significant in C-Al containing composites, that showed the maximum load-bearing ability. In contrast to what observed in previous investigations with virgin HDPE [6], in uncompatibilized materials peak stress did not decrease monotonically with increasing MC content. A possible reason is that the recycled PE matrix used in this work has developed, during service life and reprocessing, a degree of oxidation that could have slightly increased its compatibility with polar fillers. The presence of oxidation was confirmed by FTIR analysis that revealed the presence of carbonyl groups (Figure S5 in the Supplementary Materials). The overall mechanical response of composites containing up to 60 wt% of MC falls in the range of common polyolefin composites containing lignocellulosic fillers [22,23]. For these materials, then, possible applications similar to wood-polymer and cellulose-polymer composites, which are increasingly used in the transport and construction sectors, can be foreseen [7,24].

As in the case of elastic modulus, materials at high (80, 90 wt%) MC content show a different mechanical response if compared to the low (20–60 wt%) MC content materials, with a sharp decrease of strength and a negligible effect of MAPE. At high MC load the fibrous nature of the filler dominates the structure of the material and samples at 80 and 90 wt% MC can be more effectively compared to fibrous systems with polymeric binders (such as particleboards and fiberboards) or even to cellulosic sheets (heavy paperboard) than to polymer-matrix composites. In fact, mechanical properties obtained on highly filled materials were found to be similar to those recorded on fibrous and wood-based board [25,26], and on some type of paperboard [27]. The limited effect of MAPE on the mechanical response of highly filled materials can be a consequence of the highly fibrous morphology and to the insufficient adhesion evidenced by SEM analyses.

Impact tests were carried out on notched specimens by means of an instrumented Charpy pendulum, with the recorded impact resilience reported in Figure 4g,h. Interestingly, the impact performances of pristine PE were not much affected or, in some case, slightly increased by the addition of MC (at least for contents up to 60 wt%). The recycled PE used in this work has high stiffness and low deformability, thus resulting in a low impact resilience. In these conditions, the presence of a fibrous filler effectively deviates the propagating fracture front, thus increasing the energy required to break the sample (even in the case of a weak adhesion to the polymer). The presence of MAPE generally increased the impact resilience up to 60 wt% M by increasing the energy required to separate the polar filler from the polymeric phase during fracture [28]. Materials with high MC content showed a different behavior, with a decrease of impact resilience only marginally mitigated by the coupling agent as a consequence of the fibrous morphology of these samples.

3.4. Water Absorption

The water absorption behavior of the realized materials was reported in terms of water uptake as function of MC content during water immersion up to 1000 h (Figure 5). Curves of weight uptake vs. immersion time of each sample are reported in the Supplementary Materials Figure S6. The water absorption recorded increases with increasing MC content, as expected due to the highly hydrophilic nature of cellulose. Water absorption was lower for materials containing C-Al at all compositions, with an uptake value that is about 14% for CAl90 compared to the 25% of C90. This large difference cannot be justified only by the lower water uptake capacity of aluminum foil compared to cellulose, but is probably also related to a barrier effect offered by the aluminum fragments at the surface of tested samples (see Figure S7 in the Supplementary Materials). Absorption values of samples up to 60 wt% of C/C-Al are slightly lower in the presence of MAPE: this finding was attributed to the formation of a polymeric layer onto fiber surfaces, promoted by MAPE (as evidenced in Section 3.1), thus reducing their water binding tendency. At high MC content, the effect of MAPE is negligible due to high cellulose content. It is worth noting that, even at the highest

MC content, water absorption kinetics and the equilibrium uptake values are generally lower than observed in typical fiberboards [29,30], especially for the composites containing C-Al, thus making these materials interesting candidates for the substitution of fiberboards in wet environments.

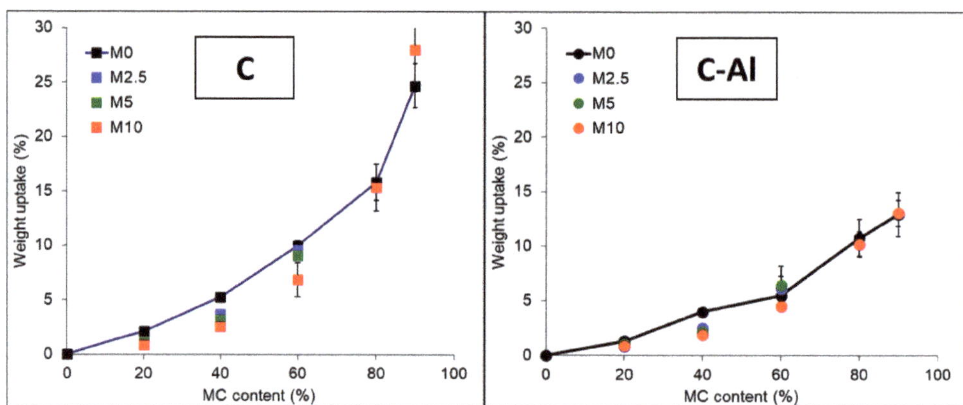

Figure 5. Water absorption after 1000 h of water immersion for the prepared composites. Lines connecting points of the M0Cx-CAlx systems are reported as a guide for the eye.

4. Conclusions

In this paper, a valorization strategy for the mechanical recycling of multilayer cartons (MCs) is reported. Industrial scraps of MCs with and without aluminum were employed in combination with recycled polyethylene to realize sustainable composites in a wide compositional range. In particular, MCs were tested as cellulosic or cellulosic/aluminum-based filler leading to the realization of thermoplastic composite materials. Good processability and formability by compression molding was demonstrated in the whole compositional range explored. Morphological analyses revealed a partial destructuration of the cellulosic component during processing, with the coexistence of both single cellulosic fibers and residual compact, paper-like cellulose clusters, and Al-foil fragments. The mechanical properties of the realized composites range from the typical values of lightly-filled polymer/wood or polymer/cellulose composites to highly fibrous materials (i.e., particleboards, fiberboards) at the highest MC content (80–90 wt%). Water absorption tests revealed a very good water resistance, in particular for the C-Al systems. Maleic anhydride modified polyethylene (MAPE) added as coupling agent during processing, was effective to improve the mechanical properties of the composites containing up to 60 wt% of MC. At higher filler content, the effect of MAPE was negligible due to the highly fibrous nature of the materials realized.

Supplementary Materials: The following are available online at https://www.mdpi.com/article/10.3390/polym14040807/s1. Figure S1. Particle size distribution of milled C and C-Al cartons as measured by sieving; Figure S2. Optical micrographs (transmitted light) of films of the M0 C40 (a–d) and M0 CAl40 (e–h) samples, highlighting the presence of single cellulose fibers, along with residual paper-like aggregates; Figure S3. SEM micrographs of impact fracture surfaces of samples containing C-Al cartons, without (M0 CAl20, M0 CAl90) and with (M10 CAl20, M10 CAl90) coupling agent, at different magnification levels; Figure S4. DSC thermograms showing the crystallization and melting peaks of M0 C-Al (a,b) and M10 C-Al (c,d) systems, respectively. Arrows indicate increasing MC content.; Figure S5. ATR-FTIR spectra of the recycled polyethylene used in this work (PE) compared to a virgin HDPE taken as a reference; Figure S6. Water absorption as a function of water immersion time for the prepared composites; Figure S7. Optical micrographs of the surface of high MC content materials, showing the presence of aluminum foil fragments at the surface in the M0 CAl80, and M0 CAl90 samples.

Author Contributions: Conceptualization, I.B., R.A., M.C., M.E.E. and G.G.; methodology, I.B., M.E.E., R.A., R.C. (Roberta Capuano) and R.C. (Rachele Castaldo); investigation, I.B., R.A., R.C. (Roberta Capuano) and R.C. (Rachele Castaldo); writing—original draft preparation, I.B., R.A. and M.E.E.; supervision, M.E.E., G.G. and M.C. All authors have read and agreed to the published version of the manuscript.

Funding: This work was supported by the project "SIRIMAP—Detection Systems of marine plastic pollution and subsequent recovery-recycling", identification code ARS01_01183, funded by the Italian Ministry of Research and Education under the National Operational Program (PON) on Research and Innovation 2014-2020–European Structural and Investment Funds ERDF/ESF.

Institutional Review Board Statement: Not applicable.

Informed Consent Statement: Not applicable.

Data Availability Statement: Not applicable.

Conflicts of Interest: The authors declare no conflict of interest.

References

1. ACE UK. What Is a Beverage Carton? Available online: http://www.ace-uk.co.uk/what-is-a-carton/ (accessed on 20 November 2021).
2. Zero Waste Europe. *Recycling of Multilayer Composite Packaging: The Beverage Carton.* Available online: https://zerowasteeurope.eu/library/recycling-of-multilayer-composite-packaging-the-beverage-carton/ (accessed on 20 November 2021).
3. Robertson, G. Recycling of Aseptic Beverage Cartons: A Review. *Recycling* **2021**, *6*, 20. [CrossRef]
4. Gesellschaft für Verpackungsmarktforschung (GVM). *Substitution of Plastic Packaging by Paper-Based Composites.* Available online: https://newsroom.kunststoffverpackungen.de/wp-content/uploads/2021/03/2021_03_23_Bericht_Substitution-durch-Papierverbunde.pdf (accessed on 20 November 2021).
5. Directive (EU) 2019/904 of the European Parliament and of the Council of 5 June 2019 on the Reduction of the Impact of Certain Plastic Products on the Environment. Available online: https://www.legislation.gov.uk/eudr/2019/904 (accessed on 20 November 2021).
6. Avella, M.; Avolio, R.; Bonadies, I.; Carfagna, C.; Errico, M.E.; Gentile, G. Recycled Multilayer Cartons as Cellulose Source in HDPE-Based Composites: Compatibilization and Structure-Properties Relationships. *J. Appl. Polym. Sci.* **2009**, *114*, 2978–2985. [CrossRef]
7. Spear, M.J.; Eder, A.; Carus, M. Wood Polymer Composites. In *Wood Composites*; Ansell, M.P., Ed.; Woodhead Publishing Ltd.: Sawston, UK, 2015; pp. 195–249.
8. Castaldo, R.; De Falco, F.; Avolio, R.; Bossanne, E.; Cicaroni Fernandes, F.; Cocca, M.; Di Pace, E.; Errico, M.E.; Gentile, G.; Jasiński, D.; et al. Critical Factors for the Recycling of Different End-of-Life Materials: Wood Wastes, Automotive Shredded Residues, and Dismantled Wind Turbine Blades. *Polymers* **2019**, *11*, 1604. [CrossRef] [PubMed]
9. Rao, J.; Zhou, Y.; Fan, M. Revealing the Interface Structure and Bonding Mechanism of Coupling Agent Treated WPC. *Polymers* **2018**, *10*, 266. [CrossRef] [PubMed]
10. Avella, M.; Cocca, M.; Errico, M.; Gentile, G. Biodegradable PVOH-Based Foams for Packaging Applications. *J. Cell. Plast.* **2011**, *47*, 271–281. [CrossRef]
11. Di Lorenzo, M.L.; Avella, M.; Avolio, R.; Bonadies, I.; Carfagna, C.; Cocca, M.; Errico, M.E.; Gentile, G. Isothermal and Nonisothermal Crystallization of HDPE Composites Containing Multilayer Carton Scraps as Filler. *J. Appl. Polym. Sci.* **2012**, *125*, 3880–3887. [CrossRef]
12. Wunderlich, B. *Macromolecular Physics, Volume 3: Crystal Melting*; Academic Press: London, UK, 1980.
13. Woodhams, R.T.; Thomas, G.; Rodgers, D.K. Wood Fibers as Reinforcing Fillers for Polyolefins. *Polym. Eng. Sci.* **1984**, *24*, 1166–1171. [CrossRef]
14. Cocca, M.; Avolio, R.; Gentile, G.; Di Pace, E.; Errico, M.E.; Avella, M. Amorphized Cellulose as Filler in Biocomposites Based on Poly(ε-Caprolactone). *Carbohydr. Polym.* **2015**, *118*, 170–182. [CrossRef]
15. Avolio, R.; Graziano, V.; Pereira, Y.D.F.; Cocca, M.; Gentile, G.; Errico, M.E.; Ambrogi, V.; Avella, M. Effect of Cellulose Structure and Morphology on the Properties of Poly(Butylene Succinate-Co-Butylene Adipate) Biocomposites. *Carbohydr. Polym.* **2015**, *133*, 408–420. [CrossRef]
16. Lin, J.-H.; Huang, C.-L.; Liu, C.-F.; Chen, C.-K.; Lin, Z.-I.; Lou, C.-W. Polypropylene/Short Glass Fibers Composites: Effects of Coupling Agents on Mechanical Properties, Thermal Behaviors, and Morphology. *Materials* **2015**, *8*, 8279–8291. [CrossRef]
17. Avolio, R.; Gentile, G.; Avella, M.; Carfagna, C.; Errico, M.E. Polymer–Filler Interactions in PET/CaCO3 Nanocomposites: Chain Ordering at the Interface and Physical Properties. *Eur. Polym. J.* **2013**, *49*, 419–427. [CrossRef]
18. Avolio, R.; Spina, F.; Gentile, G.; Cocca, M.; Avella, M.; Carfagna, C.; Tealdo, G.; Errico, M. Recycling Polyethylene-Rich Plastic Waste from Landfill Reclamation: Toward an Enhanced Landfill-Mining Approach. *Polymers* **2019**, *11*, 208. [CrossRef] [PubMed]

19. Hameed, T.; Hussein, I.A. Effect of Short Chain Branching of LDPE on Its Miscibility with Linear HDPE. *Macromol. Mater. Eng.* **2004**, *289*, 198–203. [CrossRef]
20. Zhao, L.; Choi, P. A Review of the Miscibility of Polyethylene Blends. *Mater. Manuf. Process.* **2006**, *21*, 135–142. [CrossRef]
21. Dányádi, L.; Renner, K.; Móczó, J.; Pukánszky, B. Wood Flour Filled Polypropylene Composites: Interfacial Adhesion and Micromechanical Deformations. *Polym. Eng. Sci.* **2007**, *47*, 1246–1255. [CrossRef]
22. Adhikary, K.B.; Pang, S.; Staiger, M.P. Dimensional stability and mechanical behaviour of wood–plastic composites based on recycled and virgin high-density polyethylene (HDPE). *Compos. B Eng.* **2008**, *39*, 807–815. [CrossRef]
23. Olakanmi, E.O.; Strydom, M.J. Critical materials and processing challenges affecting the interface and functional performance of wood polymer composites (WPCs). *Mater. Chem. Phys.* **2016**, *171*, 290–302. [CrossRef]
24. Müssig, J.; Graupner, N. Test Methods for Fibre/Matrix Adhesion in Cellulose Fibre–Reinforced Thermoplastic Composite Materials: A Critical Review. In *Progress in Adhesion and Adhesives, Volume 6*; Mittal, K.M., Ed.; Scrivener Publishing LLC: Beverly, MA, USA, 2021; pp. 69–130. [CrossRef]
25. Cai, Z.; Ross, R.J. Mechanical Properties of Wood-Based Composite Materials. In *Wood Handbook: Wood as an Engineering Material*; General technical report FPL; U.S. Department of Agriculture, Forest Service, Forest Products Laboratory: Madison, WI, USA, 2010; pp. 12.1–12.12.
26. Amazio, P.; Avella, M.; Emanuela Errico, M.; Gentile, G.; Balducci, F.; Gnaccarini, A.; Moratalla, J.; Belanche, M. Low Formaldehyde Emission Particleboard Panels Realized through a New Acrylic Binder. *J. Appl. Polym. Sci.* **2011**, *122*, 2779–2788. [CrossRef]
27. Kibirkštis, E.; Kabelkaitė, A. Research of Paper/Paperboard Mechanical Characteristics. *Mechanika* **2006**, *59*, 34–41.
28. Keener, T.; Stuart, R.; Brown, T. Maleated Coupling Agents for Natural Fibre Composites. *Compos. A Appl. Sci. Manuf.* **2004**, *35*, 357–362. [CrossRef]
29. Shi, S.Q. Diffusion Model Based on Fick's Second Law for the Moisture Absorption Process in Wood Fiber-Based Composites: Is It Suitable or Not? *Wood Sci. Technol.* **2007**, *41*, 645–658. [CrossRef]
30. Garcia, R.A.; Cloutier, A.; Riedl, B. Dimensional Stability of MDF Panels Produced from Fibres Treated with Maleated Polypropylene Wax. *Wood Sci. Technol.* **2005**, *39*, 630–650. [CrossRef]

polymers

Article

Oil Production by Pyrolysis of Real Plastic Waste

Laura Fulgencio-Medrano [1], Sara García-Fernández [1], Asier Asueta [1], Alexander Lopez-Urionabarrenechea [2,*], Borja B. Perez-Martinez [2] and José María Arandes [3]

[1] Gaiker Technology Center, Basque Research and Technology Alliance (BRTA), Parque Tecnológico de Bizkaia, Edificio 202, 48170 Zamudio, Spain; fulgencio@gaiker.es (L.F.-M.); garciasa@gaiker.es (S.G.-F.); asueta@gaiker.es (A.A.)
[2] Chemical and Environmental Engineering Department, Faculty of Engineering of Bilbao, University of the Basque Country (UPV/EHU), Plaza Ingeniero Torres Quevedo 1, 48013 Bilbao, Spain; borjabperez@gmail.com
[3] Department of Chemical Engineering, University of the Basque Country (UPV/EHU), P.O. Box 644, 48080 Bilbao, Spain; josemaria.arandes@ehu.eus
* Correspondence: alex.lopez@ehu.eus

Abstract: The aim of this paper is for the production of oils processed in refineries to come from the pyrolysis of real waste from the high plastic content rejected by the recycling industry of the Basque Country (Spain). Concretely, the rejected waste streams were collected from (1) a light packaging waste sorting plant, (2) the paper recycling industry, and (3) a waste treatment plant of electrical and electronic equipment (WEEE). The influence of pre-treatments (mechanical separation operations) and temperature on the yield and quality of the liquid fraction were evaluated. In order to study the pre-treatment effect, the samples were pyrolyzed at 460 °C for 1 h. As pre-treatments concentrate on the suitable fraction for pyrolysis and reduce the undesirable materials (metals, PVC, PET, inorganics, cellulosic materials), they improve the yield to liquid products and considerably reduce the halogen content. The sample with the highest polyolefin content achieved the highest liquid yield (70.6 wt.% at 460 °C) and the lowest chlorine content (160 ppm) among the investigated samples and, therefore, was the most suitable liquid to use as refinery feedstock. The effect of temperature on the pyrolysis of this sample was studied in the range of 430–490 °C. As the temperature increased the liquid yield increased and solid yield decreased, indicating that the conversion was maximized. At 490 °C, the pyrolysis oil with the highest calorific value (44.3 MJ kg^{-1}) and paraffinic content (65% area), the lowest chlorine content (128 ppm) and more than 50 wt.% of diesel was obtained.

Keywords: chemical recycling; plastic waste; industrial rejected streams; pyrolysis oil; pyrolysis; secondary raw materials; alternative fuels

Citation: Fulgencio-Medrano, L.; García-Fernández, S.; Asueta, A.; Lopez-Urionabarrenechea, A.; Perez-Martinez, B.B.; Arandes, J.M. Oil Production by Pyrolysis of Real Plastic Waste. *Polymers* **2022**, *14*, 553. https://doi.org/10.3390/polym 14030553

Academic Editor: Cristiano Varrone

Received: 4 January 2022
Accepted: 19 January 2022
Published: 29 January 2022

Publisher's Note: MDPI stays neutral with regard to jurisdictional claims in published maps and institutional affiliations.

1. Introduction

Nowadays, the huge growth in plastic production has resulted in a massive generation of this kind of waste. Despite not being considered hazardous waste, plastic waste causes cumulative and long-term environmental impacts due to its long lifespan [1,2]. In order to reduce adverse effects presented by plastic waste, a recent European Directive 2018/851 was renewed to promote the recovery of plastic waste for recycling, avoiding the deposition in landfills [3]. Nevertheless, the amount of waste that ends up in landfills is still very high. According to a recently published report, in Spain, landfill is the most recurrent measure to get rid of post-consume plastic waste (46%) [4]. Increasing the recycling rate and reducing the landfill disposal only through conventional mechanical recycling routes is sometimes complicated and not an economically viable alternative, since there are a lot of plastic waste streams that are composed by a wide and intermingled variety of materials, especially those that came from industrial recovery processes [5,6]. Therefore, new recycling alternatives are required, and pyrolysis, recently catalogued as TRL 9 (technology readiness level), seems to be a promising option [7].

The pyrolysis process consists of the thermal degradation of organic materials under an inert atmosphere. During the pyrolysis of plastics, the long carbon chains are thermally broken down into useful fractions that can serve as fuels or sources of chemicals. Typically, a liquid product, a gaseous product and a solid product are formed [8]. The solid product is usually made up of the inorganic elements of the waste (including the charges in plastics), together with the so-called "char", a carbonaceous product typical of the thermal decomposition of some polymers. Due to its heterogeneity, the solid product is not usually easy to valorise. On the contrary, the gaseous fraction normally meets the standards of a gaseous fuel, but its economic value is not sufficient to be the exclusive product of the process. Consequently, the economic success of the pyrolysis of plastic waste depends on the characteristics of the liquid product, which in principle can be largely assimilated to certain refinery streams [9]. In fact, given the petrochemical origin of plastics, returning them to the refineries when they have reached the end of life should be their circular route, provided that they cannot be mechanically recycled. In such a scenario, the pyrolysis of plastic waste allows for two benefits: the reduction of landfill disposal and the recovery of valuable hydrocarbons [10].

The characteristics and yields of the products depend to a great extent on various parameters of the pyrolysis process: temperature, residence time, reactor type, pressure, type and rate of fluidizing gas, heating rate, type of catalyst and type of feedstock [11–13]. As pyrolysis is a thermal process, the temperature is the major operational factor since it controls the cracking reactions of the polymer chains. It was reported that temperatures of the 300–500 °C range favoured conversion into liquid products [10,14]. Even though pyrolysis can tolerate mixtures of different types of plastics [5,15], polyolefins have turned out to be the most appropriate, since they produce liquid oils with low octane numbers, which are comparable to conventional fuel [15–17]. There are many references in the literature about the pyrolysis of virgin plastic and prepared plastic waste mixtures in order to achieve liquid fuel. However, few authors have analysed the pyrolysis of real waste samples which results in different liquid products in terms of composition and quality, owing to its great complexity [5,18,19]. Some undesirable materials usually present in real waste streams (PVC, metals, PET, inert materials and cellulose-based materials) deteriorate the quality of the pyrolysis products obtained. On the one hand, chlorine from PVC is detrimental since chlorinated compounds can be formed in the liquid product decreasing its quality and limiting its application [5,20,21]. The metals contained in the initial samples might remain unaltered during the pyrolysis process and could be recovered from solid product [20], but it might also produce an undesired catalytic effect [18,22–26] and of course, as part of the solid fraction, they do decrease the yield of liquids and gases. PET and cellulosic materials favour the formation of char and an aqueous phase in the pyrolysis liquid [27–30]. Thus, the source and the previous treatment of these waste streams influence the properties of the final products. Nonetheless, there are no publications analysing the effect of treatments applied to the waste stream prior to pyrolysis in order to improve the quality of the liquid obtained. Hence, in this study, the waste stream composition to pyrolize is another parameter to be studied.

In this research, three real samples were collected from different plastic-rich waste streams rejected from industrial operations and whose final disposal is normally landfill. These samples were used as feedstock in the pyrolysis process to evaluate the production and quality of the liquid products, in order to be considered for their application in refineries. The samples were processed as received and after using different pre-treatments to separate the non-desired components that could downgrade the pyrolysis oil quality. Once the effect of the pre-treatment was studied, the sample producing the most appropriate pyrolysis oil to be used as feedstock for refineries was selected. This sample was employed to investigate the effect of temperature on the production of pyrolysis oil.

2. Materials and Methods

2.1. Origin of the Samples

The samples used in this research were provided by three different recycling companies of the Basque Country (Spain). The origin and type of such waste streams, together with the annual amount generated in the Basque Country in 2017 are summarised in Table 1. The first sample was collected from a light packaging waste classification plant; there, the main components of the light packaging selectively collected in Bizkaia (a region of Basque Country) are separated for their subsequent use as raw material in recycling companies. Although more than 70% of the collected packaging waste is properly classified, there is a rejected stream with non-separated materials that is incinerated or deposited in landfill. This sample mainly consisted of PE bags and films caked with dirt, as can be seen in the picture in Table 1. The sample was named "Film sample". The second sample was collected from a company devoted to the production of newsprint paper from wastepaper recovered in street containers. As a consequence of the separation processes, a plastic containing rejected stream is also generated in this plant, mainly consisting of polyolefins and cellulosic materials. In this case, the sample was named "Paper sample". The third sample was a rejected stream coming from waste of electrical and electronic equipment (WEEE) and taken from a company devoted to dismantling and shredding WEEE to obtain high-quality metal fractions for its commercialization. This sample was named "WEEE sample".

Table 1. Annual production, industrial activity and aspect of the three samples used.

Sample	Film	Paper	WEEE
Annual production (t/year)	3491	24,341	13,228
Activity	Separation of light packaging	Recycling of paper	WEEE treatment
Aspect			

Representative samples were obtained by quartering method according to the C702 and D75 ASTM Standards. Afterward, the composition was qualitatively and quantitatively determined. First, a manual separation based on visual identification was carried out in order to separate the materials into macroscopic components (plastics, wood, textiles and inert materials). Next, the specific composition of the plastic fraction was determined by infrared spectroscopy and flame test.

2.2. Pre-Treatment Techniques

According to the composition and the specific characteristics of each waste stream, different mechanical separation technologies were applied to reduce the non-desired components in each case. For the Film sample, the separation method used was flotation (sink/float), as it takes advantage of the difference between the density of PVC and the main plastics present in the waste, i.e., polyolefins and styrene polymers (ABS and PS). Paper sample was previously deagglomerated in a jaw shredder (Oliver&Battle SOPAC-100, Badalona, Spain) to improve materials separation. After a previous screening of pre-treatment technologies for the paper-based stream, the optical separation was the selected method, as it showed the highest reduction in PVC concentration. For this purpose, automatic identification and sorting pilot line (UNISORT PX800F, RTT Systemtechnik GmbH, Zittau, Germany), based on a near-infrared (NIR) spectrophotometer (4000–10,000 cm^{-1} spectral range) and an air ejection, was employed. In this equipment, the waste flow placed on a conveyor belt passes under the measuring module (KUSTA 4004M20, LLA Instruments

GmbH, Berlin, Germany) and is irradiated with IR light, which is partly absorbed. The reflected light is captured by the sensor and conducted to the spectrophotometer, obtaining the characteristic infrared spectrum of each material. The most suitable technology employed for the pre-treatment of WEEE sample was the densimetric table since it was capable of separating PVC wires from other particles taking advantage of their different morphology. The equipment used (PETKUS KD50, Palencia, Spain) combines the movement of the table with the air-flow generated by the fans, which makes the materials slide on its surface and enables the effective separation of wires, among other undesired elements.

2.3. Pyrolysis Experiments

For pyrolysis experiments, typically 85 g of crushed samples (dp < 8 mm) were placed in a 2 L unstirred stainless steel autoclave (4570 model of Parr Instruments (Moline, IL, USA), see Figure 1). Prior to the experiments, the system was purged for 20 min with an N_2 stream, which was kept constant during reaction (80 mL min^{-1}). Then, the reactor was heated to the selected experiment temperature (430–490 °C) at a rate of 15 °C min^{-1}. As the vapours were generated, they left the reactor passing through a water-refrigerated condenser where the condensable liquids were collected. After an isothermal holding time of 1 h, the reaction system was cooled down to ambient temperature. The solid residue collected inside the reactor and the condensed liquids were weighted, and their yields were calculated according to Equation (1). The gas product yield was calculated by difference.

$$\text{Product yield (wt.\%)} = \frac{M_{\text{product}}(g)}{M_{\text{feed}}(g)} \cdot 100 \tag{1}$$

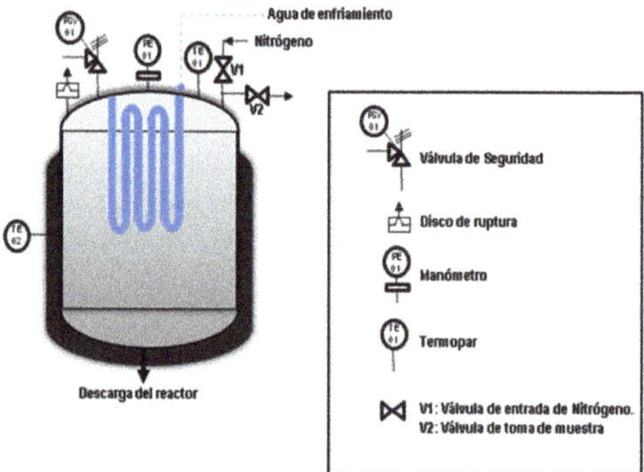

Figure 1. Pyrolysis rector.

2.4. Analytical Techniques

Both raw and pre-treated waste samples were thoroughly characterized using the following analytical techniques. Thermogravimetric analyses (TGA) of the waste samples were carried out in a Mettler-Toledo (Columbus, OH, USA) thermobalance (TGA/DSC1 Stare System). Approximately 10 mg of sample were introduced into the thermobalance and heated to 800 °C at 20 °C min^{-1} rates under a constant N_2 flow (50 mL min^{-1}). The mass loss was continuously monitored as a function of temperature. The derivative thermogravimetric curve (DTG) was calculated to determine the range of temperatures in which the greatest thermal degradation took place. Furthermore, proximate analysis (moisture, volatile matter, fixed carbon, ash) was carried out according to D3173-85 and D3174-82 ASTM standards in LECO TGA-701 equipment (St. Joseph, MI, USA). The C, H, N and S contents (ultimate analysis) were measured by a CHN-S automatic analyser (LECO TrueSpec CHN and TrueSpec S, St. Joseph, MI, USA). The content of chlorine was measured following the UNE 15408 standard, which consists of combusting the samples in a calorimetric bomb (1356 Parr Instrument, Moline, IL, USA) with pure oxygen and absorbing the combustion gases generated in a basic solution of KOH (0.2 M). The concentration of chloride anions present in the solution was then quantified by high-performance liquid chromatography (HPLC) using an ion chromatograph (ICS-1000 DIONEX, Sunnyvale, CA, USA). Waste samples were also digested with H_2SO_4 at 180 °C for 15 min and subsequently with an oxidizing mixture of H_2O_2 and HNO_3 at 200 °C for 20 min in order to determine the metal content. Metals were then analysed in aqueous phase by inductively coupled plasma-optical emission spectrometry (ICP-OES) using Optima 2100 DV Perkin-Elmer equipment (Waltham, MA, USA. At last, bulk density of samples was determined by weighting the mass occupied in a measured volume.

Concerning pyrolysis oils, a gas chromatograph (GC 6890N), equipped with an HP-5MS capillary column (30 m × 0.25 mm × 0.25 μm) and coupled to a mass spectrometer detector (MS 5975), both from Agilent Technologies (Santa Clara, CA, USA), was used to determine their composition. The higher heating value (HHV) was measured by using the 1356 Parr calorimetric bomb, which at the same time was also used to determine the halogen content (F^-, Cl^- and Br^-) following the aforementioned UNE 15408 standard. Metal content of oils was established by digestion with an oxidizing mixture of H_2O_2 and HNO_3 at 200 °C for 20 min followed by ICP-OES. Finally, simulated distillation analyses were carried out according to the ASTM D2887 standard, using an Agilent Technologies 6890 GC System (Santa Clara, CA, USA), equipped with: (i) an FID detector; and (ii) a DB-2887 semi capillary column (length, 10 m; internal diameter, 0.53 mm; thickness, 3 μm).

3. Results

3.1. Influence of the Pre-Treatment Techniques

3.1.1. Properties of Waste Samples

Table 2 shows the material composition of the collected waste samples, both in the "raw" condition (as received) and after pre-treatment. Bulk density of samples is also included in Table 2, together with two calculated ratios that allow quantifying the effectiveness of the pre-treatment techniques (see Equations (2) and (3)): "total material recovery" (TMR) and "recovery of material suitable for pyrolysis" (RMSP). While TMR refers to the amount of material obtained after pre-treatment (separation of unwanted materials), the term RMSP refers to the concentration of plastics in the recovered fraction that is appropriate for the pyrolysis process in order to obtain high liquid yields. In this research, polyolefins and styrenics were considered as the most suitable plastics for such an objective.

$$\text{TMR (wt.\%)} = \frac{\text{recovered material (kg)}}{\text{initial material (kg)}} \cdot 100 \tag{2}$$

$$\text{RMSP (wt.\%)} = \frac{\text{suitable recovered plastics (kg)}}{\text{recovered material (kg)}} \cdot 100 \tag{3}$$

The raw Film sample is the sample that showed the highest plastic content, mainly composed of polyolefins (75 wt.%). In addition, its high content of PVC, PET and inorganic matter, the last formed by aluminium cans that were trapped inside the PE bags, was remarkable. In the flotation process, most of the polyolefins, whose density is less than 1.0 g cm^{-3}, floated to the surface while other polymers such as PVC and PET, and inorganic materials, whose densities are greater, sank to the bottom. Hence, the content of such undesirable components was significantly reduced during the pre-treatment. So, flotation was enabled to recover the 93.0 wt.% of the MSP, mainly formed by polyolefins (increase from 75.0 to 93.1 wt.%) with a high TMR (78.5 wt.%). It is reported that other authors employing flotation methods to separate plastics were used wetting agents in the process. Pongstabodee et al. used 30% *w/v* calcium chloride solution to separate PP and HDPE from a mixed post-consumer plastic waste (PET, PVC, PS and ABS) [31] and Guo et al. employed a solution with 70 mg L^{-1} of sodium dodecyl sulphate to separate PS from a mixture of

PET, PVC and PC from light packaging waste [32]. However, in this case, the employment of wetting agents was not necessary to obtain high TMR and RMSP rates.

Table 2. Material composition (wt.%) and bulk density (kg m^{-3}) of raw and pre-treated samples.

Material	Film Sample		Paper Sample		WEEE Sample	
	Raw	Pre-Treated	Raw	Pre-Treated	Raw	Pre-Treated
Polyolefins (PP, PE)	75.0	93.1	36.1	68.0	14.6	19.0
Styrenics (PS, ABS)	1.7	1.0	8.8	3.6	39.2	47.9
PVC	4.5	0.0	2.8	1.5	16.3 [2]	4.8 [2]
PET	3.4	0.0	5.3	2.7	1.4	1.0
Other thermoplastics [1]	0.1	0.0	0.0	0.0	23.4	23.4
Multimaterial	3.4	4.5	1.2	1.4	2.3 [3]	1.3 [3]
Other organic	0.8	1.1	0.0	0.0	0.0	0.0
Inorganic matter	5.4	0.0	13.4	2.5	2.3	1.4
Celluloses	5.5	0.1	31.5	18.3	0.5	1.2
Textile	0.2	0.2	0.9	2.0	0.0	0.0
Bulk density	0.093	0.036	0.253	0.050	1.620	0.510
TMR	-	78.5	-	27.4	-	67.2
RMSP	-	93.0	-	43.7	-	83.7

[1] PMMA, PUR, PC, PA, PBT, POM; [2] Including electric wires; [3] PCB + rubber.

The raw Paper sample presented an important content of cellulosic materials (31.5 wt.%), principally paper and paperboard, which was expected because of the origin of the sample. Nevertheless, polyolefins constituted again the main fraction (36.1 wt.%). Moreover, it is important to highlight the high content of inorganic materials (13.4 wt.%) as well as the non-desired plastics, PVC (2.8 wt.%) and PET (5.3 wt.%) in the raw sample. In this case, the optical separation equipment achieved the removal of cellulosic material, reducing its content up to 18.3 wt.%. This resulted in a lower concentration of such oxygenated polymers, which might also improve the quality of the oil. It is also remarkable the strong reduction of inorganics (from 13.4 to 2.5 wt.%) and to a lesser extent that of PVC (from 2.8 to 1.5 wt.%) and PET (from 5.3 to 2.7 wt.%). However, compared to the other treatments, this method showed the lowest percentage of RMSP (43.7 wt.%). The separation difficulty of this sample lay in the fact that the paper was very intermingled with plastic and other materials and, in spite of the previous sample deagglomeration, the optical separation equipment could not properly identify and separate the desired polymers. In this case, the incorporation of a previous wet stage with some agent could have resulted in a better separation of polyolefins and paper. In fact, the dissolution of adhesive resins of polyolefins with the aim of separating polyolefins from post-consumer recycled paper was previously reported [33].

The WEEE sample contained plastics of diverse nature, as can be observed from the high percentages of "other thermoplastics" (23.4 wt.%), which includes many different materials, and styrenics (39.2 wt.%), formed by ABS and PS. Additionally noteworthy was the high percentage of PVC, which in this case corresponded to electric wires. After passing through the densimetric table, the electrical wires were strongly reduced (from 16.3 to 4.8 wt.%) allowing to obtain 83.7 wt.% of RMSP and 67.2 wt.% of TMR. Hiosta et al. also applied this technique to separate electric wires from WEEE [34]. Dodbiba et al. used the densimetric table to separate PP from PET/PVC fraction and concluded that the densimetric table was effective when the density difference between particles was at least 450 kg m^{-3} [35].

Concerning bulk densities (after being crushed to dp < 8 mm) Table 2 shows that WEEE samples presented the highest values, whereas Film and Paper ones had extremely low densities. That means that Film and Paper samples could present more difficulties when stored, transported or fed to the reactor. The value of the bulk densities of the three samples decreased with the pre-treatments, mainly due to the removal of inorganics. In the WEEE sample, the difference was greater, probably owing to the decrease in the number of wires. In view of Table 2, it can be said that, in general, the pre-treatment techniques employed have proved to be effective for concentrating the plastics, specifically the polyolefins, and reducing PVC, metal and other inorganic materials.

The ultimate and proximate analyses of both raw and pre-treated samples are presented in Tables 3 and 4, respectively. Table 3 also includes the HHV of the samples. As far as the ultimate analysis is concerned, the percentages of C, H and N corresponded adequately to the composition shown in Table 2. The two samples with the highest plastic content (film and WEEE) showed the highest values of carbon while the sample with high paper content showed the typical carbon values of cellulosic materials. Concerning the H/C ratio, this was in accordance with the nature of the predominant polymers they contain, being the highest for polyolefin-rich samples (Film) and the lowest for styrene plastic-rich samples (WEEE) [18,22]. With regard to nitrogen, the high percentage of this element in the WEEE sample must be noted, which is directly related to its high content of nitrogenous polymers such as ABS, PUR or PA. Finally, it is worth noting the high percentage of chlorine in the Film and WEEE samples, which must mostly come from the PVC they contain. The WEEE sample has a much higher PVC content than the Film sample, and yet both have a chlorine content of around 4 wt.%. The explanation is that, as mentioned above, the PVC counted in the WEEE sample includes electric wires, i.e., it is not only PVC but also copper. After pre-treatment, an increase in C and H is generally appreciated due to the elimination of inorganic materials [36], together with a noticeable decrease in chlorine, related to PVC elimination. This is a very important result in terms of producing pyrolysis oils with low chlorine content. At last, the Film sample showed the highest HHV as a consequence of its high polyolefinic content [37], followed by the WEEE sample and the Paper sample, respectively. In all cases, the pre-treatment techniques caused an increase in HHV, as expected from the elimination of inorganic and low-HHV materials.

Table 3. Ultimate analysis (wt.%) and HHV (MJ kg^{-1}) of raw and pre-treated waste samples (as received).

Sample	C	H	N	S	Cl	H/C	HHV
Film	70.5	11.2	0.4	<0.1	4.1	1.91	36.3
PT-Film	75.6	12.3	0.5	n.d. [1]	0.2	1.95	38.0
Paper	46.8	6.8	0.3	0.2	1.6	1.74	22.8
PT-Paper	55.9	8.2	0.4	n.d. [1]	0.9	1.76	27.0
WEEE	64.4	7.0	1.2	<0.1	4.4	1.30	26.9
PT-WEEE	74.7	7.8	1.9	n.d.[1]	1.2	1.25	33.9

[1] Not determined.

Table 4. Proximate analysis of raw and pre-treated waste samples (wt.%, as received).

Sample	Moisture	Volatile Matter	Fixed Carbon [1]	Ash
Film	0.7	91.1	1.6	6.6
PT-Film	0.3	93.1	0.4	6.2
Paper	3.5	77.9	7.0	11.6
PT-Paper	2.4	82.9	5.3	9.4
WEEE	0.0	76.6	1.4	22.0
PT-WEEE	0.0	88.7	2.4	8.9

[1] By difference.

Regarding the proximate analysis, all samples were mainly composed of volatile matter, as expected in this type of plastic and paper-rich waste. This is a desirable property because it is from this volatile matter that the pyrolysis oils are formed. Otherwise, it can be seen that the paper samples contained higher moisture and fixed carbon than the rest, as expected from a sample rich in cellulosic material. In addition, the WEEE (raw) sample showed a significant amount of ash, probably coming from the PVC wires and inorganic fillers that may be contained in the plastics of this waste. The ash content was significantly reduced with pre-treatment (also in the other two samples), which increased the amount of volatile matter in the waste, a circumstance that would possibly improve the yield of pyrolysis oils, as mentioned above.

Table 5 shows the metal content of the samples. When analysing this table, the uncertainty associated with the multi-stage analysis of these complex samples must be taken into account. It can be seen that the major metals were calcium, titanium, aluminium and, in the case of the WEEE sample, copper from electric wires. It was surprising that the highest amount of iron was present in the paper sample, as this is an unsuitable material for paper/cardboard waste collection, although it is used in paper and printing ink applications [38]. If such iron came from steel, it is possible that there were no magnetic separators in the waste paper and paperboard sorting plant, and this iron ended up in the rejected fraction under study in this work. Regarding heavy metals (Ni, Pb, Cd, Cr, As, Cu, Co, Tl, Sb, Sn, Hg, Mn, Zn), zinc was the most present, with the exception of copper in the case of the WEEE sample. On this occasion, no clear effect of the pre-treatment could be established for the three samples, although the reduction of the amount of copper in the WEEE sample was evident, which was in agreement with results obtained in previous characterizations.

Table 5. Metal content in raw and pre-treated samples (ppm, as received).

Metal	Film	PT-Film	Paper	PT-Paper	WEEE	PT-WEEE
Zn	86.5	107	114	72	457	222
Sb	7.7	3.1	7.7	5.0	<1	621
P	169	133	60.4	236	269	818
Pb	<1	7.2	8.6	3.5	<1	91.1
Co	<1	4.5	3.2	38.8	9.3	7.9
Cd	<1	<1	<1	<1	10.6	22.4
Ni	<1	< 1	20.2	8.3	65.5	58.7
Fe	399	323	4257	949	498	632
B	<1	<1	<1	<1	33.3	16.4
Si	175	118	320	500	270	246
Mn	10.2	9.1	31.7	18.5	115	174
Cr	3.8	4.1	32.2	6.4	16.3	15.1
Mg	182	183	559	334	763	550
Ca	14,620	13,890	12,060	17,390	13,820	8022
Cu	22.0	32.4	76.5	24.7	48620	6337
Ti	6546	8173	1520	2993	5842	6264
Al	8463	3620	17430	6960	25,580	27,820
Na	253	192	843	421	75	68.8

Concentration of Sn, Tl, As, Mo, Ba, V and Ag was <1 ppm.

The TGA profiles of all the samples are illustrated in Figure 2, where it can be seen that temperatures slightly higher than 500 °C were needed for the total conversion of the three samples. In view of these results and taking into account that an isotherm of 1 h would be used in the pyrolysis experiments, a lower temperature (460 °C) was selected for the initial experiments, in order to avoid gas formation. As far as the decomposition phenomena occurring in the different samples are concerned, different behaviour can be observed between them. The Film sample showed a decomposition that took place practically in a single step at temperatures close to 500 °C, which is usual in samples whose main content is polyolefins [39]. After pre-treatment, it seemed that decomposition happened in a lower temperature range (narrower DTG peak), which is a consequence of the removal of polymers that can start to decompose at lower temperatures than polyolefins (styrenics, PVC, etc.).

The thermogravimetric profile of the Paper sample showed three main stages of decomposition. (1) The first one close to 100 °C, corresponding to moisture loss, (2) another one around 350 °C, which is related to the decomposition of the cellulosic materials, and the last one (3), at temperatures similar to those observed for the Film sample, corresponding to the cracking of the polyolefins [39]. After pre-treatment, the third DTG peak was higher, as a consequence of the polyolefin concentration resulting from pre-treatment.

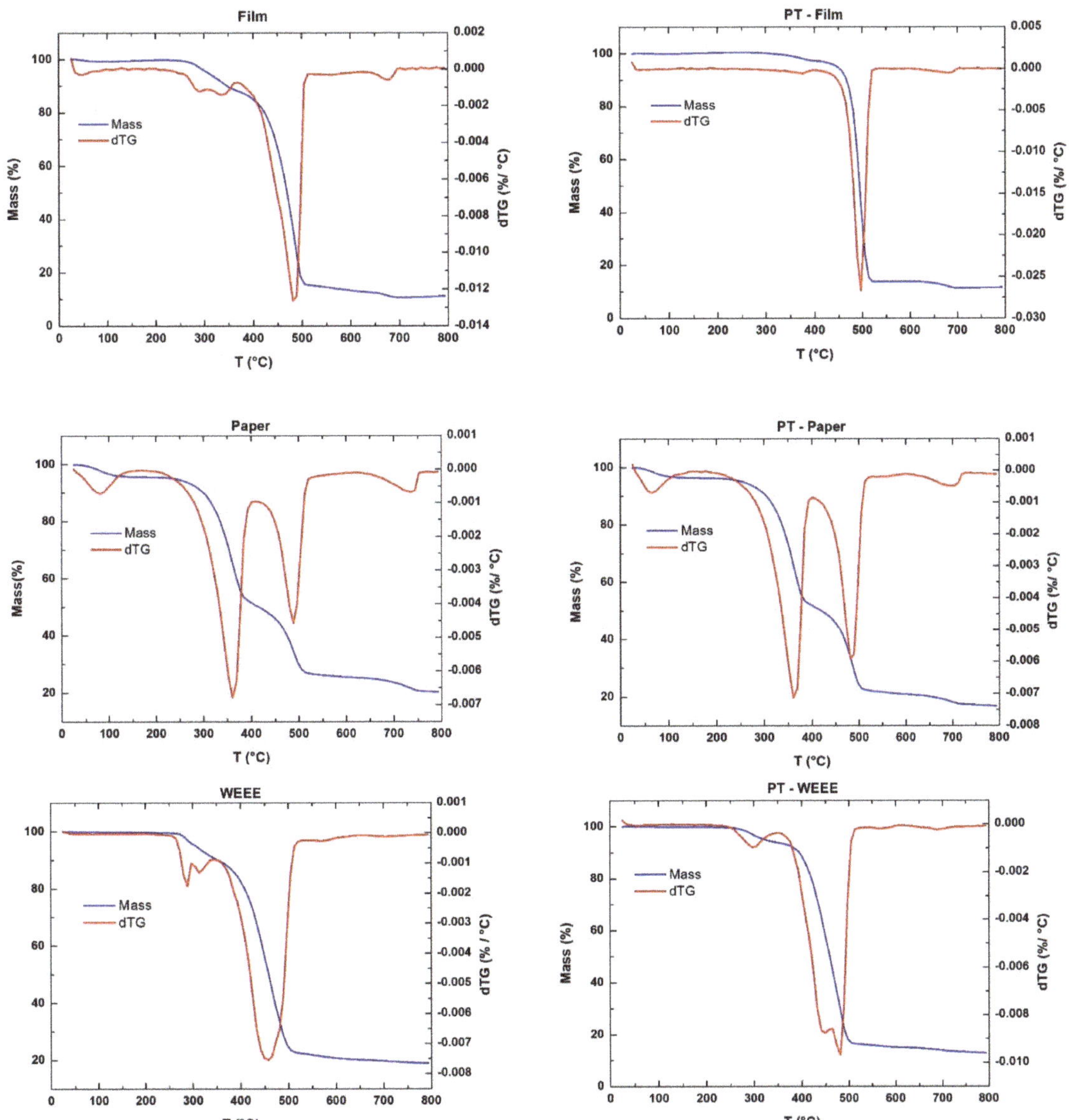

Figure 2. Thermogravimetric profiles of the raw and pre-treated waste samples.

Finally, the WEEE sample showed the classical decomposition phenomenon of PVC at 300 °C and the subsequent decomposition of the rest of the plastics, in this case in a wider temperature range than in the case of Film sample, due to the early decomposition of styrene plastics, compared to polyolefins [20]. In fact, in the pre-treated sample, a decoupling at the peak of the main decomposition can be seen, due to the higher percentage of styrenics compared to the raw sample. A smaller peak size can also be observed at 300 °C, due to the lower PVC content.

3.1.2. Pyrolysis Process

The yields obtained in the pyrolysis of the three samples, raw and pre-treated, at 460 °C, are shown in Table 6. Regarding the film-rich samples (raw and pre-treated), it can be seen that pyrolysis oils were the main product, followed by gas and solid. These results are directly attributed to the high polyolefin content in the initial sample (Table 2). In particular, the liquid yield of the pre-treated film sample reached 70.6 wt.% owing to the 93.0% of RMSP present in the feedstock. These results are in accordance with those obtained in previous papers. Lopez et al. obtained 65 wt.% of liquid yield at 500 °C using a sample that contained 92.3 wt.% of plastic [5]. Yan et al. reported the pyrolysis of PP and LDPE waste at 460 °C, reaching the 65.4 wt.% and 77.1 wt.% liquid yields, respectively [40]. Regarding the effect of pre-treatment, the increase in the yield of liquids can be related to the decrease in the yield of solids. Such a decrease in solid yield can be explained by the elimination of inorganic compounds and polymers that have a tendency to carbonize (PVC, PET, cellulose) during the pre-treatment.

Table 6. Pyrolysis yields of the raw and pre-treated samples (wt.%).

Sample	Oils		Gas [1]	Solid
	Organic	Aqueous		
Film	61.0	0.0	14.3	24.7
PT-Film	70.6	0.0	12.8	16.6
Paper	17.8	19.9	21.7	40.6
PT-Paper	42.5	10.9	20.0	26.6
WEEE	51.6	0.0	19.9	28.5
PT-WEEE	60.1	0.0	20.8	19.1

[1] By difference.

In the case of the Paper sample, the main fraction was the solid one, followed by liquids and gases. Such performance in solids can be explained by the high amount of inorganics contained in this sample (13.4 wt.%), together with a large amount of polymers with a tendency to carbonize, mainly cellulosic materials (31.5 wt.%). It is remarkable that this sample produced an aqueous liquid phase. This is explained by the presence of cellulosic-based materials rich in -OH and =O groups [5,20]. In the case of this sample, pre-treatment reduced the total solid yield by half and increased more than twice the organic liquid yield (from 17.8 to 42.5 wt.%). This fact might be explained by the significant effect of the pre-treatment on the reduction of (1) inorganic content (13.4 to 2.5 wt.%), and (2) char precursor materials, PET (5.3 to 2.7 wt.%) and especially cellulose (from 31.5 to 18.3 wt.%). However, the aqueous liquid phase could not be completely removed by the pre-treatment.

Finally, the WEEE sample generated also liquids as the main product, followed by solids and gases. Concerning the high solid yield, this sample did not contain a lot of inorganic material as such (2.3 wt.%), but it is necessary to remember the aforementioned issue of electric wires; in fact, the ash content determined by proximate analysis was high (22.0 wt.%, Table 4). Furthermore, the group constituted by "other thermoplastics" contained polymers with a tendency to carbonize and within "multimaterial" there were some inorganic elements coming from circuit printed boards. After pre-treatment, the higher liquid yield was observed and, at the same time, the solid yield decreased, as a consequence of the removal of PVC, inorganics and multimaterial. The liquid yield obtained from these two WEEE samples was similar to those obtained by Caballero et al. when investigating the pyrolysis of WEEE plastics at 500 °C. They found that landline waste (phones) generated a 58 wt.% liquid yield while mobile phones a 54 wt.% [22]. Higher values (around 70 wt.%) were obtained by Hall et al. during pyrolysis of mixed WEEE in a fixed bed reactor at 600 °C [41].

To summarize, it can be said that for the three different waste samples, the pre-treatment led to higher liquid yields and lower solid yields as compared to the pyrolysis of raw samples, while gas yields remained almost constant. This is the evidence that the pre-treatments produced the desired effect, which is the promotion of pyrolysis oils through the elimination of undesired materials of the original samples.

3.1.3. Pyrolysis Oils

Liquid products had a different appearance depending on the composition of the pyrolyzed waste sample. The liquids obtained from Film waste samples resulted in a waxy-like product instead of liquid oil, which can be ascribed to the high H/C ratio of the waste samples (see Table 3), principally explained by their great PE content [42,43]. In fact, Kiran et al. and Sharudin et al. experimented with operational blockage problems in pipelines and condenser tubes with waxes formation when pyrolyzed samples richer in PE [44,45]. Nevertheless, these waxes obtained from the polyolefins pyrolysis can serve as feedstock for FCC units of petroleum refineries [46,47]. Paper samples presented two differenced phases (organic and aqueous) in the liquid as was explained in Section 3.1.2. The organic phase of the pre-treated paper oil presented a more waxy-like appearance than the non-pre-treated one according to the promotion of polyolefins with the pre-treatment (see Table 2). By contrast, the pyrolysis of WEEE samples, with an aromatic/naphthenic nature, result in dark-brown coloured oils, which resemble petroleum fractions [5,18,22].

In order to evaluate the quality of the organic liquid products, several of their properties, such as higher heating value (HHV), halogen and metal content, and composition, were determined. First, the most limiting properties, i.e., HHV, halogen content and metal content, were analysed. This information is presented in Tables 7 and 8. Table 7 shows the HHV and the halogen content of these liquid products. The HHV of the pyrolysis oils was high (40–43 MJ kg^{-1}) and close to those of liquid fossil fuels (diesel 45 MJ kg^{-1} and heavy fuel oil 42–43 MJ kg^{-1} [43]), with the exception of Paper samples (37–39 MJ kg^{-1}). This is an important result, as it provides the possibility of using these oils as alternative fuels. Again, the pre-treatment improved the calorific properties of the pyrolysis oils, increasing the HHV in all cases due to the reduction of impurities and PET [36], and the concentration of MSP.

Table 7. HHV (MJ kg^{-1}) and halogen content (ppm) of the organic fraction of pyrolysis oils.

Sample	HHV	F$^-$	Cl$^-$	Br$^-$	% Cl$^-$ Transferred
Film	40.4	57	12,213	13	30
PT-Film	42.6	27	160	<10	8
Paper	37.4	26	1479	42	9
PT-Paper	39.2	7	894	11	10
WEEE	39.7	19	13,078	709	30
PT-WEEE	40.3	17	2076	796	17

Table 8. Metal content (ppm) in the pyrolysis oils from raw and pre-treated samples.

Metal	Film	PT-Film	Paper	PT-Paper	WEEE	PT-WEEE
Zn	8.3	5.9	8.2	6.0	<1	<1
Sb	7.7	3.1	7.7	5.0	<1	<1
P	5.5	<1	< 1	2.8	92.7	95.8
Pb	5.1	7.0	< 1	7.1	6.5	6.2
Ni	6.9	10.0	5.5	3.1	<1	<1
Fe	41.3	47.0	30.9	12.0	16.0	<1
Si	106	290	876	217	1813	567
Mn	<1	2.0	<1	<1	<1	<1
Cr	11.1	7.7	7.4	3.4	2.5	<1
Mg	<10	228	<10	<10	<10	<10
Ca	53.5	319	100	68.9	59.4	50.2
Al	21.6	4.2	6.1	4.8	5.8	3.4
Na	<10	70.8	<10	<10	20.6	<10

Concentration of Co, Cd, Cu, Sn, B, Tl, Ti, As, Mo, Ba, V and Ag was <1 ppm.

On the other hand, it is important to consider the halogen content, since they have an important and negative impact on the direct application of pyrolysis oils as fuels [20,48]. In this work, fluorine, chlorine and bromine were measured. It is clear from Table 7 that the main halogen element in the pyrolysis oils was chlorine. The fluorine and bromine values were very low, with the exception of the bromine content of the liquids from the WEEE samples, probably due to the presence of brominated flame retardants as part of the additives in the plastic materials of this waste. Liquids generated from raw waste samples presented more chlorine content, directly related to the presence of PVC, as was concluded by several authors [20,49,50]. Although all pre-treatments achieved a reduction in the halogen content, pyrolysis liquids still showed relatively high chlorine content. PT-Film sample registered the lowest chlorine content (160 ppm). This value, although low, is higher than the value established for use in existing petrochemical plants (3–10 ppm), as stated by some authors [51–53] However, these pyrolysis oils could probably be blended with other refinery streams before usage and most likely could be used as alternative fuels in cement kilns, where the required chlorine concentration is not usually so low. In the conditions of this work, about 10–30 wt.% of the chlorine content present in the waste samples was transferred to the liquid product. These transfer ratios can be reduced in several ways: using solid catalysts or adsorbents [20,49,54] or by the application of stepwise pyrolysis (two-stage pyrolysis) [55] but this was out of the scope of this paper.

Table 8 shows the metal content of the pyrolysis oils. In this case, two issues need to be considered: the heavy metal content, which could lead to environmental problems, and the presence of metals that could act as poisons in catalysts used in petrochemical processes. With regard to heavy metals (in bold in the table), zinc, antimony, lead, nickel, manganese and chromium were detected, all in concentrations below 8 ppm in the oils from the pre-treated samples. This means that all these oils were free of heavy metals such as cadmium, copper, arsenic, cobalt, thallium, tin and mercury, or at least the concentration of these metals was below 1 ppm. Among the metals that can cause problems in catalysts, the presence of silicon was particularly noticeable in the oils from the WEEE samples. For this reason, it is important to take it into account when processing oil in the refinery, since requirements are usually established to avoid its presence and prevent damage to the catalysts. The limits of the metals will depend on each refinery, the processing unit in which the oil is included and the degree of dissolution that the oil presents along with the conventional feed used.

As the liquid fractions coming from the pre-treated samples showed better quality, the characterization by the GC/MS was carried out only in these oils. Figure 3 shows the compounds identified by this method grouped according to their nature in paraffinic, naphthenic, olefinic and aromatic compounds. Only those compounds with areas > 1% and an identification quality degree > 90% were included in such groups. The oil from the pre-treated Film sample was composed mainly of paraffins (59.9% area) and olefins (30.2% area), due to the high content of PE presented in the original sample (see Table 2). It was proven by other authors that during the degradation of PE, free radical fragments are formed and react with hydrogen chains, giving rise to alkanes and alkenes [5,56]. According to Das et al., olefins are the precursors of many industrial organic chemicals such as vinyl acetate, acetaldehyde and vinyl chloride, therefore, the concentration of olefins in the pyrolytic oil could be used in numerous industrial applications [57].

On the other hand, pre-treated WEEE sample oil consisted of more than 97% area of aromatic compounds, with small quantities of naphthenes (2.4% area). The high quantity of aromatics is attributed to the great styrene content and the low content of polyolefins in the original sample (see Table 2). In previous investigations, 80% of aromatic hydrocarbons were obtained in the pyrolysis of PS [58]. Since a high concentration of aromatics is desired for gasoline production [10], this could be the most appropriate application for PT-WEEE oil provided the chlorine content is reduced. The major compounds in the PT-Paper sample oil included paraffins (45.8% area), naphthenes (16.9% area) and aromatics (34.5% area). This wide distribution is related to the composition of the sample. As was previously mentioned, polyolefins generate paraffinic and olefinic compounds while styrenics favour aromatic content in the liquid oils. Moreover, other fractions such as PET could also favour the formation of the former compounds [15].

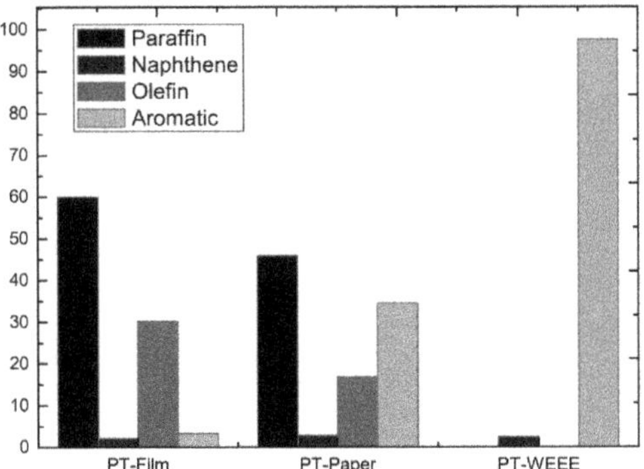

Figure 3. Composition of pyrolysis oils coming from pre-treated samples.

3.2. Effect of Temperature in the Production of Oils Coming from PT-Film Sample

At this point of the investigation, it was considered that the PT-Film sample was the most suitable sample to deepen the possibilities of pyrolysis oil production. This decision was based on the fact that the composition of these liquids allowed them to be considered a priori as feedstock for refineries or as a source of olefins, and the HHV and halogen content enabled its use as an alternative fuel. In addition, it was the sample that generated the greatest amount of liquids. Therefore, this sample was selected to study the effect of the cracking temperature, which is the most significant variable in the pyrolysis process, showing a critical influence in the conversion and product distribution [59]. The pyrolysis experiments were run at three different temperatures: 430, 460 and 490 °C. The yields are presented in Table 9.

Table 9. Pyrolysis yields of PT-Film sample at different temperatures (wt.%).

Temperature (°C)	Oils		Gas [1]	Solid
	Organic	Aqueous		
430	48.8	0.0	12.5	38.7
460	70.6	0.0	12.8	16.6
490	78.0	0.0	14.2	7.8

[1] By difference.

As it can be seen in Table 9, solid and liquid yields were strongly affected by temperature, while the gas formation did not show such a wide variation. The liquid was the main product and its yield rose with the increase in temperature from 48.8 wt.% at 430 °C to 78.0 wt.% at 490 °C. Equally, an important decrease in solid yield was observed in the same temperature range (from 38.7 to 7.8 wt.%). This fact indicates that pyrolysis was incomplete until 490 °C, that is, organic matter was still remained for cracking in the experiments carried out at lower temperatures. This phenomenon was previously reported in pyrolysis tests carried out at temperatures below 500 °C with similar samples [16]. As far as gas yield is concerned, the most common thing is to observe a trend of higher gas yields as the temperature increases, due to the stronger breaking of the polymer chains that happens at high temperatures, as happened in this work [28,48].

The temperature effect was also investigated in the properties of pyrolysis oils. HHV and halogen content are presented in Table 10. Concerning HHV, a slight increase in the HHV was produced as the temperature rose, ranging 44.3 MJ kg^{-1} at 490 °C (Heavy fuel oil: 42–43 MJ kg^{-1}). The same tendency was found in other works [36]. As was mentioned before, halogen content, especially chlorine, is limited by the requirements of refineries. In general, no significant effect of the temperature on the halogen content was found. Concerning chlorine content, higher temperatures led to a slightly lower presence of chlorine in the liquid products. However, previously published papers concluded that there is usually a chlorine increase with temperature (from 460 °C to 600 °C) as a result of the quicker interactions between radical fragments and HCl released from PVC [16]. Anyway, this depends to a large extent on the operating conditions and the design of the pyrolysis plant. Moreover, the differences in this work were not very significant (units are in ppm), and could be part of the intrinsic error of experimentation and analytics.

Table 10. HHV (MJ kg^{-1}) and halogen content (ppm) of the pyrolysis oils of PT-Film pyrolyzed at different temperatures.

Temperature (°C)	HHV	F$^-$	Cl$^-$	Br$^-$
430	42.2	14	245	<10
460	42.6	27	160	<10
490	44.3	12	128	<10

Concerning composition, in spite of the increase in the cracking temperature, all liquids were wax-like products that solidified at ambient temperature and easily re-melted above 40 °C. Nevertheless, it was reported that higher cracking temperatures can decrease the viscosity of the liquids. This effect is observed at operating temperatures above 600 °C when waxes chains are broken down to lighter components due to the higher thermal cracking produced [15,27]. However, there is no evidence of this effect at the temperature range studied in this research. Figure 4 shows the distribution of the hydrocarbon's nature from the compounds identified by GC/MS. It was discussed in Section 3.1.3 that the oil coming from PT-Film mainly consisted of paraffinic and olefinic compounds due to the original composition of the sample. Now, as the temperature increased, the paraffin content raised whereas aromatic distribution was reduced. This is due to the fact that the temperature favours the intramolecular hydrogen transfer, generating a more paraffinic fraction [57,60,61]. Nevertheless, other authors experimented with the reverse trend, Onwudili et al., who pyrolyzed LDPE and PS in a batch reactor from 300 °C to 500 °C, concluded that higher temperatures and higher residence times favoured the aromatic proportion in pyrolytic oils due to the cyclization and aromatization at 500 °C [62] The explanation for the difference between these results can lie in the different designs of reactors and reaction systems, which have relevant importance in the routes and mechanisms of reaction that take place. In any case, the removal of aromatics in these pyrolysis oils is a good result, as this means a purer stream of paraffins and olefins.

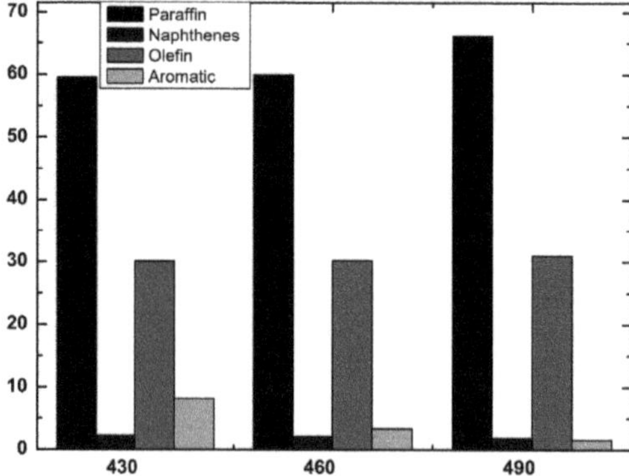

Figure 4. Composition of pyrolysis oils of PT-Film pyrolyzed at different temperatures.

At last, the results of simulated distillation are presented in Figure 5 and Table 11. The raw distillation curve (presented in Figure 5) showed that the final boiling point at T95% was 506.6 °C for the oil obtained at 490 °C, 500 °C for the oil obtained at 460 °C and 485 °C for the oil obtained at 430 °C. Moreover, the hydrocarbon fractions were classified based on their boiling temperature: naphtha (T < 216 °C), middle distillates (216 °C < T < 343 °C) and heavy diesel (T > 343 °C). Attending to the results shown in Table 11, a temperature effect can be observed: rising temperature reduced the light fraction (naphtha) while the heavy fraction (heavy diesel) was increased. Other authors reported the same effect [16,60].

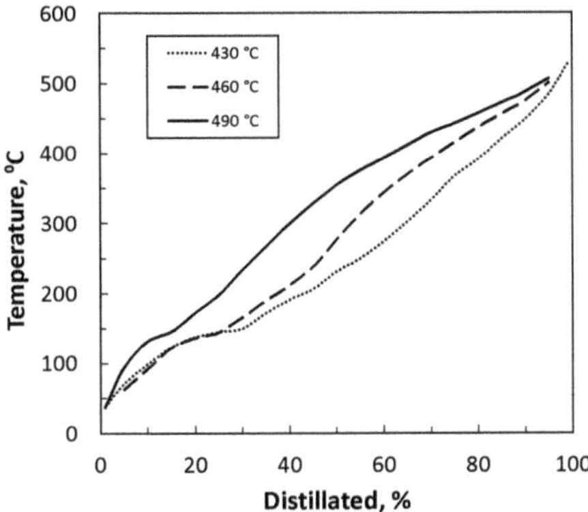

Figure 5. Simulated distillation curves of pyrolysis oils of PT-Film pyrolyzed at different temperatures.

Table 11. Distillation fractions of pyrolysis oils of PT-Film pyrolyzed at different temperatures.

Temperature (°C)	Naphtha	Middle Distillates	Heavy Diesel
430	45.8	25.4	28.8
460	39.9	20.0	40.1
490	26.3	21.8	51.9

4. Conclusions

Pyrolysis appears as an attractive alternative for recycling rejected streams with high plastic content. The idea is to obtain pyrolysis oils that can be used for petrochemical processes or as alternative fuels. However, industrial rejected streams present different natures depending on their origin and this decisively influences the production of pyrolysis oils. After analysing rejected streams from sorting plants for packaging waste (Film sample), paper/cardboard waste (Paper sample) and waste from the electrical and electronics sector (WEEE sample), it was found that they contain significant quantities of materials that can reduce the quantity and quality of pyrolysis oils. These materials are mainly PVC, PET and cellulosic materials, and inorganic matter such as metals, which lead to the generation of chlorinated oils (PVC), aqueous phases in the oils (PET and cellulosic materials) and high quantities of pyrolysis solids in detriment of liquids (inorganic matter as metals).

These samples were subjected to mechanical separation processes (pre-treatments) and all pre-treatments were effective in concentrating materials suitable for pyrolysis (mainly polyolefins and styrenic plastics). Flotation and densimetric separation achieved a high recovery rate for the Film and WEEE samples, respectively. By contrast, a great deal of material mixture in the Paper sample made the separation by NIR spectroscopy less effective. Nevertheless, all the pre-treated samples achieved higher liquid and lower solid yields compared with raw samples. Regarding the quality of pyrolysis oils, the higher heating value of the oils coming from pre-treated Film and WEEE samples were similar to heavy fuel oil, showing its potential application as fuel. Moreover, the oil from the pre-treated Film sample was mainly composed of olefins and paraffins, whereas the oil coming from the pre-treated WEEE sample was based on aromatic compounds. The halogen content was considerably reduced in the oils after pre-treatment; however, a significant proportion of chlorine was transferred to oils, limiting its application.

The temperature effect was also studied in the 430–490 °C range, using the pre-treated Film sample. The temperature favoured the formation of liquid products (from 48.8 to 78.0 wt.%) and solid yield decreased (from 38.7 to 7.8 wt.%). In addition to increasing liquid production, at 490 °C, an oil with very low chlorine concentration (128 ppm), high HHV (44.3 MJ kg^{-1}) and high paraffin content was produced. The results presented in this work demonstrate that the implementation of mechanical material separation processes can be an interesting option as a preliminary step to pyrolysis processes, with the aim of producing more quantities of pyrolysis oils with improved properties. This information should be taken into account when designing recycling processes for complex waste by pyrolysis.

Author Contributions: Conceptualization, A.A. and S.G.-F.; methodology, A.A., S.G.-F. and L.F.-M.; validation, S.G.-F. and L.F.-M.; formal analysis, A.A., S.G.-F. and A.L.-U.; investigation, L.F.-M., B.B.P.-M. and J.M.A.; resources, A.A.; data curation, B.B.P.-M. and J.M.A.; writing—original draft preparation, L.F.-M.; writing—review and editing, A.L.-U.; visualization, A.L.-U. and J.M.A.; supervision, A.A. and S.G.-F.; project administration, A.A. and S.G.-F.; funding acquisition, A.A. All authors have read and agreed to the published version of the manuscript.

Funding: This research was funded by the Basque Government through the project with reference KK-2020/00107 (ELKARTEK program) and through the support of the SUPREN group (GIC10/31, GIC15/13, S-PE13UN126 (SAI13/190)).

Institutional Review Board Statement: Not applicable.

Informed Consent Statement: Not applicable.

Acknowledgments: The authors thank the recycling companies for providing the waste samples.

Conflicts of Interest: The authors declare no conflict of interest. The funders had no role in the design of the study; in the collection, analyses, or interpretation of data; in the writing of the manuscript, or in the decision to publish the results.

References

1. Moharir, R.V.; Kumar, S. Challenges associated with plastic waste disposal and allied microbial routes for its effective degradation: A comprehensive review. *J. Clean. Prod.* **2019**, *208*, 65–76. [CrossRef]
2. Emadian, S.M.; Onay, T.T.; Demirel, B. Biodegradation of bioplastics in natural environments. *Waste Manag.* **2017**, *59*, 526–536. [CrossRef] [PubMed]
3. European Union. *DIRECTIVE (EU) 2018/851 of the European Parliament and of the Council 2008/98/EC on Waste*; EU Publications: Luxembourg, 2018; pp. 109–140.
4. Plastics Europe. *Plásticos—Situación en 2017*; Plastics Europe: Brussels, Belgium, 2017; p. 50.
5. López, A.; de Marco, I.; Caballero, B.M.; Laresgoiti, M.F.; Adrados, A. Pyrolysis of municipal plastic wastes: Influence of raw material composition. *Waste Manag.* **2010**, *30*, 620–627. [CrossRef] [PubMed]
6. Esposito, L.; Cafiero, L.; De Angelis, D.; Tuffi, R.; Vecchio Ciprioti, S. Valorization of the plastic residue from a WEEE treatment plant by pyrolysis. *Waste Manag.* **2020**, *112*, 1–10. [CrossRef] [PubMed]
7. Solis, M.; Silveira, S. Technologies for chemical recycling of household plastics—A technical review and TRL assessment. *Waste Manag.* **2020**, *105*, 128–138. [CrossRef] [PubMed]
8. Kunwar, B.; Cheng, H.N.; Chandrashekaran, S.R.; Sharma, B.K. Plastics to fuel: A review. *Renew. Sustain. Energy Rev.* **2016**, *54*, 421–428. [CrossRef]
9. Ragaert, K.; Delva, L.; Van Geem, K. Mechanical and chemical recycling of solid plastic waste. *Waste Manag.* **2017**, *69*, 24–58. [CrossRef]
10. Demirbas, A. Pyrolysis of municipal plastic wastes for recovery of gasoline-range hydrocarbons. *J. Anal. Appl. Pyrolysis* **2004**, *72*, 97–102. [CrossRef]
11. Sipra, A.T.; Gao, N.; Sarwar, H. Municipal solid waste (MSW) pyrolysis for bio-fuel production: A review of effects of MSW components and catalysts. *Fuel Process. Technol.* **2018**, *175*, 131–147. [CrossRef]
12. Sharuddin, S.D.A.; Abnisa, F.; Daud, W.M.A.W.; Aroua, M.K. Pyrolysis of plastic waste for liquid fuel production as prospective energy resource. *IOP Conf. Ser. Mater. Sci. Eng.* **2018**, *334*, 012001. [CrossRef]
13. Diaz Silvarrey, L.S.; Phan, A.N. Kinetic study of municipal plastic waste. *Int. J. Hydrogen Energy* **2016**, *41*, 16352–16364. [CrossRef]

14. Ciliz, N.K.; Ekinci, E.; Snape, C.E. Pyrolysis of virgin and waste polypropylene and its mixtures with waste polyethylene and polystyrene. *Waste Manag.* **2004**, *24*, 173–181. [CrossRef] [PubMed]
15. de Marco, I.; Caballero, B.M.; López, A.; Laresgoiti, M.F.; Torres, A.; Chomón, M.J. Pyrolysis of the rejects of a waste packaging separation and classification plant. *J. Anal. Appl. Pyrolysis* **2009**, *85*, 384–391. [CrossRef]
16. López, A.; de Marco, I.; Caballero, B.M.; Laresgoiti, M.F.; Adrados, A. Influence of time and temperature on pyrolysis of plastic wastes in a semi-batch reactor. *Chem. Eng. J.* **2011**, *173*, 62–71. [CrossRef]
17. Kumar, S.; Panda, A.K.; Singh, R.K. A review on tertiary recycling of high-density polyethylene to fuel. *Resour. Conserv. Recycl.* **2011**, *55*, 893–910. [CrossRef]
18. De Marco, I.; Caballero, B.M.; Chomón, M.J.; Laresgoiti, M.F.; Torres, A.; Fernández, G.; Arnaiz, S. Pyrolysis of electrical and electronic wastes. *J. Anal. Appl. Pyrolysis* **2008**, *82*, 179–183. [CrossRef]
19. Adrados, A.; de Marco, I.; Caballero, B.M.; Laresgoiti, M.F. Pyrolysis behavior of different type of materials contained in the rejects of packaging waste sorting plants. *Waste Manag.* **2013**, *33*, 52–59. [CrossRef]
20. López, A.; De Marco, I.; Caballero, B.M.; Laresgoiti, M.F.; Adrados, A. Dechlorination of fuels in pyrolysis of PVC containing plastic wastes. *Fuel Process. Technol.* **2011**, *92*, 253–260. [CrossRef]
21. Scheirs, J. *Overview of Commercial Pyrolysis Processes for Waste Plastics*; John Wiley & Sons, Inc.: Hoboken, NJ, USA, 2006. ISBN 9780470021545.
22. Caballero, B.M.; de Marco, I.; Adrados, A.; Solar, J.; Gastelu, N. Possibilities and limits of pyrolysis for recycling plastic rich waste streams rejected from phones recycling plants. *Waste Manag.* **2016**, *57*, 226–234. [CrossRef]
23. Ateş, F.; Miskolczi, N.; Borsodi, N. Comparision of real waste (MSW and MPW) pyrolysis in batch reactor over different catalysts. Part I: Product yields, gas and pyrolysis oil properties. *Bioresour. Technol.* **2013**, *133*, 443–454. [CrossRef]
24. Akubo, K.; Nahil, M.A.; Williams, P.T. Aromatic fuel oils produced from the pyrolysis-catalysis of polyethylene plastic with metal-impregnated zeolite catalysts. *J. Energy Inst.* **2019**, *92*, 195–202. [CrossRef]
25. Uemichi, Y.; Makino, Y.; Kanazuka, T. Degradation of polyethylene to aromatic hydrocarbons over metal-supported activated carbon catalysts. *J. Anal. Appl. Pyrolysis* **1989**, *14*, 331–344. [CrossRef]
26. Sharypov, V.I.; Marin, N.; Beregovtsova, N.G.; Baryshnikov, S.V.; Kuznetsov, B.N.; Cebolla, V.L.; Weber, J.V. Co-pyrolysis of wood biomass and synthetic polymer mixtures. Part I: Influence of experimental conditions on the evolution of solids, liquids and gases. *J. Anal. Appl. Pyrolysis* **2002**, *64*, 15–28. [CrossRef]
27. Williams, E.A.; Williams, P.T. The pyrolysis of individual plastics and a plastic mixture in a fixed bed reactor. *J. Chem. Technol. Biotechnol.* **1997**, *70*, 9–20. [CrossRef]
28. Singh, R.K.; Ruj, B. Time and temperature depended fuel gas generation from pyrolysis of real world municipal plastic waste. *Fuel* **2016**, *174*, 164–171. [CrossRef]
29. Muhammad, C.; Onwudili, J.A.; Williams, P.T. Thermal degradation of real-world waste plastics and simulated mixed plastics in a two-stage pyrolysis-catalysis reactor for fuel production. *Energy Fuels* **2015**, *29*, 2601–2609. [CrossRef]
30. Lopez-Urionabarrenechea, A.; de Marco, I.; Caballero, B.M.; Adrados, A.; Laresgoiti, M.F. Empiric model for the prediction of packaging waste pyrolysis yields. *Appl. Energy* **2012**, *98*, 524–532. [CrossRef]
31. Pongstabodee, S.; Kunachitpimol, N.; Damronglerd, S. Combination of three-stage sink-float method and selective flotation technique for separation of mixed post-consumer plastic waste. *Waste Manag.* **2008**, *28*, 475–483. [CrossRef]
32. Guo, J.; Li, X.; Guo, Y.; Ruan, J.; Qiao, Q.; Zhang, J.; Bi, Y.; Li, F. Research on Flotation Technique of Separating PET from Plastic Packaging Wastes. *Procedia Environ. Sci.* **2016**, *31*, 178–184. [CrossRef]
33. Mumbach, G.D.; de Sousa Cunha, R.; Machado, R.A.F.; Bolzan, A. Dissolution of adhesive resins present in plastic waste to recover polyolefin by sink-float separation processes. *J. Environ. Manag.* **2019**, *243*, 453–462. [CrossRef]
34. Hiosta, J.; Zurovec, D.; Kratochivil, M.; Botula, J.; Zegzuika, J. WEEE sorting process and separation of copper wires with support of DEM modeling. *Inz. Mineriana* **2017**, *1*, 159–164.
35. Dodbiba, G.; Fujita, T. Air Tabling-A Dry Gravity Solid-Solid Separation Technique. *Prog. Filtr. Sep.* **2015**, 527–555.
36. Sogancioglu, M.; Ahmetli, G.; Yel, E. A Comparative Study on Waste Plastics Pyrolysis Liquid Products Quantity and Energy Recovery Potential. *Energy Procedia* **2017**, *118*, 221–226. [CrossRef]
37. Paolo, L.M.F. Polymer mechanical recycling: Downcycling or upcycling? *Prog. Rubber Plast. Recycl. Technol.* **2004**, *20*, 11–24.
38. Miskolczi, N.; Ateş, F.; Borsodi, N. Comparison of real waste (MSW and MPW) pyrolysis in batch reactor over different catalysts. Part II: Contaminants, char and pyrolysis oil properties. *Bioresour. Technol.* **2013**, *144*, 370–379. [CrossRef] [PubMed]
39. López, A.; de Marco, I.; Caballero, B.M.; Laresgoiti, M.F.; Adrados, A.; Torres, A. Pyrolysis of municipal plastic wastes II: Influence of raw material composition under catalytic conditions. *Waste Manag.* **2011**, *31*, 1973–1983. [CrossRef]
40. Yan, G.; Jing, X.; Wen, H.; Xiang, S. Thermal cracking of virgin and waste plastics of PP and LDPE in a semibatch reactor under atmospheric pressure. *Energy Fuels* **2015**, *29*, 2289–2298. [CrossRef]
41. Hall, W.J.; Williams, P.T. Analysis of products from the pyrolysis of plastics recovered from the commercial scale recycling of waste electrical and electronic equipment. *J. Anal. Appl. Pyrolysis* **2007**, *79*, 375–386. [CrossRef]
42. Miandad, R.; Barakat, M.A.; Aburiazaiza, A.S.; Rehan, M.; Ismail, I.M.I.; Nizami, A.S. Effect of plastic waste types on pyrolysis liquid oil Kingdom of Saudi Arabia. *Int. Biodeterior. Biodegrad.* **2016**, *119*, 2239–2252.
43. Lee, K.H. Effects of the types of zeolites on catalytic upgrading of pyrolysis wax oil. *J. Anal. Appl. Pyrolysis* **2012**, *94*, 209–214. [CrossRef]

44. Anuar Sharuddin, S.D.; Abnisa, F.; Wan Daud, W.M.A.; Aroua, M.K. Energy recovery from pyrolysis of plastic waste: Study on non-recycled plastics (NRP) data as the real measure of plastic waste. *Energy Convers. Manag.* **2017**, *148*, 925–934. [CrossRef]
45. Kiran, N.; Ekinci, E.; Snape, C.E. Recyling of plastic wastes via pyrolysis. *Resour. Conserv. Recycl.* **2000**, *29*, 273–283. [CrossRef]
46. Butler, E.; Devlin, G.; McDonnell, K. Waste polyolefins to liquid fuels via pyrolysis: Review of commercial state-of-the-art and recent laboratory research. *Waste Biomass Valorization* **2011**, *2*, 227–255. [CrossRef]
47. Arandes, J.M.; Torre, I.; Castaño, P.; Olazar, M.; Bilbao, J. Catalytic cracking of waxes produced by the fast pyrolysis of polyolefins. *Energy Fuels* **2007**, *21*, 561–569. [CrossRef]
48. Miandad, R.; Barakat, M.A.; Aburiazaiza, A.S.; Rehan, M.; Nizami, A.S. Catalytic pyrolysis of plastic waste: A review. *Process Saf. Environ. Prot.* **2016**, *102*, 822–838. [CrossRef]
49. Bhaskar, T.; Kaneko, J.; Muto, A.; Sakata, Y.; Jakab, E.; Matsui, T.; Uddin, M.A. Pyrolysis studies of PP/PE/PS/PVC/HIPS-Br plastics mixed with PET and dehalogenation (Br, Cl) of the liquid products. *J. Anal. Appl. Pyrolysis* **2004**, *72*, 27–33. [CrossRef]
50. Bhaskar, T.; Uddin, M.A.; Murai, K.; Kaneko, J.; Hamano, K.; Kusaba, T.; Muto, A.; Sakata, Y. Comparison of thermal degradation products from real municipal waste plastic and model mixed plastics. *J. Anal. Appl. Pyrolysis* **2003**, *70*, 579–587. [CrossRef]
51. Kaminsky, W. Chemical recycling of mixed plastics of pyrolysis. *Adv. Polym. Technol.* **1995**, *14*, 337–344. [CrossRef]
52. Kaminsky, W.; Kim, J.S. Pyrolysis of mixed plastics into aromatics. *J. Anal. Appl. Pyrolysis* **1999**, *51*, 127–134. [CrossRef]
53. Kusenberg, M.; Eschenbacher, A.; Djokic, M.R.; Zayoud, A.; Ragaert, K.; De Meester, S.; Van Geem, K.M. Opportunities and challenges for the application of post-consumer plastic waste pyrolysis oils as steam cracker feedstocks: To decontaminate or not to decontaminate? *Waste Manag.* **2022**, *138*, 83–115. [CrossRef]
54. Lopez-Urionabarrenechea, A.; De Marco, I.; Caballero, B.M.; Laresgoiti, M.F.; Adrados, A. Upgrading of chlorinated oils coming from pyrolysis of plastic waste. *Fuel Process. Technol.* **2015**, *137*, 229–239. [CrossRef]
55. Lopez-Urionabarrenechea, A.; De Marco, I.; Caballero, B.M.; Laresgoiti, M.F.; Adrados, A. Catalytic stepwise pyrolysis of packaging plastic waste. *J. Anal. Appl. Pyrolysis* **2012**, *96*, 54–62. [CrossRef]
56. Singh, R.K.; Ruj, B.; Sadhukhan, A.K.; Gupta, P. Thermal degradation of waste plastics under non-sweeping atmosphere: Part 1: Effect of temperature, product optimization, and degradation mechanism. *J. Environ. Manag.* **2019**, *239*, 395–406. [CrossRef] [PubMed]
57. Das, P.; Tiwari, P. The effect of slow pyrolysis on the conversion of packaging waste plastics (PE and PP) into fuel. *Waste Manag.* **2018**, *79*, 615–624. [CrossRef] [PubMed]
58. Ramli, A.; Bakar, D.R.A. Effect of calcination method on the catalytic degradation of polystyrene using A12O3 supported Sn and Cd catalysts. *J. Appl. Sci.* **2011**, *11*, 1346–1350. [CrossRef]
59. Aguado, J.; Serrano, D. Thermal Processes. In *Feedstock Recycling of Plastic Wastes*; Royal Society of Chemistry: London, UK, 1999; pp. 73–124.
60. Marcilla, A.; Beltrán, M.I.; Navarro, R. Evolution of products during the degradation of polyethylene in a batch reactor. *J. Anal. Appl. Pyrolysis* **2009**, *86*, 14–21. [CrossRef]
61. Singh, B.; Sharma, N. Mechanistic implications of plastic degradation. *Polym. Degrad. Stab.* **2008**, *93*, 561–584. [CrossRef]
62. Onwudili, J.A.; Insura, N.; Williams, P.T. Composition of products from the pyrolysis of polyethylene and polystyrene in a closed batch reactor: Effects of temperature and residence time. *J. Anal. Appl. Pyrolysis* **2009**, *86*, 293–303. [CrossRef]

Review

Valorization of Spent Coffee Grounds as Precursors for Biopolymers and Composite Production

Anne Shayene Campos de Bomfim [1], Daniel Magalhães de Oliveira [1], Herman Jacobus Cornelis Voorwald [1], Kelly Cristina Coelho de Carvalho Benini [1], Marie-Josée Dumont [2] and Denis Rodrigue [3,*]

1   Fatigue and Aeronautical Materials Research Group, Department of Materials and Technology, UNESP-São Paulo State University, Guaratinguetá 12516-410, São Paulo, Brazil; anne.shayene@unesp.br (A.S.C.d.B.); daniel.m.oliveira@unesp.br (D.M.d.O.); h.voorwald@unesp.br (H.J.C.V.); kcccbenini@gmail.com (K.C.C.d.C.B.)
2   Bioresource Engineering Department, McGill University, 21111 Lakeshore Road, Ste-Anne-de-Bellevue, QC H9X 3V9, Canada; marie-josee.dumont@mcgill.ca
3   Department of Chemical Engineering and CERMA, Université Laval, Quebec, QC G1V0A6, Canada
*   Correspondence: denis.rodrigue@gch.ulaval.ca

Abstract: Spent coffee grounds (SCG) are a current subject in many works since coffee is the second most consumed beverage worldwide; however, coffee generates a high amount of waste (SCG) and can cause environmental problems if not discarded properly. Therefore, several studies on SCG valorization have been published, highlighting its waste as a valuable resource for different applications, such as biofuel, energy, biopolymer precursors, and composite production. This review provides an overview of the works using SCG as biopolymer precursors and for polymer composite production. SCG are rich in carbohydrates, lipids, proteins, and minerals. In particular, carbohydrates (polysaccharides) can be extracted and fermented to synthesize lactic acid, succinic acid, or polyhydroxyalkanoate (PHA). On the other hand, it is possible to extract the coffee oil and to synthesize PHA from lipids. Moreover, SCG have been successfully used as a filler for composite production using different polymer matrices. The results show the reasonable mechanical, thermal, and rheological properties of SCG to support their applications, from food packaging to the automotive industry.

Keywords: spent coffee grounds; biopolymer precursors; polysaccharides; composites

check for
updates

Citation: Bomfim, A.S.C.d.; Oliveira, D.M.d.; Voorwald, H.J.C.; Benini, K.C.C.d.C.; Dumont, M.-J.; Rodrigue, D. Valorization of Spent Coffee Grounds as Precursors for Biopolymers and Composite Production. *Polymers* **2022**, *14*, 437. https://doi.org/10.3390/polym14030437

Academic Editor: Sheila Devasahayam

Received: 15 December 2021
Accepted: 19 January 2022
Published: 22 January 2022

Publisher's Note: MDPI stays neutral with regard to jurisdictional claims in published maps and institutional affiliations.

1. Introduction

Coffee, as a beverage, is consumed all over the world and is now considered a commodity. In 2018, coffee production was evaluated at 9.5 million tons [1,2]. After brewing, SCG are generated as waste, corresponding to about 90% of the initial coffee beans [3]. Thus, SCG waste is an environmental concern, since it can be toxic when not properly discarded because of the caffeine, tannin, and polyphenol emissions during coffee fermentation [4]. This is why efforts to valorize and reuse this waste via environmentally friendly pathways, ones that do not cause any kind of pollution, have been proposed. SCG, which are a lignocellulose-rich waste, have been applied in different fields, such as biofuel production, energy production, and in polymer precursors and composites [5–7].

It is well established that SCG are rich in carbohydrates, lipids, proteins, and minerals. Much attention has been given to the extraction and valorization of the individual fractions of SCG. Carbohydrates represent half of the total coffee beans, being mainly polysaccharides of hemicellulose (30–40 wt.%) and cellulose (8–15 wt.%) [8,9] (Figure 1). These polysaccharides can be hydrolyzed to obtain fermentable sugars, such as mannose, glucose, galactose, and arabinose, which can then be converted by microbial fermentation into lactic acid, acetic acid, succinic acid, polyhydroxyalkanoate (PHA), and other molecules of interest [10–12]. The SCG oil fraction (7–21 wt.%) is also a valuable resource that is reported in PHA synthesis, biosurfactant production, biodiesel and bioethanol synthesis, as well as in sunscreen formulations [13,14]. The residues of SCG can be directly used as fillers in composites (mainly polymer matrices), or after chemical modifications (the extraction of the molecules of interest). Typical examples of polymer matrices are synthetic polymers, such as polyurethane [15] and polypropylene [16–18], or biopolymers, such as polylactic acid (PLA) [19–21]. The resulting materials have been proposed for various applications, such as for packaging (mostly food), disposable products (single-use products/composting), and 3D printing [15,21,22]. The development of SCG composites is proposed as a sustainable pathway because it reduces the amount of SCG waste in the environment [22] while creating added-value products, which can even accelerate the degradation process of a biodegradable matrix [23].

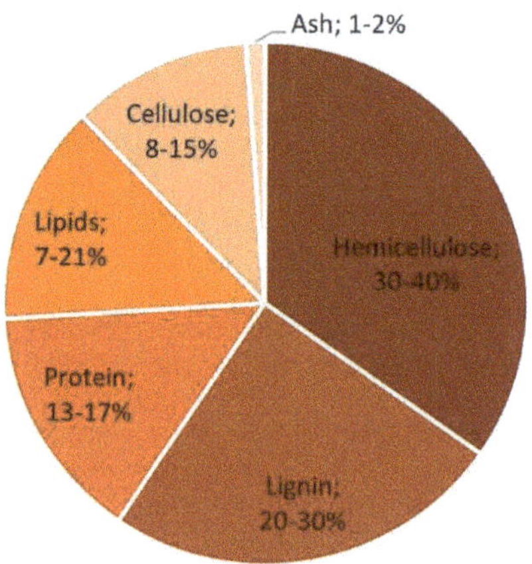

Figure 1. SCG chemical compounds after brewing.

Several reviews report the possible valorization of SCG and other coffee byproducts [2,3,8,11,24]. Nevertheless, none of them discuss in detail, with specific applications, their use in the plastics industry. This work aims to fill this gap by presenting the state-of-the-art possibilities of SCG valorization in the plastics field over the last 20 years (2000–2021). In particular, there is a focus on polymer precursors, biopolymer production, and composites. The first part reports information on the composition of SCG, the possibility of using SCG as green polymer precursors, and their application in different polymeric matrices. Then, future prospects are discussed, highlighting the significance of this subject for more academic/scientific research, as well as for commercial/industrial developments.

2. Polysaccharide Extraction from SCG

Various polysaccharides can be extracted from SCG. As stated, these polysaccharides mainly consist of hemicelluloses, such as galactomannan and arabinogalactan, as well as cellulose, galactan, mannan, and arabinan, in lower amounts [25–27]. Mussatto et al. (2011) identified a hemicellulose composition with mannose as the main monosaccharide present (21.2 g/100 g), followed by galactose (13.8 g/100 g), and arabinose (1.7 g/100 g); however, no xylose was found [25].

Alkali extraction is the main technique used to extract polysaccharides from SCG, and it allows access to the molecules still present after infusion. The technique consists of soaking the SCG in hot water, followed by a series of soakings at room temperature with the progressive addition of a base, such as potassium hydroxide or sodium hydroxide, which leads to the precipitation of polysaccharides [26–28]. The extraction sequence was: 2 L of distilled water (90 °C for 1 h), followed by 2 L and 0.5 M of imidazole (70 °C for 1 h) and 1-L NaOH extractions (0.05, 1, and 4 M), three times at room temperature for 2 h; however, an inert atmosphere is required for the last extractions to avoid alkaline oxidation and the peeling reactions of the polysaccharides [26]. The extract can then be filtered, followed by the suspension and dialysis of the precipitate containing the polysaccharides, followed by centrifugation. The defatting of the SCG can be considered prior to alkali extraction [29]. It is also possible to do a roasting pretreatment, leading to an increase in the polysaccharide extraction yield [30]. Following their extraction, the polysaccharides can be hydrolyzed to separate the monosaccharides by enzymatic or acid hydrolyses [31]. It should be noted that hydrolysis via the anaerobic digestion of polysaccharides is also possible [32]. Another method of extracting the polysaccharides from SCG is through the application of an alkali extraction with hydrogen and peroxide in the solution [33]. Finally, dilute acid hydrolysis is another option [25]. The main advantage of the latter method is that the monosaccharides forming the polysaccharides are released and separated, thus avoiding the need of the extra hydrolysis step for the other extraction methods. Dilute acid hydrolysis consists of soaking the SCG in sulfuric acid at a high temperature (100–180 °C) in a batch reactor. At the end of the reaction, the reactor is cooled down in an ice bath, and the solid material can be separated by filtration, with the filtrate containing the polysaccharide hydrolysates. The optimal conditions for hydrolysis were found to be at 100 mg of acid/g of dry matter, and a 10 g/g liquid-to-solid ratio at 163 °C for 45 min. Following the hydrolysis step, various compounds are formed, including hydroxymethylfurfural (HMF), acetic acid, phenols, and heavy metals, which interfere with the bioconversion processes, e.g., the fermentation of the hydrolysate, and inhibit them [34]. To purify and separate the hydrolysate from the compounds of these inhibitors, different methods can be used, such as biological methods, involving enzymes; physical methods, involving the evaporation of the volatile compounds; and chemical methods, involving precipitation or ionization. These methods can be used alone or in any combination.

Autohydrolysis is a new method proposed for the extraction of polysaccharides that does not require any chemical agent, as it only involves water. In an autohydrolysis process, SCG are mixed with water at a very high temperature (160–200 °C) in a batch reactor. The temperature, the solid-to-liquid ratio, and the extraction time play crucial roles in the extraction performance, and the optimal conditions are 160 °C, 15 mL/g of SCG, and 10 min, respectively. Following extraction, the reactor is cooled down in an ice bath. The liquid phase is then mixed with ethanol to precipitate the polysaccharides, which can then be recovered by centrifugation [5,35].

It should be noted that pretreatments can be performed on the SCG before the polysaccharide extraction via water hydrolysis, ultrasound, microwave, or supercritical carbon dioxide (SC-CO_2). Some parameters can be set to compare the yields of the SCG after each pretreatment, such as the extraction temperature, pressure, and time. The optimal conditions were found to be 180 °C, 40 bar, and 10 min, respectively. On the basis of these results, it was found that the ultrasound pretreatment led to higher yields (18 wt.%) and it was shown to be a more successful method for extracting polysaccharides from SCG [36]. The microwave-assisted pretreatment can work at a lower operation time and is influenced by the temperature, the microwave radiation exposure time, and the water or alkali solution in the suspended sample. The microwave pretreatment followed by water hydrolysis showed that arabinogalactan needs a higher temperature to be extracted (up to 170 °C), while galactomannan can be extracted at lower temperatures (from 140 °C) [37,38].

3. Conversion of Polysaccharides and Their Hydrolysates into Biopolymer Precursors

3.1. 2,3-Butanediol Synthesis

Butanediol (BD) is a chemical compound obtained via microbiological fermentation, with an alternative bioprocess for several industrial applications, such as polyesters, rubbers, fertilizers, cosmetics, pharmaceuticals, and food additives [39]. BD is synthesized by different microorganisms in a mixed acid fermentation process that produces other coproducts, such as ethanol, acetoin, acetate, lactate, and succinate, depending on the microorganism used, the pH value, and the oxygen level. The synthesis of 2,3-BD occurs from pyruvate with three main enzymes (α-acetolactate synthase, α-acetolactate decarboxylase, and butanediol dehydrogenase), as reported in Figure 2 [40]. Pyruvate is generated from some monosaccharides, such as glucose, xylose, or other sugars. From glucose, pyruvate is produced by glycolysis. It is then converted to α-acetolactate via decarboxylation, followed by acetoin formation via α-acetolactate decarboxylase. In the presence of oxygen, acetoin can also be formed by diacetyl reductase after a spontaneous formation of diacetyl. Both types of acetoin are converted to 2,3-BD via butanediol dehydrogenase. It is important to know that 2,3-BD can be formed with different isomers because of the distinct acetoin formation [40–42].

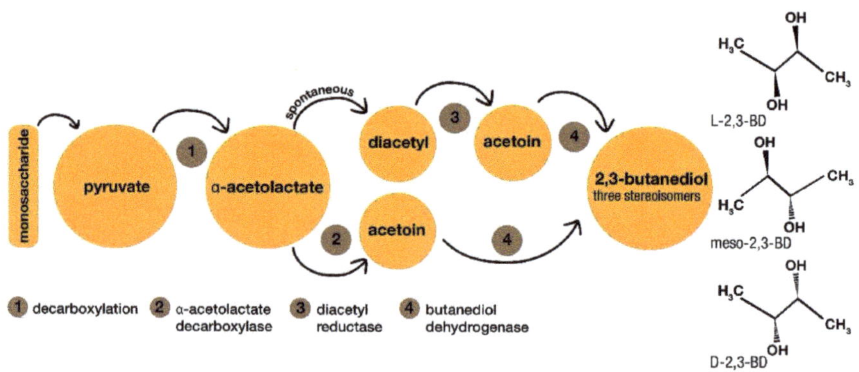

Figure 2. The synthesis steps leading to 2,3-butanediol.

The oxygen supply, the pH, and the temperature should be considered when starting the fermentation process, which is different between laboratory and industrial conditions. The yield of the 2,3-BD production, and the effectiveness of the aerobic conditions, are mostly related to the oxygen level. An appropriate initial pH (6–7) and pH setting (5–8) during the whole fermentation process directly affect the production. Moreover, bacteria inoculums only grow under specific temperatures (30–37 °C), but selecting an appropriate temperature directly affects the yield of 2,3-BD. After the fermentation process, 2,3-BD purification should be evaluated and optimized, but its high boiling point (180–184 °C) and hydrophilic behavior make the separation and purification more difficult [42–44].

Despite the biological and nontoxic process, the yield and the cost of 2,3-BD synthesis is a current issue that stands against the use of petroleum-based chemicals. This is why there have been some attempts to enhance 2,3-BD synthesis, in terms of higher yields, lower costs, and an efficient substrate feedstock [39]. Therefore, waste materials rich in lignocellulose are attractive for 2,3-BD production because they are inexpensive, abundant, and nontoxic. In the last few years, lignocellulosic biomass has been used as a substrate for 2,3-BD production, since cellulose and hemicellulose can be hydrolyzed to recover 2,3-BD [42,45]. Several works present good results using microorganisms and lignocellulosic biomasses to produce 2,3-BD (Table 1). Moreover, Hazeena et al. (2020) report a list of lignocellulosic biomasses that have already been used in the literature for 2,3-BD production, e.g., corn stover, corncobs, sugarcane bagasse, rice, kenaf, and soybean hull [46].

There is no work on the direct production of 2,3-BD from SCG. However, it has been reported that 2,3-BD was detected after the fermentation of green coffee beans from *C. cellulans*, with a spontaneous fermentation using mesophilic and lactic acid bacteria to improve the fermentation process [47], and from *Rhizopus oligosporus*, with valine biosynthetic fermentation, followed by diacetyl formation and acetoin formation aimed at the modulation of the coffee aroma [48]. These two works detected 2,3-BD, but this was not the objective of their work. Nonetheless, SCG could also be used directly for 2,3-BD production, upgrading this lignocellulose-rich waste, since 2,3-BD has several valuable coproducts and can be converted to polymer precursors, as will be described next.

Table 1. Examples of the lignocellulosic biomasses used for 2,3-BD production.

Lignocellulosic Biomass	Production Method	Type of Sugar	2,3-BD Yield	References
Mixed biomass	Hydrolyses and flask fermentation by *S. cerevisiae*	Xylose	0.27 g/g	[49]
Sorghum biomass and wood	Hydrolyses and shaken flask, followed by bioreactor fermentation by *B. licheniformis*	Glucose and Xylose	0.45 g/g 0.40 g/g	[50]
Corncob	Alkali pretreatment, hydrolyses, and batch/fed-batch fermentation by *E. cloacae*	Glucose and Xylose	0.42 g/g	[51]
Kenaf core	Calcium peroxide pretreatment, hydrolyses, and batch fermentation by *K. pneumoniae*	Glucose and Xylose	0.38 g/g	[52]
Sunflower and pine tree	Hydrolyses and shaken flask fermentation by *K. oxytoca*	Glucose, Xylose, Galactose, and Mannose	0.29 g/g 0.22 g/g	[53]
Sugar cane bagasse	Hydrolyses and fed-batch fermentation by *E. ludwigii*	Xylose	0.38 g/g	[54]
Brewer's spent grain	Microwave-assisted alkali pretreatment, hydrolyses, and shaken flask fermentation by *E. ludwigii*	Glucose	0.48 g/g	[55]

3.1.1. Conversion of 2,3-BD into Biopolymer Precursors

It is well established that 2,3-BD can produce polymer precursors and other derivatives, such as methyl ethyl ketone (MEK), 1,3-butadiene (BD), acetoin, diacetyl, 2,3-butanediol diester, and polyurethane precursor. From dehydration, 2,3-BD can be converted into 1,3-BD using different catalysts, which are commonly applied in elastomer production, such as styrene-butadiene rubbers (SBR) [42,46,56]. Nguyen et al. (2019) produced 1,3-BD through the one-step catalytic dehydration of 2,3-BD with a rare earth orthophosphate-based catalytic reaction. They showed that this process could reach the industrial scale, with an operating temperature of 300 °C. The results also show that it was possible to recover 58 wt.% of 1,3-BD with MEK and methyl propanal (MPA) as coproducts [57]. Sun et al. (2020) reviewed several works about the extraction of C4 alcohols from biomass to produce 1,3-BD. They concluded that one of the most efficient and competitive methods for recovering 1,3-BD is through 2,3-BD production, mainly from lignocellulosic biomasses [58].

MEK is also produced from the dehydration of 2,3-BD, a fuel additive that can be used in the polymer industry to produce resins, paints, and solvents. From polymerization, 2,3-BD can be used as a polyurethane precursor, i.e., for polyol and polymeric isocyanates [42,59].

3.1.2. Valorization of the Coproducts from 2,3-Butanediol Synthesis

The microbial synthesis of 2,3-BD results in the formation of coproducts, such as ethanol, acetoin, acetate, lactate, succinate, and formate [41,45]. Rehman et al. (2021) produced 2,3-BD from pure glucose and xylose-oil-palm-derived fermentation using *K. pneumoniae*, and they identified the formation of small amounts of ethanol, lactic acid, and succinic acid [43]. Narisetty et al. (2021) synthesized 2,3-BD from sugarcane-bagasse-rich xylose with *E. ludwigii*, and they obtained acetic acid as a coproduct [54]. Liakou et al. (2018) used various fruits, such as plums, apples, and pears, as well as vegetables, such as broccoli, cabbage, lettuce, fresh beans, corn salad, carrots, peppers, and eggplants, to extract fructose, sucrose, glucose, xylose, galactose, and arabinose, which were fermented by *E. ludwigii* to produce succinic acid, ethanol, and lactic acid as coproducts from 2,3-BD synthesis [60]. Furthermore, all these coproducts from 2,3-BD synthesis are valuable compounds that can also be directly extracted from the fermentation of the lignocellulosic biomass.

Bioethanol, produced from biomass sources, is biodegradable and less polluting than conventional ethanol, and can be used as a fuel for transport. From ethanol dehydration, it is possible to produce ethylene with a low investment. Ethylene has applications in the polymer field, e.g., in biobased polyethylene [61]. Previous research shows that bioethanol could be effectively produced from lignocellulosic biomass fermented by *S. cerevisiae* [62] and *Antarctic psychrophilic* [63]. Mussatto et al. (2012) evaluated the ethanol production from SCG sugars (glucose, arabinose, galactose, and mannose) fermented with three different strains (*S. cerevisiae*, *Pichia stipitis*, and *Kluyveromyces fragilis*). The results show that *S. cerevisiae* was the most effective for ethanol production: 11.7 g/L with a yield of 0.26 g/g [64]. Rocha et al. (2014) produced ethanol from the hydrolysis of SCG oil-free extract by ultrasound-assisted extraction, fermented by *S. cerevisiae* [65]. In addition, SCG oil was used with brewer's spent grain to produce high amounts of ethanol (30–55 wt.%), for which the authors evaluated the relation between the increasing yield (more than 100%, compared to SCG oil alone) and the decreasing cost (from EUR 9.31/kg of SCG oil, to EUR 3.89/kg of SCG/brewer's spent grain) for the ethanol produced [66].

Acid compounds, such as succinic, acetic, and lactic acids, are usually found from the beginning until the end of the SCG fermentation process [67]. Succinic acid is largely used in the production of 1,4-butanediol, tetrahydrofuran, and biodegradable polymers, such as poly(ester amides) [68]. Acetic acid is applied in vinyl acetate production, which is of interest for the production of vinyl plastics, adhesives, textile, and latex paints [69]. It also works as a solvent for the precipitation polymerization of poly(divinylbenzene) (PDVB) [70]. Lactic acid has been used in the pharmaceutical, cosmetic, chemical, and food industries for several years. Today, its main application is for poly(lactic acid) (PLA) production, a well-known biodegradable and biocompatible polyester [71,72].

Some of the coproducts from SCG fermentation are listed in Table 2. Liu et al. (2021) investigated the fermentation of SCG hydrolysates by *S. cerevisiae* and *Lachancea thermotolerans*, with and without yeast extracts. The results show that both strains produced succinic acid, acetic acid, lactic acid, and volatile compounds, but that the addition of yeast extract improved the succinic acid yield [12]. In another work, Liu et al. (2021) used SCG hydrolysates fermented by *Oenococcus oeni* and *L. thermotolerans* to produce volatile and nonvolatile compounds. The authors identified succinic, acetic, and lactic acids on SCG hydrolysates, before and after the fermentation process, with higher yields at the end of the fermentation process [73]. The specific production of lactic acid from SCG hydrolysates fermented by *Lactobacillus rhamnosus* was conducted and was aimed at industrial and economic demands [10]. Moreover, it was reported that 1000 g of SCG could produce about 100 g of lactic acid through the fermentation of SCG hydrolysates by *Lactobacillus brevis* and *Lactobacillus parabuchneri* [74]. Another work investigated lactic acid production via acid-pretreated SCG and washed pretreated SCG fermented by *S. cerevisiae*. The lactic acid yield was four times higher than the SCG pretreated with acid [75].

Table 2. Examples of coproducts from SCG fermentation.

Coproducts	Production Method	Yield [1]	References
Bioethanol	Hydrolysis of SCG fermented by *S. cerevisiae*	0.26 g/g	[64]
Bioethanol	Hydrolysis of SCG oil extracted by ultrasound-assisted extraction fermented by *S. cerevisiae*	0.5 g/g	[65]
Bioethanol	Hydrolysis of SCG oil and brewer's spent grain oil, extracted by Soxhlet extraction, fermented by *S. cerevisiae*	57.3%	[66]
Succinic acid, acetic acid, and lactic acid	Hydrolysis of SCG fermented by *S. cerevisiae* with yeast extract	2.6 g/L 0.8 g/L 0.2 g/L	[12]
Succinic acid, acetic acid, and lactic acid	Hydrolysis of SCG fermented by *O. oeni* coinoculated with *L. thermotolerans*	16.4 g/L 5.2 g/L 22.4 g/L	[73]
Lactic acid	Hydrolysis of SCG fermented by *L. rhamnosus*	98%	[10]
Lactic acid	Hydrolysis of alkali-treated SCG fermented by *L. brevis* (Lb) and *L. parabuchneri* (Lp)	40.1% (Lb) 55.8% (Lp)	[74]
Lactic acid	SCG pretreated with sulfuric acid whole slurry (s) and washed (w) and fermented by *S. cerevisiae*	11.2 g/L (s) 3.4 g/L (w)	[75]

[1] When many different samples were found, only the highest yields are shown.

3.2. Polyhydroxyalkanoate (PHA) Synthesis

PHA is a bioplastic that is produced as an alternative to petroleum-based polymers with respect to sustainable industrial development [76]. A microbial fermentation process is used for the synthesis of this polyester, but it has a higher cost than petroleum-based polymers (USD 3.50/kg compared to USD 1.30/kg). Several efforts have been devoted to the use of waste biomass as a substrate source for PHA production, as the substrate represents around 28–50% of the entire PHA production costs [77,78]. Since PHA is biodegradable and biocompatible, it is used in several fields, such as in the production of biomedical devices (surgical pins and bone screws), as well as in electronics, construction, and the automotive industry, and in packaging, wrapping films, and bottles [77].

PHA can be found with short-chain lengths (3–5 carbon atoms) and medium-chain lengths (6–14 carbon atoms) [77,79]. The former is more crystalline and has thermoplastic properties, while the latter has more elastomeric properties. Short-chain-length PHA is the most used and competes with poly(lactic acid) (PLA) in several fields, such as food packaging. Its functional groups can be replaced by 3-hydroxybutyrate or 3-hydroxyvalerate, forming poly(3-hydroxybutyrate) P(3HB), poly(4-hydroxybutyrate) P(4HB), poly(3-hydroxyvalerate) P(3HV), and a copolymer of P(3HB–co–3HV) [77,79]. PHA can be synthesized via three pathways: the acetyl-CoA to 3-hydroxybutyryl-CoA pathway; the fatty acid degradation by β-oxidation pathway; and the fatty acid synthesis pathway (Figure 3). Sugars (glucose, sucrose, and saccharose) and fatty acids (myristic, palmitic, and oleic) can be used as substrates for PHA production [77]. In the first pathway, the sugars or fatty acids are converted to acetyl-CoA. Two molecules of acetyl-CoA are converted to acetoacetyl-CoA by β-ketoacyl-CoA thiolase. In the next stage, acetoacetyl-CoA is reduced to 3-hydroxybutyryl-CoA by acetoacetyl-CoA reductase. Finally, PHB or P(3HB) is produced by polymerizing 3-hydroxybutyryl-CoA with PHA polymerase (PhaC). The second pathway starts from the fatty acid and β–oxidation cycle, which can be performed through 2-enoyl-CoA by hydratase, through 3-hydroxyacyl-CoA by epimerase, or through 3-ketoacyl-CoA by reductase. Then, 3-hydroxyacyl-CoA is obtained and converted to the PHA medium-chain length by PHA polymerase (PHaC). In the third pathway, sugars or fatty acids are converted to 3-hydroxyacyl-ACP. Through the enzyme, 3-hydroxyacyl-ACP-CoA transferase, 3-hydroxyacyl-ACP is converted to 3-hydroxyacyl-CoA, and this is followed by the synthesis of PHA polymerized by PHAC [11,76]. It should be noted that PHA production can be improved with an enhanced microbial fermentation process using specific conditions, such as fed-batch fermentation, the feast–famine strategy, solid-state fermentation, and continuous fermentation [79].

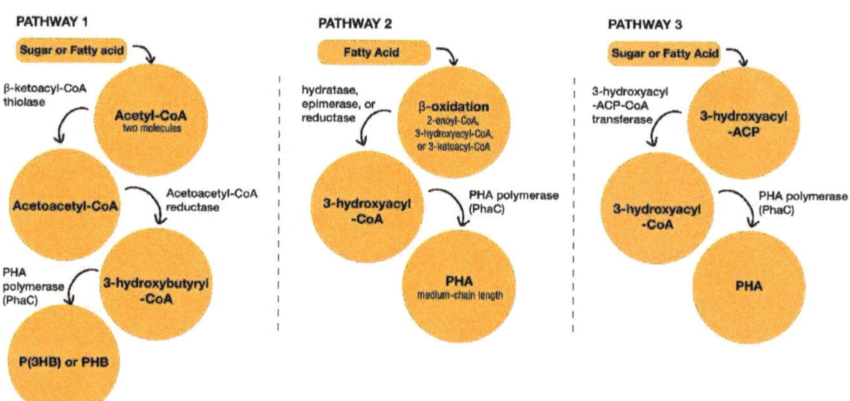

Figure 3. The three pathways leading to PHA synthesis.

The 3-hydroxybutyrate (PHB) homopolymer can be produced from SCG oil extracted by supercritical carbon dioxide and fermented by *Cupriavidus necator*. The PHB yield can reach 0.77 kg per kg of SCG oil. Moreover, the mechanical properties of PHB films show a tensile strength of 16.0 MPa, a Young's modulus of 1.0 GPa, and an elongation at break of 1.3% [80]. Kovalcik et al. (2018) produced PHB after the detoxification of SCG hydrolysates (the extracts of coffee oil and phenolics) by *Halomonas halophila* fermentation [81]. Obruca et al. (2014) also tested the detoxification of SCG hydrolysates to produce PHA by *Burkholderia cepacia*. It was found that polyphenol removal by ethanol was the most suitable process since it improved the PHA yield up to 25%. Therefore, copolymers of 3-hydroxybutyrate and 3-hydroxyvalerate were produced [82]. Although several works report using cellulose and hemicellulose sugars from lignocellulosic waste for PHA production [78,79,83], lignin is also an effective substrate source since acetyl-CoA can be produced from these aromatic compounds, followed by fatty acid synthesis to produce PHA [77,84].

Several works focus on producing higher yields of PHA from waste biomass. However, effective extraction methods and the quality of the biopolymer properties require more attention. Moreover, waste biomass is not only used to reduce the PHA production costs; it is also the key to a circular economy because it closes the loop of material consumption [85]. The challenges in producing PHA from SCG are highlighted in Figure 4, while PHA production from SCG oil will be detailed in Section 4 with regard to the extraction and application of SCG oil.

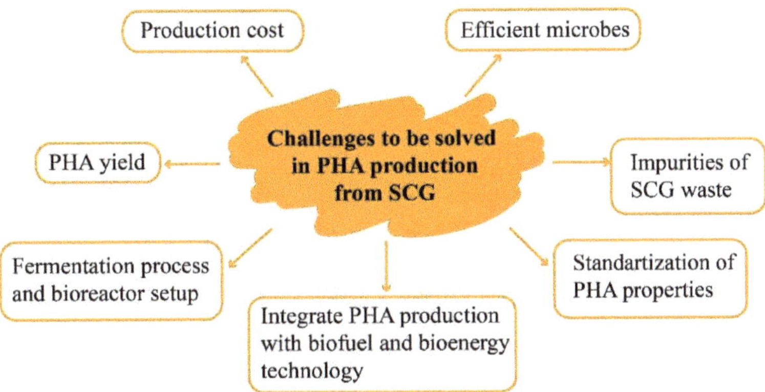

Figure 4. Challenges related to PHA production from SCG.

3.3. Other Biopolymer Precursors

Other polymers can be produced from sugars, such as cyclic monomers, with or without the functional groups of the original sugar structure [86], and synthetic polymers, with sugar units integrated into their main chains [87]. Lactic acid is a cyclic monomer generally produced from glucose and sucrose fermentation, which can be polymerized into poly(lactic acid) (PLA) through polycondensation or ring-opening polymerization (ROP). Another example is the cyclic monomer, ε-caprolactone (ε-CL), produced from fructose, glucose, or mannose by 5-(hydroxymethyl)furfural. ε-CL can react with ammonia to synthesize Nylon-6 (ε-caprolactam), or by ROP to produce poly(ε-caprolactone) (ε-PCL) [86]. Polyesters, polycarbonates, polyamides, polyesteramides, polyurethanes, and polyureas can all be synthesized using sugar-based monomers, such as alditols, aldonic acids and lactones, aldaric acids (polyamides and polyesters), and amino sugars (polyamides, polyurethanes, and polyureas) [87]. Moreover, some sugar-based polymers are biodegradable, biocompatible, and nonimmunogenic. This is why they are mostly used in biomedical applications [88].

It was reported that aliphatic polyesters could be produced from sugar-based bicyclic alditol monomers from D-glucose (glux-diol and isosorbide) and D-mannose (manx-diol) by polycondensation. Nevertheless, glux-diol is the most suitable monomer in terms of the molecular weight (46,500 g/mol), the higher glass transition temperature (more than 100% increasing), and the higher mechanical properties (a 300% increase in tensile strength) [89]. Moreover, polycarbonate was effectively synthesized from the cyclic carbonate monomer obtained from D-mannose, showing good thermal properties for biomedical applications [90]. Polyols were also effectively produced from sugars and were used in polyurethane reactions. Fructose presented the most appropriate chemical properties for a fructose–isocyanate reaction, and resulted in a stable and low-shrinkage sugar-based polyurethane [91]. All these works confirm the possibility of producing a sugar-based polymer from different lignocellulosic sources. In particular, SCG were used in polyol production through the acid-liquefaction process, followed by polyurethane (PU) production. It should be noted that the optimal parameters for polyol acid liquefaction from SCG were 160 °C, 4 wt.% of sulfuric acid, and a reaction time of 80 min, which resulted in a yield of 70 wt.% of polyol [7]. Another work produced 74 wt.% of polyol from SCG, and then polymerized a PU foam (10–50 wt.% of SCG polyol) with a dodecahedron-cell structure and undamaged cell windows, similar to conventional PU foams for thermal insulation [92]. PU foam from SCG polyol has already been tested for thermal insulation applications and it demonstrates optimum thermal properties. Moreover, the foam exhibited great elasticity properties, since foams that were compressed to up to 70% could return to their initial state [93]. Polyols from SCG were also polymerized into PU foams for sound-absorption applications. The results show sound-absorption coefficient improvements at low and high frequencies, compared to the foams derived from glycerol, because of the increase in the cells size and the low stiffness of SCG, respectively [94].

Finally, a biofilm from the sugars of SCG was reported, using SCG galactomannans treated by alkaline and enzymatic treatments. The mechanical results show tensile strengths of 0.30 MPa and 0.21 MPa, Young's moduli of 4.8 MPa and 8.5 MPa, and elongations at break of 13.2% and 5.26%, for the alkaline treatment and the enzymatic treatment, respectively. The water vapor permeability was also investigated, resulting in 2.3 g/m s Pa for the alkaline treatment, and 9.35 g/m s Pa for the enzymatic treatment. The authors conclude that both biofilms can be used for food packaging, vibration damping, and gas separation membranes [95].

4. SCG Oil

SCG oil has been extracted for applications in several fields, such as biodiesel production, polymer production, and cosmetics [81]. Coffee oil (corresponding to 7–21 wt.% of SCG) is rich in lipids/fatty acids (linoleic, palmitic, oleic, polyphenolic, etc.) and other valuable compounds, such as tocopherol and polyphenolic compounds. The oil can be recovered by solvent extraction using hexane, ethanol, and methanol, or by supercritical conditions with CO_2. Supercritical CO_2 extraction is more sustainable than solvent extraction and it can recover a higher oil yield (15–20%, instead of 6–27%) [13]. The combination of the two-phase solvent (methanol and hexane), assisted with ultrasound, can more effectively extract the coffee oil (higher yield), compared to the solvent extraction alone [96]. Nonthermal plasma technology and ultrasound-assisted methods were used with Soxhlet extraction (solvent) to investigate their influence on the oil yields. The results show that plasma technology recovered the highest amount of oil (19%), compared to ultrasound (14%) and Soxhlet extraction alone (9.4%) [97]. Couto et al. (2009) obtained coffee oil by supercritical CO_2 under different temperatures, pressures, and time conditions. It was shown that 50 °C, 25 MPa, and 3 h were the optimal parameters, leading to a yield of 15%, which represented 85% of the total amount of oil [98]. Araújo et al. (2019) used ethanol as a solvent with supercritical CO_2 to enhance coffee oil extraction. The results show a yield of 16%, but a shorter extraction time and a lower amount of organic solvent were used, compared to supercritical CO_2 alone [99].

Several efforts have been made to produce PHA from SCG oil and to optimize the process to increase the PHA yield in a shorter time. One study shows that the homopolymer, 3-hydroxybutyrate, could be obtained from SCG oil extracted by supercritical CO_2 [80]. The optimal parameters of the oil extraction were: 50 °C; 25 MPa; a solvent flow rate of 10 kg/h; a CO_2:coffee mass ratio of 35:1; and 1.5 h of extraction time. These conditions enabled the recovery of 90% of the total coffee oil. Then, the coffee oil was fermented using *Cupriavidus necator* in a batch reactor, allowing a PHA conversion of 78% (w/w), with a molecular weight of 2.34×10^5 g/mol, a melting temperature of 172 °C, a glass transition of 8 °C, and a crystallinity degree of 58% [80].

Obruca et al. (2014) used SCG oil fermented by *C. necator* in a batch and fed-batch reactor for PHB production. The fed-batch mode produced a higher PHB yield (49.4 g/L vs. 26.5 g/L), increased productivity (1.33 g/h vs. 0.66 g/h), and it produced a higher molecular mass (4.74×10^5 vs. 4.27×10^5 g/mol), compared to the batch mode. The authors also discuss the advantages of reusing SCG as a carbohydrate source for PHB production, since millions of tons of waste with toxic compounds are discarded in the environment (caffeine and tannins) [100].

Bhatia et al. (2018) obtained coffee oil from four different solvents (hexane, isobutanol, methanol, and ethanol), and fermented the solvent with the highest oil yield (coffee oil from hexane) with *Ralstonia eutropha* to produce a copolymer of poly(3-hydroxybutyrate-co-3-hydroxyhexanoate) (P(HB-co-HHx)). Using a β-oxidation pathway and coffee oil as a carbon source, it was possible to produce the copolymer precursor (HB-co-HHx). Then, 69% (w/w) of P(HB-co-HHx) was obtained, with 78 mol% of HB, and 22 mol% of HHx [101].

Ingram and Winterburn (2021) used coffee oil and sunflower seed oil as carbon sources for poly(3-hydroxybutyrate-co-3-hydroxyvalerate) (P(3HB-co-3HV)) production, fermented by *C. necator*. After 72 h, the coffee oil and the sunflower oil were produced, with 89% (*w*/*w*) and 88% (*w*/*w*) P(3HB-co-3HV) yields, respectively, showing that P(3HB-co-3HV) can be produced without any traditional 3HV precursor. The authors claim that it was the first time that P(3HB-co-3HV) was produced via *C. necator* from coffee oil, which will have a large influence on future works in this field [102].

5. SCG Polymer Composites

Polymers may be classified as thermoplastics, thermosets, and elastomers, on the basis of their responses to temperature. Above their melting temperature, thermoplastics become soft, presenting a fluid-like behavior, and they can then be molded. However, heating a thermoset eventually leads to its degradation [103]. Elastomers can have thermoplastic characteristics, or chains developing a network via covalent crosslinks, formed in a separate post-polymerization stage known as "vulcanization" [104]. In this study, some of the polymers used as matrices for the composites of SCG are presented and discussed.

5.1. Polythylene (PE) Composites

The most common synthetic polymer matrices filled with SCG are polyethylene (PE) and polypropylene (PP), because of their widespread applications in packaging, their low costs, and their overall good properties [2,105]. PE can be classified into several types, but high-density polyethylene (HDPE) and low-density polyethylene (LDPE) are the main commercial ones. Worldwide, PE is one of the most widely used thermoplastic polyolefins for blow and injection molding. Because of its high toughness, its ease of processing, its low electrical conductivity, and its chemical inertness, PE is used in different applications, such as in pipes, sheets, containers, and other similar products. Compared to LDPE, HDPE presents excellent impact strength and electrical insulation properties [106–108]. As a thermoplastic matrix for composites, HDPE can be found in packaging and automotive parts, as well as in biomedical and space applications [107,109].

Mendes et al. (2021) obtained HDPE/SCG composites (10–30 wt.%) via extrusion and injection molding. They reported that the incorporation of SCG particles in the HDPE preserved, or improved, the physicomechanical properties of the composites, without treatment or the use of a coupling agent. Adding SCG fibers increased the stiffness (elastic modulus) by 49% but decreased the tensile strength by 21%. Nevertheless, a 13% increase in the elasticity was observed with 10 wt.% of SCG. The incorporation of SCG did not modify the thermal properties, so the same processing conditions were used as in the neat HDPE matrix [110].

Alkali-treated SCG were used as fillers for the oxo-biodegradable HDPE composites, using different volume fractions (5, 10, 15, and 20%). The alkaline-treated SCG composites presented better structural, thermal, and mechanical properties than the untreated ones. The FTIR spectra show that the alkali treatment eliminated amorphous contents and impurities, improving the filler/matrix interface adhesion, as is confirmed by the SEM images. The degree of crystallinity increased by 5% after the treatment, and the thermal stabilities for both the untreated and treated composites were similar. In terms of the mechanical properties, at 10 wt.% of SCG, improvements of 25% for the tensile strength, and 24% for the tensile modulus, were observed, compared to the untreated composite. Using 15 wt.% of SCG led to a 6% improvement in the impact resistance. The authors indicate that this composite has high potential for several engineering applications, such as automotive, packaging, and lightweight furniture applications [111].

Cestari and Mendes (2013) elaborated the HDPE/SCG composites using four different types of SCG (integral, extracted, large-size, and small-size) in order to study the effects of the particle sizes and the soluble extraction on HDPE properties. The composites were prepared with 10 wt.% of filler. The results show that integral SCG presented similar properties to the small-size sample, and that it was superior to the extracted sample. The thermal properties show that no degradation occurred for the processing temperature range investigated (160–190 °C) for HDPE. HDPE/integral SCG and HDPE/small-size SCG composites degraded similarly, at higher temperatures than HDPE/extracted SCG and HDPE/large SCG. The melting temperature of all of the composites was similar to the neat polymer (134 °C). Except for the large SCG composites, a slight reduction in crystallinity was observed [112].

The effect of the SCG content (0 to 60 wt.%) on the crystallization kinetics of recycled HDPE composites was studied using differential scanning calorimetry (DSC). The results show that no degradation occurred within the processing temperature range (from −220 °C to 178 °C) of recycled HDPE. Moreover, the melting temperature of the composites was similar to the neat polymer (131 °C) [113].

Recycled HDPE compounds were elaborated from 0 to 60 wt.% of SCG using an extruder, and the samples were injection molded. A thermal analysis shows that the composites degraded in two steps. The first degradation step occurred at around 250 °C, and it was followed by another step at around 350 °C. The DSC curves show that the melting temperature of the composites was similar to that of the recycled HDPE, which presented a second peak at around 160 °C, which was probably due to the presence of PP as a typical contamination. The compressive moduli of the composites were similar to the neat polymer. The composite with 30 wt.% of SCG had the same compressive modulus as the neat polymer. However, the composites with 40 and 50 wt.% fillers had 11% lower compressive moduli, while at a 60 wt.%, the loss was 29% [114].

Biochar from SCG pyrolysis was used as a filler for the HDPE composites produced via melt mixing. A rheological characterization was performed for different flow fields (linear and nonlinear dynamic shear flows). A decrease in the relaxation dynamics of HDPE macromolecules was observed, which was due to the porous structure of the filler. Stress relaxation measurements revealed pseudo-solid-like behavior for the composites containing high amounts of filler. The biochar improved the thermo-oxidative stability of the composites and modified their melting enthalpies, which decreased the polymer crystallinity, compared to the neat matrix. The authors claim that this decrease in the polymer crystallinity was reported early in the literature because of the reduction in the polymer chain mobility in contact with the natural particle surface [115].

5.2. Polypropylene (PP) Composites

Polypropylene is a low-cost polymer and a well-known general-purpose commodity thermoplastic used in different applications, including in packaging, films, fibers, and automotive parts. PP can be used alone, or as a matrix for composites with different reinforcements (particles, fibers, etc.), especially biobased materials, such as wood [116], sugarcane bagasse [117], hemp [118], and rice straw [119]. Nevertheless, recent works report that SCG are a promising filler for PP composites, leading to different PP structures. It was reported that PP homopolymer and copolymer were successfully filled with 10, 20, and 30 wt.% of SCG. Overall, the mechanical properties decreased with the increasing SCG content, compared to the neat PP. However, the impact strength was improved by 77% (10 wt.% SCG), compared to the PP homopolymer. The authors conclude that SCG must have a surface treatment in order to improve the interfacial adhesion with the matrix and to improve the tensile strength for future works. These composites, based on SCG and PP homopolymer, could be used as substitutes for virgin PP in some applications, where the impact strength and the sustainable principles are needed [16].

PP composites can be improved by using treated fillers to increase the interfacial adhesion [16,120]. Coupling agents are generally used to create a chemical bond between the matrix and the reinforcement. The most common ones are based on maleic anhydride-grafted polypropylene (MAPP), and they yield effective results. However, a direct surface treatment of the fiber can be performed using palmitoyl chloride, silane, and MAPP to improve the properties of PP reinforced with SCG (20 wt.%). The flexural properties of the composite were not affected by either treatment (palmitoyl chloride and silane) or

MAPP (2 wt.%). However, palmitoyl led to a significant improvement (54%) in the impact strength. The thermal stability of the untreated SCG composite was higher than the neat PP, but the addition of MAPP and the surface treatments decreased it. Moreover, palmitoyl provided an effective hydrophobic behavior and showed a better SCG particle dispersion in the matrix, with the authors of the study concluding that palmitoyl-chloride-modified SCG was the best filler studied (20 wt.% SCG) [17]. MAPP (1, 3, and 5 wt.%) was also reported with PP homopolymer and copolymer, reinforced with 15 wt.% of SCG [121]. The PP copolymer, with and without MAPP, presented the best mechanical results, which were related to a better interaction between the chemical structure of the copolymer (propylene and ethylene) with the MAPP and the OH groups of the filler surface. The MAPP content was found to directly affect the mechanical properties, but 3 wt.% of MAPP showed a balance between all the results [121]. The other coupling agents investigated were silane and styrene–ethylene–butene–styrene grafted with maleic anhydride (SEBS-g-MA) for the PP composites filled with 15 wt.% of SCG. The coupling agents were compared with the bleached SCG to enhance the interfacial adhesion between the matrix/filler by removing the amorphous components, which leads to an improvement in the surface roughness of the filler, for which, in this case, bleached SCG caused no significant mechanical properties (less than 10%). Both coupling agents had good mechanical properties, but SEBS-g-MA was the only one that slightly improved the tensile strength (6%) without losing ductility, and while maintaining rigidity (Young's modulus) [18].

Wu et al. (2016) report that the oil extracted from SCG could improve the properties of PP composites. The oil extraction was performed by ultrasonication, followed by the production of composites with 40 wt.% of SCG, 2.5 wt.% of MAPP, and 2.5 wt.% of stearic acid. The composites with oil-extracted SCG presented lower water absorption, combined with higher mechanical strength (12%). These results validate that this additive is efficient at improving PP composites [120]. However, the oil-extracted SCG can also be used in other applications, such as biopolymer or biodiesel production [122]. Another interesting study was reported on SCG and coffee chaff as fillers for PP composites. Although SCG are more widely known and available, coffee chaff led to better thermal stability (an increase of 35 °C), better tensile (16%) and flexural (21%) strengths, as well as better tensile (43%) and flexural (52%) moduli. However, the coffee chaff produced a more brittle composite (lower elongation at yield (45%) and at break (55%), with a similar impact strength). Furthermore, coffee chaff has a dense fibrous morphology, while SCG present a granular porous morphology that is directly related to the better mechanical improvement of coffee chaff [123].

One of the most significant advantages of PP is its thermoplastic behavior, allowing for the possibility of using a recycled resin as the matrix to produce composites, thereby developing a more environmentally friendly material, with properties similar to those of virgin PP. Recycled PP (rPP) was applied in several works, with different fibers [124–126]. However, only one work was found on combining rPP with SCG, where the rPP was obtained from waste espresso coffee capsules and was reinforced with 20 and 30 wt.% of espresso SCG, with and without MAPP (10 wt.% of the reinforcement amount). The mechanical properties and the thermal stability results show that the composites based on rPP had properties similar to those of the virgin PP. Moreover, the effect of MAPP led to increases in the tensile strength (18%) and the Young's modulus (20%), but to decreases in the elongation at break (more than 100%) and the impact strength (23%). However, these composites were developed for a specific application on home composters. The authors conclude that the composite with 30 wt.% of SCG and without MAPP was the most suitable for this application, which was aiming for an environmentally friendly product [127].

5.3. Polyurethane (PU) Composites

Polyurethane is a versatile polymer that was created to replace rubbers by reacting a diisocyanate with a polyol. It has been used in several fields, such as in insulators, rigid foams, coating, adhesives, and elastomers [128]. Because of environmental concerns, some efforts are devoted to producing polyol from a natural source, and to producing PU composites with natural or waste-based fillers. So far, a very limited number of works are available on using SCG as a filler in PU composites.

Funabashi et al. (2003) compared different fillers in rigid PU foam composites, including SCG [129]. Hatakeyama and Hatakeyama (2010) produced PU from lignin polyol filled with SCG (50–80 wt.%). The results show increases in the flexural strength and the modulus, as the amount of filler increased with the constant density [130]. More recently, SCG (10–40 wt.%) were added to a viscoelastic PU foam composite, leading to a reduction in the foam growth time (11% for 10 wt.% SCG), and an increase in the foam density with increasing filler content (24% for 30 wt.% of SCG). The compression tests after 75% and 90% of the original deformation show that PU/SCG composites presented lower values of permanent deformation (about 2–3%), compared to neat PU (10% and 85%), which were below the acceptable limit of 10% [15].

5.4. Poly(Lactic Acid) (PLA) Composites

Biodegradable polymers have been reported to be more costly. Therefore, PLA composites with lignocellulosic wastes have been produced to reduce the PLA content, consequently becoming more affordable for the different applications for replacing synthetic polymers. In the last ten years, several works have reported on SCG valorization as a filler for PLA composites to reduce the amount of SCG in the environment. Baek et al. (2013) studied the effect of the SCG content (10, 20, 30, and 40 wt.%) in PLA and 4,4-methylene diphenyl diisocyanate (MDI) as a coupling agent. The mechanical strength decreased with the increasing filler content, but MDI was shown to improve the values at 30 wt.% by creating an urethane bond between the PLA and SCG [131]. Furthermore, the tensile strength of the 30 wt.% of the PLA/SCG composite without MDI (27.5 MPa) [131] was slightly higher than for the 30 wt.% of the PP copolymer/SCG composite (26.0 MPa) [16]. This comparison between the PLA and PP composites highlights the possibility of using PLA as a substitute for synthetic polymers, without using a nonenvironmentally friendly coupling agent, such as MDI. Arrigo et al. (2020) investigated two alternative processing routes to produce PLA filled with SCG biochar (1, 2.5, 5, and 7.5 wt.%): melt mixing and solvent casting. Both processes successfully produced PLA/biochar composites. However, a rheological characterization study suggests a poor particle dispersion for the solvent casting process, while polymer degradation (lower molar mass) was observed for the melt mixing process because of the higher temperature [21].

The main advantage of PLA reinforced with natural materials is the biodegradability of both materials, resulting in environmentally friendly compounds with low carbon footprints and easy end-of-life disposal (composting). The biodegradation rate was investigated for PLA/SCG composites, with and without coupling agents. After 60 days of incubation, the results show that the PLA filled with 20 wt.% of SCG had a higher mass loss than the composites with a coupling agent and the neat PLA. This biodegradation rate change was supported by a microscopy analysis, which showed more disruptions with larger voids for the PLA/SCG without a coupling agent [23]. It was also reported that a photodegradation test could precede the biodegradation test, accelerating the degradation process in PLA/SCG composites. The results show a decrease in the crystallinity degree and impact strength, but an increase in the water absorption, compared to the biodegradation test alone. Moreover, SEM micrographs show that the photodegradation followed by biodegradation caused a roughening of the material surfaces, producing more cracks, voids, and erosions [20].

Some efforts have been devoted to the food packaging field, mainly through the development of PLA biofilm composites. SCG were reported to act as a plasticizer and a lubricant in biofilms. The results show a more flexible PLA behavior (higher elongation at break) after the addition of SCG [132]. It was also shown that SCG could act as an oxygen barrier in PLA reinforced with diatomite, improving their possible use as food packaging [133]. Songtipya et al. (2019) investigated the use of PLA with polybutylene adipate terephthalate (PBAT) and SCG for food packaging, using toluene 2,4-diisocyanate (TDI) as a coupling agent. The composite presented good mechanical properties, with overall migration values (0.03–0.28 mg/dm^2) lower than 10 mg/dm^2, which is the limit for chemical compounds on the surfaces of food packaging. This indicates that these PLA/SCG biofilms can be used for food packaging, with a minimum number of affinities among the chemical compounds of the polymer and the food [19].

Besides food packaging, PLA has been widely used as a filament for 3D printing. Recently, researchers were looking to produce lower-cost PLA filaments by using different fillers, such as lignocellulosic wastes. PLA filled with oil-extracted SCG was effectively produced, and they presented an important toughness increase (419%), with 20 wt.% of SCG, compared to the neat PLA filament, which also led to a higher impact strength [134]. In another work, SCG were decolorized by bleaching and were mixed with PLA to change the final printing color of the PLA/SCG filament from brown (SCG) to other colors, depending on the pigments used. The decolorized composites showed similar mechanical strengths to the neat PLA, resulting in colored filaments with high melt flows and good printing quality [135].

SCG oil was also proposed as a plasticizer for PLA composites, using 40 wt.% of recycled coffee cups (paper). The results show that the SCG oil plasticizer (30 wt.%) improved the hydrophobicity and decreased the brittle behavior of the composite (an 86% increase in the elongation at break). The authors state that this composite had balanced mechanical properties and a nontoxic behavior for several food applications, especially for the coffee beverage industry [136]. SCG were also investigated to produce luminescent quantum dots (QD) as a filler for PLA. A composite with 1 wt.% of QD showed good UV shielding and important transmission to the visible light, coupled with significantly improved mechanical properties, compared to the neat PLA, i.e., 69% and 67% increases in the tensile strength and the elastic modulus, respectively. The authors also propose the use of this material as a high-performance nanocomposite for applications involving transparency and UV protection [137]. Another interesting work reported good results by using PP and lignin as a compatibilizer for PLA/SCG composites. PP and lignin were combined to improve the mechanical, thermal, and morphological properties, compared to those of neat PP or lignin alone, and they showed some synergy and a compatibilization efficiency for PLA/SCG composites [138].

5.5. Poly(Butylene Adipate-Co-Terephthalate) (PBAT) Composites

PBAT is a biodegradable polymer that is mainly used in packaging and the biomedical fields. Because of its low thermomechanical properties and high production cost, PBAT was used as a matrix for lignocellulosic composites [139]. Coffee waste is a current material for PBAT composites filled with coffee husks [140,141] and coffee silver skins [142,143].

Moustafa et al. (2017) published two works on PBAT/SCG composites. The first one investigated the effect of a plasticizer (polyethylene glycol, PEG). The results show a good interaction between the SCG, PBAT, and the plasticizer, which led to a higher tensile strength, good SCG dispersion, higher hydrophobicity, and higher thermal stability [144]. In their second work, the effect of torrefied SCG on the hydrophobicity of PBAT composites was studied. It was shown that coffee torrefaction at 250 °C and 270 °C was efficient to improve the composite hydrophobicity by more than 20%, while also improving the tensile strength of 10 wt.% of torrefied SCG by 27% and 63%, compared to the neat PBAT and the nontorrefied SCG composite, respectively [145].

5.6. Polyvinyl Alcohol (PVA) Composites

Polyvinyl alcohol is a biodegradable, biocompatible, water-soluble, and hydrophilic synthetic polymer that is used in several fields, such as biomedical and food packaging [146]. PVA fully biodegradable composites filled with SCG have been reported in different studies, mainly for their adsorbent properties. Lessa et al. (2018) investigated PVA filled with SCG and chitosan to adsorb pharmaceutical contaminants from water. They observed a substantial improvement (from 10 to 40%) in the absorption properties with a 5 wt.% of SCG, compared to the PBAT/chitosan composite. It was also possible to remove acetylsalicylic acid, caffeine, acetaminophen, and metamizole from the water [147]. Another study developed PBAT composites on the basis of Fe_3O_4 for Pb(II) ion adsorption. The optimal adsorption conditions were at a pH of 5, 24 h of contact time, room temperature, and an $SCG:Fe_3O_4$ ratio of 4:1. It was shown that the $PBAT/SCG/Fe_3O_4$ maintained a 78% adsorption efficiency after five cycles [148]. Minh and Thuan (2021) also produced $PBAT/SCG/Fe_3O_4$ composites for the adsorption of methylene blue, congo red, and tannic acid from an aqueous solution. The composite was analyzed by adsorption kinetics and adsorption thermodynamics, and the results present a high adsorption capacity, as the process was characterized as spontaneous and endothermic, with a blend of physisorption (electrostatic interaction, internal and external pore diffusion) and chemisorption (load shared or transferred from the organic molecules to the surface functional groups of the sorbent to create a chemical bond) adsorption mechanisms [149].

Cellulose nanofibers extracted from SCG were also studied as fillers for PVA composites as a source of non-wood cellulose material [150]. It was reported that nanoparticles from SCG can be used as a filler for PVA composites, leading to a higher tensile strength (from 80 MPa to 125 MPa) and a better deodorization performance, which means the removal of compounds that causes undesirable odors (from 98.9% to 89.5%), compared to neat PVA and PVA/carbon black composites [151]. Another interesting work used antioxidants extracted from SCG and citric acid as an active compound for PVA/starch films. The authors report that these antioxidants were effective at improving the antioxidant and antimicrobial properties of the films for food packaging [152].

5.7. Epoxy Composites

The effect of the SCG addition (5, 10, 15, 20, 25, and 30 wt.%) on the mechanical properties of compression-molded epoxy composites was studied. Compared to the neat resin, better mechanical properties for SCG/epoxy composites were found, especially for 25 wt.% of SCG. The fracture toughness increased with the SCG content, and a uniform distribution in the epoxy matrix enabled good stress distribution and better interface bonding. The 30 wt.% of SCG presented highly reduced wettability, with the epoxy matrix leading to lower mechanical properties, e.g., a decrease of 50% in the ultimate stress, and a decrease of 23.8% in the toughness [153].

Chemically treated SCG with NaOH were mixed with an epoxy resin at different weight contents (30, 40, 50, and 60 wt.%) [154]. The "30 wt.% of SCG" composites presented the most suitable properties, with better compatibility, compared to the other concentrations. A tensile strength of 45 MPa, a flexural strength of 80 MPa, a compressive strength of 112 MPa, and an Izod impact strength of 8 kJ/m^2 were reported. Their flame-retardant properties showed that the oxygen index was 20%, and the burning rate, according to the UL94HB, was 27 mm/min. Adding 30 wt.% of SCG with glass fiber led to the production of an epoxy hybrid composite. The morphological structure of the SCG/fiberglass/epoxy hybrid composite showed that the interface was strongly bonded and interactive, but no effect on the flame-retardant properties was observed, as only the epoxy resin and part of the SCG were burned, while the glass remained intact. The mechanical properties of the SCG/fiberglass/epoxy hybrid composite were similar to those of the epoxy-based composites reinforced with SCG alone.

Biochar derived from SCG (1 and 3 wt.%) was also used to obtain thermoset-based composites for 3D printing [155]. Their particles presented a nanostructured morphology. The rheological results show that the addition of SCG biochar increased the resin viscosity. Nevertheless, the 3D printed samples with lower SCG biochar contents (1 wt.%) had improved mechanical properties. The storage modulus increased by 27%, while the flexural modulus and strength increased by 55% and 43%, respectively. Unfortunately, adding 3 wt.% of SCG biochar substantially decreased the viscoelastic and flexural properties, which is due to agglomeration and the improper crosslinking between the chains. Moreover,

SCG biochar (15 and 20 wt.%) was also blended with epoxy for electrical purposes [156]. The results show that composites with 20 wt.% of SCG biochar produced a higher electrical conductivity (four times) than carbon black composites, and a higher tensile strength (18%), compared to the carbon black composite and neat epoxy.

SCG oil was successfully extracted from SCG and then the extracted SCG were blended (10 wt.%) with an epoxy resin [157]. In general, the composites presented lower mechanical properties relative to the neat epoxy. However, improvements were observed for the extracted SCG composites, compared to the SCG composites: an increased tensile strength, from 20.9 to 23.4 MPa; an increased flexural modulus, from 2.09 to 3.02 GPa; and increased flexural strength, from 33.0 to 42.9 MPa. Another recent work extracted the oil from SCG to fill epoxy composites (35 wt.%), and they claimed that extracted SCG provided enhanced tensile strength (more than 200%) and toughness (more than 100%), compared to neat epoxy and the SCG composite. The curing kinetics were also investigated and showed that extracted SCG composites cured faster at room temperature [158].

A novel work used SCG treated with phosphorus as a flame retardant in epoxy composites (5, 15, and 30 wt.%) [159]. The results show that the sample with 30 wt.% of SCG presented a 40% decrease in the peak of the heat release rate compared to neat epoxy, which confirmed the flame-retardant behavior of the epoxy/SCG composite. Furthermore, the burning tests reveal the self-extinguishing behavior of this composite after 40 s of ignition, proving the efficiency of phosphorus treatment on SCG particles, which caused the flame-retardant behavior and formed a compact char.

5.8. Rubber Composites

For elastomers or rubbers as a matrix, SCG, PP, and PLA particles were used as fillers instead of the commonly used carbon black (CB) [160]. The effects of the filler additions on the vulcanization characteristics of the rubber compounds, as well as on the physical, mechanical, and dynamic mechanical properties, were analyzed. Compared to the reference sample, the minimum and maximum torque values of the PP, PLA, and SCG composites were lower, while the optimum vulcanization time for PP and SCG were slightly higher. This indicates that these alternative fillers lead to lower vulcanization rates. The tensile strengths of the PP and SCG composites were similar and slightly higher, compared to the reference. However, the hardness and storage moduli of the PLA and SCG composites decreased.

The properties of natural rubber (NR) filled with various amounts of SCG, and the surface treated by a silane coupling agent (TESPT) and liquid epoxidized natural rubber (LENR) was studied [161]. The incorporation of SCG resulted in a faster cure (50%) and a higher curing efficiency. However, it did not provide adequate reinforcement and it retarded the vulcanization process. The surface treatment improved the rubber properties, which is due to the better rubber–filler interaction and the higher cure. TESPT-SCG provided a composite with a higher crosslink density (21%), hardness (6%), and modulus (13%), compared to LENR-SCG, producing the highest mechanical properties, followed by LENR-SCG and untreated SCG, respectively.

A more recent work used SCG treated via pyrolysis as a filler for styrene–butadiene rubber (5, 10, 15, and 20 wt.%) [162]. The pyrolysis treatment (700 °C and 900 °C) decreased the average particle size of the SCG and improved the surface roughness, leading to a better interaction with the rubber than untreated SCG/rubber composites. The SCG treatment behaved as an activator for vulcanization, decreasing the cure time, increasing the crosslink density, and increasing the mechanical properties by 30% (the tensile strength and the modulus), compared to the untreated SCG/rubber composite.

6. SCG Reuse Routes

SCG have been widely used in other fields since their high amounts of generated waste are a concern for the environment; however, they are a nutrient-rich material, containing polysaccharides, lipids, proteins, and minerals. A recent review lists the potential

SCG applications, which include their use as an antioxidant source, as well as in energy production, soil fertilizers, dietary fiber, adsorbents, biogas production, and microbial biotechnology [163]. In the present review, SCG reuse is focussed on two areas: biopolymer precursors and composite production. These two areas are different, but they can be complementary, since the composite's matrix can be produced from a polymer based on SCG as a precursor.

As a raw material, SCG can be a filler for composites and a biofertilizer for the soil. Composites filled with SCG can initiate a new lifecycle as a plastic product, decreasing the amount of synthetic material, and they are biodegradable/compostable, especially when using polylactic acid (PLA) or polyhydroxyalkanoates (PHA) as matrices. Moreover, composites filled with lignocellulosic materials are being reported for food packaging [2,19,164], automotive parts [111,165], and household furniture [111,157].

On the other hand, SCG can be chemically treated to remove valuable compounds, generating other applications using their sugars or oil fractions. Sugars are directly related to the production of biopolymer precursors, as well as bioethanol, biogas, and biodiesel [13,22]. From coffee oil, it is possible to produce PHA, as well as pharmaceutical, food, and cosmetics products [13].

7. Conclusions

Several works report the use of SCG in the plastics field as an environmentally friendly material. As a starting point, several works investigate and discuss the possibility of extracting SCG polysaccharides, which can be fermented to produce polymer precursors, such as lactic acid and polyol, or that can directly produce a biopolymer, such as PHA. These works show that microbial fermentation is effective in extracting the main compounds from SCG. In particular, the oil fraction was found to be a valuable resource, not only for biofuel production, but also for PHA synthesis. SCG particles can also be included in different polymer matrices, such as PP, PE, PU, PLA, epoxy, and rubbers, to produce composites with suitable properties for numerous applications, such as food packaging, automotive parts, 3D printing, and UV shielding. Moreover, SCG can be used as a plasticizer.

Although a great deal of effort was devoted to combining SCG and different polymer matrices, some gaps in the literature were found and could be the subjects for future research. For example, 2,3-BD synthesis was reported as a valuable source of polymer precursors and coproducts, but nothing was found on the direct synthesis of 2,3-BD from SCG. Moreover, lactic acid, succinic acid, and other organic compounds could be directly extracted from SCG for novel applications in the biopolymer field, with the aim of developing new green polymers. Coffee oil, as a valuable fraction of SCG, could be further studied in the plastics field, not only for biopolymer synthesis, but also as biofillers for the production of composites and/or bioadditives to modify their properties. More works on polymer composites must be performed for a wider range of matrices, both synthetic and biobased. For example, PU made from SCG polyol, PLA made from SCG lactic acid, and PHA made from SCG could also be filled with SCG. In fact, nothing was found on PHA composites filled with SCG. Finally, recycled matrices should be further investigated, as well as the possibility of recycling these composites after their end-of-life.

Author Contributions: Conceptualization and methodology, A.S.C.d.B. and D.R.; validation, K.C.C.d.C.B. and H.J.C.V.; formal analysis, D.R.; investigation, A.S.C.d.B.; resources, D.R.; data curation, A.S.C.d.B. and D.M.d.O.; writing—original draft preparation, A.S.C.d.B. and D.M.d.O.; writing—review and editing, D.R., K.C.C.d.C.B., M.-J.D. and H.J.C.V.; visualization, A.S.C.d.B. and D.R.; supervision, D.R. and H.J.C.V.; project administration, D.R.; funding acquisition, A.S.C.d.B. and D.M.d.O. All authors have read and agreed to the published version of the manuscript.

Funding: This research was funded by Coordenação de Aperfeiçoamento de Pessoal de Nível Superior-Brazil (CAPES), finance code, 001, and grant number, 88887.495399/2020-00, and the Fundação de Amparo à Pesquisa do Estado de São Paulo (FAPESP), grant: 2019/02607-6 and 2020/02361-4.

Polymers **2022**, *14*, 437

Acknowledgments: The authors would like to acknowledge Université Laval and São Paulo State University for their administrative and technical support.

Conflicts of Interest: The authors declare no conflict of interest.

References

1. Wu, C.T.; Agrawal, D.C.; Huang, W.Y.; Hsu, H.C.; Yang, S.J.; Huang, S.L.; Lin, Y.S. Functionality Analysis of Spent Coffee Ground Extracts Obtained by the Hydrothermal Method. *J. Chem.* **2019**, *2019*, 4671438. [CrossRef]
2. Garcia, C.V.; Kim, Y.T. Spent Coffee Grounds and Coffee Silverskin as Potential Materials for Packaging: A Review. *J. Polym. Environ.* **2021**, *29*, 2372–2384. [CrossRef]
3. Kourmentza, C.; Economou, C.N.; Tsafrakidou, P.; Kornaros, M. Spent coffee grounds make much more than waste: Exploring recent advances and future exploitation strategies for the valorization of an emerging food waste stream. *J. Clean. Prod.* **2018**, *172*, 980–992. [CrossRef]
4. Mussatto, S.I.; Machado, E.M.S.; Martins, S.; Teixeira, J.A. Production, Composition, and Application of Coffee and Its Industrial Residues. *Food Bioprocess Technol.* **2011**, *4*, 661–672. [CrossRef]
5. Zhang, S.; Yang, J.; Wang, S.; Rupasinghe, H.P.V.; He, Q. (Sophia) Experimental exploration of processes for deriving multiple products from spent coffee grounds. *Food Bioprod. Process.* **2021**, *128*, 21–29. [CrossRef]
6. Atabani, A.E.; Mercimek, S.M.; Arvindnarayan, S.; Shobana, S.; Kumar, G.; Cadir, M.; Al-Muhatseb, A.H. Valorization of spent coffee grounds recycling as a potential alternative fuel resource in Turkey: An experimental study. *J. Air Waste Manag. Assoc.* **2017**, *68*, 196–214. [CrossRef]
7. Soares, B.; Gama, N.; Freire, C.S.R.; Barros-Timmons, A.; Brandão, I.; Silva, R.; Neto, C.P.; Ferreira, A. Spent coffee grounds as a renewable source for ecopolyols production. *J. Chem. Technol. Biotechnol.* **2015**, *90*, 1480–1488. [CrossRef]
8. Kovalcik, A.; Obruca, S.; Marova, I. Valorization of spent coffee grounds: A review. *Food Bioprod. Process.* **2018**, *110*, 104–119. [CrossRef]
9. Ballesteros, L.F.; Teixeira, J.A.; Mussatto, S.I. Chemical, Functional and Structural Properties of Spent Coffee Grounds and Coffee Silverskin. *Food Bioprocess. Technol.* **2014**, *7*, 3493–3503. [CrossRef]
10. Hudeckova, H.; Neureiter, M.; Obruca, S.; Frühauf, S.; Marova, I. Biotechnological conversion of spent coffee grounds into lactic acid. *Lett. Appl. Microbiol.* **2018**, *66*, 306–312. [CrossRef]
11. Saratale, G.D.; Bhosale, R.; Shobana, S.; Banu, J.R.; Pugazhendhi, A.; Mahmoud, E.; Sirohi, R.; Kant Bhatia, S.; Atabani, A.E.; Mulone, V.; et al. A review on valorization of spent coffee grounds (SCG) towards biopolymers and biocatalysts production. *Bioresour. Technol.* **2020**, *314*, 123800. [CrossRef] [PubMed]
12. Liu, Y.; Yuan, W.; Lu, Y.; Liu, S.Q. Biotransformation of spent coffee grounds by fermentation with monocultures of Saccharomyces cerevisiae and Lachancea thermotolerans aided by yeast extracts. *LWT* **2021**, *138*, 110751. [CrossRef]
13. Battista, F.; Barampouti, E.M.; Mai, S.; Bolzonella, D.; Malamis, D.; Moustakas, K.; Loizidou, M. Added-value molecules recovery and biofuels production from spent coffee grounds. *Renew. Sustain. Energy Rev.* **2020**, *131*, 110007. [CrossRef]
14. Kwon, E.E.; Yi, H.; Jeon, Y.J. Sequential co-production of biodiesel and bioethanol with spent coffee grounds. *Bioresour. Technol.* **2013**, *136*, 475–480. [CrossRef]
15. Auguścik-Królikowska, M.; Ryszkowska, J.; Ambroziak, A.; Szczepkowski, L.; Oliwa, R.; Oleksy, M. The structure and properties of viscoelastic polyurethane foams with fillers from coffee grounds. *Polimery* **2020**, *65*, 708–718. [CrossRef]
16. Sohn, J.S.; Ryu, Y.; Yun, C.S.; Zhu, K.; Cha, S.W. Extrusion compounding process for the development of eco-friendly SCG/PP composite pellets. *Sustainability* **2019**, *11*, 1720. [CrossRef]
17. García-García, D.; Carbonell, A.; Samper, M.D.; García-Sanoguera, D.; Balart, R. Green composites based on polypropylene matrix and hydrophobized spend coffee ground (SCG) powder. *Compos. Part B Eng.* **2015**, *78*, 256–265. [CrossRef]
18. Essabir, H.; Raji, M.; Laaziz, S.A.; Rodrique, D.; Bouhfid, R.; Qaiss, A.E.K. Thermo-mechanical performances of polypropylene biocomposites based on untreated, treated and compatibilized spent coffee grounds. *Compos. Part B* **2018**, *149*, 1–11. [CrossRef]
19. Songtipya, L.; Limchu, T.; Phuttharak, S.; Songtipya, P.; Kalkornsurapranee, E. Poly(lactic acid)-based Composites Incorporated with Spent Coffee Ground and Tea Leave for Food Packaging Application: A Waste to Wealth. *IOP Conf. Ser. Mater. Sci. Eng.* **2019**, *553*. [CrossRef]
20. da Silva, A.P.; de Pereira, M.P.; Passador, F.R.; Montagna, L.S. PLA/Coffee Grounds Composites: A Study of Photodegradation and Biodegradation in Soil. *Macromol. Symp.* **2020**, *394*, 1–9. [CrossRef]
21. Arrigo, R.; Bartoli, M.; Malucelli, G. Poly(lactic Acid)–Biochar Biocomposites: Effect of Processing and Filler Content on Rheological, Thermal, and Mechanical Properties. *Polymers* **2020**, *12*, 892. [CrossRef]
22. McNutt, J.; He, Q. (Sophia) Spent coffee grounds: A review on current utilization. *J. Ind. Eng. Chem.* **2019**, *71*, 78–88. [CrossRef]
23. Wu, C.S. Renewable resource-based green composites of surface-treated spent coffee grounds and polylactide: Characterisation and biodegradability. *Polym. Degrad. Stab.* **2015**, *121*, 51–59. [CrossRef]
24. Campos-Vega, R.; Loarca-Piña, G.; Vergara-Castañeda, H.A.; Dave Oomah, B. Spent coffee grounds: A review on current research and future prospects. *Trends Food Sci. Technol.* **2015**, *45*, 24–36. [CrossRef]
25. Mussatto, S.I.; Carneiro, L.M.; Silva, J.P.A.; Roberto, I.C.; Teixeira, J.A. A study on chemical constituents and sugars extraction from spent coffee grounds. *Carbohydr. Polym.* **2011**, *83*, 368–374. [CrossRef]

26. Simões, J.; Madureira, P.; Nunes, F.M.; do Rosário Domingues Domingues, M.; Vilanova, M.; Coimbra, M.A. Immunostimulatory properties of coffee mannans. *Mol. Nutr. Food Res.* **2009**, *53*, 1036–1043. [CrossRef]
27. Fischer, M.; Reimann, S.; Trovato, V.; Redgwell, R.J. Polysaccharides of green Arabica and Robusta coffee beans. *Carbohydr. Res.* **2001**, *330*, 93–101. [CrossRef]
28. Simões, J.; Nunes, F.M.; Domingues, M.; do Rosário Domingues, M.; Coimbra, M.A. Structural features of partially acetylated coffee galactomannans presenting immunostimulatory activity. *Carbohydr. Polym.* **2010**, *79*, 397–402. [CrossRef]
29. Ballesteros, L.F.; Cerqueira, M.A.; Teixeira, J.A.; Mussatto, S.I. Characterization of polysaccharides extracted from spent coffee grounds by alkali pretreatment. *Carbohydr. Polym.* **2015**, *127*, 347–354. [CrossRef]
30. Simões, J.; Nunes, F.M.; Domingues, M.R.; Coimbra, M.A. Extractability and structure of spent coffee ground polysaccharides by roasting pre-treatments. *Carbohydr. Polym.* **2013**, *97*, 81–89. [CrossRef]
31. Wyman, C.E.; Decker, S.R.; Himmel, M.E.; Brady, J.W.; Skopec, C.E.; Viikari, L. *Acid Hydrolysis of Cellulose and Hemicellulose*, 2nd ed.; Marcel Dekker: New York, NY, USA, 2005.
32. Dos Santos, L.C.; Adarme, O.F.H.; Baêta, B.E.L.; Gurgel, L.V.A.; de Aquino, S.F. Production of biogas (methane and hydrogen) from anaerobic digestion of hemicellulosic hydrolysate generated in the oxidative pretreatment of coffee husks. *Bioresour. Technol.* **2018**, *263*, 601–612. [CrossRef] [PubMed]
33. Batista, M.J.P.A.; Ávila, A.F.; Franca, A.S.; Oliveira, L.S. Polysaccharide-rich fraction of spent coffee grounds as promising biomaterial for films fabrication. *Carbohydr. Polym.* **2020**, *233*, 115851. [CrossRef] [PubMed]
34. Mussatto, S.I.; Roberto, I.C. Acid hydrolysis and fermentation of brewer's spent grain to produce xylitol. *J. Sci. Food Agric.* **2005**, *85*, 2453–2460. [CrossRef]
35. Ballesteros, L.F.; Teixeira, J.A.; Mussatto, S.I. Extraction of polysaccharides by autohydrolysis of spent coffee grounds and evaluation of their antioxidant activity. *Carbohydr. Polym.* **2017**, *157*, 258–266. [CrossRef]
36. Getachew, A.T.; Cho, Y.J.; Chun, B.S. Effect of pretreatments on isolation of bioactive polysaccharides from spent coffee grounds using subcritical water. *Int. J. Biol. Macromol.* **2018**, *109*, 711–719. [CrossRef]
37. Passos, C.P.; Rudnitskaya, A.; Neves, J.M.M.G.C.; Lopes, G.R.; Evtuguin, D.V.; Coimbra, M.A. Structural features of spent coffee grounds water-soluble polysaccharides: Towards tailor-made microwave assisted extractions. *Carbohydr. Polym.* **2019**, *214*, 53–61. [CrossRef]
38. Passos, C.P.; Coimbra, M.A. Microwave superheated water extraction of polysaccharides from spent coffee grounds. *Carbohydr. Polym.* **2013**, *94*, 626–633. [CrossRef]
39. Ji, X.J.; Huang, H.; Ouyang, P.K. Microbial 2,3-butanediol production: A state-of-the-art review. *Biotechnol. Adv.* **2011**, *29*, 351–364. [CrossRef]
40. Hakizimana, O.; Matabaro, E.; Lee, B.H. The current strategies and parameters for the enhanced microbial production of 2,3-butanediol. *Biotechnol. Rep.* **2020**, *25*, e00397. [CrossRef]
41. Celińska, E.; Grajek, W. Biotechnological production of 2,3-butanediol—Current state and prospects. *Biotechnol. Adv.* **2009**, *27*, 715–725. [CrossRef]
42. Song, C.W.; Park, J.M.; Chung, S.C.; Lee, S.Y.; Song, H. Microbial production of 2,3-butanediol for industrial applications. *J. Ind. Microbiol. Biotechnol.* **2019**, *46*, 1583–1601. [CrossRef] [PubMed]
43. Rehman, S.; Khairul Islam, M.; Khalid Khanzada, N.; Kyoungjin An, A.; Chaiprapat, S.; Leu, S.Y. Whole sugar 2,3-butanediol fermentation for oil palm empty fruit bunches biorefinery by a newly isolated Klebsiella pneumoniae PM2. *Bioresour. Technol.* **2021**, *333*, 125206. [CrossRef] [PubMed]
44. Jiang, B.; Li, Z.G.; Dai, J.Y.; Zhang, D.J.; Xiu, Z.L. Aqueous two-phase extraction of 2,3-butanediol from fermentation broths using an ethanol/phosphate system. *Process Biochem.* **2009**, *44*, 112–117. [CrossRef]
45. Syu, M.J. Biological production of 2,3-butanediol. *Appl. Microbiol. Biotechnol.* **2001**, *55*, 10–18. [CrossRef] [PubMed]
46. Hazeena, S.H.; Sindhu, R.; Pandey, A.; Binod, P. Lignocellulosic bio-refinery approach for microbial 2,3-Butanediol production. *Bioresour. Technol.* **2020**, *302*, 122873. [CrossRef]
47. Ribeiro, L.S.; da Cruz Pedrozo Miguel, M.G.; Martinez, S.J.; Bressani, A.P.P.; Evangelista, S.R.; Batista, C.F.S.E.; Schwan, R.F. The use of mesophilic and lactic acid bacteria strains as starter cultures for improvement of coffee beans wet fermentation. *World J. Microbiol. Biotechnol.* **2020**, *36*, 1–15. [CrossRef]
48. Lee, L.W.; Cheong, M.W.; Curran, P.; Yu, B.; Liu, S.Q. Modulation of coffee aroma via the fermentation of green coffee beans with Rhizopus oligosporus: I. Green coffee. *Food Chem.* **2016**, *211*, 916–924. [CrossRef]
49. Kim, S.J.; Seo, S.O.; Park, Y.C.; Jin, Y.S.; Seo, J.H. Production of 2,3-butanediol from xylose by engineered Saccharomyces cerevisiae. *J. Biotechnol.* **2014**, *192*, 376–382. [CrossRef]
50. Guragain, Y.N.; Chitta, D.; Karanjikar, M.; Vadlani, P.V. Appropriate lignocellulosic biomass processing strategies for efficient 2,3-butanediol production from biomass-derived sugars using Bacillus licheniformis DSM 8785. *Food Bioprod. Process.* **2017**, *104*, 147–158. [CrossRef]
51. Ling, H.Z.; Cheng, K.K.; Ge, J.P.; Ping, W.X. Corncob Mild Alkaline Pretreatment for High 2,3-Butanediol Production by Spent Liquor Recycle Process. *Bioenergy Res.* **2017**, *10*, 566–574. [CrossRef]
52. Saratale, R.G.; Shin, H.S.; Ghodake, G.S.; Kumar, G.; Oh, M.K.; Saratale, G.D. Combined effect of inorganic salts with calcium peroxide pretreatment for kenaf core biomass and their utilization for 2,3-butanediol production. *Bioresour. Technol.* **2018**, *258*, 26–32. [CrossRef] [PubMed]

53. Cha, J.W.; Jang, S.H.; Kim, Y.J.; Chang, Y.K.; Jeong, K.J. Engineering of Klebsiella oxytoca for production of 2,3-butanediol using mixed sugars derived from lignocellulosic hydrolysates. *GCB Bioenergy* **2020**, *12*, 275–286. [CrossRef]

54. Narisetty, V.; Amraoui, Y.; Abdullah, A.; Ahmad, E.; Agrawal, D.; Parameswaran, B.; Pandey, A.; Goel, S.; Kumar, V. Bioresource Technology High yield recovery of 2,3-butanediol from fermented broth accumulated on xylose rich sugarcane bagasse hydrolysate using aqueous two-phase extraction system. *Bioresour. Technol.* **2021**, *337*, 125463. [CrossRef]

55. Amraoui, Y.; Prabhu, A.A.; Narisetty, V.; Coulon, F.; Kumar Chandel, A.; Willoughby, N.; Jacob, S.; Koutinas, A.; Kumar, V. Enhanced 2,3-Butanediol production by mutant Enterobacter ludwigii using Brewers' spent grain hydrolysate: Process optimization for a pragmatic biorefinery loom. *Chem. Eng. J.* **2022**, *427*, 130851. [CrossRef]

56. van Haveren, J.; Scott, E.L.; Sanders, J. Bulk chemicals from biomass. *Biofuels Bioprod. Biorefining* **2007**, *2*, 41–57. [CrossRef]

57. Nguyen, N.T.T.; Matei-Rutkovska, F.; Huchede, M.; Jaillardon, K.; Qingyi, G.; Michel, C.; Millet, J.M.M. Production of 1,3-butadiene in one step catalytic dehydration of 2,3-butanediol. *Catal. Today* **2019**, *323*, 62–68. [CrossRef]

58. Sun, D.; Li, Y.; Yang, C.; Su, Y.; Yamada, Y.; Sato, S. Production of 1,3-butadiene from biomass-derived C4 alcohols. *Fuel Process. Technol.* **2020**, *197*, 106193. [CrossRef]

59. Tinôco, D.; Borschiver, S.; Coutinho, P.L.; Freire, D.M.G. Technological development of the bio-based 2,3-butanediol process. *Biofuels Bioprod. Biorefining* **2021**, *15*, 357–376. [CrossRef]

60. Liakou, V.; Pateraki, C.; Palaiogeorgou, A.M.; Kopsahelis, N.; Machado de Castro, A.; Guimarães Freire, D.M.; Nychas, G.J.E.; Papanikolaou, S.; Koutinas, A. Valorisation of fruit and vegetable waste from open markets for the production of 2,3-butanediol. *Food Bioprod. Process.* **2018**, *108*, 27–36. [CrossRef]

61. Morschbacker, A. Bio-ethanol based ethylene. *Polym. Rev.* **2009**, *49*, 79–84. [CrossRef]

62. Domínguez-Bocanegra, A.R.; Torres-Muñoz, J.A.; López, R.A. Production of Bioethanol from agro-industrial wastes. *Fuel* **2015**, *149*, 85–89. [CrossRef]

63. Alvarez-Guzmán, C.L.; Balderas-Hernández, V.E.; De Leon-Rodriguez, A. Coproduction of hydrogen, ethanol and 2,3-butanediol from agro-industrial residues by the Antarctic psychrophilic GA0F bacterium. *Int. J. Hydrogen Energy* **2020**, *45*, 26179–26187. [CrossRef]

64. Mussatto, S.I.; Machado, E.M.S.; Carneiro, L.M.; Teixeira, J.A. Sugars metabolism and ethanol production by different yeast strains from coffee industry wastes hydrolysates. *Appl. Energy* **2012**, *92*, 763–768. [CrossRef]

65. Rocha, M.V.P.; de Matos, L.J.B.L.; de Lima, L.P.; da Silva Figueiredo, P.M.; Lucena, I.L.; Fernandes, F.A.N.; Gonçalves, L.R.B. Ultrasound-assisted production of biodiesel and ethanol from spent coffee grounds. *Bioresour. Technol.* **2014**, *167*, 343–348. [CrossRef]

66. Barampouti, E.M.; Grammatikos, C.; Stoumpou, V.; Malamis, D.; Mai, S. Emerging Synergies on the Co-treatment of Spent Coffee Grounds and Brewer's Spent Grains for Ethanol Production. *Waste Biomass Valoriz.* **2021**, *6*, 1–15. [CrossRef]

67. Ruta, L.L.; Farcasanu, I.C. Coffee and yeasts: From flavor to biotechnology. *Fermentation* **2021**, *7*, 9. [CrossRef]

68. Bechthold, I.; Bretz, K.; Kabasci, S.; Kopitzky, R.; Springer, A. Succinic acid: A new platform chemical for biobased polymers from renewable resources. *Chem. Eng. Technol.* **2008**, *31*, 647–654. [CrossRef]

69. Merli, G.; Becci, A.; Amato, A.; Beolchini, F. Acetic acid bioproduction: The technological innovation change. *Sci. Total Environ.* **2021**, *798*, 149292. [CrossRef]

70. Yan, Q.; Zhao, T.; Bai, Y.; Zhang, F.; Yang, W. Precipitation polymerization in acetic acid: Study of the solvent effect on the morphology of poly(divinylbenzene). *J. Phys. Chem. B* **2009**, *113*, 3008–3014. [CrossRef]

71. Abdel-Rahman, M.A.; Tashiro, Y.; Sonomoto, K. Recent advances in lactic acid production by microbial fermentation processes. *Biotechnol. Adv.* **2013**, *31*, 877–902. [CrossRef]

72. Eş, I.; Mousavi Khaneghah, A.; Barba, F.J.; Saraiva, J.A.; Sant'Ana, A.S.; Hashemi, S.M.B. Recent advancements in lactic acid production-a review. *Food Res. Int.* **2018**, *107*, 763–770. [CrossRef] [PubMed]

73. Liu, Y.; Seah, R.H.; Abdul Rahaman, M.S.; Lu, Y.; Liu, S.Q. Concurrent inoculations of Oenococcus oeni and Lachancea thermotolerans: Impacts on non-volatile and volatile components of spent coffee grounds hydrolysates. *LWT* **2021**, *148*, 111795. [CrossRef]

74. Lee, K.H.; Jang, Y.W.; Lee, J.; Kim, S.; Park, C.; Yoo, H.Y. Statistical optimization of alkali pretreatment to improve sugars recovery from spent coffee grounds and utilization in lactic acid fermentation. *Processes* **2021**, *9*, 494. [CrossRef]

75. Kim, J.W.; Jang, J.H.; Yeo, H.J.; Seol, J.; Kim, S.R.; Jung, Y.H. Lactic Acid Production from a Whole Slurry of Acid-Pretreated Spent Coffee Grounds by Engineered Saccharomyces cerevisiae. *Appl. Biochem. Biotechnol.* **2019**, *189*, 206–216. [CrossRef]

76. Adeleye, A.T.; Odoh, C.K.; Enudi, O.C.; Banjoko, O.O.; Osiboye, O.O.; Toluwalope Odediran, E.; Louis, H. Sustainable synthesis and applications of polyhydroxyalkanoates (PHAs) from biomass. *Process Biochem.* **2020**, *96*, 174–193. [CrossRef]

77. Li, M.; Wilkins, M.R. Recent advances in polyhydroxyalkanoate production: Feedstocks, strains and process developments. *Int. J. Biol. Macromol.* **2020**, *156*, 691–703. [CrossRef]

78. Obruca, S.; Benesova, P.; Marsalek, L.; Marova, I. Use of lignocellulosic materials for PHA production. *Chem. Biochem. Eng. Q.* **2015**, *29*, 135–144. [CrossRef]

79. Saratale, R.G.; Cho, S.K.; Saratale, G.D.; Kadam, A.A.; Ghodake, G.S.; Kumar, M.; Bharagava, R.N.; Kumar, G.; Kim, D.S.; Mulla, S.I.; et al. A comprehensive overview and recent advances on polyhydroxyalkanoates (PHA) production using various organic waste streams. *Bioresour. Technol.* **2021**, *325*, 124685. [CrossRef]

80. Cruz, M.V.; Paiva, A.; Lisboa, P.; Freitas, F.; Alves, V.D.; Simões, P.; Barreiros, S.; Reis, M.A.M. Production of polyhydroxyalkanoates from spent coffee grounds oil obtained by supercritical fluid extraction technology. *Bioresour. Technol.* **2014**, *157*, 360–363. [CrossRef]

81. Kovalcik, A.; Kucera, D.; Matouskova, P.; Pernicova, I.; Obruca, S.; Kalina, M.; Enev, V.; Marova, I. Influence of removal of microbial inhibitors on PHA production from spent coffee grounds employing Halomonas halophila. *J. Environ. Chem. Eng.* **2018**, *6*, 3495–3501. [CrossRef]

82. Obruca, S.; Benesova, P.; Petrik, S.; Oborna, J.; Prikryl, R.; Marova, I. Production of polyhydroxyalkanoates using hydrolysate of spent coffee grounds. *Process Biochem.* **2014**, *49*, 1409–1414. [CrossRef]

83. Sirohi, R.; Prakash Pandey, J.; Kumar Gaur, V.; Gnansounou, E.; Sindhu, R. Critical overview of biomass feedstocks as sustainable substrates for the production of polyhydroxybutyrate (PHB). *Bioresour. Technol.* **2020**, *311*, 123536. [CrossRef]

84. Tomizawa, S.; Chuah, J.A.; Matsumoto, K.; Doi, Y.; Numata, K. Understanding the limitations in the biosynthesis of polyhydroxyalkanoate (PHA) from lignin derivatives. *ACS Sustain. Chem. Eng.* **2014**, *2*, 1106–1113. [CrossRef]

85. Khatami, K.; Perez-Zabaleta, M.; Owusu-Agyeman, I.; Cetecioglu, Z. Waste to bioplastics: How close are we to sustainable polyhydroxyalkanoates production? *Waste Manag.* **2021**, *119*, 374–388. [CrossRef]

86. Gregory, G.L.; Lopez-Vidal, E.M.; Buchard, A. Polymers from sugars: Cyclic monomer synthesis, ring-opening polymerisation, material properties and applications. *Chem. Commun.* **2017**, *53*, 2198–2217. [CrossRef]

87. Galbis, J.A.; García-Martín, M.D.G.; De Paz, M.V.; Galbis, E. Synthetic Polymers from Sugar-Based Monomers. *Chem. Rev.* **2016**, *116*, 1600–1636. [CrossRef]

88. Zhang, Y.; Chan, J.W.; Moretti, A.; Uhrich, K.E. Designing polymers with sugar-based advantages for bioactive delivery applications. *J. Control. Release* **2015**, *219*, 355–368. [CrossRef]

89. Zakharova, E.; De Ilarduya, A.M.; León, S.; Muñoz-Guerra, S. Sugar-based bicyclic monomers for aliphatic polyesters: A comparative appraisal of acetalized alditols and isosorbide. *Des. Monomers Polym.* **2017**, *20*, 157–166. [CrossRef]

90. Gregory, G.L.; Jenisch, L.M.; Charles, B.; Kociok-Köhn, G.; Buchard, A. Polymers from sugars and CO_2: Synthesis and polymerization of a d-mannose-based cyclic carbonate. *Macromolecules* **2016**, *49*, 7165–7169. [CrossRef]

91. Lu, M.Y.; Surányi, A.; Viskolcz, B.; Fiser, B. Molecular design of sugar-based polyurethanes. *Croat. Chem. Acta* **2018**, *91*, 299–307. [CrossRef]

92. Sendijarevic, I.; Pietrzyk, K.W.; Schiffman, C.M.; Sendijarevic, V.; Kiziltas, A.; Mielewski, D. Polyol from spent coffee grounds: Performance in a model pour-in-place rigid polyurethane foam system. *J. Cell. Plast.* **2020**, *56*, 630–645. [CrossRef]

93. Gama, N.V.; Soares, B.; Freire, C.S.R.; Silva, R.; Neto, C.P.; Barros-Timmons, A.; Ferreira, A. Bio-based polyurethane foams toward applications beyond thermal insulation. *Mater. Des.* **2015**, *76*, 77–85. [CrossRef]

94. Gama, N.; Silva, R.; Carvalho, A.P.O.; Ferreira, A.; Barros-Timmons, A. Sound absorption properties of polyurethane foams derived from crude glycerol and liquefied coffee grounds polyol. *Polym. Test.* **2017**, *62*, 13–22. [CrossRef]

95. Coelho, G.O.; Batista, M.J.A.; Ávila, A.F.; Franca, A.S.; Oliveira, L.S. Development and characterization of biopolymeric films of galactomannans recovered from spent coffee grounds. *J. Food Eng.* **2020**, *289*. [CrossRef]

96. Abdullah, M.; Koc, A.B. Oil removal from waste coffee grounds using two-phase solvent extraction enhanced with ultrasonication. *Renew. Energy* **2013**, *50*, 965–970. [CrossRef]

97. Cubas, A.L.V.; Machado, M.M.; Bianchet, R.T.; da Costa Hermann, K.A.; Bork, J.A.; Debacher, N.A.; Lins, E.F.; Maraschin, M.; Coelho, D.S.; Moecke, E.H.S. Oil extraction from spent coffee grounds assisted by non-thermal plasma. *Sep. Purif. Technol.* **2020**, *250*, 117171. [CrossRef]

98. Couto, R.M.; Fernandes, J.; da Silva, M.D.R.G.; Simões, P.C. Supercritical fluid extraction of lipids from spent coffee grounds. *J. Supercrit. Fluids* **2009**, *51*, 159–166. [CrossRef]

99. Araújo, M.N.; Azevedo, A.Q.P.L.; Hamerski, F.; Voll, F.A.P.; Corazza, M.L. Enhanced extraction of spent coffee grounds oil using high-pressure CO_2 plus ethanol solvents. *Ind. Crops Prod.* **2019**, *141*, 111723. [CrossRef]

100. Obruca, S.; Petrik, S.; Benesova, P.; Svoboda, Z.; Eremka, L.; Marova, I. Utilization of oil extracted from spent coffee grounds for sustainable production of polyhydroxyalkanoates. *Appl. Microbiol. Biotechnol.* **2014**, *98*, 5883–5890. [CrossRef]

101. Bhatia, S.K.; Kim, J.H.; Kim, M.S.; Kim, J.; Hong, J.W.; Hong, Y.G.; Kim, H.J.; Jeon, J.M.; Kim, S.H.; Ahn, J.; et al. Production of (3-hydroxybutyrate-co-3-hydroxyhexanoate) copolymer from coffee waste oil using engineered *Ralstonia eutropha*. *Bioprocess Biosyst. Eng.* **2018**, *41*, 229–235. [CrossRef]

102. Ingram, H.R.; Winterburn, J.B. Anabolism of poly(3-hydroxybutyrate-co-3-hydroxyvalerate) by Cupriavidus necator DSM 545 from spent coffee grounds oil. *New Biotechnol.* **2021**, *60*, 12–19. [CrossRef] [PubMed]

103. Pascault, J.; Sautereau, H.; Verdu, J.; Williams, R.J.J. *Thermosetting Polymers*, 1st ed.; Marcel Dekker: New York, NY, USA, 2002; ISBN 0824706706.

104. Franck, A.J. *Understanding Rheology of Thermosets*; TA Instruments: New Castle, DE, USA, 2004.

105. Hejna, A. Potential applications of by-products from the coffee industry in polymer technology—Current state and perspectives. *Waste Manag.* **2021**, *121*, 296–330. [CrossRef] [PubMed]

106. Peacock, A.J. *Handbook of Polyethylene: Structures, Properties, and Applications*, 1st ed.; CRC Press: Boca Raton, FL, USA, 2000; ISBN 9780429180774.

107. Khanam, P.N.; AlMaadeed, M.A.A. Processing and characterization of polyethylene-based composites. *Adv. Manuf. Polym. Compos. Sci.* **2015**, *1*, 63–79. [CrossRef]

108. Shih, Y.F.; Kotharangannagari, V.K.; Chen, R.M. Green composites based on high density polyethylene and recycled coffee gunny: Morphology, thermal, and mechanical properties. *Mod. Phys. Lett. B* **2020**, *34*, 1–6. [CrossRef]
109. León, L.D.V.E.; Escocio, V.A.; Visconte, L.L.Y.; Junior, J.C.J.; Pacheco, E.B.A.V. Rotomolding and polyethylene composites with rotomolded lignocellulosic materials: A review. *J. Reinf. Plast. Compos.* **2020**, *39*, 459–472. [CrossRef]
110. Mendes, J.F.; Martins, J.T.; Manrich, A.; Luchesi, B.R.; Dantas, A.P.S.; Vanderlei, R.M.; Claro, P.C.; de Sena Neto, A.R.; Mattoso, L.H.C.; Martins, M.A. Thermo-physical and mechanical characteristics of composites based on high-density polyethylene (HDPE) e spent coffee grounds (SCG). *J. Polym. Environ.* **2021**, *29*, 2888–2900. [CrossRef]
111. Tan, M.Y.; Nicholas Kuan, H.T.; Lee, M.C. Characterization of Alkaline Treatment and Fiber Content on the Physical, Thermal, and Mechanical Properties of Ground Coffee Waste/Oxobiodegradable HDPE Biocomposites. *Int. J. Polym. Sci.* **2017**, *2017*, 6258151. [CrossRef]
112. Cestari, S.P.; Mendes, L.C. Thermal properties and morphology of high-density polyethylene filled with coffee dregs. *J. Therm. Anal. Calorim.* **2013**, *114*, 1–4. [CrossRef]
113. Cestari, S.P.; Mendes, L.C.; Altstädt, V.; Mano, E.B.; Da Silva, D.F.; Keller, J.H. Crystallization kinetics of recycled high density polyethylene and coffee dregs composites. *Polym. Polym. Compos.* **2014**, *22*, 541–549. [CrossRef]
114. Cestari, S.P.; Mendes, L.C.; da Silva, D.F.; Chimanowsky, J.P., Jr.; Altstädt, V.; Demchuk, V.; Lang, A.; Leonhardt, R.G.; Keller, J.H. Properties of Recycled High Density Polyethylene and Coffee Dregs Composites. *Polimeros* **2013**, *23*, 733–737. [CrossRef]
115. Arrigo, R.; Jagdale, P.; Bartoli, M.; Tagliaferro, A.; Malucelli, G. Structure-property relationships in polyethylene-based composites filled with biochar derived from waste coffee grounds. *Polymers* **2019**, *11*, 1336. [CrossRef] [PubMed]
116. Aydemir, D.; Alsan, M.; Can, A.; Altuntas, E.; Sivrikaya, H. Accelerated weathering and decay resistance of heat-treated wood reinforced polypropylene composites. *Drv. Ind.* **2019**, *70*, 279–285. [CrossRef]
117. Luz, S.M.; Gonçalves, A.R.; Del'Arco, A.P. Mechanical behavior and microstructural analysis of sugarcane bagasse fibers reinforced polypropylene composites. *Compos. Part A Appl. Sci. Manuf.* **2007**, *38*, 1455–1461. [CrossRef]
118. Badji, C.; Beigbeder, J.; Garay, H.; Bergeret, A.; Bénézet, J.C.; Desauziers, V. Correlation between artificial and natural weathering of hemp fibers reinforced polypropylene biocomposites. *Polym. Degrad. Stab.* **2018**, *148*, 117–131. [CrossRef]
119. Karina, M.; Syampurwadi, A.; Satoto, R.; Irmawati, Y.; Puspitasari, T. Physical and Mechanical Properties of Recycled Polypropylene Composites Reinforced with Rice Straw Lignin. *BioResources* **2017**, *12*, 5801–5811. [CrossRef]
120. Wu, H.; Hu, W.; Zhang, Y.; Huang, L.; Zhang, J.; Tan, S.; Cai, X.; Liao, X. Effect of oil extraction on properties of spent coffee ground–plastic composites. *J. Mater. Sci.* **2016**, *51*, 10205–10214. [CrossRef]
121. Chitra, N.J.; Vasanthakumari, R.; Amanulla, S. Preliminary Studies of the Effect of Coupling Agent on the Properties of Spent Coffee Grounds Polypropylene Bio-Composites. *Int. J. Eng. Res. Technol.* **2014**, *7*, 9–16.
122. Karmee, S.K. A spent coffee grounds based biorefinery for the production of biofuels, biopolymers, antioxidants and biocomposites. *Waste Manag.* **2018**, *72*, 240–254. [CrossRef]
123. Zarrinbakhsh, N.; Wang, T.; Rodriguez-Uribe, A.; Misra, M.; Mohanty, A.K. Characterization of Wastes and Coproducts from Coffee Industry for Composite Material Production. *BioResources* **2016**, *11*, 7637–7653. [CrossRef]
124. Zhang, X.; Bo, X.; Cong, L.; Wei, L.; McDonald, A.G. Characteristics of undeinked, alkaline deinked, and neutral deinked old newspaper fibers reinforced recycled polypropylene composites. *Polym. Compos.* **2018**, *39*, 3537–3544. [CrossRef]
125. Ibrahim, I.D.; Jamiru, T.; Sadiku, R.E.; Kupolati, W.K.; Agwuncha, S.C. Dependency of the Mechanical Properties of Sisal Fiber Reinforced Recycled Polypropylene Composites on Fiber Surface Treatment, Fiber Content and Nanoclay. *J. Polym. Environ.* **2017**, *25*, 427–434. [CrossRef]
126. Moreno, D.D.P.; de Camargo, R.V.; dos Santos Luiz, D.; Branco, L.T.P.; Grillo, C.C.; Saron, C. Composites of Recycled Polypropylene from Cotton Swab Waste with Pyrolyzed Rice Husk. *J. Polym. Environ.* **2020**, *29*, 350–362. [CrossRef]
127. De Bomfim, A.S.C.; Voorwald, H.J.C.; de Benini, K.C.C.C.; de Oliveira, D.M.; Fernandes, M.F.; Cioffi, M.O.H. Sustainable application of recycled espresso coffee capsules: Natural composite development for a home composter product. *J. Clean. Prod.* **2021**, *297*. [CrossRef]
128. Akindoyo, J.O.; Beg, M.D.H.; Ghazali, S.; Islam, M.R.; Jeyaratnam, N.; Yuvaraj, A.R. Polyurethane types, synthesis and applications-a review. *RSC Adv.* **2016**, *6*, 114453–114482. [CrossRef]
129. Funabashi, M.; Hirose, S.; Hatakeyama, T.; Hatakeyama, H. Effect of filler shape on mechanical properties of rigid polyurethane composites containing plant particles. *Macromol. Symp.* **2003**, *197*, 231–242. [CrossRef]
130. Hatakeyama, H.; Hatakeyama, T. Lignin Structure, Properties, and Applications. In *Advance in Poymer Science*; Springer: Berlin/Heidelberg, Germany, 2010; Volume 232, pp. 1–63. ISBN 978-3-642-13629-0.
131. Baek, B.S.; Park, J.W.; Lee, B.H.; Kim, H.J. Development and Application of Green Composites: Using Coffee Ground and Bamboo Flour. *J. Polym. Environ.* **2013**, *21*, 702–709. [CrossRef]
132. Suaduang, N.; Ross, S.; Ross, G.M.; Pratumshat, S.; Mahasaranon, S. Effect of spent coffee grounds filler on the physical and mechanical properties of poly(lactic acid) bio-composite films. *Mater. Today Proc.* **2019**, *17*, 2104–2110. [CrossRef]
133. Cacciotti, I.; Mori, S.; Cherubini, V.; Nanni, F. Eco-sustainable systems based on poly(lactic acid), diatomite and coffee grounds extract for food packaging. *Int. J. Biol. Macromol.* **2018**, *112*, 567–575. [CrossRef]
134. Chang, Y.C.; Chen, Y.; Ning, J.; Hao, C.; Rock, M.; Amer, M.; Feng, S.; Falahati, M.; Wang, L.J.; Chen, R.K.; et al. No Such Thing as Trash: A 3D-Printable Polymer Composite Composed of Oil-Extracted Spent Coffee Grounds and Polylactic Acid with Enhanced Impact Toughness. *ACS Sustain. Chem. Eng.* **2019**, *7*, 15304–15310. [CrossRef]

135. Li, S.; Shi, C.; Sun, S.; Chan, H.; Lu, H.; Nilghaz, A.; Tian, J.; Cao, R. From brown to colored: Polylactic acid composite with micro/nano-structured white spent coffee grounds for three-dimensional printing. *Int. J. Biol. Macromol.* **2021**, *174*, 300–308. [CrossRef]
136. Gama, N.; Ferreira, A.; Evtuguin, D.V. New poly(lactic acid) composites produced from coffee beverage wastes. *J. Appl. Polym. Sci.* **2021**, *139*, 51434. [CrossRef]
137. Xu, H.; Xie, L.; Li, J.; Hakkarainen, M. Coffee Grounds to Multifunctional Quantum Dots: Extreme Nanoenhancers of Polymer Biocomposites. *ACS Appl. Mater. Interfaces* **2017**, *9*, 27972–27983. [CrossRef] [PubMed]
138. Lee, H.J.; Lee, H.K.; Lim, E.; Song, Y.S. Synergistic effect of lignin/polypropylene as a compatibilizer in multiphase eco-composites. *Compos. Sci. Technol.* **2015**, *118*, 193–197. [CrossRef]
139. Ferreira, F.V.; Cividanes, L.S.; Gouveia, R.F.; Lona, L.M.F. An overview on properties and applications of poly(butylene adipate-co-terephthalate)–PBAT based composites. *Polym. Eng. Sci.* **2019**, *59*, E7–E15. [CrossRef]
140. Lule, Z.C.; Kim, J. Properties of economical and eco-friendly polybutylene adipate terephthalate composites loaded with surface treated coffee husk. *Compos. Part A Appl. Sci. Manuf.* **2021**, *140*, 106154. [CrossRef]
141. Lule, Z.C.; Wondu, E.; Kim, J. Highly rigid, fire-resistant, and sustainable polybutylene adipate terephthalate/polybutylene succinate composites reinforced with surface-treated coffee husks. *J. Clean. Prod.* **2021**, *315*, 128095. [CrossRef]
142. Sarasini, F.; Tirillò, J.; Zuorro, A.; Maffei, G.; Lavecchia, R.; Puglia, D.; Dominici, F.; Luzi, F.; Valente, T.; Torre, L. Recycling coffee silverskin in sustainable composites based on a poly(butylene adipate-co-terephthalate)/poly(3-hydroxybutyrate-co-3-hydroxyvalerate) matrix. *Ind. Crops Prod.* **2018**, *118*, 311–320. [CrossRef]
143. Sarasini, F.; Luzi, F.; Dominici, F.; Maffei, G.; Iannone, A.; Zuorro, A.; Lavecchia, R.; Torre, L.; Carbonell-Verdu, A.; Balart, R.; et al. Effect of different compatibilizers on sustainable composites based on a PHBV/PBAT matrix filled with coffee silverskin. *Polymers* **2018**, *10*, 1256. [CrossRef]
144. Moustafa, H.; Guizani, C.; Dufresne, A. Sustainable biodegradable coffee grounds filler and its effect on the hydrophobicity, mechanical and thermal properties of biodegradable PBAT composites. *J. Appl. Polym. Sci.* **2017**, *134*, 1–11. [CrossRef]
145. Moustafa, H.; Guizani, C.; Dupont, C.; Martin, V.; Jeguirim, M.; Dufresne, A. Utilization of torrefied coffee grounds as reinforcing agent to produce high-quality biodegradable PBAT composites for food packaging applications. *ACS Sustain. Chem. Eng.* **2017**, *5*, 1906–1916. [CrossRef]
146. Nagarkar, R.; Patel, J. Polyvinyl Alcohol: A Comprehensive Study. *Acta Sci. Pharm. Sci.* **2019**, *3*, 34–44.
147. Lessa, E.F.; Nunes, M.L.; Fajardo, A.R. Chitosan/waste coffee-grounds composite: An efficient and eco-friendly adsorbent for removal of pharmaceutical contaminants from water. *Carbohydr. Polym.* **2018**, *189*, 257–266. [CrossRef] [PubMed]
148. Le, V.T.; Pham, T.M.; Doan, V.D.; Lebedeva, O.E.; Nguyen, H.T. Removal of Pb(ii) ions from aqueous solution using a novel composite adsorbent of Fe3o4/PVA/spent coffee grounds. *Sep. Sci. Technol.* **2019**, *54*, 3070–3081. [CrossRef]
149. Minh, P.T.; Thuan, L. Van Investigation of sorption mechanism of methylene blue, congo red and tannic acid from aqueous solutions onto magnetic composite sorbent obtained from alkaline pretreated spent coffee grounds. *BIO Web Conf.* **2021**, *30*, 02008. [CrossRef]
150. Kanai, N.; Honda, T.; Yoshihara, N.; Oyama, T.; Naito, A.; Ueda, K.; Kawamura, I. Structural characterization of cellulose nanofibers isolated from spent coffee grounds and their composite films with poly(vinyl alcohol): A new non-wood source. *Cellulose* **2020**, *27*, 5017–5028. [CrossRef]
151. Lee, H.K.; Park, Y.G.; Jeong, T.; Song, Y.S. Green nanocomposites filled with spent coffee grounds. *J. Appl. Polym. Sci.* **2015**, *132*, 2–7. [CrossRef]
152. Ounkaew, A.; Kasemsiri, P.; Kamwilaisak, K.; Saengprachatanarug, K.; Mongkolthanaruk, W.; Souvanh, M.; Pongsa, U.; Chindaprasirt, P. Polyvinyl Alcohol (PVA)/Starch Bioactive Packaging Film Enriched with Antioxidants from Spent Coffee Ground and Citric Acid. *J. Polym. Environ.* **2018**, *26*, 3762–3772. [CrossRef]
153. Muniappan, A.; Srinivasan, R.; Sai Sandeep, M.V.V.; Senthilkumar, N.; Senthiil, P.V. Mode-1 fracture toughness analysis of coffee bean powder reinforced polymer composite. *Mater. Today Proc.* **2020**, *21*, 537–542. [CrossRef]
154. Nguyen, T.A.; Nguyen, Q.T. Hybrid Biocomposites Based on Used Coffee Grounds and Epoxy Resin: Mechanical Properties and Fire Resistance. *Int. J. Chem. Eng.* **2021**, *2021*, 1919344. [CrossRef]
155. Alhelal, A.; Mohammed, Z.; Jeelani, S.; Rangari, V.K. 3D printing of spent coffee ground derived biochar reinforced epoxy composites. *J. Compos. Mater.* **2021**, *55*, 3651–3660. [CrossRef]
156. Giorcelli, M.; Bartoli, M. Development of coffee biochar filler for the production of electrical conductive reinforced plastic. *Polymers* **2019**, *11*, 1916. [CrossRef] [PubMed]
157. Leow, Y.; Yew, P.Y.M.; Chee, P.L.; Loh, X.J.; Kai, D. Recycling of spent coffee grounds for useful extracts and green composites. *RSC Adv.* **2021**, *11*, 2682–2692. [CrossRef]
158. Tellers, J.; Willems, P.; Tjeerdsma, B.; Sbirrazzuoli, N.; Guigo, N. Spent Coffee Grounds as Property Enhancing Filler in a Wholly Bio-Based Epoxy Resin. *Macromol. Mater. Eng.* **2021**, *306*, 1–10. [CrossRef]
159. Vahabi, H.; Jouyandeh, M.; Parpaite, T.; Saeb, M.R.; Ramakrishna, S. Coffee wastes as sustainable flame retardants for polymer materials. *Coatings* **2021**, *11*, 1021. [CrossRef]
160. Pajtášová, M.; Ondrušová, D.; Janík, R.; Mičicová, Z.; Pecušová, B.; Labaj, I.; Kohutiar, M.; Moricová, K. Using of alternative fillers based on the waste and its effect on the rubber properties. *MATEC Web Conf.* **2019**, *254*, 04010. [CrossRef]

161. Siriwong, C.; Boopasiri, S.; Jantarapibun, V.; Kongsook, B.; Pattanawanidchai, S.; Sae-Oui, P. Properties of natural rubber filled with untreated and treated spent coffee grounds. *J. Appl. Polym. Sci.* **2018**, *135*, 1–9. [CrossRef]
162. Boopasiri, S.; Sae-Oui, P.; Lundee, S.; Takaewnoi, S.; Siriwong, C. Reinforcing Efficiency of Pyrolyzed Spent Coffee Ground in Styrene-Butadiene Rubber. *Macromol. Res.* **2021**, *29*, 597–604. [CrossRef]
163. Stylianou, M.; Agapiou, A.; Omirou, M.; Vyrides, I.; Ioannides, I.M.; Maratheftis, G.; Fasoula, D. Converting environmental risks to benefits by using spent coffee grounds (SCG) as a valuable resource. *Environ. Sci. Pollut. Res.* **2018**, *25*, 35776–35790. [CrossRef]
164. Thiagamani, S.M.K.; Nagarajan, R.; Jawaid, M.; Anumakonda, V.; Siengchin, S. Utilization of chemically treated municipal solid waste (spent coffee bean powder) as reinforcement in cellulose matrix for packaging applications. *Waste Manag.* **2017**, *69*, 445–454. [CrossRef]
165. Nguyen, D.M.; Nhung, V.T.; Le Do, T.C.; Ha-Thuc, C.N.; Perre, P. Effective Synergistic Effect of Treatment and Modification on Spent Coffee Grounds for Sustainable Biobased Composites. *Waste Biomass Valoriz.* **2021**, *13*, 1339–1348. [CrossRef]

Correction

Correction: Devasahayam, S. Decarbonising the Portland and Other Cements—Via Simultaneous Feedstock Recycling and Carbon Conversions Sans External Catalysts. *Polymers* 2021, *13*, 2462

Sheila Devasahayam

Department of Chemical Engineering, Faculty of Science and Engineering, Monash University, Melbourne 3800, Australia; sheiladevasahayam@gmail.com

The author wishes to make the following correction to the above paper: [1]. In the original version of the published article, there is a mistake in Equation (16) on Page 14. It should be:

$$CH_4 + 2H_2O \rightarrow CO_2 + 4H_2 \ (\Delta H \ 298 \ K = +165 \ kJ/mol) \tag{16}$$

The author apologises for any inconvenience caused to the readers by these changes. These changes have no material impact on the conclusions of our paper. The original article has been updated.

Reference

1. Devasahayam, S. Decarbonising the Portland and other Cements—Via Simultaneous Feedstock Recycling and Carbon Conversions Sans External Catalysts. *Polymers* **2021**, *13*, 2462. [CrossRef] [PubMed]

Citation: Devasahayam, S. Correction: Devasahayam, S. Decarbonising the Portland and Other Cements—Via Simultaneous Feedstock Recycling and Carbon Conversions Sans External Catalysts. *Polymers* 2021, *13*, 2462. *Polymers* **2022**, *14*, 281. https://doi.org/10.3390/polym14020281

Received: 30 November 2021
Accepted: 8 December 2021
Published: 11 January 2022

Publisher's Note: MDPI stays neutral with regard to jurisdictional claims in published maps and institutional affiliations.

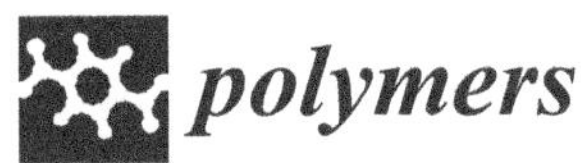

Review

Composite Polymers from Leather Waste to Produce Smart Fertilizers

Daniela Simina Stefan [1], Magdalena Bosomoiu [1,*], Rodica Roxana Constantinescu [2] and Madalina Ignat [2]

1   Department of Analytical Chemistry and Environmental Engineering, Faculty of Applied Chemistry and Materials Science, University Politehnica of Bucharest, 1-7 Polizu Street, 011061 Bucharest, Romania; daniela.stefan@upb.ro
2   Leather and Footwear Research Institute (ICPI) Division, National Research & Development Institute for Textiles and Leather, 93 Ion Minulescu Street, 031215 Bucharest, Romania; rodica.constantinescu@icpi.ro (R.R.C.); madalina.ignat@icpi.ro (M.I.)
*   Correspondence: mbosomoiu@yahoo.com

Abstract: The leather industry is facing important environmental issues related to waste disposal. The waste generated during the tanning process is an important resource of protein (mainly collagen) which can be extracted and reused in different applications (e.g., medical, agricultural, leather industry). On the other side, the utilization of chemical fertilizers must be decreased because of the negative effects associated to an extensive use of conventional chemical fertilizers. This review presents current research trends, challenges and future perspectives with respect to the use of hide waste to produce composite polymers that are further transformed in smart fertilizers. Hide waste contains mostly protein (collagen that is a natural polymer), that is extracted to be used in the cross-linking with water soluble copolymers to obtain the hydrogels which are further valorised as smart fertilizers. Smart fertilizers are a new class of fertilizers which allow the controlled release of the nutrients in synchronization with the plant's demands. Characteristics of hide and leather wastes are pointed out. The fabrication methods of smart fertilizers and the mechanisms for the nutrients release are extensively discussed. This novel method is in agreement with the circular economy concepts and solves, on one side, the problem of hide waste disposal, and on the other side produces smart fertilizers that can successfully replace conventional chemical fertilizers.

Keywords: bio-polymer; hide waste; circular economy

Citation: Stefan, D.S.; Bosomoiu, M.; Constantinescu, R.R.; Ignat, M. Composite Polymers from Leather Waste to Produce Smart Fertilizers. *Polymers* **2021**, *13*, 4351. https://doi.org/10.3390/polym13244351

Academic Editor: Sheila Devasahayam

Received: 15 October 2021
Accepted: 7 December 2021
Published: 12 December 2021

Publisher's Note: MDPI stays neutral with regard to jurisdictional claims in published maps and institutional affiliations.

1. Introduction

It has been demonstrated that crop quality and yield is closely related to the type and concentration and release mode of fertilizers used. Nitrogen, carbon and phosphorous are essential nutrients for the growth of plants. Over the years, it has been evidenced that most of the chemical synthetic fertilizers have reduced efficiency in time, because of volatilization, leaching due to their good mobility all together with the disadvantage of necessity to apply large quantities frequently. This creates environmentally related problems regarding water, air and soil pollution, such as water contamination (especially when these substances penetrate below the plant's roots and pollute the ground water), eutrophication, soil erosion, food contamination and effective hazardous emissions. Nitrate and phosphate leaching has also been reported, due to an excess of nutrients release, which are further transported from soils to water, causing the eutrophication [1–3]. Over long-term the extensive use of synthetic fertilizers could cause even a more reduced soil fertility, because of the increased need of food quantities. Another problem associated with inappropriate fertilization practices include low disease resistance of crops, causing a decrease of the productivity and poor-quality crops [4].

To reduce the environmental impact, chemical fertilizer substitutes, can be applied (e.g., organic fertilizers, biofertilizers). Work has been done in developing new formula fertilizers that allow slower and controlled nutrient release in accordance with the plant life cycle [5].

Organic fertilizers such as animal manure or sewage sludge are used to increase the soil fertility in nitrogen, carbon and phosphorous nutrients [4,6], but this brings a series of problems related to the risk of accumulation of heavy metals and organic pollutants (phthalate esters) [7,8]. That is why some countries (e.g., Switzerland) have prohibited the use of such fertilizers [9].

Biological fertilizers contain different types of microorganisms that convert the main nutrients from an inaccessible to an accessible form, during biological processes, and lead to the development of root systems and better seed germination [10]. The mechanisms involved are complex and depend on the microorganism type; it has been found that bacteria Pseudomonas and Azotobacter combined with organic manures (vermicompost and farm yard manure) enhanced plant growth and determinate early flowering for strawberry [11].

To overcome all the above mentioned disadvantages, a new class of fertilizers is needed, which allows the controlled release of the nutrients in synchronization with the plants demands. In turn will enhance the efficiency of fertilizers use and optimize the fertilizers application, thus reducing the costs associated to this operation [12]. This new type of fertilizers is called smart fertilizers [13].

Another alternative to conventional fertilizers, that has recently received great attention is constituted by the hydrogels, because of their properties of water/aqueous solutions absorption and retention, as well as their slow release of the nutrients together with the absorbed water, when the soil humidity decreases [14]. These properties are highly influenced by the concentration and pH of the aqueous solution, and temperature [15]. The precursor of hydrogel, called superabsorbent, can absorb large amounts of solutions containing the nutrients (e.g., urea); the nutrients release is controlled by the concentration gradient between the hydrogel and the environment around the hydrogel (soil), which corresponds to the plant demand in nutrients [16]. Hydrogels have been initially used in agriculture only as an alternative water resource, because of their capacity to absorb water [17]. Therefore, the use of hydrogels contributes also to a better management of water resources by reducing the irrigation frequency and preventing water loss through evaporation.

There are many studies dedicated to developing synthetic hydrogels [18–23].

2. Characteristics of Hide and Leather Wastes

The problematic disposal of hide waste has received a particular interest because of the high quantities that are generated from leather industry and its negative impact on the environment. Hide and skins are by-products in the meat industry, and raw material in the leather industry [24,25]. Over the last 20 years, a continuously increasing number of raw hides and skins has been seen, passing from about 470 thousand tons in 1999 to 574 thousand tons in 2014 for heavy leather, respectively from 11,978 million square feet in 1999 to 14,540 million square feet in 2014, for light leather [26]. This means that the quantity of hide waste generated is also increasing.

More than 99% of the world leather production comes from the processing of raw hides and skins from animals raised mainly for milk and/or meat production. The leather industry produces solid, liquid and gaseous phases waste. About 20% of the raw hide is transformed in finished leather, the rest being lost during the manufacturing process (20 kg of leather can be obtained from 100 kg of raw hide) [27,28]. The solid waste consists of hair, trimmings, flesh, keratin [29–31]. A detailed presentation of the steps in the tanning process is given by Sundar et al. (2011) [27].

During the tanning process, the so called "wet blue", a stable and inert polynuclear chromium-collagen complex, is formed. The next step of the fabrication process is to equalize the thickness of "wet blue" and to cut the uneven parts. In this way, large amounts of material from shavings and trimming are produced, around 40% of product turned into waste by this stage [32].

The circular economy concept has gathered substantial regional and worldwide interest. According to this concept, the materials and resources must be recovered and reintegrated in the system at the end of their life cycles, by optimizing their potential use (Figure 1). This is done by recycling, reusing, repairing, considering that any residual stream can be used either in the same process, or to make a new product. The major obstacles encountered in developing a circular economy for the leather industry are: (1) significant environmental and social impacts of waste leather landfilling operations; (2) continuous increase of production of leather-based products, and, implicitly, of hide waste quantity, especially in developing countries and; (3) availability of very few alternative disposal methods to waste landfill.

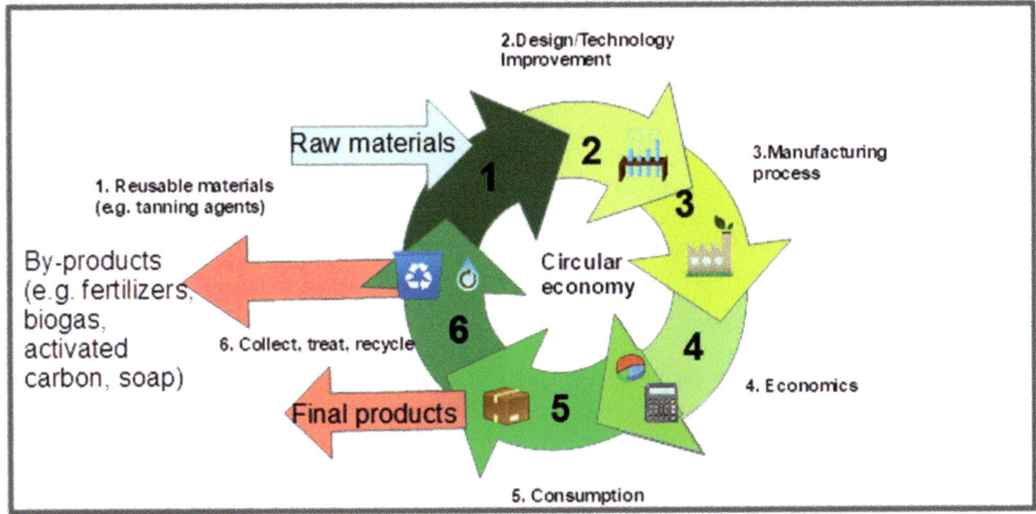

Figure 1. Circular economy in leather industry.

The influence of the tannery process on the environment can be measured in significant changes of parameters like chemical oxygen demand (COD), total dissolved solids (TDS), chlorides, sulphates and heavy metal pollution. The chemical substances discharged in the aquatic systems accumulate and generate polluted sediments and rivers salinization. Regarding the negative impact reduction on environment by tannery processes, there are two main directions: first is related to optimizing the process technologies to decrease the load of streams in toxic compounds, the second one consists in recovery, treatment and reuse of wastes generated during the tanning process [33]. Although a lot of progress has been made regarding chrome recovery in the tanning step, there is still more to achieve regarding the use of hide waste and development of a good strategy that can be applied at large scale. The leather industry can be a resource of by-products obtained by the recovery and use of different generated raw materials such as chromium, nutrients, collagen hydrolysate, fats, biogas and anaerobic digestate, which can be reused in other applications (e.g., in agriculture as fertilizers, for energy generation as biofuels, and in pharmacy and cosmetics) [34].

Another aspect to be taken into account regards the new indications of the European Commission on the reduction by 30% of the use of fertilizers from non-renewable resources. This can be accomplished by the valorisation of wastes that are suitable for production of smart fertilizers. The use of hide waste is a practical solution to recover valuable fertilizer components (namely proteins like collagen). This involves the construction of small installations for fertilizer production at the site of waste generation, which will solve both the problem of waste transport and sanitary hazards [35].

Kilic et al. (2018) [36] analysed the case of a tannery in Turkey, and provided recommendations for the process improvement related to energy consumption reduction, use of renewable energy resources, waste management and the reduction of water consumption.

Vidaurre Arbizu et al. (2021) [37] studied the case of a tannery located in Navarra (North of Spain) following the concepts of circular economy. This tannery was producing around 2 tons/day of leather shavings and leather dust, and 10 tons/day of discarded hair. Three different types of solid wastes (chromium free tanned shavings, chromium buffing dust, and discarded hair) were analysed, in an attempt to recycle them. Two directions were proposed for the reuse of the weekly tons of leftovers (both shavings and hair) produced by the tannery, instead of the usual composting in an external composting plant, or by landfilling. The first direction was to use the discarded material to obtain biomass for the company's thermal production plant. For that, the calorific value of the discarded hair and shavings was estimated, to see if they are of competitive value in comparison with actual biomass products (e.g., wood pellets). The second approach involved the use of tannery-generated waste in the construction sector, as acoustic panels.

Various uses or disposal methods have been reported for tannery waste, in the attempt to reduce its impact on the environment, and to create efficient models of circular economy in the leather industry [34,38]; among these methods there are: pyrolysis [39–41], biotransformation [42–44], use as adsorbent after transformation in activated carbon [28,45,46], biodiesel production [47], transformation into composite sheets [48] or doped nanocarbon [49], landfilling [50] etc. Alibardi and Cossu (2016) [50] proposed a sustainable method for landfilling of tannery sludge generated after the tannery wastewater treatment. The applied pretreatment processes consist of aerobic stabilization, compaction and drying, produced a reduction of volume, mass and biodegradability of treated sludge, demonstrating a reduced leachability of organic and inorganic compounds from the treated sludge. However, landfilling in the case of hide waste has numerous disadvantages related to the leaching of Cr (III) by the acid rains to the groundwater, soil contamination and high cost, on one side, and, on the other, that all the raw materials contained in the hide waste are not recovered (e.g., Cr (III) and proteins). Pyrolysis also presents a series of disadvantages regarding the gaseous emissions (HCH, NH_3, as nitrogen is present in the form of amino acids in the leather), the ashes that contain Cr (III) or Cr (VI), depending on pH. It has been found that in the pH ranges 6.3–11.5, the dominant chromium species is Cr (III), whilst for pH above 11.5 the dominant species is Cr (VI) [51]. Torres-Filho et al. (2016) [52] studied the pyrolysis of leather wastes from tanning to obtain carbonized leather residue that is further used in metallurgical processes. Tang et al. (2021) [53] used non-tanned hide wastes to produce an efficient adsorbent for dye removal from tannery wastewater. In Figure 2, the possible utilizations of wastes generated during the tanning process are schematically presented.

One of the first steps in the tanning process is to remove the hairs from the hide, hairs that end up in the sludge, after the water treatment. Untanned skin waste can be transformed to produce organic derivatives, such as glue and gelatine [29]. Keratin hydrolysate and fleshing hydrolysate (after a chemical modification) can be used in the retanning process [29,54]. Fleshing wastes can also be used to produce glue, gelatine. Chrome and buffing dust are used to produce tanning agents, fertilizers etc. [29,55]. Fat and other tissues resulting from leather can be a source of biogas after an anaerobic treatment [56–60]. Puhazhselvan et al. (2017) [61] developed a method for the extraction in the presence of enzymes (Bacillus subtilis) of lipids from tannery fleshing waste, allowing the reducing of solvent consumption by 1.9 to 7 times corresponding to the production of 1 kg lipids, compared to conventional methods.

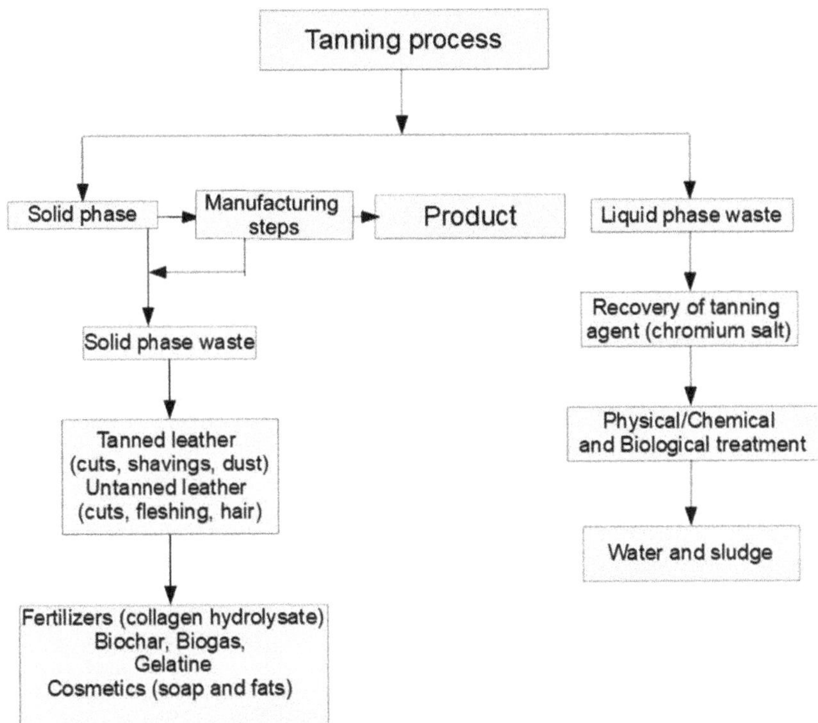

Figure 2. Tanning wastes and alternative ways for wastes valorisation.

Pecha et al. (2021) [44] built a mathematical model based on experimental data obtained from enzymatic hydrolysis of chrome-tanned solid waste. This study includes experimental verification of reaction kinetics of alkali-enzymatic hydrolysis of chrome shavings and their dependence on enzyme concentration, time of reaction. The simulations showed the existence of a restraint range of optimal reaction conditions, corresponding to the minimum of unit operating costs.

Al-Jabari et al. (2021) [31] studied the case of a tannery implemented in Palestine, to better understand the possible directions to follow for a cleaner production, and to propose resource recovery technologies. To do that, the processes involved in the leather tannery have been reviewed, based on raw materials, processing parameters and effluent composition, changes being proposed at different stages (e.g., fleshing after liming, counter-current soak, use of new materials: enzymes for unhairing, and carbon dioxide and/or organic acid, for deliming and pickling, waste recycling: protein recycling from fleshing, salt collection and reuse, for hide preservation and waste recycling).

Smart fertilizers produced from hide waste protein hydrolysate (such as collagen) offer the advantages of valorising a large quantity of the wastes generated by the meat and leather industry. At the same time it determinates the obtaining of higher crop yields, with a lower cost, contributes to the conservation of the soil fertility (by not using the conventional chemical fertilizers) and combats environmental challenges related to waste disposal [62].

3. Hide and Leather Waste Processing

Sometimes, the hide waste needs to be processed for hair and flesh removal prior to its use for collagen recovery. The process uses lime and sodium sulphide to pulp the hair while the flesh is removed mechanically [49]. The processing of hides and skins involves multiple operations to achieve their conversion to the final products. Therefore, hide and skins trimming wastes are less contaminated by chemicals compared to the trimming generated by tanned and finished leathers. Tanning is the process by which the leather is given more stability and resistance to the chemical, thermal and biological degradation, by stabilizing the protein (collagen). This is done by chromium tanning which consists in the cross linking of collagen free carboxyl groups with the chromium ions [63]. It was found that around 15–30% of proteinous solid wastes generated from tanneries are chrome contaminated shavings, produced when the tanned hide is shaved to a uniform thickness [29,55]. As reported by Tahiri et al. (2007) [64], the chromium oxide content in chromium tanned leather shaving is about 4.4%. Other authors have reported slightly lower chromium contents, namely between 2 and 4% [65,66]. Therefore, the main toxic compound found in the tanned leather waste is chromium which can be recovered by extraction and used again in the tanning process of leather. El Boushy et al. (1991) [67] utilized a method consisting of several washing steps (alternating alkaline-acid wash) of waste leather in order to decrease the chromium content. At the end the authors obtained a material rich in hide protein (74.9%), with a fibrous texture consisting mainly of collagen and having a digestibility of 98%. The chromium content was reported to be 0.2% [67]. A recent method consists of two steps: leaching with H_2SO_4 and ion exchange step using cation exchange resins, allowed to reduce the Cr (III) content at ppm level (14 ppm) [68]. However, this same process of leather stabilization will generate problems at the moment of waste leather disposal.

It has been evidenced that initially the cow hide has about 60–70% water, 30–35% proteins, 0.5–2% lipids and 0.35–0.5% mineral compounds. The tanning process determinates a decrease in the water content, which gives in the end a content of about 70% proteins (mostly collagen and small amounts of elastin) [69,70].

The main protein that is encountered in hide wastes is the collagen that has 28 different types that can vary in abundance, distribution and functionality within a tissue; the most abundant is collagen type I that can form up to 90% of the connective matrix. A collagen molecule consists of three polypeptide chains assembled into a triple helix structure and a repeating amino acid sequence is responsible for the helical arrangement. [71]. Walters and Stegemann (2014) [72], described the collagen complex structure from the nano to macroscale emphasizing the role of collagen at each scale (Figure 3).

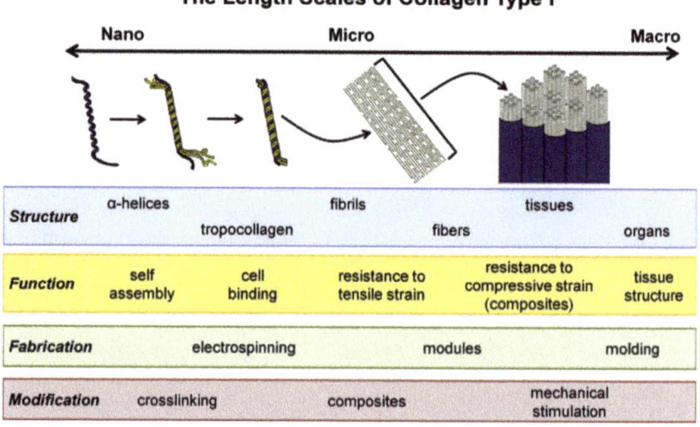

Figure 3. Representation of collagen type I structure at different scales [72].

Two of the three polypeptide chains are identical, the third has a distinct sequence of amino acids. The α-chains are composed of repeating sequences of three amino acids that have a glycine at every third interval. Glycine amino acid allows the rotational freedom that leads to the helical structure. Other amino acids are responsible for stability, rigidity, biochemical and physical characteristics of collagen [72,73].

4. Preparation of Polymer Based Smart Fertilizer

A good fertilizer releases the nutrients in time and at the same time is biodegradable. The release of nutrients is based on the concentration gradient that exists between soil and fertilizer matrix. The fertilizer biodegradability allows an advanced release of the nutrients once the surface nutrients are consumed and at the same time provides the carbon necessary for the plant to grow.

Hydrogels are one type of polymer materials that have the advantage of absorbing and retaining high quantities of water, while they do not dissolve in contact with the water. This ability is given by the numerous functional hydrophilic groups (carboxylic acids, alcohols, amides and amines) attached to the polymeric chain, while the resistance to water dissolution is the result of cross-linked chains forming a three-dimensional network [74]. The high quantity of water retained, facilitates the diffusion of the nutrients through the polymer structure. Different types of hydrogels can be synthesized depending on the protein that is employed: collagen, gelatine, fibrin, silk, elastin, keratin [75]. Composite materials synthesized by hide waste hydrolysis are nontoxic compounds that have two major areas of utilization (Figure 4): (a) in medical applications and cosmetic products [75–79] and (b) in agriculture as fertilizers or as additive for animal feed [22,67,80,81]. In recent years, the use of collagen recovered from leather waste as food additive for animals feeding has been forbidden in the European Union [80]. The difference between the two major utilizations is that for the medical use the entire gelatine pelt is used while in the leather industry the gelatine pelts are processed (cutting, tanning) to obtain the final product and the waste is reused to extract the collagen.

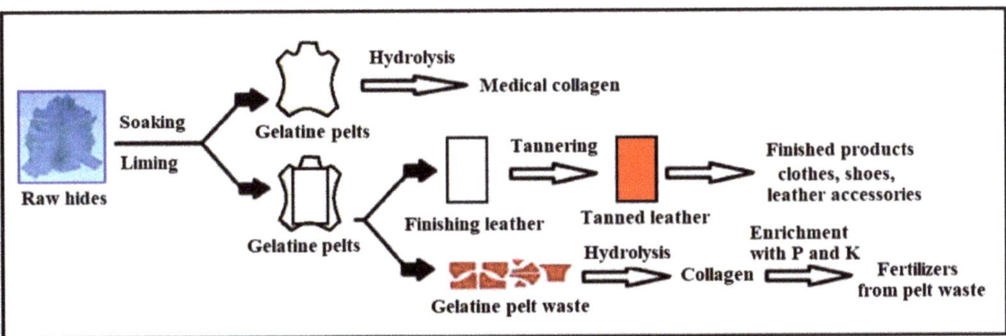

Figure 4. Differences in the preparation of medical collagen and agricultural collagen, readapted from [80].

To produce the biofertilizer, the first step is to recover the collagen from the hide waste. This is made by extraction via the hydrolysis process. A partial hydrolysis process gives rise to gelatine whilst a more advanced hydrolysis generates the collagen hydrolysate, a liquid that contains low molecular weight peptides [82,83]. The collagen hydrolysate is further modified by chemical cross-linking which consists in reactions between collagen's reactive groups and the functional groups of a water-soluble copolymer. This step is necessary in order to overcome the disadvantages related to the liquid state of collagen hydrolysate which limits its utilization as fertilizer due to the odour, risk of microbial development and difficulty to be applied on the soils [84].

Masilamani et al. (2016) [85] presented a method for the extraction of collagen from trimming waste using acetic or propionic acid. Both acetic acid and propionic acid were effective in the extraction of collagen from trimming waste but propionic acid gives relatively higher amount of collagen extracted.

The steps of a smart fertilizer synthesizing starting from leather wastes, are depicted in Figure 5. In the first stage, the collagen matrix is obtained by hydrolysing the hide waste. The hydrolysis can be either using base [86] or acid chemicals [84]. The collagen hydrolysate is a liquid that contrary to medical applications, cannot be used as a fertilizer in this form. Therefore, a stabilization step is necessary by cross-linking with water soluble polymers. The resulted copolymers will incorporate more easily the nutrients, and release them later according to the plants' needs.

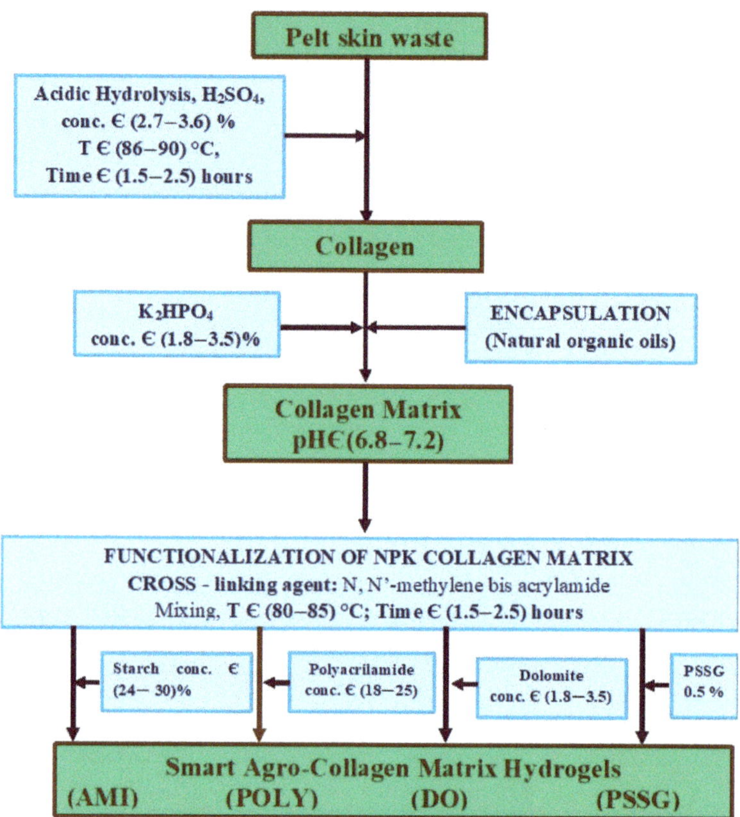

Figure 5. Technology scheme for obtaining smart-fertilizers by using acid hydrolysis, readapted from [80] (CH—collagen hydrolysate, Ref—CH—collagen hydrolysate with nutrients encapsulated as reference sample, PSSG—Ref—CH functionalized with P(SSNa—co—GMAx) copolymer, POLY—Ref—CH functionalized with polyacrylamide, AMI—Ref—CH functionalized with starch, AMI—Ref—CH functionalized with dolomite).

Figure 6 shows images of different intermediary products obtained during the synthesis of a smart fertilizer starting from wet blue wastes (resulted by leather tanning with chromium salt) [87].

Tzoumani et al. (2019) [84] selected poly (sodium 4-styrenesulfonate-co-glycidyl methacrylate) (P(SSNa-co-GMAx)) as water -soluble copolymer, because the behaviour of a charged polyelectrolyte combined with the reactive epoxy groups will be used in the cross-linking process. Different ratios of monomers have been used in the preparation of copolymers named P(SSNa-co-GMAx). The collagen hydrolysate was modified with P(SSNa-co-GMAx) or starch. To confirm the cross-linking between collagen hydrolysate and the epoxy groups, ATR-FTIR analysis was used. The authors have compared the variation of the release degree for oxidable compounds in water, in time, and found a controlled release in the case of enriched collagen functionalized with synthetic polymer and starch, compared to the un-functionalized enriched collagen (functionalization of collagen results in slowing their release capacity).

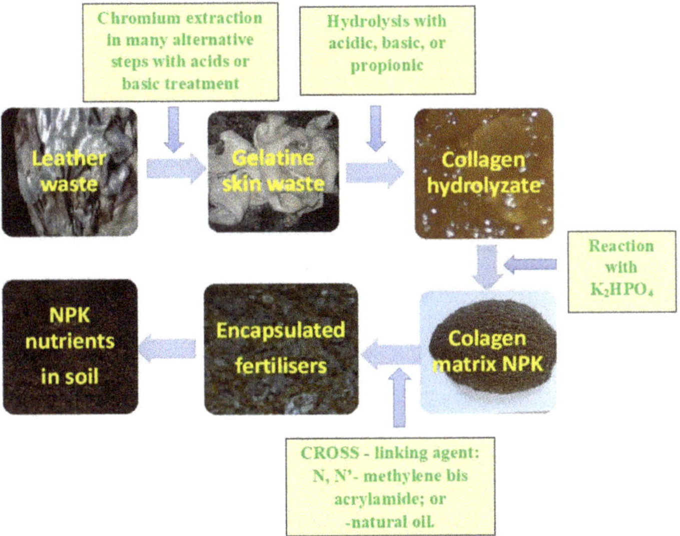

Figure 6. From waste leather to bio-polymer fertilizer—Intermediary products [87].

Hu et al. (2021) [16] described the preparation of a new fertilizer using leather waste as source of collagen, having the abilities of nutrient controlled release and heavy metal adsorption. The ability of heavy metal removal was evaluated under different environmental parameters (pH, coexisting ions), to demonstrate the potential application of this fertilizer as a green and sustainable agrochemical. The hydrogel was obtained by hydrolysis of leather waste, using KOH solution, at 60 °C under stirring. The protein hydrolysate is functionalized by adding a mixture of neutralized acrylic acid, ammonized maleic anhydride, N, N-methylene-bis-acrylamide, ammonium persulfate, and sodium bisulphite, at the weight ratio of 1000:100:0.55:2.2:1.1, at 60 °C. The product was swollen, washed, filtered and dewatered, in excess ethanol. Afterward, tests to evaluate the swelling characteristics, nutrient release ratios and biodegradation, were made.

In Tables 1 and 2 a comparison among the different characteristics of the hydrogel-based fertilizers is presented. Higher pore diameter values determinate higher surface area to volume ratio which produces a fertilizer with enhanced swelling rate and biodegradability [16,81].

Table 1. Comparison among different types of biopolymer-based fertilizers.

Hydrogel	Water Absorption Capacity	Cr (III) Adsorption Capacity
Biofertilizer (collagen-nitrogen and potassium) [16]	2208 g H_2O/g	149.3 g Cr (III)/g
Biofertilizer (source P nutrient) Collagen-g-poly(acrylic acid-co-2-acrylamido-2-methyl-1-propane sulfonic acid)–iron(III) [81]	2595 g H_2O/g	Not tested
Collagen-polyacrylic acid-co-2-acrylamido-2-methyl-1-propane sulfonic acid) [81]	3578 g H_2O/g	Not tested

Table 2. Characteristics of hydrogels.

Hydrogel	Pore Diameter	Days of Controlled Nutrient Release
Biofertilizer (collagen-nitrogen and potassium) [16] Biofertilizer (source P nutrient)	1.26–6.73 μm	More than 40
Collagen-g-poly(acrylic acid-co-2-acrylamido-2-methyl-1-propane sulfonic acid)–iron(III) [81]	4–9 μm	More than 30

Some researchers combined the use of wet blue leather, as a source of nitrogen, with other materials (poultry bone meal and water hyacinth ash, as a source of phosphorous, and potassium, respectively) for the production of N-P-K (Nitrogen-Phosphorous-Potassium) enriched organic fertilizer [88]. The chromium was extracted from wet blue leather by basic hydrolysis, followed by acidic hydrolysis. The collagenic material was further mixed with the poultry bone meal, and potassium enriched water hyacinth ash, the resulted organic N-P-K based manure was then tested as a nutrient source for Catharanthus roseus (Madagascar Periwinkle). Results were compared with plant growth on soils without fertilizers, and on soils containing a conventional chemical fertilizer. It was found that the release of nutrients was controlled for the polymer-based fertilizer with a sustained plant growth over time, while the chemical fertilizer dissolved faster in the soil moisture and gave extra plant growth in the initial stage, but slower afterwards. In a more recent study, two organic ingredients, namely chromium-free collagen of wet blue leather (WBL) waste-as nitrogen source, and potato peel biochar-as potassium-phosphorus source, were used for the synthesis of NPK rich bio-fertilizer [89]. The chromium was extracted from wet blue leather by treatment with H_3PO_4 [90,91]; the so treated wet blue leather waste was washed with distilled water, and tested for residual chromium, both in the washing solution, and WBL, using atomic absorption spectrophotometer. The degree of chromium removal at the final washing step was of 90.38%. The potato peel biochar was washed with water to remove the residual dust, dried at 80 °C for humidity decrease, crushed and then carbonized in an electric furnace at 450 °C for 1 h. The two resulted materials WBL (as a nitrogen source) and potato peel biochar (as potassium and phosphorus source) were mixed in the ratio of 1:2.5. The mixture was dried at 30 °C for 24 h, and crushed, to obtain the bio-organic NPK fertilizer powder, which is easy to spread on the agricultural fields. The final product was checked by analytical characterization by SEM, EDS and FT-IR, and the final composition was: nitrogen 13.10%, phosphorus 2.41%, potassium 20.20% and magnesium 1.16%, carbon 33.74% and chromium 0.23%, indicating that all major nutrients are present as required by any commercial fertilizer. The biofertilizer was compared with a chemical fertilizer regarding their nutrients release in time: the chemical fertilizer released the nutrients more rapidly in time, causing a faster growth during the initial stage of the plant growth, whereas bio-organic NPK fertilizer loses nutrients in a controlled way, and determinates the plant growth uniformly in time.

The influence of chromium was also discussed. The analysis of the soil composition before using the biofertilizer, indicated a concentration of 0.055 mg Cr/kg soil. The chromium content of the bio-organic fertilizer (0.71 mg Cr/kg) increased the soil chromium content to 0.765 mg Cr/kg soil, which is well below the maximum allowable limit of chromium in soils, which as recommended by WHO is 100 mg Cr/kg [92]. However, the authors proposed an advanced washing of the WBL with H_3PO_4, in order to further decrease the chromium content of the biofertilizer.

Constantinescu et al. (2015) [93], presented the development of biocomposite fertilizers and their application in agriculture for plant growth (namely soybean crop), and remediation of soil content in required nutrients. For this purpose, the authors have used untanned waste provided by a local leather processing company, and the fertilizer was synthesized by alkaline hydrolysis of raw hide leather. Dipotassium phosphate was added, to improve the nutritional characteristics as regarding K and P. Results obtained on soils treated with this biofertilizer were compared to untreated soils, and showed that application of biofertilizer stimulated the plant growth, and the production increase.

Zainescu et al. (2018) [94], synthesized a hydrogel based on collagen hydrolysate cross-linked with acrylamide synthetic polymer. Acrylamide was chosen because it offers several advantages: it is chemically inert, transparent and stable in a wide range of pH and temperature. The presence of cross-linking between collagen and polyacrylamide in the molecular structure of hydrogel was confirmed by optical microscopy and IR analysis.

Collagen recovered from wet blue leather wastes was used as adsorbent for K and P, in order to obtain an NPK-fertilizer [95]. The adsorption of P and K takes place in a multilayer at the surface of the adsorbent, and the process was very well described by Freundlich models. The resulted fertilizer was applied as a source of nutrients for promoting the growth of rice plants with promising results.

5. Biodegradation of Polymers Extracted from Hide and Leather Waste

The collagen is a natural polymer that by itself is enzymatically degradable [96]. Soil humidity and temperature are influencing factors that stimulate the biodegradation of hydrogels, under the influence of proteolytic bacteria [94]. The biodegradation time of hydrogels vary from one up to six months [80,97,98]. Generally, the enzymatic biodegradation takes place through the action of enzymes and/or chemical deterioration associated with living organisms.

Biodegradability depends on the polymer chemical structure and the environmental degrading conditions (pH, water availability, temperature, light) [99].

The biodegradation mechanism of polymer-coated controlled-release PC-CRT fertilizers in soils involves several steps (Figure 7) [100]:

(1) swelling: the ionic functional groups like hydroxyl, carboxyl, and amino can form the hydrogen bounds with water and more easily swell resulting in a porous network. This behaviour is specific only for hydrogels that have more functional ionic groups. The swollen porous structure increases the pore size, allowing the release of incorporated fertilizers such as urea, phosphates, etc.

(2) biodeterioration: polymer fragmentation into lower molecular mass species in abiotic reactions (oxidation, photodegradation, hydrolysis).

(3) biofragmentation: the polymer is fragmentated in biotic reactions, i.e., hydrolysis of macromolecules in oligomer, dimer or monomer, the mediators being microorganisms.

(4) assimilation: the monomer can be absorbed by the microorganisms and degraded for example by deamination or decarboxylation, resulting ammonia or nitrate, acids and alcohols etc.

(5) mineralization: is the process of degradation of organic compounds in aerobic and anaerobic conditions to mineral compounds (nitrate, carbon dioxide, hydrogen, methane).

In leather waste, the collagen is crosslinked with the tanning agent, which gives stability to biodegradation. In order to assess the suitability of using leather waste as fertilizers, and to evaluate their behaviour and potentially adverse environmental effects, studies of leather biodegradability have been conducted. Stefan et al. (2012) [70], made an experimental study on the identification of microorganisms that are suitable to be used in the improvement of waste leather biodegradation. Their investigation consisted in the isolation, selection and characterization of microorganisms that produce the extracellular protease and lipase. The inoculum of microorganisms was taken from an old waste storage dump leather. The Bacillus species showed higher extracellular proteolytic and lipolytic activity, the maximum production being obtained after 48 h.

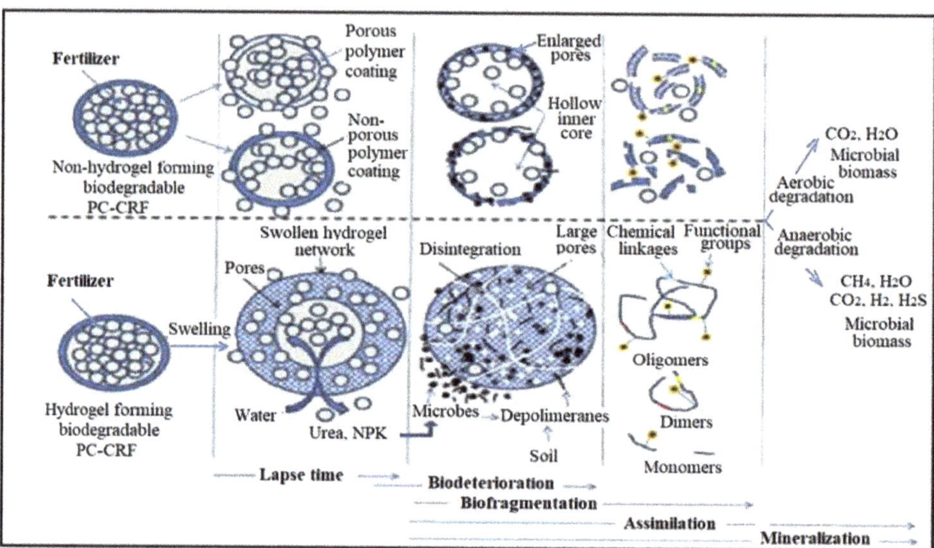

Figure 7. Biodegradation in soils of hydrogel compared with non-hydrogel polymers [100].

Because of the environmental issues related to the use of chromium-based salts in the leather tanning process, in recent years, methods employing vegetal materials as tanning agents have been developed [101–105]. A comparative study of the biodegradation in aqueous solution in aerobic condition of leather wastes, generated in the tanning process using chromium or vegetal compounds, for a period of 100 days, was made by Stefan et al. (2011) [106]. To reproduce the effect of nutrients in fertilizers, a mixture of nutritive salts containing K, N, P, Na, Ca was added, and the leather was used as carbon source for the crop growth, while as inoculum, the liquid extract of compost generated from leather waste disposal was used. Through this system, air free of carbon dioxide was bubbled at a flow rate between 1.5 ÷ 2.5 L/s, at constant temperature 20 ± 0.5 °C. Several parameters were monitored, in order to account for leather biodegradation process: carbon dioxide resulted, respectively pH, conductivity, TOC (total organic carbon) and TON (total organic nitrogen) of the liquid phase.

The biodegradability of untanned, chrome tanned and vegetal tanned leather under anaerobic conditions was investigated, and the effect of untanning on the leather biodegradability was reported [107]. As inoculum anaerobic microorganisms isolated from anaerobic sludge were used, generated by wastewater treatment plant employed for the treatment of tannery wastewater, and also from the sludge obtained from the sewage treatment plant treating domestic wastewater. The results showed that biodegradation of chrome tanned leather waste is possible using anaerobic sludge, and that in certain conditions, the degradation degree is higher than for the vegetable tanned leather waste (when the untanning process is introduced before the biodegradation process, the resulted material is less stable).

Zainescu et al. (2018) [94], studied the biodegradability of the collagen hydrogel with encapsulated nutrients by two methods: in soil, by measuring the weight loss, and in aerobic conditions in aqueous medium (according to SR EN ISO 14852/2005). The second method is recommended by the authors for evaluating the hydrogel biodegradability due to its accuracy.

Although not so high as the activated sewage sludge, soil has still an important biological activity, being rich in species that could be a source of inoculum. Stefan et al. (2020) [80] performed biodegradability tests for several fertilizers, in water in aerobic conditions, and in composting conditions. The experiments realized in composting conditions aim to study the fertilizer biodegradability in conditions similar to that encountered in agricultural environment. The tests were performed according to the standard procedures respectively SR EN ISO 14852/2005, for determination of the aerobic biodegradability in aqueous medium of plastic materials, and SR EN ISO 14855-1/2008, for determination of the final aerobic biodegradability in composting-controlled conditions. Both standards evaluate biodegradability based on the measuring of CO_2 quantity released during the polymer consumption. Experiments evidenced the existence of four regions corresponding to the steps of biodegradation process:

(1) stagnant zone: 0 ÷ 6 days, the biodegradation process is initiated and the process rate is small.
(2) acceleration zone: 2 ÷ 50 days, the biodegradability is linearly increasing.
(3) slowing zone: 14 ÷ 56 days, the biodegradability rate is decreasing.
(4) stationary zone: 41 ÷ 75 days, biodegradation reaches its maximum degree and the biodegradability rate goes to zero.

The results indicated that the biodegradability degree is higher for the biopolymer-based products, and smaller in the case of compounds functionalized with synthetic polymers (polyacrylamide and P(SSNa—co—GMAx) copolymer, noted PSSG). As regarding the environment, biodegradation is slower in composting medium, than in water medium. The evolution of biodegradability in time, in water and in composting conditions, for the studied fertilizers, compared with collagen hydrolysate is shown in Table 3. As expected, the collagen hydrolysate (CH and Ref CH) has the highest biodegradability, followed by collagen functionalized with starch while the functionalized fertilizers with synthetic polymers have the lowest biodegradability (POLY and PSSG).

Table 3. Biodegradation degree evolution in time * [80].

Time, Days	Biodegradation Degree							
	CH	Ref CH		AMI		POLY		PSSG
	W	W	C	W	C	W	C	W
10	42	35	17	37	33	15	12	20
20	58	48	33	50	48	27	21	35
75	99	74	50	80	64	62	40	63

* CH—collagen hydrolysate, Ref-CH—collagen hydrolysate with nutrients encapsulated as reference sample, PSSG—Ref—CH functionalized with P(SSNa—co—GMAx) copolymer, POLY—Ref—CH functionalized with poly-acrylamide, AMI—Ref—CH functionalized with starch, W—water, C—composting conditions.

In literature it is reported that 15–30% of the fertilizers remained unreleased from PC-CRFs due to the concentration gradient difference across the polymer coatings [108]. It can be seen that the amount of fertilizers remained unreleased in the case of gelatine-based fertilizers functionalized with natural polymers is in the range of 20–25% in time, and for those functionalized with synthetic polymers in the range of 32–34%.

There have been numerous studies performed on starch-based hydrogels functionalized with natural and synthetic polymers for example: chitosan, ethyl cellulose, polyacrylic acid, clay, lignin and polyurethane, polyvinyl alcohol, polybutylene succinates and other variants [19,108–112].

The unreleased amounts of fertiliser starch-based hydrogels functionalized with natural and synthetic polymers varied between 10–25%. Fertilizers obtained from both natural collagen-based and starch-based polymers release fertilizers more easily than synthetic ones. The degree of biodegradation is also higher in the case of fertilizers obtained from natural polymers. Furthermore, the degradation compounds obtained from fertilizers obtained from natural polymers are less toxic [100].

6. Release Mechanism of Nutrients

For the release mechanism of the nutrients, several stages were proposed: in a first step, organic and inorganic matter having high solubility (e.g., peptides with short chains, soluble polymers (starch), amino acids, glucides (mono- and disaccharides), nitrate salts, acids) is released. Afterwards, the release of oxidable compounds decreased for all fertilizers corresponding to organic compounds with low solubility (proteins) [80,84].

Du et al. (2006) [97], studied the influence of different parameters on the release of nutrients for two polymer coated compound fertilizers having the same core composition but different ratio in the nutrients of the coating. It was found that the nutrients release from CRF was mainly controlled by the diffusion mechanism. In this case, among the parameters that influence the diffusion, the temperature and coating thickness had an important role: lower temperature and thicker membrane determinate a lower diffusion coefficient of coated membrane, which slowed the nutrients release rate. Furthermore, nutrients release rate was different, depending on the diffusion medium, the fastest release rate was in water, then water saturated sand, and the last was in sand at field capacity. Nutrients release profile over 70 days indicate three steps: lag period, linear stage and decay period. The lag period of P was significantly longer than of other nutrients indicating that there is a strong bond between P and the core composition.

6.1. Hydrogel Modelling—Kinetic Modelling of Nutrients Release

There are few studies dedicated to modelling of hydrogel materials regarding the release of different chemicals. The first model used for fitting the experimental data of substance release from porous hydrophilic polymers is Korsmeyer-Peppas model [113–116]. The Nernst -Planck equation, accounting for the fluxes of mobile ions in the hydrogel structure and in external diffusion layer has been used by Pareek et al. (2017) [117] and Goh et al. (2017) [118]. Assuming that the hydrogel pores are narrow enough so that the diffusion dominates the transport across the hydrogel and that the nutrients are uniformly dispersed throughout the matrix, unsteady-state nutrient diffusion in a one-dimensional direction can be described using Fick's second law of diffusion [119]. Mass conservation balance is expressed in a mathematical equation that accounts for diffusion of species inside and outside hydrogels, electrostatic interactions, hydrophobic associations, cleavable covalent linkage and degradation [80]. An empirical equation developed by Peppas et al. (2000) [120] assumes a time-dependent power law function for the released quantity of nutrient:

$$R_f = k\, t^n$$

where R_f represents the release factor,

k—kinetic constant dependent on material

n—exponent depending on type of transport, hydrogel geometry and polymer polydispersity; values of n close to 0.5 indicate that the controlling step in the transport mechanism is the diffusion while for n values close to 1, the surface deterioration is the controlling step [80].

The kinetic parameters, as reported by Stefan et al. (2020) [80], are given in Table 4, being calculated from experimental data for the leaching degree of oxidable compounds for the tested fertilizers, during almost one month (27 days). The leaching test consisted in the determination of chemical oxygen demand with $KMnO_4$ (CODMn).

Table 4. Kinetic parameters values for compounds release * [80].

Fertilizer Type	k	*n*	R2
CH	0.1370	0.6693	0.9159
REF-CH	0.0480	0.9514	0.9240
AMI	0.0880	0.7995	0.9212
POLY	0.0560	0.9059	0.9212
PSSG	0.0710	0.8644	0.9097

* CH—collagen hydrolysate, Ref-CH—collagen hydrolysate with nutrients encapsulated as reference sample, PSSG—Ref—CH functionalized with P(SSNa—co—GMAx) copolymer, POLY—Ref—CH functionalized with poly-acrylamide, AMI—Ref—CH functionalized with starch.

The values of the exponent n are above 0.5, indicating that the transport mechanism is controlled by several steps. The kinetic parameters were recalculated using a logarithmic form, and two linear regions were identified corresponding to: (1) the initial step days 3–10, $n > 1$, (2) the second step n is around 0.5. In the first zone, the controlling step is attributed to surface deterioration by $KMnO_4$, while in the second zone, the controlling step is the diffusion of oxidation compounds.

6.2. Hydrogel Modelling—Swelling Process

The model mentioned above considers only the nutrients transport across the matrix. More complex models include also the water transport and the hydrogel swelling/ shrinkage [121]. The hydrogel swelling depends on a series of parameters such as temperature, pH, solvent and hydrogel structure, because the swelling process is thermodynamically controlled by hydrogel-solvent interactions [15]. The equilibrium state for the swelling process of the smart hydrogel in solvent is reached when the solvent inside the hydrogels is in thermodynamic equilibrium with that outside [15,122].

Hydrogel swelling ratio is defined as follows:

$$S.R. = (m_{hydrated} - m_{dehydrated})/m_{dehydrated}$$

where S.R. is the swelling ration and m is the mass of either hydrated or dehydrated hydrogel.

Activity coefficients for species in bulk volume or in hydrogel phase were calculated using either UNIQUAC or NRTL models [123–125]. Recent papers have reported, based on experimental-modelling data comparisons, that NRTL model is more suitable to be used, as UNIQUAC model fails to accurately account for liquid phase-hydrogel interaction in cross linked polymers [15].

Sheth et al. (2019) [126] developed an 1D (one-dimensional) computational model for the diffusion and swelling, that accounts for time-varying of diffusivity and geometry to predict profiles of substances released from degradable hydrogels. Time snapshots of diffusivity and hydrogel geometry data measured experimentally were used as inputs in the computational model, which predicted the components profiles.

A three-phase complex model (a solid matrix, the hydrogel and the liquid solvent) was developed by Sauerwein and Steeb (2020), model that has been validated by experimental data obtained for hydrogel swelling in different solvents [122]. The governing equations of the model for mass, momentum and electrostatic charges, were written assuming isothermal conditions [127]. However, their model has to be adapted when used for the fertilizers case, because the fertilizers do not include the matrix phase.

Three steps are influencing the swelling behaviour of a hydrogel: the local polymer and ionic liquid phase concentrations, the elasticity of the polymer network (given by the crosslinking degree), and the behaviour of the ions at the hydrogel-liquid phase interface governed by Donnan equilibrium [128]. As regarding the swelling kinetics, this is mainly limited by the diffusion step [129]. However, work still needs to be done regarding the approximation of swelling/deswelling time, which are not equal.

7. Future Perspectives of the Production of Smart Fertilizers from Leather Waste

In order to identify the nutritional status and, consequently the needed fertilizer dose, it is necessary to know the initial state of soil fertility (defined by the agrochemical characteristics of the soil), the species and variety of plants to be cultivated, and the type of the ecological zone (i.e., if there are underground waters that are susceptible to be polluted by the levigated nutrients).

The choice among the different types of smart fertilizers obtained using collagen hydrolysate obtained by acidic, basic or enzymatic hydrolysis is made after the characterisation of soils. The soil composition regarding the initial content in nutrients (mainly N, P, K) and acidity must be known prior to the decision whether to use, for example, a P-enriched smart fertilizer, or a more acid, or a more basic one. During the synthesis steps, the ratio between the nutrients N, P, K and the pH of the final product can be adjusted to correspond to soil deficiencies and to the crop demand.

The use of enzymatic hydrolysis is highly recommended, when the final fertilizer is intended to be applied on soils with high salinity, and for which it is not recommended a further increase in the salt content.

As chromium containing components are dangerous for human health and for the environment, chromium content should be carefully checked when the fertilizer is synthesised by using tanned wastes as its content must be in agreement with the imposed regulations. This step will not add a high extra cost to the fertilizers final price, because the tanneries employing chromium in their processing are already recovering Cr containing compounds before waste disposal to landfilling.

8. Conclusions

The scope of this review was to highlight the potential of hide waste as a source of collagen-based biopolymers. The characteristics of hide leather wastes have been presented and, as well as the problems generated, if these wastes are left untreated. The methods for the synthesis of smart fertilizers have been detailed, together with the advantages generated by reducing the impact on the environment (by recovering resources from wastes and by offering a bio-fertilizer as an alternative to the chemical fertilizer).

The biggest advantage of this type of fertilizers is that the synthesis method can be easily adapted to the beneficiary demands, namely it can be adjusted to soil characteristics (pH and nutrients ratio), as well as to crop necessity in nutrients.

The main advantage that arises from replacing the conventional chemical fertilizers with bio-polymer containing fertilizers is related to the gradual release of nutrients over a longer period of time, which leads to increasing crop production and improving the quality of plants by controlled release of nutrients. On the contrary, in the case of chemically fertilizers, the release of nutrients is immediate, and if combined with the meteorological conditions (e.g., rains), it can lead to the nutrients' leachability in the deeper layers of the soil and aquifer causing major pollution problems.

The smart fertilizers obtained by enzymatic hydrolysis of leather wastes are indicated for soils with high salinity. This ensures the reduction of stress caused by increasing further the soil salinity, or by applying treatments with phytosanitary products. This type of smart fertilizer improves the beneficial activity of microorganisms in the soil, and increases the permeability of cell membranes from the root system, favouring the nutrients absorption and retention.

In the case of alkaline soils (which are the worst soils for plants growth), collagen-based fertilizers act as a naturally chelating agent for micronutrients, favouring their accessibility to the plant.

The capacity to retain high quantities of water and to gradually release it reduces the irrigation frequency, and prevents water loss by evaporation.

The recovery and reuse of high quantities of waste generated by the leather industry by extracting a valuable raw material: the collagen, ensures the compliance of leather industry with more strict regulations related to waste disposal and in agreement with the principles of circular economy.

Author Contributions: Conceptualization, D.S.S. and M.B.; methodology, R.R.C. and M.I.; validation, D.S.S. and R.R.C.; investigation, M.B. and M.I.; resources, D.S.S. and M.B.; data curation, D.S.S., R.R.C. and M.I.; writing—original draft preparation, M.B.; writing—review and editing, D.S.S.; supervision, D.S.S. and M.B. All authors have read and agreed to the published version of the manuscript.

Funding: This research received no external funding.

Institutional Review Board Statement: Not applicable.

Informed Consent Statement: Not applicable.

Conflicts of Interest: The authors declare no conflict of interest.

References

1. Helaly, P.M.; Abo-Elela, S.I. Protection of surface water from eutrophication via controlled release of phosphate fertilizer. *J. Control. Release* **1990**, *12*, 39–44. [CrossRef]
2. Huang, J.; Xu, C.C.; Ridoutt, B.; Wang, X.C.; Ren, P.A. Nitrogen and phosphorus losses and eutrophication potential associated with fertilizer application to cropland in China. *J. Clean. Prod.* **2017**, *159*, 171–179. [CrossRef]
3. Ashitha, A.; Rakhimol, K.R.; Jyothis, M. Fate of the conventional fertilizers in environment. In *Controlled Release Fertilizers for Sustainable Agriculture*, 1st ed.; Lewu, F.B., Volova, T., Thomas, S., Rakhimol, K.R., Eds.; Academic Press: London, UK, 2020; pp. 25–39.
4. Li, P.; Kong, D.; Zhang, H.; Xu, L.; Li, C.; Wu, M.; Jiao, J.; Li, D.; Xu, L.; Li, H.; et al. Different regulation of soil structure and resource chemistry under animal-and plant-derived organic fertilizers changed soil bacterial communities. *Appl. Soil Ecol.* **2021**, *165*, 104020. [CrossRef]
5. Naghdi, A.A.; Piri, S.; Khaligi, A.; Morad, P.J. Enhancing the qualitative and quantitative traits of potato by biological, organic, and chemical fertilizers. *Saudi Soc. Agric. Sci.* **2021**, in press. [CrossRef]
6. Hui, K.; Tang, J.; Cui, Y.; Xi, B.; Tan, W. Accumulation of phthalates under high versus low nitrogen addition in a soil-plant system with sludge organic fertilizers instead of chemical fertilizers. *Environ. Pollut.* **2021**, *291*, 118193. [CrossRef] [PubMed]
7. Gao, D.; Li, Z.; Wang, H.; Liang, H. An overview of phthalate acid ester pollution in China over the last decade: Environmental occurrence and human exposure. *Sci. Total Environ.* **2018**, *645*, 1400–1409. [CrossRef] [PubMed]
8. Seleiman, M.; Santanen, A.; Mäkelä, P. Recycling sludge on cropland as fertilizer–Advantages and risks. *Resour. Conserv. Recycl.* **2020**, *155*, 104647. [CrossRef]
9. Bigalke, M.; Ulrich, A.; Rehmus, A.; Keller, A. Accumulation of cadmium and uranium in arable soils in Switzerland. *Environ. Pollut.* **2017**, *221*, 85–93. [CrossRef] [PubMed]
10. Mishra, D.J.; Singh, R.; Mishra, U.K.; Shahi, S.K. Role of Bio-Fertilizer in Organic Agriculture: A Review. *Res. J. Recent Sci.* **2013**, *2*, 39–41.
11. Negi, Y.K.; Sajwan, P.; Uniyal, S.; Mishra, A.C. Enhancement in yield and nutritive qualities of strawberry fruits by the application of organic manures and biofertilizers. *Sci. Hortic.* **2021**, *283*, 110038. [CrossRef]
12. Lubkowski, K.; Grzmil, B. Controlled release fertilizers. *Polish, J. Chem. Technol.* **2007**, *9*, 81–84. [CrossRef]
13. Vejan, P.; Khadiran, T.; Abdullah, R.; Ahma, N. Controlled release fertilizer: A review on developments, applications and potential in agriculture. *J. Control. Release* **2021**, *339*, 321–334. [CrossRef] [PubMed]
14. León, O.; Soto, D.; Antúnez, A.; Fernández, R.; González, J.; Piña, C.; Muñoz-Bonilla, A.; Fernandez-García, M. Hydrogels based on oxidized starches from different botanical sources for release of fertilizers. *Int. J. Biol. Macromol.* **2019**, *136*, 813–822. [CrossRef] [PubMed]
15. Rashedul Islam, M.; Tanveer, S.; Chen, C.C. Modeling swelling behavior of hydrogels in aqueous organic solvent. *Chem. Eng. Sci.* **2021**, *242*, 116744. [CrossRef]
16. Hu, Z.Y.; Chen, G.; Yi, S.H.; Wang, Y.; Liu, Q.; Wang, R. Multifunctional porous hydrogel with nutrient controlled-release and excellent biodegradation. *J. Environ. Chem. Eng.* **2021**, *9*, 106146. [CrossRef]
17. Kazanskii, K.S.; Dubrovskii, S.A. Chemistry and physics of "agricultural" hydrogels. *Adv. Polym. Sci.* **1992**, *104*, 97–133.
18. Azeem, B.; Ku Shaari, K.; Man, Z.B.; Basit, A.; Thanh, T.H. Review on materials & methods to produce controlled release coated urea fertilizer. *J. Control. Release* **2014**, *181*, 11–21. [PubMed]
19. Qiao, D.; Liu, H.; Yu, L.; Bao, X.; Simon, G.P.; Petinakis, E.; Chen, L. Preparation and characterization of slow-release fertilizer encapsulated by starch-based superabsorbent polymer. *Carbohydr. Polym.* **2016**, *147*, 146–154. [CrossRef] [PubMed]
20. Essawy, H.A.; Ghazy, M.B.; El-Hai, F.A.; Mohamed, M.F. Superabsorbent hydrogels via graft polymerization of acrylic acid from chitosan-cellulose hybrid and their potential in controlled release of soil nutrients. *Int. J. Biol. Macromol.* **2016**, *89*, 144–151. [CrossRef]
21. Hu, Z.; Yi, S.; Hu, W.; Tang, Y.; Wang, R. Synthesis and urea-loading of a novel biosuperabsorbent polymer based on leather waste. *J. Soc. Leather Technol. Chem.* **2015**, *99*, 51–57.
22. Senna, A.M.; Botaro, V.R. Biodegradable hydrogel derived from cellulose acetate and EDTA as a reduction substrate of leaching NPK compound fertilizer and water retention in soil. *J. Control. Release* **2017**, *260*, 194–201. [CrossRef] [PubMed]
23. Zhou, T.; Wang, Y.; Huang, S.; Zhao, Y. Synthesis composite hydrogels from inorganic-organic hybrids based on leftover rice for environment-friendly controlled-release urea fertilizers. *Sci. Total Environ.* **2018**, *615*, 422–430. [CrossRef]
24. Langmaier, F.; Kolozmik, K.; Sukop, S.; Mladek, M. Products of enzymatic decomposition of chrome- tanned leather waste. *J. Soc. Leather Technol. Chem.* **1999**, *83*, 187–195.
25. Ockerman, H.W.; Basu, L. BY-PRODUCTS Hides and Skins in *Encyclopedia of Meat Sciences*, 2nd ed.; Michael Dikeman, M., Devine, C., Eds.; Academic Press: London, UK, 2014; pp. 112–124.
26. Mascianà, P. World Statistical Compendium for Raw Hides and Skins, Leather and Leather footwear 1999–2015. Available online: fao.org (accessed on 1 October 2021).
27. Sundar, V.J.; Gnanamani, A.; Muralidharan, C.; Chandrababu, N.K.; Mandal, A.B. Recovery and utilization of proteinous wastes of leather making: A review. *Rev. Environ. Sci. Biotechnol.* **2011**, *10*, 151–163. [CrossRef]

28. Cabrera-Codony, A.; Ruiz, B.; Gil, R.R.; Popartan, L.A.; Santos-Clotas, E.; Martín, M.; Fuente, E. From biocollagenic waste to efficient biogas purification: Applying circular economy in the leather industry. *Environ. Technol. Innov.* **2021**, *21*, 101229. [CrossRef]

29. Kanagaraj, J.; Velappan, K.C.; Chandra Babu, N.K.; Sadulla, S. Solid wastes generation in the leather industry and its utilization for cleaner environment: A review. *J. Sci. Ind. Res.* **2006**, *65*, 541–548. [CrossRef]

30. Brugnoli, F.; Král, I. Life Cycle Assessment, Carbon Footprint. In Proceedings of the Leather Processing, Eighteenth Session of the Leather and Leather Products Industry Panel, Eighteenth Session of the Leather and Leather Products Industry, Shanghai, China, 1–5 September 2012; Available online: open.unido.org (accessed on 29 September 2021).

31. Al-Jabari, M.; Sawalha, H.; Pugazhendhi, A.; Rene, E.R. Cleaner production and resource recovery opportunities in leather tanneries: Technological applications and perspectives. *Bioresour. Technol. Rep.* **2021**, in press. [CrossRef]

32. Sreeram, K.J.; Saravanabhavan, S.; Raghava Rao, J.; Unni Nair, B. Use of chromium-collagen wastes for the removal of tannins from wastewater. *Ind. Eng. Chem. Res.* **2004**, *43*, 5310–5317. [CrossRef]

33. Dixit, S.; Yadav, A.; Dwivedi, P.D.; Das, M. Toxic hazards of leather industry and technologies to combat threat: A review. *J. Clean. Prod.* **2015**, *87*, 39–49. [CrossRef]

34. Chojnacka, K.; Skrzypczak, D.; Mikula, K.; Witek-Krowiak, A.; Izydorczyk, G.; Kuligowski, K.; Bandrow, P.; Kułazynski, M. Progress in sustainable technologies of leather wastes valorization as solutions for the circular economy. *J. Clean. Prod.* **2021**, *313*, 127902. [CrossRef]

35. Chojnacka, K.; Moustakas, K.; Witek-Krowiak, A. Bio-based fertilizers: A practical approach towards circular economy. *Bioresour. Technol.* **2020**, *295*, 122223. [CrossRef]

36. Kiliç, E.; Puig, R.; Zengin, G.; Zengin, C.A.; Fullana-i-Palmer, P. Corporate carbon footprint for country Climate Change mitigation: A case study of a tannery in Turkey. *Sci. Total Environ.* **2018**, *635*, 60–69. [CrossRef] [PubMed]

37. Vidaurre-Arbizu, M.; Perez-Bou, S.; Zuazua-Ros, A.; Martín-Gomez, C. From the leather industry to building sector: Exploration of potential applications of discarded solid wastes. *J. Clean. Prod.* **2021**, *291*, 125960. [CrossRef]

38. Hu, J.; Xiao, Z.; Zhou, R.; Deng, W.; Wang, M.; Ma, S. Ecological utilization of leather tannery waste with circular economy model. *J. Clean. Prod.* **2011**, *19*, 221–228. [CrossRef]

39. Yılmaz, O.; Cem Kantarli, I.; Yuksel, M.; Saglam, M.; Yanik, J. Conversion of leather wastes to useful products. *Resour. Conserv. Recycl.* **2007**, *49*, 436–448. [CrossRef]

40. Gil, R.R.; Girón, R.P.; Lozano, M.S.; Ruiz, B.; Fuente, E. Pyrolysis of biocollagenic wastes of vegetable tanning. Optimization and kinetic study. *J. Anal. Appl. Pyrolysis* **2012**, *98*, 129–136. [CrossRef]

41. Amdouni, S.; Hassen Trabelsi, A.B.; Mabrouk Elasmi, A.; Chagtmi, R.; Haddad, K.; Jamaaoui, F.; Khedhira, H.; Cherif, C. Tannery fleshing wastes conversion into high value-added biofuels and biochars using pyrolysis process. *Fuel* **2021**, *294*, 120423. [CrossRef]

42. Bhatia, S.K.; Joo, H.S.; Yang, Y.H. Biowaste-to-bioenergy using biological methods—A mini-review. *Energy Convers. Manag.* **2018**, *177*, 640–660. [CrossRef]

43. Simioni, T.; Borges Agustini, C.; Dettmer, A.; Gutterres, M. Nutrient balance for anaerobic co-digestion of tannery wastes: Energy efficiency, waste treatment and cost-saving. *Bioresour. Technol.* **2020**, *308*, 123255. [CrossRef] [PubMed]

44. Pecha, J.; Barinova, M.; Kolomaznik, K.; Nguyen, T.N.; Dao, A.T.; Le, V.T. Technological-economic optimization of enzymatic hydrolysis used for the processing of chrome-tanned leather waste. *Process. Saf. Environ. Prot.* **2021**, *152*, 220–229. [CrossRef]

45. Cem Kantarli, I.; Jale, Y. Activated carbon from leather shaving wastes and its application in removal of toxic materials. *J. Hazard. Mater.* **2010**, *179*, 348–356. [CrossRef] [PubMed]

46. Kong, J.; Yue, Q.; Gao, B.; Li, Q.; Wang, Y.; Ngo, H.H.; Guo, W. Porous structure and adsorptive properties of hide waste activated carbons prepared via potassium silicate activation. *J. Anal. Appl. Pyrolysis* **2014**, *109*, 311–314. [CrossRef]

47. Sanek, L.; Pecha, J.; Kolomaznik, K.; Barinova, M. Biodiesel production from tannery fleshings: Feedstock pretreatment and process modeling. *Fuel* **2015**, *148*, 16–24. [CrossRef]

48. Ashokkumar, M.; Thanikaivelan, P.; Krishnaraj, K.; Chandrasekaran, B. Transforming chromium containing collagen wastes into flexible composite sheets using cellulose derivatives: Structural, thermal and mechanical investigations. *Polym. Compos.* **2011**, *32*, 1009–1017. [CrossRef]

49. Ashokkumar, M.; Narayanan, N.T.; Mohana Reddy, A.L.; Gupta, B.K.; Chandrasekaran, B.; Talapatra, S.; Ajayan, P.M.; Thanikaivelan, P. Transforming collagen wastes into doped nanocarbons for sustainable energy applications. *Green Chem.* **2012**, *14*, 1689–1695. [CrossRef]

50. Alibardi, L.; Cossu, R. Pre-treatment of tannery sludge for sustainable landfilling. *Waste Manag.* **2016**, *52*, 202–211. [CrossRef]

51. Rai, D.; Eary, L.E.; Zachara, J.M. Environmental chemistry of chromium. *Sci. Total Environ.* **1989**, *86*, 15–23. [CrossRef]

52. Tôrres Filho, A.; Lange, L.C.; Caldeira Bandeira de Melo, G.; Praes, G.E. Pyrolysis of chromium rich tanning industrial wastes and utilization of carbonized wastes in metallurgical process. *Waste Manag.* **2016**, *48*, 448–456. [CrossRef]

53. Tang, Y.; Zhao, J.; Zhang, Y.; Zhou, J.; Shi, B. Conversion of tannery solid waste to an adsorbent for high-efficiency dye removal from tannery wastewater: A road to circular utilization. *Chemosphere* **2021**, *263*, 127987. [CrossRef]

54. Sathish, M.; Madhan, B.; Rao, J.R. Leather solid waste: An eco-benign raw material for leather chemical preparation—A circular economy example. *Waste Manag.* **2019**, *87*, 357–367. [CrossRef] [PubMed]

55. Pati, A.; Chaudhary, R.; Subramani, S. A review on management of chrome-tanned leather shavings: A holistic paradigm to combat the environmental issues. *Environ. Sci. Pollut. Res.* **2014**, *21*, 11266–11282. [CrossRef] [PubMed]

56. Shanmugam, P.; Horan, N.J. Optimising the biogas production from leather fleshing waste by co-digestion with MSW. *Bioresour. Technol.* **2009**, *100*, 4117–4120. [CrossRef] [PubMed]

57. Priebe, G.P.S.; Kipper, E.; Gusmao, A.L.; Marcilio, N.R.; Gutterres, M. Anaerobic digestion of chrome-tanned leather waste for biogas production. *J. Clean. Prod.* **2016**, *129*, 410–416. [CrossRef]

58. Agustini, C.; da Costa, M.; Gutterres, M. Biogas production from tannery solid wastes—Scale-up and cost saving analysis. *J. Clean. Prod.* **2018**, *187*, 158–164. [CrossRef]

59. Agustini, C.B.; Spier, F.; da Costa, M.; Gutterres, M. Biogas production for anaerobic co-digestion of tannery solid wastes under presence and absence of the tanning agent. *Resour. Conserv. Recycl.* **2018**, *130*, 51–59. [CrossRef]

60. Diamantis, V.; Eftaxias, A.; Stamatelatou, K.; Noutsopoulos, C.; Vlachokostas, C.; Aivasidis, A. Bioenergy in the era of circular economy: Anaerobic digestion technological solutions to produce biogas from lipid-rich wastes. *Renew. Energy* **2021**, *168*, 438–447. [CrossRef]

61. Puhazhselvan, P.; Aparna, R.; Ayyadurai, N.; Gowthaman, M.K.; Saravanan, P.; Kamini, N.R. Enzyme based cleaner process for enhanced recovery of lipids from tannery fleshing waste. *J. Clean. Prod.* **2017**, *144*, 187–191. [CrossRef]

62. Stefan, D.S.; Manea-Saghin, A.M.; Meghea, A.; Stefan, M. Biodegradation of composite fertilizers in aerobic aqueous and composting conditions. In Proceedings of the 19th International Multidisciplinary Scientific GeoConference SGEM, Albena, Bulgaria, 30 June–6 July 2019; Volume 19, pp. 49–56.

63. Ockerman, H.W.; Hansen, E.I. *Animal By-Product Processing*, 1st ed.; VCH: Wienheim, Germany, 1988; pp. 89–131.

64. Tahiri, S.; Albizane, A.; Messaoudi, A.; Azzi, M.; Bennazha, J.; Younssi, S.A.; Bouhria, M. Thermal behaviour of chrome shavings and of sludges recovered after digestion of tanned solid wastes with calcium hydroxide. *Waste Manag.* **2007**, *27*, 89–95. [CrossRef] [PubMed]

65. Cabeza, L.F.; Taylor, M.M.; DiMaio, G.L.; Brown, E.M.; Marmer, W.N.; Carrió, R.; Celma, P.J.; Cot, J. Processing of leather waste: Pilot scale studies on chrome shavings. Isolation of potentially valuable protein products and chromium. *Waste Manag.* **1998**, *18*, 211–218. [CrossRef]

66. Famielec, S. Chromium Concentrate Recovery from Solid Tannery Waste in a Thermal Process. *Materials* **2020**, *13*, 1533. [CrossRef] [PubMed]

67. El Boushy, A.R.; van der Poel, A.F.B.; Koene, J.I.A.; Dieleman, S.H. Tanning waste by-product from cattle hides, its suitability as a feedstuff. *Bioresour. Technol.* **1991**, *35*, 321–323. [CrossRef]

68. Wang, L.; Li, J.; Jin, Y.; Chen, M.; Luo, J.; Zhu, X.; Zhang, Y. Study on the removal of chromium(III) from leather waste by a two-step method. *J. Ind. Eng. Chem.* **2019**, *79*, 172–180. [CrossRef]

69. Teles, F.R.R.; Cabral, J.M.S.; Santos, J.A.L. Enzymatic degreasing of a solid waste from the leather industry by lipases. *Biotechnol. Lett.* **2001**, *23*, 1159–1163. [CrossRef]

70. Stefan, D.S.; Dima, R.; Pantazi, M.; Ferdes, M.; Meghea, A. Identifying Microorganisms Able to Perform Biodegradation of Leather Industry Waste. *Mol. Cryst. Liq. Cryst.* **2012**, *556*, 301–308. [CrossRef]

71. Ricard-Blum, S. The Collagen Family. *Cold Spring Harb. Perspect. Biol.* **2011**, *3*, 1–19. [CrossRef] [PubMed]

72. Walters, B.D.; Stegemann, J.P. Strategies for directing the structure and function of three-dimensional collagen biomaterials across length scales. *Acta Biomater.* **2014**, *10*, 1488–1501. [CrossRef]

73. Shoulders, M.D.; Raines, R.T. Collagen Structure and Stability. *Annu. Rev. Biochem.* **2009**, *78*, 929–958. [CrossRef] [PubMed]

74. Ho, E.; Lowman, A.; Marcolongo, M. Synthesis and Characterization of an Injectable Hydrogel with Tunable Mechanical Properties for Soft Tissue Repair. *Biomacromolecules* **2006**, *7*, 3223–3228. [CrossRef]

75. Rutz, A.L.; Shah, R.N. *Protein Based Hydrogels in Polymeric Hydrogels as Smart Biomaterials*, 1st ed.; Kalia, S., Ed.; Springer International Publishing: Cham, Switzerland, 2016; pp. 73–104.

76. Ramakrishna, S.; Mayer, J.; Wintermantel, E.; Leong, K.W. Biomedical applications of polymer-composite materials: A review. *Compos. Sci. Technol.* **2001**, *61*, 1189–1224. [CrossRef]

77. Horue, M.; Berti, I.R.; Cacicedo, M.L.; Castro, G.R. Microbial production and recovery of hybrid biopolymers from wastes for industrial applications—A review. *Bioresour. Technol.* **2021**, *340*, 125671. [CrossRef]

78. Sionkowska, A. Collagen blended with natural polymers: Recent advances and trends. *Prog. Polym. Sci.* **2021**, *122*, 101452. [CrossRef]

79. Ucar, B. Natural biomaterials in brain repair: A focus on collagen. *Neurochem. Int.* **2021**, *146*, 105033. [CrossRef]

80. Stefan, D.S.; Zainescu, G.; Manea-Saghin, A.M.; Triantaphyllidou, I.E.; Tzoumani, I.; Tatoulis, T.I.; Syriopoulos, G.T.; Meghea, A. Collagen-Based Hydrogels Composites from Hide Waste to Produce Smart Fertilizers. *Materials* **2020**, *13*, 4396. [CrossRef] [PubMed]

81. Xu, S.; Li, X.; Wang, Y.; Hu, Z.; Wang, R. Characterization of slow-release collagen-g-poly(acrylic acid-co-2-acrylamido-2-methyl-1-propane sulfonic acid)–iron(III) superabsorbent polymer containing fertilizer. *J. Appl. Polym. Sci.* **2019**, *136*, 47178. [CrossRef]

82. Teramoto, N.; Hayashi, A.; Yamanaka, K.; Sakiyama, A.; Nakano, A.; Shibata, M. Preparation and Mechanical Properties of Photo-Crosslinked Fish Gelatin/Imogolite Nanofiber Composite Hydrogel. *Materials* **2012**, *5*, 2573–2585. [CrossRef]

83. Liguori, A.; Uranga, J.; Panzavolta, S.; Guerrero, P.; de la Caba, K.; Focarete, M.L. Electrospinning of Fish Gelatin Solution Containing Citric Acid: An Environmentally Friendly Approach to Prepare Crosslinked Gelatin Fibers. *Materials* **2019**, *12*, 2808. [CrossRef] [PubMed]

84. Tzoumani, I.; Lainioti, G.C.; Aletras, A.J.; Zainescu, G.; Stefan, S.; Meghea, A.; Kallitsis, J.K. Modification of Collagen Derivatives with Water-Soluble Polymers for the Development of Cross-Linked Hydrogels for Controlled Release. *Materials* **2019**, *12*, 4067. [CrossRef] [PubMed]

85. Masilamani, D.; Madhan, B.; Shanmugam, G.; Palanivel, S.; Narayan, B. Extraction of collagen from raw trimming wastes of tannery: A waste to wealth approach. *J. Clean. Prod.* **2016**, *113*, 338–344. [CrossRef]

86. Gousterova, A.; Nustorova, M.; Christov, P.; Nedkov, P.; Neshev, G.; Vasileva-Tonkova, E. Development of a biotechnological procedure for treatment of animal wastes to obtain inexpensive biofertilizer. *World J. Microbiol. Biotechnol.* **2008**, *24*, 2647–2652. [CrossRef]

87. Constantinescu, R.R.; (National Research & Development Institute for Textiles and Leather, Bucharest, Romania); Ignat, M.; (National Research & Development Institute for Textiles and Leather, Bucharest, Romania). Personal communication, 2013.

88. Majee, S.; Halder, G.; Mandal, T. Formulating nitrogen-phosphorous-potassium enriched organic manure from solid waste: A novel approach of waste valorization. *Process. Saf. Environ. Prot.* **2019**, *132*, 160–168. [CrossRef]

89. Majee, S.; Halder, G.; Mandal, D.D.; Tiwari, O.N.; Mandal, T. Transforming wet blue leather and potato peel into an eco-friendly bio-organic NPK fertilizer for intensifying crop productivity and retrieving value-added recyclable chromium salts. *J. Hazard. Mater.* **2021**, *411*, 125046. [CrossRef] [PubMed]

90. Lima de Oliveira, D.Q.; Gomes Carvalho, K.T.; Ribeiro Bastos, A.R.; Alves de Oliveira, L.C.; João José Granate de Sá e Melo Marques, J.J.; Severina de Melo Pereira do Nascimento, R. Use of leather industry residues as nitrogen sources for elephantgrass. *Rev. Bras. Ciênc. Solo* **2008**, *32*, 417–424.

91. Lima, D.Q.; Oliveira, L.C.A.; Bastos, A.R.R.; Carvalho, G.S.; Marques, J.G.S.M.; Carvalho, J.G.; de Souza, G.A. Leather Industry Solid Waste as Nitrogen Source for Growth of Common Bean Plants. *Appl. Environ. Soil Sci.* **2010**, *2010*, 703842. [CrossRef]

92. Chiroma, T.M.; Ebewele, R.O.; Hymore, F.K. Comparative Assessment of Heavy Metal Levels In Soil, Vegetables and Urban Grey Waste Water Used for Irrigation in Yola And Kano. *Int. Ref. J. Eng. Sci.* **2014**, *3*, 1–9.

93. Constantinescu, R.R.; Zainescu, G.; Stefan, D.S.; Sirbu, C.; Voicu, P. Protein biofertilizer development and application on soybean cultivated degraded soil. *Leather Footwear, J.* **2015**, *15*, 169–178. [CrossRef]

94. Zainescu, G.; Albu, L.; Constantinescu, R.R. Study of Collagen Hydrogel Biodegradability over Time. *Rev. Chim.* **2018**, *69*, 101. [CrossRef]

95. Nogueira, F.G.E.; do Prado, N.T.; Oliveira, L.C.A.; Bastos, A.R.R.; Lopes, J.H.; Carvalho, J.G. Incorporation of mineral phosphorus and potassium on leather waste (collagen): A new NcollagenPK-fertilizer with slow liberation. *J. Hazard. Mater.* **2010**, *176*, 374–380. [CrossRef] [PubMed]

96. Nair, L.S.; Laurencin, C.T. Biodegradable polymers as biomaterials. *Progr. Polym. Sci.* **2007**, *32*, 762–798. [CrossRef]

97. Du, C.W.; Zhou, J.M.; Shaviv, A. Release Characteristics of Nutrients from Polymer-coated Compound Controlled Release Fertilizers. *J. Polym. Environ.* **2006**, *14*, 223–230. [CrossRef]

98. Zainescu, G.; Constantinescu, R.R.; Sirbu, C. Smart Hydrogels with Collagen Structure Made of Pelt Waste. *Rev. Chim.* **2017**, *68*, 393. [CrossRef]

99. Vroman, I.; Tighzert, L. Biodegradable Polymers. *Materials* **2009**, *2*, 307–344. [CrossRef]

100. Majeed, Z.; Ramli, N.K.; Mansor, N.; Man, Z. A comprehensive review on biodegradable polymers and their blends used in controlled release fertilizer processes. *Rev. Chem. Eng.* **2015**, *31*, 69–95. [CrossRef]

101. Zuriaga-Agustí, E.; Galiana-Aleixandre, M.V.; Bes-Pia, A.; Mendoza-Roca, J.A.; Risueno-Puchades, V.; Segarra, V. Pollution reduction in an eco-friendly chrome-free tanning and evaluation of the biodegradation by composting of the tanned leather wastes. *J. Clean. Prod.* **2015**, *87*, 874–881.

102. China, C.R.; Hilonga, A.; Nyandoro, S.S.; Schroepfer, M.; Kanth, S.V.; Meyer, M.; Njau, K.N. Suitability of selected vegetable tannins traditionally used in leathermaking in Tanzania. *J. Clean. Prod.* **2020**, *251*, 119687. [CrossRef]

103. Carsote, C.; Sendrea, C.; Micu, M.C.; Adams, A.; Badea, E. Micro-DSC, FTIR-ATR and NMR MOUSE study of the dose-dependent effects of gamma irradiation on vegetable-tanned leather: The influence of leather thermal stability. *Radiat. Phys. Chem.* **2021**, *189*, 109712. [CrossRef]

104. Pradeep, S.; Sundaramoorthy, S.; Sathish, M.; Jayakumar, G.C.; Rathinam, A.; Madhan, B.; Saravanan, P.; Rao, J.R. Chromium-free and waterless vegetable-aluminium tanning system for sustainable leather manufacture. *Chem. Eng. J. Adv.* **2021**, *7*, 100108. [CrossRef]

105. Shi, J.; Zhang, R.; Mi, Z.; Lyu, S.; Ma, J. Engineering a sustainable chrome-free leather processing based on novel lightfast wet-white tanning system towards eco-leather manufacture. *J. Clean. Prod.* **2021**, *282*, 124504. [CrossRef]

106. Stefan, D.S.; Meghea, I.; Apetroaei, M.R. Study of biodegradation of leather tanning with chromium and vegetal compounds. In Proceedings of the 11th International Multidisciplinary Scientific GeoConference SGEM2011, Albena, Bulgaria, 20–25 June 2011; Volume 3, pp. 221–228.

107. Dhayalan, K.; Fathima, N.N.; Gnanamani, A.; Rao, J.R.; Nair, B.U.; Ramasami, T. Biodegradability of leathers through anaerobic pathway. *Waste Manag.* **2001**, *27*, 760–767.

108. Ge, J.; Yu, H.; Zhong, W.; Li, W.; Yu, T. Study on the utilization of biodegradable polyurethane material from the bark of Acacia mearnsii (I). Coating material of controlled slow-release fertilizer. *J. Funct. Polym.* **1998**, *11*, 478–482.

109. Chen, L.; Xie, Z.; Zhuang, X.; Chen, X.; Jing, X. Controlled release of urea encapsulated by starch-g-poly(l-lactide). *Carbohydr. Polym.* **2008**, *72*, 342–348. [CrossRef]

110. Han, X.; Chen, S.; Hu, X. Controlled-release fertilizer encapsulated by starch/polyvinyl alcohol coating. *Desalination* **2009**, *240*, 21–26. [CrossRef]
111. Celli, A.; Sabaa, M.W.; Jyothi, A.N.; Kalia, S. Chitosan and Starch-Based Hydrogels Via Graft Copolymerization. In *Polymeric Hydrogels as Smart Biomaterials*; Springer Series on Polymer and Composite Materials; Kalia, S., Ed.; Springer: Cham, Switzerland, 2016; pp. 189–234.
112. Mahkam, M. Synthesis and Characterization of Polymer/Silica Composite for Colon-Specific Drug Delivery. *Int. J. Polym. Mater.* **2011**, *60*, 456–468. [CrossRef]
113. Korsmeyer, R.W.; Gumy, R.; Doelker, E.; Buri, P.; Peppas, N.A. Mechanisms of solute release from porous hydrophilic polymers. *Int. J. Pharm.* **1983**, *15*, 25–35. [CrossRef]
114. Lee, P.I. Kinetics of drug release from hydrogel matrices. *J. Control. Release* **1985**, *2*, 277–288. [CrossRef]
115. Li, J.; Mooney, D.J. Designing hydrogels for controlled drug delivery. *Nat. Rev. Mater.* **2016**, *1*, 16071. [CrossRef] [PubMed]
116. Nešovic, K.; Jankovic, A.; Peric-Grujic, A.; Vukašinovic-Sekulic, M.; Radetic, T.; Živkovic, L.; Park, S.J.; Rheed, K.Y.; Miškovic-Stankovic, V. Kinetic models of swelling and thermal stability of silver/poly(vinyl alcohol)/chitosan/graphene hydrogels. *J. Ind. Eng. Chem.* **2019**, *77*, 83–96. [CrossRef]
117. Pareek, A.; Maheshwari, S.; Cherlo, S.; Thavva, R.S.R.; Runkana, V. Modeling drug release through stimuli responsive polymer hydrogels. *Int. J. Pharm.* **2017**, *532*, 502–510. [CrossRef] [PubMed]
118. Goh, K.B.; Li, H.; Lam, K.Y. Development of a multiphysics model to characterize the responsive behavior of urea-sensitive hydrogel as biosensor. *Biosens. Bioelectron.* **2017**, *91*, 673–679. [CrossRef] [PubMed]
119. Lin, C.C.; Metters, A.T. Hydrogels in controlled release formulations: Network design and mathematical modeling. *Adv. Drug Deliv. Rev.* **2006**, *58*, 1379–1408. [CrossRef]
120. Peppas, N.A.; Bures, P.; Leobandung, W.; Ichikawa, H. Hydrogels in pharmaceutical formulations. *Eur. J. Pharm. Biopharm.* **2000**, *50*, 27–46. [CrossRef]
121. Caccavo, D. An overview on the mathematical modeling of hydrogels' behavior for drug delivery systems. *Int. J. Pharm.* **2019**, *560*, 175–190. [CrossRef] [PubMed]
122. Sauerwein, M.; Steeb, H. Modeling of dynamic hydrogel swelling within the pore space of a porous medium. *Int. J. Eng. Sci.* **2020**, *155*, 103353. [CrossRef]
123. Chun, S.W.; Kim, J.D. Swelling and Deswelling Transition of Water-Soluble Poly(N-isopropylacrylamide) by a Method of Blob Rescaling. *Korean J. Chem. Eng.* **2002**, *19*, 803–807. [CrossRef]
124. Poschlad, K.; Enders, S. Thermodynamics of aqueous solutions containing poly (N-isopropylacrylamide). *J. Chem. Thermodynamics* **2011**, *43*, 262–269. [CrossRef]
125. Althans, D.; Langenbach, K.; Enders, S. Influence of different alcohols on the swelling behaviour of hydrogels. *Mol. Phys.* **2012**, *110*, 1391–1402. [CrossRef]
126. Sheth, S.; Barnard, E.; Hyatt, B.; Rathinam, M.; Zustiak, S.P. Predicting Drug Release from Degradable Hydrogels Using Fluorescence Correlation Spectroscopy and Mathematical Modeling. *Front. Bioeng. Biotechnol.* **2019**, *7*, 410. [CrossRef]
127. Sauerwein, M.; Steeb, H. A modified effective stress principle for chemical active multiphase materials with internal mass exchange. *Geomech. Energy Environ.* **2018**, *15*, 19–34. [CrossRef]
128. Zhi, D.; Huang, Y.; Hu, S.; Liu, H.; Hu, Y. Molecular thermodynamic model for swelling behavior and volume phase transition of multi-responsive hydrogels. *Fluid Phase Equilibria.* **2011**, *312*, 106–115. [CrossRef]
129. Bayat, M.R.; Dolatabadi, R.; Baghani, M. Transient swelling response of pH-sensitive hydrogels: A monophasic constitutive model and numerical implementation. *Int. J. Pharm.* **2020**, *577*, 119030. [CrossRef]

Article

Development of Oil and Gas Stimulation Fluids Based on Polymers and Recycled Produced Water

Mustafa AlKhowaildi [1], Bassam Tawabini [2,*], Muhammad Shahzad Kamal [2,*], Mohamed Mahmoud [2,*], Murtada Saleh Aljawad [2] and Mohammed Bataweel [1]

[1] Advanced Research Center, Saudi Aramco, Dhahran 31311, Saudi Arabia; mustafa.alkhowaildi@aramco.com (M.A.); batawema@yahoo.com (M.B.)

[2] College of Petroleum Engineering and Geosciences, King Fahad University of Petroleum and Minerals, Dhahran 31261, Saudi Arabia; mjawad@kfupm.edu.sa

* Correspondence: bassamst@kfupm.edu.sa (B.T.); shahzadmalik@kfupm.edu.sa (M.S.K.); mmahmoud@kfupm.edu.sa (M.M.)

Abstract: Freshwater scarcity is a highly pressing and accelerating issue facing our planet. Therefore, there is a great incentive to develop sustainable solutions by reusing wastewater or produced water (PW), especially in places where it is generated abundantly. PW represents the water produced as a by-product during oil and gas extraction operations in the petroleum industry. It is the largest wastewater stream within the industry, with hundreds of millions of produced water barrels per day worldwide. This research investigates a reuse opportunity for PW to replace freshwater utilization in well stimulation applications. Introducing an environmentally friendly chelating agent (GLDA) allowed formulating a PW-based fluid system that has similar rheological properties in fresh water. This work aims at evaluating the rheological properties of the developed stimulation fluid. The thickening profile of the fluid was controlled by chelation chemistry and varying different design parameters. The experiments were carried out using a high-pressure, high-temperature (HPHT) viscometer. Variables such as polymer concentration and pH have a great impact on the viscosity, while temperature and concentration of the chelating agents are shown to control the thickening profile, as well as its stability and breakage behaviors. Furthermore, 50 pptg of carboxymethyl hydroxypropyl guar (CMHPG) polymer in 20 wt.% chelating solution was shown to sustain 172 cP viscosity for nearly 2.5 h at 150 °F and 100 S^{-1} shear rate. The newly developed fluid system, solely based on polymer, chelating agent, and PW, showed great rheological capabilities to replace the conventional stimulation fluids based on fresh water. The newly developed fluid can also have economic value realization due to fewer additives, compared with conventional fluids.

Keywords: polymers; stimulation fluid; oilfield produced water; chelating agents; water sustainability

Citation: AlKhowaildi, M.; Tawabini, B.; Kamal, M.S.; Mahmoud, M.; Aljawad, M.S.; Bataweel, M. Development of Oil and Gas Stimulation Fluids Based on Polymers and Recycled Produced Water. *Polymers* 2021, 13, 4017. https://doi.org/10.3390/polym13224017

Academic Editor: Antonio Zuorro

Received: 15 October 2021
Accepted: 15 November 2021
Published: 20 November 2021

Publisher's Note: MDPI stays neutral with regard to jurisdictional claims in published maps and institutional affiliations.

1. Introduction

Water scarcity and depletion of freshwater resources are a global concern and among the most predominant environmental challenges of the 21st century. One of the key challenges is the enormous water consumption of humans. As the hydrological cycle is tightly connected with climate change, these changes will significantly affect the quality and availability of water. Concerns over the impact and consequences of climate change often dominate discussions on water challenges by both scientists and policymakers. The widespread water crisis is mostly linked to growing populations and the extensive consumption of water [1]. Water scarcity is a highly pressing issue, as highlighted by the World Economic Forum in its Global Risks 2019 report. It is thought to have one of the highest impacts and most likely risks facing the planet. Water stress is defined as the ratio of total water withdrawals for different consumption purposes by a country relative to the available renewable surface water [2]. The severe water-scarcity threshold set by the United Nations is 500 cubic meters per capita per year [3]. According to the World Resources

Institute, 17 countries, mostly in the Middle East and North Africa (MENA) region, face the risk of extremely high water stress. All Gulf Cooperation Council (GCC) countries (Saudi Arabia, Kuwait, United Arab Emirates, Oman, Qatar, and Bahrain) are classified as water-scarce nations. Due to the severe scarcity of water resources, the MENA region will be the most liable to climate effects on water resources [4].

In the oil and gas industry, freshwater consumption is rising across different productivity enhancement operations such as fracking [5,6]. Each well treated could consume 0.5 million to 6 million gallons of fresh water, depending on the well type and extent of treatment. These amounts are usually extracted from nearby groundwater aquifer wells. An alternative source is to recycle the enormous amount of generated produced water from the oil and gas industry to suffice some of the industry's own water-needing operations. In exploration and production (E&P), some of the processes associated with hydrocarbon recovery require a massive quantity of fresh water. For example, the water usage in hydraulic fracturing operations in 2010 in the US was estimated to be between 70 to 140 billion gallons of water [7]. Thus, reusing produced water in well stimulation operations has emerged as a win–win proposition, transforming the industry's biggest waste product into a resource, with added benefits of reducing the environmental footprint [8]. However, this process involves some technical challenges that need to be addressed to formulate a fluid system that meets the performance requirements. Some of these challenges include sustaining high viscosity at certain shear rates (i.e., 200 cP at a shear rate of 100 1/s for a minimum of 2 h), having no precipitation or suspended solids, and the ability to carry fracturing sands (elasticity properties).

Natural gas, an alternative energy source with a low carbon footprint, is often trapped inside pores of extremely low permeable rocks, which require formation stimulation/treatment for commercial production. Over the past decade, technologies such as horizontal drilling and hydraulic fracturing have enabled the excessive growth of the natural gas industry (i.e., shale gas) [9]. Well stimulation techniques are categorized into two main types: hydraulic fracturing and matrix stimulation. The hydraulic fracturing process involves an injection of pressurized fluid into the wellbore to create cracks within subsurface rock formation through which natural gas flows freely. Matrix stimulation is a process that involves pumping acid nearby the wellbore region to dissolve minerals that could hinder the well's productivity. These fluids are generally water based, comprising 99% water and 1% chemical additives to meet the required properties of the fluid [10]. As mentioned before, these stimulation fluids must have the following characteristics: high proppant carrying capacity (viscosity), low pipe friction, low fluid loss (fluid efficiency), easy preparation, and easy removal from the treated reservoir (clean up).

Fluid's viscosity can be increased using a gelling agent, guar gum derivatives are usually used, such as hydroxypropyl guar (HPG) and carboxymethyl hydroxypropyl guar (CMHPG). Less residue and faster hydration are achieved using these gelling agents; however, they are sensitive to salts and solids content in the source water [11]. Enormous amounts of fresh water are normally used to prepare these fluids; however, finding an alternative (produced water) to fresh water needs the introduction of new chemicals. Chelating agents have been used for a variety of different applications across the oil and industry; however, their effect in thickening and breaking the stimulation fluid is in its infancy. This paper provided a fluid formulation exhibiting optimum rheology for stimulation applications in oil and gas. Replacing fresh water with produced water (PW) containing high total dissolved solids (TDS) will introduce technical challenges, one of which is achieving the required fluid rheology (i.e., 200 cP viscosity under $100 \, \text{s}^{-1}$ shear rate, for 2 h, under high-pressure, high-temperature (HPHT) conditions). This paper investigates the overall effect of polymeric gel (CMHPG) when mixed with a chelating agent L-glutamic acid-N,N-diacetic acid (GLDA) to determine the optimum concentrations of different components. GLDA chelating agent is preferred over other types of chelating agents because it has a wide range of solubility in different waters at different pH values. Compared with ethylenediaminetetraacetic acid (EDTA) and diethylenetriaminepentaacetic acid (DTPA) chelating agents, GLDA is very soluble in acidic medium, and it is the most stable chelating agent in high salinity water. GLDA is readily biodegradable and environmentally friendly, compared with other chelating agents, because it has one nitrogen atom, which is responsible for biodegradation [12]. Previous work conducted with EDTA and DTPA chelating agents showed that both chelates do not have the capability of breaking the polymers because they formed very stable components. GLDA has the ability to thicken the polymer viscosity, and at the same time, it can break the polymer backbone based on its ph and concentration at different temperatures [13]. DTPA and EDTA chelating agents are very common in oil and gas industry applications, but they have limited solubility at low ph values and cannot handle high salinity water such as produced water [14].

Produced water accounts for approximately 98% of the total generated waste volume by oilfield operations in the United States. According to the American Petroleum Institute (API), around 18 billion barrels of produced water were generated in 1995 by US onshore operations alone. Additional large volumes of produced water are generated by US offshore operations and from thousands of additional wells in other countries worldwide [15]. Khatib et al. [16] estimated that around 77 billion bbl of produced water was generated worldwide in 1999. Dickhout et al. [17] estimated that more than 70 billion bbl of produced water was generated in 2009, of which the United States generates 21 billion bbl. PW has a complex chemical characteristic that consists of many inorganic and organic compounds. Table 1 summarizes a general range of constituent concentrations found in produced water gathered from the literature [18].

Table 1. Generic PW Composition [18]. Reprinted with permission from ref [18]. 2019 Elsevier.

Parameter	Concentration (mg/L)
Major Parameters	
TDS (Total dissolved solids)	100–400,000
TSS (Total suspended solids)	1.2–1000
COD (Chemical oxygen demand)	1220–2600
TOC (Total organic content)	0–1500
Total organic acids	0.001–10,000
Chemicals Additives	
Glycol	7.7–2000
Corrosion inhibitor	0.3–10
Scale inhibitor	0.2–30
BTEX	
Benzene	0.032–778.51
Toluene	0.058–5.86
Ethylbenzene	0.026–399.84
Xylene	0.01–1.29
Other pollutants	
Saturated hydrocarbons	17–30
Total oil and grease	2–560
Phenol	0.001–10,000
Metals	
Na	0–150,000
Sr	0–6250
Zn	0.01–35

Table 1. *Cont.*

Parameter	Concentration (mg/L)
Li	0.038–64
Al	0.4–410
As	0.002–11
Ba	0–850
Cr	0.002–1.1
Fe	0.1–1100
Mn	0.004–175
K	24–4300
Other ions	
B	5–95
Ca^{2+}	0–74,000
$SO4^{2-}$	0–15,000
Mg^{2+}	8–6000
$HCO3^{-}$	77–3990
Cl^{-}	0–270,000

Produced water contains various species of salts; the amount of ionic composition differs from one source of produced water to another. Examining the rheological performance then becomes important [19]. Several chelating agents are known for their compatibility in stimulation fluid to capture ions; however, their stability under harsh conditions and impact on the formation is crucial. In our previous work, we found that polymer dissolved in seawater and chelating agent proved extremely effective in leaving minimal formation damage and exhibiting favorable characteristics in fracking fluids, such as requiring no breaker and showing excellent stability under high temperatures [13,20]. This paper will instead focus on the produced water as means for recycling, which, as found in the literature above, can have a distinctively different composition than seawater.

2. Materials and Methods

2.1. Materials

Sodium hydroxide (NaOH) in the form of solid pellets was received from Sigma Aldrich. Low pH GLDA (3–4) with an active agent concentration of 40 wt % was manufactured by Nouryon, The Netherlands. High pH GLDA (11–12) with an active agent concentration of 47 wt% was also manufactured by Nouryon, The Netherlands. The GLDA used as a chelating agent in this study is readily biodegradable and environmentally friendly. The chemical structure is shown in Figure 1a.

Carboxymethyl hydroxypropyl guar (CMHPG) polymer was used as a gelling agent and supplied by a service provider company. CMHPG is a biopolymer, a guar gum derivative that can be sustainable at high-temperature conditions (300 °F), and it is widely used in the industry. The selection of this polymer type was based on practicality and ease of deployment in industrial applications. The chemical structure is shown in Figure 1b.

2.2. Simulated Produced Water

It is a widely accepted notion that PW composition varies considerably from a geographic place to another, from one field to another, and, in some cases, from one well to another. Table 2 shows a generic representation of PW composition, consisting of 70,000 ppm total dissolved ions. The produced water used in this study was synthesized using this composition.

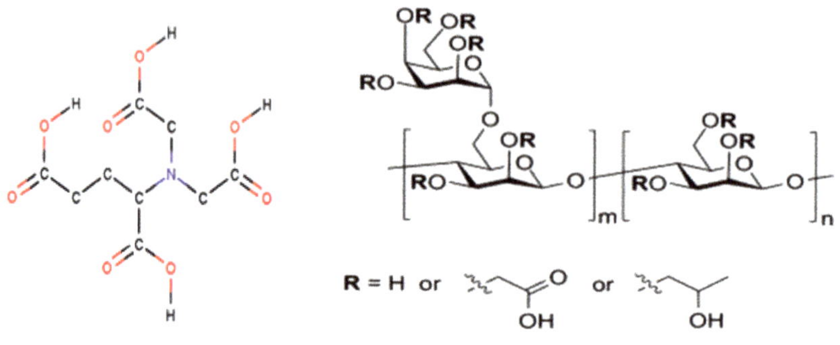

(a)   (b)

Figure 1. Chemical structure of low pH GLDA (**a**) and chemical structure of CMHPG (**b**).

Table 2. Salt composition in synthesized PW.

Salts	g/L
NaCl, g/L	48.6
$CaCl_2 \cdot 2H_2O$, g/L	22
$MgCl_2 \cdot 6H_2O$, g/L	8.4

2.3. Viscosity Measurement

Rheological measurements were conducted using two different viscometer apparatuses. Fann Model 35 was used to measure fluid viscosity at ambient conditions. Chandler Model 5550 HPHT viscometer was used to assess the rheological profile of the developed fluid at elevated conditions.

The main objective of this work was to characterize the rheological properties of the developed stimulation fluid at different conditions. It is crucial to understand the effect of the various parameters controlling the rheological profiles of the fluid and its effectiveness in carrying out specific application functions, which is stimulation of oil and gas wells in this study. The fluid was formulated using different concentrations of the chelating agent (GLDA) and polymer (CMHPG) diluted in synthesized produced water. The fluid was then subjected to HPHT conditions at various shear rates to study the stability and rheological properties of the developed fluid. The effect of pH, additives concentration, shear rates, and temperature was studied. For the pH, three different systems were tested to investigate the fluid's versatility under different pH systems: acidic, neutral, and alkaline (pH = 4, 8, and 12). The conditions for shear rate were chosen at high and low shear rate values to capture the fluid's movement inside the wellbore of the well (low shear rate) and then inside the formation and fractures (high shear rate), the values of shear rate were 100, 170 and 511 s^{-1}. The additives concentration values were chosen based on the industry's current practices and reported studies in the literature covering detailed studies on each additive. The temperature values ranged from room temperature to high gas-well temperatures in common fields (300 °F).

3. Results and Discussion

3.1. Effect of Chelating Agent's Presence

Initial baseline experiments were performed to establish an understanding of how the polymer hydrates in different water systems. It is well understood that guar gum polymers hydrate better in freshwater systems and that dissolved ions tend to hinder the thickening behavior. This was reaffirmed with clear observation in the lab using our 70 k TDS PW and CMHPG polymer. In Figure 2, deionized water was mixed with 30 lb/1000 gal CMHPG, showing an apparent viscosity of 54 cP at 171 s^{-1}, while the viscosity of the polymer dissolved in PW was 42 cP at the same shear rate and polymer concentration. Approximately a 22% reduction in apparent viscosity due to the presence of TDS in the water was observed. The experiment was conducted at multiple polymer concentrations, and the reduction in apparent viscosity remained visible, however, with varying percentages. The hydration of the polymer is, therefore, a function of to what extent are ions present in the water phase. It is well understood that water freshness controls the effectiveness of gelation buildup in this polymer type. This is due to the charge screening effect when salts are added to the solution. The CMHPG polymer is anionic in nature due to the presence of carboxymethyl groups at various points of the polymer chain. In deionized water or fresh water, the negatively charged carboxymethyl groups will repel each other, which results in large hydrodynamic volume and higher viscosity. However, the addition of produced water brings cations in the solution and charge screening, resulting in less hydrodynamic volume and lower viscosity [21].

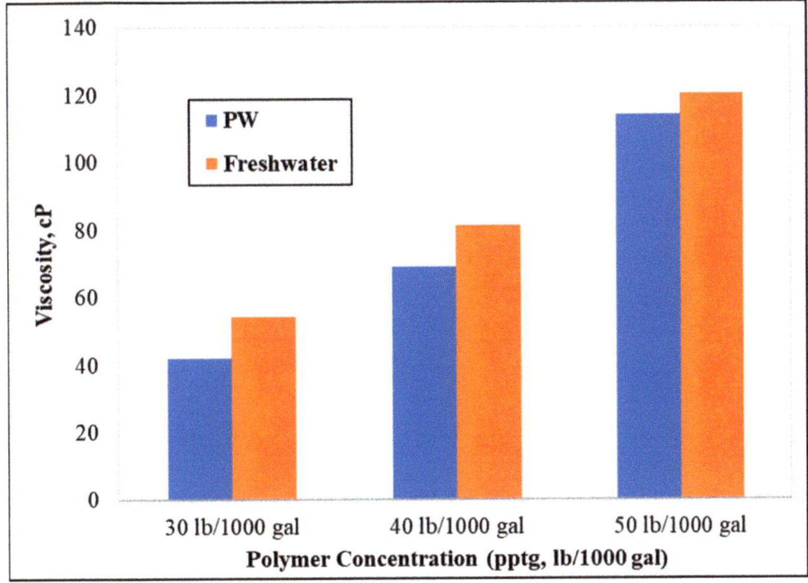

Figure 2. Viscosity measurements of CMHPG polymer in DI and PW (shear rate = 171 s^{-1}, pH = 7, temperature = 77 °F).

In another baseline experiment, 10 wt.% of GLDA was introduced to a 70 k TDS PW and hydrated with the same amount of CMHPG polymer (40 lb/1000 gal). The test was conducted at ambient conditions and a shear rate of 170 s^{-1}. The results showed an increase in the apparent viscosity in the solution containing 10 wt.% GLDA, indicating a strong effect in minimizing the TDS. As shown in Figure 3, with the 10 wt.% GLDA, the viscosity was 93 cP, a 27% higher than the solutions without GLDA. These results align with the literature in specifying GLDA as having an affinity to chelate Ca^{+2} and Mg^{+2} ions in the water, therefore allowing more free water to hydrate the polymer.

In another experiment, the effect of GLDA (10 wt.%) addition was observed at different shear rates. The test was performed at 200 °F, 500 psi, and at a shear rate of 100 s^{-1}. The addition of 10 wt.% GLDA showed a strong water-softening effect on the 70 k ppm TDS PW as well as a noticeable thickening behavior of CMHPG polymer. At elevated temperatures, the stability was different with GLDA, compared with the solution without GLDA. As shown in Figure 4, the solution containing 10 wt.% GLDA stayed stable at around 73 cP for 25 min, while the polymer solution without GLDA degraded to 64 cP at the same time.

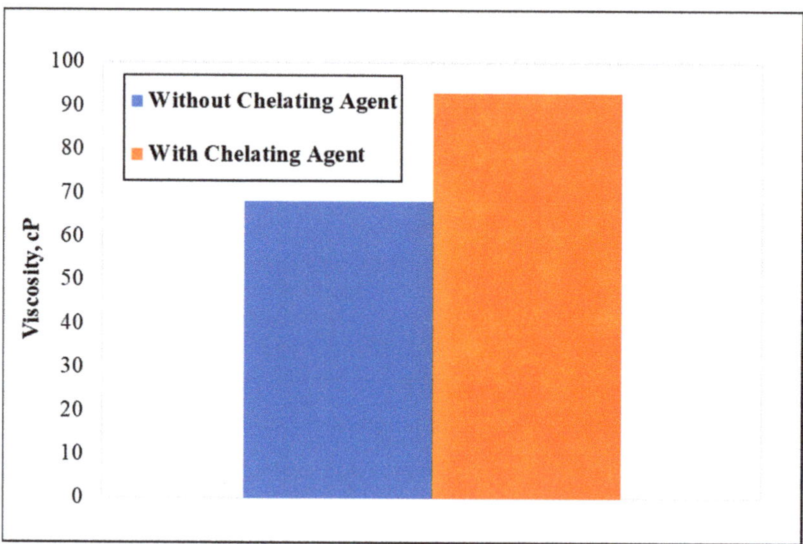

Figure 3. The effect of GLDA addition (10 wt%) on the viscosity (polymer concentration = 40 pptg, shear rate = 171 s^{-1}, temperature= 77 °F, water salinity = 70 k).

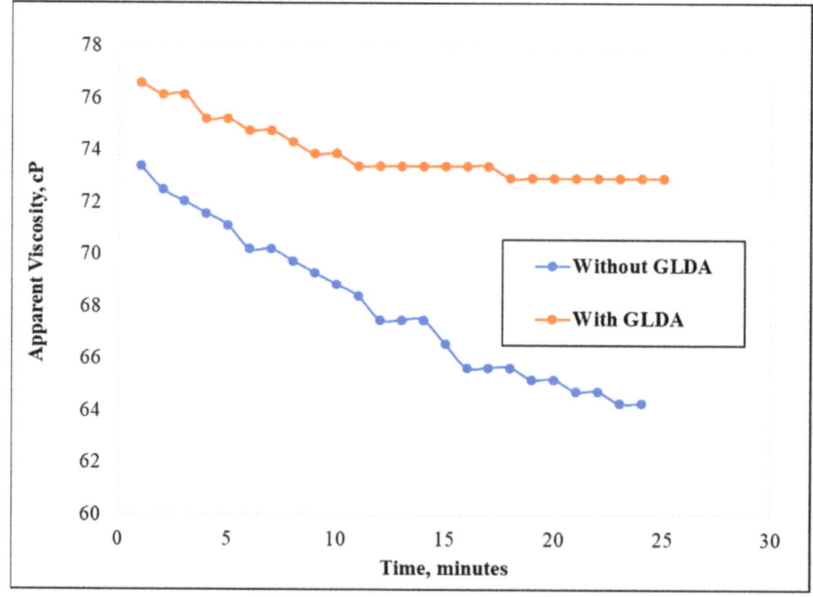

Figure 4. Viscosity measurement over time for samples with and without GLDA (temperature = 200 °F, shear rate = 100 s^{-1}, pH = 8, CMHPG concentration = 50 pptg).

3.2. *Effect of GLDA on Water Softening*

The water-softening abilities of the GLDA were evaluated by comparing the viscosity of the polymer in fresh water and the presence of different concentrations of Ca^{++}. The concentration of the GLDA was fixed at 4 wt.%. Various concentrations of calcium ions (Ca^{++}) were dissolved in the water, i.e., 4000, 6000, 8000, and 10,000 ppm, with 45 lb/1000 gal CMHPG polymer concentration. A minimum of 10 min of hydration was allowed on all samples before measuring viscosities across different shear rates at ambient conditions. The effectiveness is simply determined through the ability of water to hydrate the polymer and build a viscous gel, where it is normally hindered in high TDS water systems or produced water exhibiting high hardness profiles (presence of calcium and magnesium ions). The 4 wt.% of GLDA was shown to chelate or capture most divalent ions present in the water systems, with up to 10,000 ppm of Ca^{++} dissolved in the solution. This was observed when comparing and showing similar values of viscosity of water containing various concentrations of Ca^{++}, compared with that in deionized water (Figure 5). As shown in the graph, similar values of viscosity indicated the successful chelation effect of 4 wt.% of GLDA, preventing the interruption of thickening behavior normally seen without the presence of chelating agents.

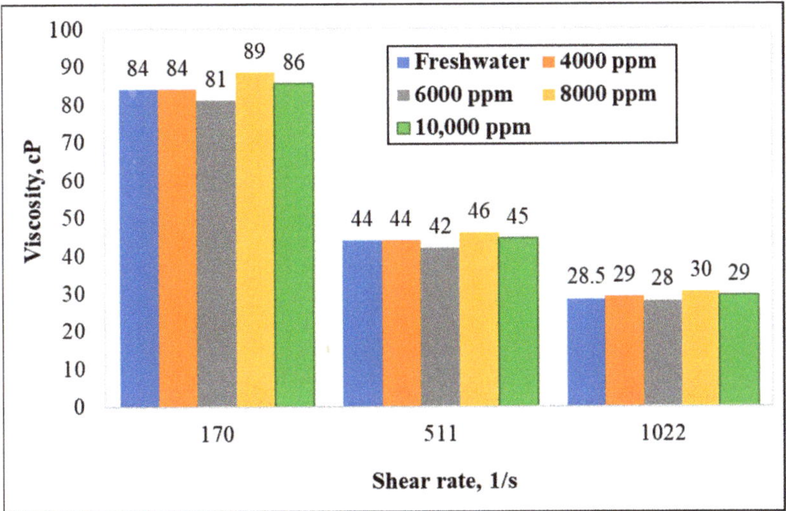

Figure 5. Viscosity measurements in the presence of 4 wt.% GLDA for different water systems (GLDA concentration = 4 wt.%, polymer concentration = 45 pptg, temperature = 77 °F).

3.3. Effect of Chelating Agent Concentration

The effect of chelating agent concentration on the viscosity was assessed using two different concentrations of chelating agents (10 wt.% and 20 wt.%). A total of 50 lb/1000 gal of CMHPG polymer was mixed in PW solution at a pH of 7.5. The apparent viscosity was measured against time at 150 °F and 500 psi pressure at a shear rate of 100 s^{-1}. The results presented in Figure 6 showed that the higher concentration of active chelating agents increased the viscosity readings. This indicates that with the additional amount of GLDA available in the system, more thickening occurs with the polymer, assuming that only a certain amount of GLDA is held to capture the system's ions.

3.4. Effect of pH

The neutral solution resulted in the best apparent viscosity slightly outperforming the acidic solution. As shown in Figure 7, the neutral pH system maintained a constant viscosity at around 112 cP, the acidic pH system maintained a slightly decreasing viscosity at around 106 cP, while the basic pH system read at 82 cP. The pH influenced the thickening behavior of the mixture and its stability, with a pH of 7–8 showing the most favorable conditions. It is worth mentioning that the basic pH systems took a long time (5–6 h) to hydrate the polymer and build up a viscous fluid, while this occurred almost instantly in other systems.

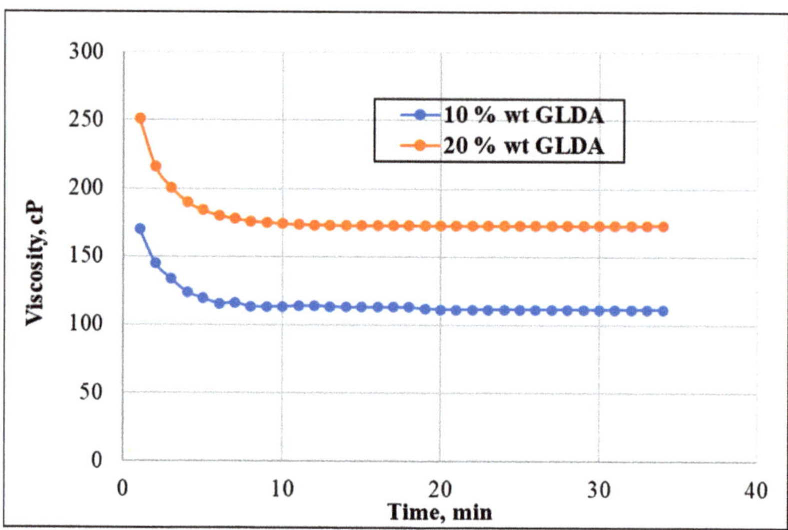

Figure 6. Effect of GLDA concentration on the viscosity (shear rate = 100 s^{-1}, temperature = 150 °F, pH = 8, and CMHPG concentration = 50 pptg).

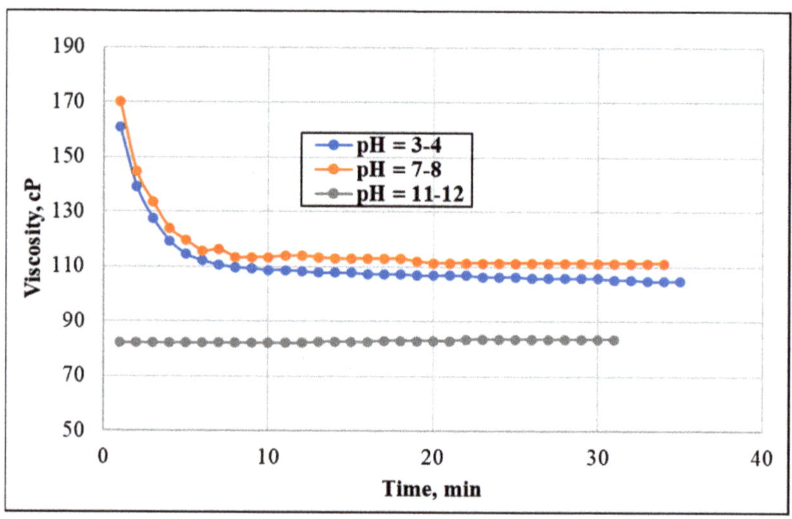

Figure 7. Effect of pH on the solution viscosity (shear rate = 100 s^{-1}, temperature = 150 °F, GLDA concentration = 10 wt.%, and CMHPG concentration = 50 pptg).

3.5. Effect of CMHPG Concentration

The effect of concentration was assessed using three different concentrations of CMHPG (30, 40, and 50 pptg). The results in Figure 8 showed that the highest CMHPG concentration resulted in the highest apparent viscosity (112 cP). The solution with a 40 pptg concentration resulted in apparent viscosity of 73 cP, while the 30 pptg solutions resulted in a viscosity of 33 cP, indicating that polymer loading directly enhances the thickening behavior in these solutions.

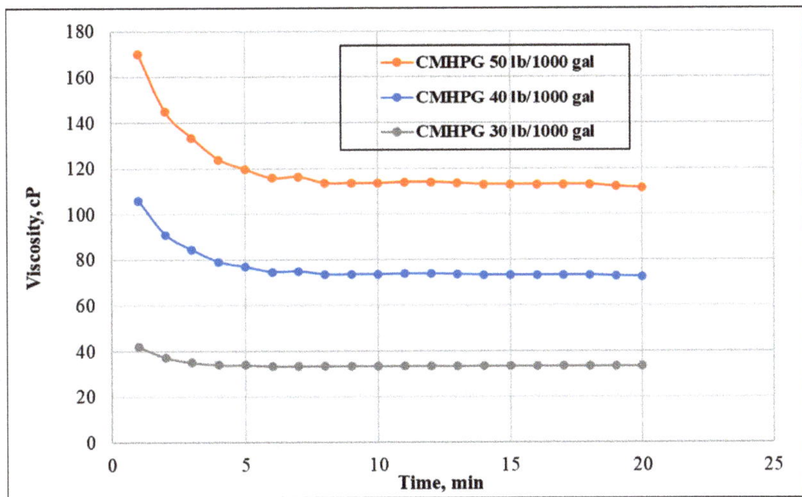

Figure 8. Effect of polymer concentration on the viscosity (shear rate = $100\ \text{s}^{-1}$, temperature = 150 °F, and pH = 8).

3.6. Stability and Breakage Behaviors

The stability of the newly developed stimulation fluid is critical to understand. To deploy in the field, a minimum of 1 h is usually needed to pump down this fluid inside a wellbore. This experiment found that stability is highly dependent on the temperature, polymer concentration, and chelating agent concentrations. In a 10 wt.% GLDA, 50 pptg CMHPG polymer concentrations, and at 150 °F temperature, the fluid stabilized for around 2 h and broke completely after a total time of 4.5 h, shown in Figure 9.

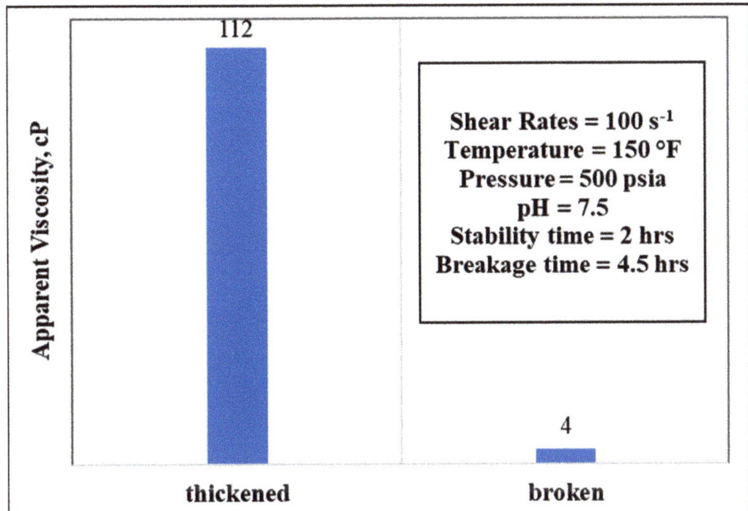

Figure 9. Thickening and breaking behaviors with respective parameters (polymer concentration = 50 pptg, GLDA concentration = 10 wt.%, water salinity = 70 k).

In a 20 wt.% GLDA, 50 pptg CMHPG polymer concentrations, and at 150 °F temperature, the fluid stabilized for around 2.4 h and broke completely after a total time of 6 h, reading higher viscosity values, as indicated in Figure 10.

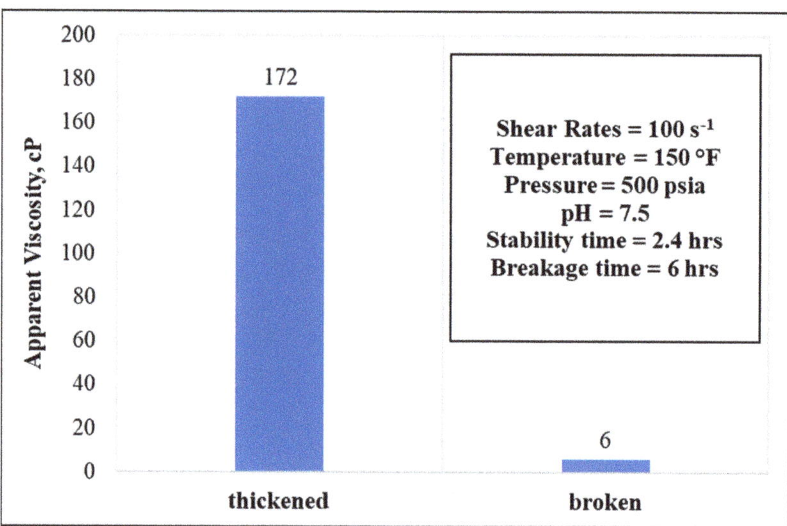

Figure 10. Thickening and breaking behaviors with respective parameters (polymer concentration = 50 pptg, GLDA concentration = 20 wt.%, water salinity = 70 k).

However, elevating the temperature while holding the remaining parameters constant clearly showed a variety of stability profiles. As shown in Figure 11, in a 10 wt.% GLDA, 50 pptg CMHPG polymer concentrations, and at 200 °F temperature, the fluid stabilized for around 1.5 h and broke completely after a total time of 4 h while also showing less apparent viscosity, at 73 cP.

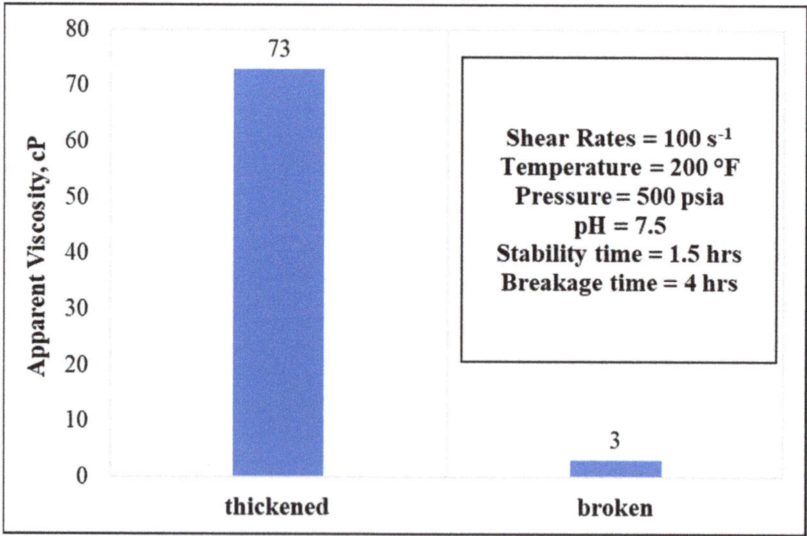

Figure 11. Thickening and breaking behaviors with respective parameters (polymer concentration = 50 pptg, GLDA concentration = 10 wt.%, water salinity = 70 k).

Finally, setting the experiment at the highest temperature (300 °F) drastically changed the stability time window. In a 10 wt.% GLDA, 50 pptg CMHPG polymer concentrations, the fluid stabilized for only 0.5 h and broke completely after a total time of 1.0 h, showing an apparent viscosity of 44 cP at the thickening stage (Figure 12).

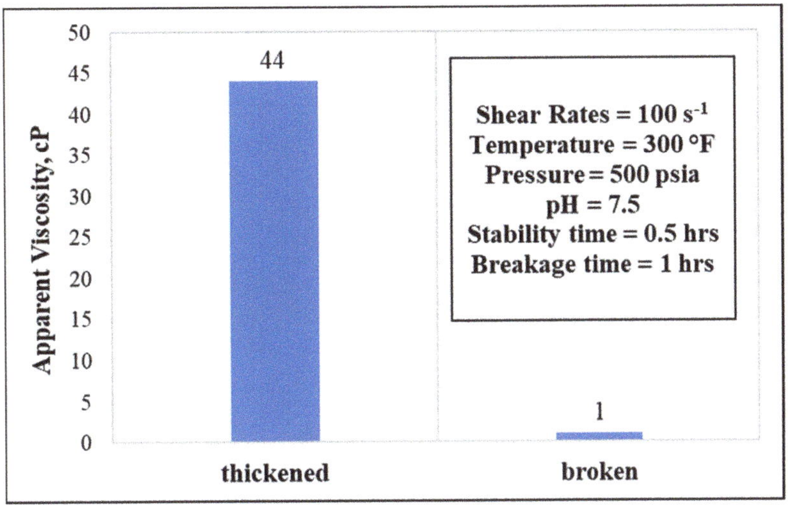

Figure 12. Thickening and breaking behaviors with respective parameters (polymer concentration = 50 pptg, GLDA concentration = 10 wt.%, water salinity = 70 k).

4. Conclusions and Recommendation

In this work, a new environmentally friendly stimulation fluid was developed to alleviate the burden on the exhausted groundwater resources, which is often available to support well stimulation jobs in the oil industry. The new simply constructed fluid was based on oilfield-produced water, introduced as a reuse opportunity that potentially eliminates environmental impacts associated with disposing and discharging such wastewater streams, thus promoting a more sustainable water use practice across a vital industry such as the petroleum industry. The newly developed fluid was composed of only chelating agents (GLDA), polymeric gel (CMHPG), and produced water as a base fluid. In comparison with conventional stimulation fluids, fewer additives were used to meet the rheological requirements for stimulation fluids, providing an environmentally sound solution as well as an economic advantage.

Rheological characterization was conducted on this newly developed fluid, studying the effects of multiple parameters such as concentrations of polymer, the concentration of the chelating agent, pH, shear rate, water chemistry, and temperature. Viscosity profile against time was investigated, in addition to assessing the thickening and breakage profiles of these fluids at different concentrations and settings.

Results showed that GLDA has excellent water-softening and thickening effects when mixed with CMHPG polymer and can break by itself without adding a breaker. The increase in GLDA concentration from 10 to 20 wt.% was shown to improve the fluid viscosity and stability time. The study showed that the most optimum concentrations of GLDA and CMHPG were 20 wt.% and 50 pptg, respectively, while the most optimum conditions were a neutral pH system of 7.5 and a temperature of 150 °F.

The highest apparent viscosity profile, using mentioned optimum concentrations and conditions, was 172 cP at 100 s^{-1} shear rate, exhibiting a stable thickened phase for nearly 150 min before breaking completely in a total of 360 min. The findings of this research can aid researchers in the oil and gas upstream business to search for new ways to develop stimulation fluids and find alternatives to using freshwater resources in stimulation applications. The use of 4 wt.% GLDA offers great water-softening capabilities in holding off major divalent ions present in PW, with up to 10 k ppm hardness; this can be utilized in multiple applications to prevent scaling. Neutral pH system with 20 wt.% GLDA concentration and 50 pptg polymer concentration result in adequate viscosity values for fracturing fluids carrying proppant such as hydraulic fracturing applications. The formulation is environmentally friendly, as GLDA can replace crosslinker, breaker, biocide, and clay stabilizer from fracturing fluid formulation.

Author Contributions: Conceptualization, B.T., M.S.K. and M.M.; methodology, M.A. and M.B.; formal analysis, M.A., M.S.A. and M.S.K.; investigation, M.A.; resources, M.S.K., M.M., B.T. and M.B.; data curation, M.A.; writing—original draft preparation, M.A. and M.S.A.; writing—review and editing, M.S.K., M.M., B.T. and M.B.; supervision, B.T. and M.M.; project administration, M.S.K. and M.B. All authors have read and agreed to the published version of the manuscript.

Funding: This research was funded by Saudi Aramco and King Fahd University of Petroleum and Minerals, Grant Number CIPR 2345, and the APC was funded by King Fahd University of Petroleum and Minerals, Dhahran Saudi Arabia.

Institutional Review Board Statement: Not applicable.

Informed Consent Statement: Not applicable.

Data Availability Statement: Not applicable.

Acknowledgments: The authors would like to thank the College of Petroleum and Geosciences, King Fahd University of Petroleum and Minerals, EXPEC ARC Saudi Aramco, for providing research opportunities to produce this paper through Project Number CIPR2345.

Conflicts of Interest: The authors declare no conflict of interest.

References

1. Hogeboom, R.J. The Water Footprint Concept and Water's Grand Environmental Challenges. *One Earth* **2020**, *2*, 218–222. [CrossRef]
2. Reig, P.; Luo, T.; Proctor, J.N. *Global Shale Gas Development: Water Availability & Business Risks*. World Resources Institute. 2014. Available online: https://www.wri.org/research/global-shale-gas-development-water-availability-business-risks (accessed on 19 November 2021).
3. Samad, N.A.; Bruno, V.L. The urgency of preserving water resources. *EnviroNews* **2013**, *21*, 3–6.
4. Sowers, J.; Vengosh, A.; Weinthal, E. Climate change, water resources, and the politics of adaptation in the Middle East and North Africa. *Clim. Chang.* **2010**, *104*, 599–627. [CrossRef]
5. Zhang, L.; Hascakir, B. A review of issues, characteristics, and management for wastewater due to hydraulic fracturing in the U.S. *J. Pet. Sci. Eng.* **2021**, *202*, 108536. [CrossRef]
6. Yao, S.; Chang, C.; Hai, K.; Huang, H.; Li, H. A review of experimental studies on the proppant settling in hydraulic fractures. *J. Pet. Sci. Eng.* **2021**, 109211. [CrossRef]
7. LeBas, R.; Lord, P.; Luna, D.; Shahan, T. Development and Use of High-TDS Recycled Produced Water for Crosslinked-Gel-based Hydraulic Fracturing. Presented at the SPE Hydraulic Fracturing Technology Conference, The Woodlands, TX, USA, 4–6 February 2013; pp. 125–133. [CrossRef]
8. Whalen, T. The Challenges of Reusing Produced Water. *J. Pet. Technol.* **2012**, *64*, 18–20. [CrossRef]
9. Zendehboudi, S.; Bahadori, A. *Shale Oil and Gas Handbook: Theory, Technologies and Challenges*; Gulf Professional Publishing: Houston, TX, USA, 2016; ISBN 9780128021002.
10. Cheremisinoff, N.P.; Davletshin, A. *Hydraulic Fracturing Operations: Handbook of Environmental Management Practices*; John Wiley & Sons: Beverly, MA, USA, 2015; ISBN 9781119099987.
11. Bonapace, J.; Giglio, M.; Moggia, J.; Krenz, A. Water Conservation: Reducing Freshwater Consumption by Using Produced Water for Base Fluid in Hydraulic Fracturing-Case Histories in Argentina. Presented at the SPE Latin America and Caribbean Petroleum Engineering Conference, Mexico City, Mexico, 16–18 April 2012; pp. 199–223. [CrossRef]
12. Mahmoud, M.A.; Nasr-El-Din, H.A.; De Wolf, C.A.; LePage, J.N.; Bemelaar, J.H. Evaluation of a New Environmentally Friendly Chelating Agent for High-Temperature Applications. *SPE J.* **2011**, *16*, 559–574. [CrossRef]
13. Kamal, M.S.; Mohammed, M.; Mahmoud, M.; Elkatatny, S. Development of chelating agent-based polymeric gel system for hydraulic fracturing. *Energies* **2018**, *11*, 1663. [CrossRef]
14. Abdelgawad, K.Z.; Mahmoud, M.A.; Elkatatny, S.M. Stimulation of High Temperature Carbonate Reservoirs using Seawater and GLDA Chelating Agents: Reaction Kinetics Comparative Study. Presented at the SPE Kuwait Oil & Gas Show and Conference, Kuwait City, Kuwait, 15–18 October 2017.

15. Veil, J.A.; Puder, M.G.; Elcock, D.; Redweik, R.J.J.; Punder, M.G.; Elcock, D.; Redweik Jr, R.J. *A White Paper Describing Produced Water from Production of Crude Oil, Natural Gas, and Coal Bed Methane*; Argonne National Laboratry: Du Page, IL, USA, 2004.
16. Khatib, Z.; Verbeek, P. Water to value—Produced water management for sustainable field development of mature and green fields. *JPTJ Pet. Technol.* **2003**, *55*, 26–28. [CrossRef]
17. Dickhout, J.M.; Moreno, J.; Biesheuvel, P.M.; Boels, L.; Lammertink, R.G.H.; de Vos, W.M. Produced water treatment by membranes: A review from a colloidal perspective. *J. Colloid Interface Sci.* **2017**, *487*, 523–534. [CrossRef] [PubMed]
18. Al-Ghouti, M.A.; Al-Kaabi, M.A.; Ashfaq, M.Y.; Da'na, D.A. Produced water characteristics, treatment and reuse: A review. *J. Water Process Eng.* **2019**, *28*, 222–239. [CrossRef]
19. Kalam, S.; Kamal, M.S.; Patil, S.; Hussain, S.M.S. Impact of Spacer Nature and Counter Ions on Rheological Behavior of Novel Polymer-Cationic Gemini Surfactant Systems at High Temperature. *Polymers* **2020**, *12*, 1027. [CrossRef] [PubMed]
20. Othman, A.; Aljawad, M.S.; Mahmoud, M.; Kamal, M.S.; Patil, S.; Bataweel, M. Chelating Agents Usage in Optimization of Fracturing Fluid Rheology Prepared from Seawater. *Polymers* **2021**, *13*, 2111. [CrossRef] [PubMed]
21. Lei, C.; Clark, P.E. Crosslinking of Guar and Guar Derivatives. *SPE J.* **2007**, *12*, 316–321. [CrossRef]

Article

Functionalization of Graphene Oxide with Polysilicone: Synthesis, Characterization, and Its Flame Retardancy in Epoxy Resin

Jiangbo Wang

School of Materials and Chemical Engineering, Ningbo University of Technology, Ningbo 315211, China; jiangbowang@nbut.edu.cn

Abstract: A novel polysilicone flame retardant (PMDA) has been synthesized and covalently grafted onto the surfaces of graphene oxide (GO) to obtain GO-PMDA. The chemical structure and morphology of GO-PMDA was characterized and confirmed by the Fourier transform infrared (FTIR) spectroscopy, X-ray photoelectron spectrometer (XPS), atomic force microscope (AFM), and thermogravimetric analysis (TGA). The results of dynamic mechanical analysis (DMA) indicated that the grafting of PMDA improved the dispersion and solubility of GO sheets in the epoxy resin (EP) matrix. The TGA and cone calorimeter measurements showed that compared with the GO, GO-PMDA could significantly improve the thermal stability and flame retardancy of EP. In comparison to pure EP, the peak heat release rate (pHRR) and total heat release (THR) of EP/GO-PMDA were reduced by 30.5% and 10.0% respectively. This greatly enhanced the flame retardancy of EP which was mainly attributed to the synergistic effect of GO-PMDA. Polysilicone can create a stable silica layer on the char surface of EP, which reinforces the barrier effect of graphene.

Keywords: graphene oxide; polysilicone; functionalization; flame retardancy; epoxy resin

Citation: Wang, J. Functionalization of Graphene Oxide with Polysilicone: Synthesis, Characterization, and Its Flame Retardancy in Epoxy Resin. *Polymers* **2021**, *13*, 3857. https://doi.org/10.3390/polym13213857

Academic Editor: Sheila Devasahayam

Received: 30 September 2021
Accepted: 4 November 2021
Published: 8 November 2021

Publisher's Note: MDPI stays neutral with regard to jurisdictional claims in published maps and institutional affiliations.

1. Introduction

Epoxy resin (EP) was widely used in various industrial fields such as coating, adhesives, laminates, and composites, etc., [1–5]. However, one of the main drawbacks of epoxy resin is its inherent flammability, which restricts its application in many fields for safety consideration. Therefore, it is an important issue to improve the flame retardancy of EP. Halogen-free flame retardants (such as hydrated alumina, aromatic phosphates, etc.) with their environment friendly property has become a new trend of replacing the original position of halogen-containing flame retardants in improving fire resistance of epoxy resin [6–8].

As the two-dimensional sp2-hybridized carbon, graphene is currently the most intensively studied material. This single-atom-thick sheet of carbon atoms arrayed in a honeycomb pattern is the world's thinnest, strongest, and stiffest material [9,10]. Until recently, most studies of graphene-based nanocomposites focused on the incorporation of graphene, modified graphene, or graphene oxide (GO) into polymer matrices. Because of their layered structures, graphene and GO can act as barriers reducing the heat released and insulating against the transfer of combustion gases into the inflammable polymer matrix [11–15].

However, due to the high surface area and strong Van der Waals force, the re-aggregating phenomenon is inclined to appear between graphene sheets, which limit its use in polymer matrix [16,17]. The problem is usually solved by covalent functionalization. After oxidation, rich oxygen-containing groups (e.g., hydroxyl, epoxide, carboxyl and carbonyl groups, etc.) are brought to the surface of graphene sheets. Through further chemically functionalizing the GO, grafting the organic molecules on GO was widely adopted in improving the dispersion and thermal stability of GO [18–21]. Polyhedral oligomeric silsesquioxane (POSS, which contains an inner inorganic framework made up of silicone and oxygen $(SiO_{1.5})_x$ and is externally covered by organic substituents.), phosphorus-containing molecule, and intumescent flame retardant have been designed and covalently grafted onto the surface of graphene sheets to obtain a novel flame retardant [22–27]. In our previous research [28,29], polysiloxane was grafted onto the surface of carbon nanotubes, which not only improved the dispersion of carbon nanotubes in matrix, but also improved the flame retardancy of the epoxy resin and thiol-ene polymer.

In this study, a novel polysilicone (PMDA), which consisted of 60 mol% phenylsiloxane, 35 mol% methylsiloxane, and 5 mol% aminosiloxane as a unit, was synthesized by the hydrolysis and polycondensation method. Subsequently, PMDA was grafted onto the surface of graphene oxide to improve the dispersion and flame retardancy of graphene in the polymer. The results of Fourier transform infrared (FTIR) spectroscopy, X-ray photoelectron spectrometer (XPS), atomic force microscope (AFM), and thermogravimetric analysis (TGA) measurements indicated that the polysilicone has been successfully attached to the surfaces of GO. It is confirmed that the covalent grafting of polysilicone onto graphene sheets improved both the flame retardancy efficiency and the dispersibility of graphene in the EP matrix, which provides a new path to obtain a new kind of graphene- based flame-retardant nanocomposites.

2. Materials and Methods

2.1. Materials

Graphite powders (spectrum pure), concentrated sulfuric acid (98%), phosphoric acid (85%), potassium permanganate, hydrogen peroxide (30%), tetramethylammonium hydroxide (TMAOH), methyltrimethoxysilane (MTMS), (3-aminopropyl)trimethoxysilane (APT), N,N'-dicyclohexylcarbodiimide (DCC) and tetrahydrofuran (THF) were all purchased from Alfa Aesar Chemical Reagent Co., Ltd. (Tewksbury, MA, USA). Phenyltrimethoxysilane (PTMS) and dimethyldimethoxysilane (DMDS) were reagent grade and purchased from Gelest Chemical Reagent Co., Ltd. (Morrisville, PA, USA). Ethyl alcohol was supplied by Sigma-Aldrich Reagent Co., Ltd. (St. Louis, MO, USA). Chloroform $(CHCl_3)$ and hydrochloric acid were supplied by Fisher Scientific Chemical Co. (Waltham, MA, USA). EPON 826 with an epoxy equivalent weight of 178–186 g was supplied by Hexion Specialty Chemicals Inc. (Columbus, OH, USA) and used as received. The hardener, Jeffamine D230, with an amine equivalent weight of 60 g, was supplied by Huntsman Corp. (Woodlands, TX, USA) and also used as received.

2.2. Synthesis of Polysilicone (PMDA)

The polysilicone PMDA was synthesized by the hydrolysis and polycondensation method as shown in Scheme 1. The raw material consisted of 60 mol% phenylsiloxane, 35 mol% methylsiloxane and 5 mol% aminosiloxane as a unit. The ratio of organic groups to silicon atoms (R/Si) was 1.2, which was used to characterize the extent of branches to a polysiloxane structure; the molecular structure of polysilicone is illustrated in Scheme 1. Distilled water (25 mL), EtOH (75 mL), and TMAOH (1 mL) were mixed in a 250 mL flask under stirring, followed by adding the mixture of PTMS, MTMS, DMDS, and APS at certainly molar ratios (0.69:0.06:0.20:0.05) and maintaining 10% weight percentage solution. The stirring was maintained for 8 h, and the resulting solution was stored at room temperature overnight. Through decantation of most clear supernatant, precipitated condensate was collected and then washed by vacuum filtration with distilled H_2O/EtOH (1/3 by volume), then washed again in pure EtOH. Finally, the acquired rinsed powder (PMDA) was thoroughly dried under vacuum for 20 h at room temperature [28].

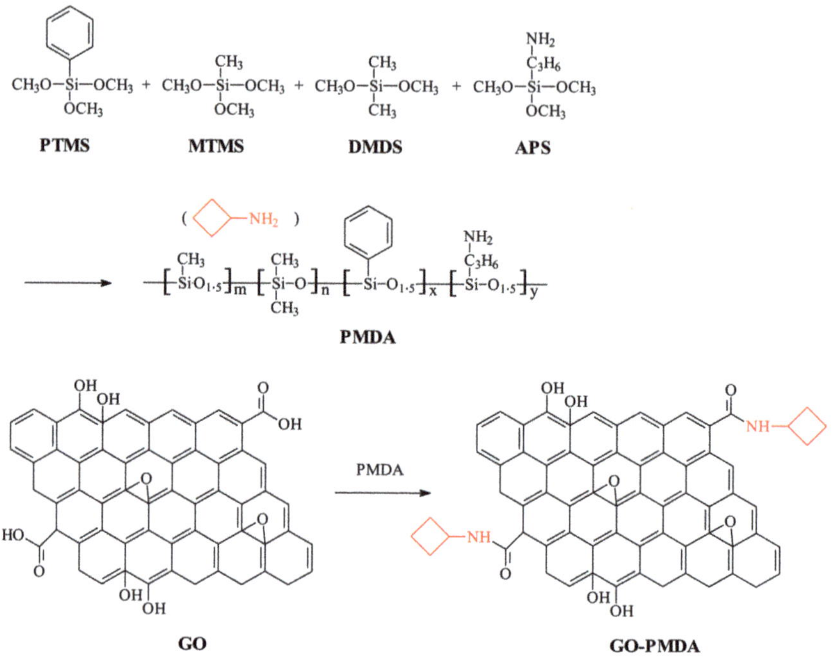

Scheme 1. Synthesis route of GO-PMDA.

2.3. Functionalization of Graphene Oxide (GO)

GO was prepared from graphite by modified Hummers' method [30]. In a 500 mL three-neck flask, the as-prepared GO (0.2 g) was first suspended in THF (200 mL) under ultrasonication for 90 min. Subsequently, the PMDA (0.8 g) and DCC (0.1 g, as cat.) were introduced into the above flask, and followed by ultrasonication for 30 min. With stirring, the mixture was heated to 66 °C and refluxed for 20 h under nitrogen atmosphere. Afterwards, the mixture was centrifuged and thoroughly washed with anhydrous THF to remove the residual PMDA. Then, the product dried in a vacuum at room temperature for 12 h to remove the solvent (Scheme 1).

2.4. Preparation of Epoxy Composite

Briefly, the EP/GO-PMDA composites were prepared as follows: The GO-PMDA (2 g) was dispersed in acetone and sonicated for 60 min to form a uniform black suspension. Then, EPON 826 (73.5 g) was added to the mixture and dispersed by a mechanical stirrer for 30 min. The mixture was heated in a vacuum oven at 50 °C for 10 h to remove the solvent. After that, D230 (24.5 g) was added into the mixture and stirring for 30 min. After degassing in vacuum for 10 min to remove any trapped air, the samples were cured at 80 °C for 2 h and post cured at 135 °C for 2 h. For comparison, pure epoxy (EP) and 2 wt% GO/epoxy (EP/GO) composites were also prepared at same processing condition.

2.5. Characterization and Measurement

The Fourier transform infrared spectroscopy (FTIR) was tested using a Digilab Scimitar FTS-2000 IR spectrometer (Digilab Inc., Hopkinton, MA, USA) at a resolution of 2 cm^{-1} with 20 scans. The samples were mixed with potassium bromide and pressed to a disc, which was used for measurement. X-ray photoelectron spectroscopy (XPS) was carried out in a Thermo Scientific ESCALAB 250Xi X-ray photoelectron spectrometer (Thermo Fisher Scientific Inc., Waltham, MA, USA) equipped with a mono-chromatic Al Kα X-ray source (1486.6 eV). AFM observation was performed on the Bruker Dimension Icon atomic force microscope (Bruker Corp., Karlsruhe, Germany) in tapping-mode. The aqueous GO suspension and DMF suspension of GO-PMDA were spin-coated onto freshly cleaved silica surfaces. Thermogravimetric analysis (TGA) was carried out on a TA instrument Q5000 thermogravimetric analyzer (TA Instrument Corp., New Castle, DE, USA). The sample was dried in an oven at 100 °C for 5 h to remove moisture, and then about 10 mg sample was heated from 50 to 700 °C at a 10 °C/min heating ramp rate in nitrogen atmosphere. Dynamic mechanical analysis (DMA) was determined using a Rheometric Scientific SR-5000 dynamic mechanical analyzer (Rheometric Scientific Inc., West Yorkshire, UK). Data were collected from 60 to 120 °C at a scanning rate of 5 °C/min. Cone calorimeter measurement was performed on an FTT cone calorimeter (Fire Testing Technology Ltd., East Grinstead, West Sussex, UK) according to ASTM E1354. The dimension of each specimen was $100 \times 100 \times 3$ mm^3. All the measurements were repeated three times and the results were averaged.

3. Results and Discussion

3.1. Structural Characterization

The dispersion of GO-PMDA is very important for the preparation of GO-PMDA polymer nanocomposites. Figure 1 shows the digital photos of dispersion stability of low-concentration GO and GO-PMDA (0.5 mg/mL) in insoluble mixtures of water and CHCl$_3$ after 24 h of static placement.

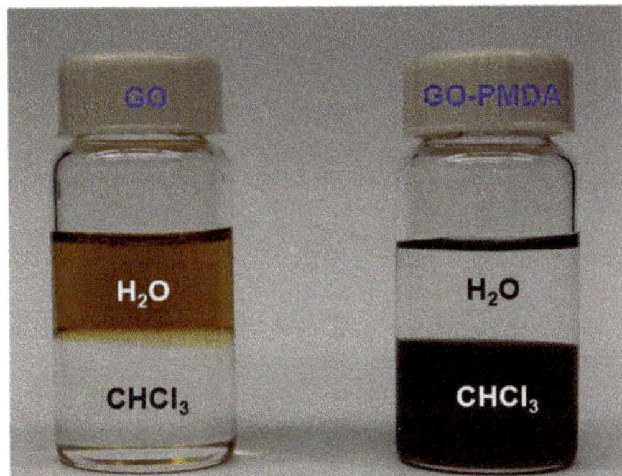

Figure 1. Solubility of GO and GO-PMDA.

As is well-known, carboxylic, epoxy, carbonyl, and hydroxide groups are present on the surface and edge of GO sheets, which make GO hydrophilic. Therefore, it can be found that the GO homogeneously disperses in water and exhibits a light yellow color. Meanwhile, the stability of the dispersion of GO-PMDA is very good in CHCl$_3$ without noticeable precipitation. The strong interaction between PMDA and CHCl$_3$ makes the dispersion of GO-PMDA in CHCl$_3$ stable. The evolution of surface functionality during reaction usually leads to a change of the graphene sheet from hydrophilic into hydrophobic. Moreover, the color of GO gradually changes to black after the one-step functionalization and reduction process, which is usually considered to be a sign of GO reduction.

FTIR spectra for the GO, PMDA, and GO-PMDA is measured in Figure 2. As expected, GO presents the typical spectra of oxygen-based functional groups: –OH at 3408 cm^{-1}, C=O at 1705 cm^{-1}, C–O at 1211 cm^{-1}. The intensities of these IR peaks decrease significantly after chemical attachment of PMDA onto the GO. In addition, the strong peak at 1200–1000 cm^{-1} characteristic of the Si-O-Si stretching vibration in PMDA is clearly seen in the FTIR spectra of GO-PMDA, indicating that PMDA has been successfully grafted onto GO.

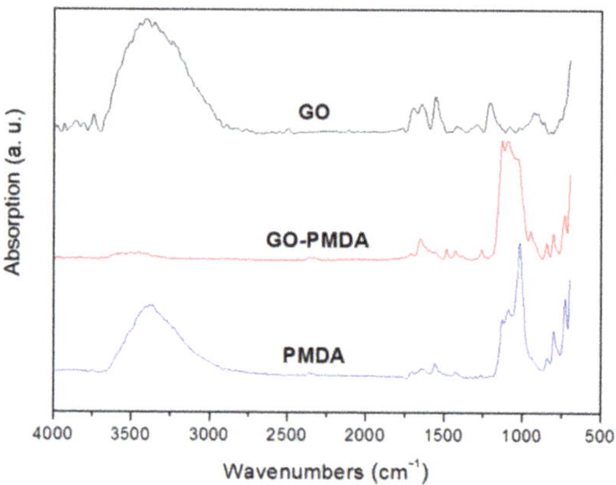

Figure 2. FTIR spectra of GO, PMDA, and GO-PMDA.

XPS measurement is performed to gain more information about chemical bonds formed on the surface of GO before and after its functionalization with PMDA. As expected, only C1s and O1s peaks are obtained in the XPS survey spectra for GO (Figure 3a). Compared to that of GO, the XPS survey spectra for GO-PMDA shows additional N1s, Si2s, and Si2p peaks, accompanied by the reduced O1s peak with respect to the C1s peak. Furthermore, it is easy to see the N1s band of GO-PMDA locates at 399.3 eV (C–N) and 402.6 eV (CO–N) in Figure 3b, which further proves that GO is modified with PMDA through the success of the covalent bonding of PMDA onto GO.

$$\text{—COOH} + \text{NH}_2\text{—} \rightarrow \text{—CO—NH—} \tag{1}$$

$$\text{—CH(O)CH}_2 + \text{NH}_2\text{—} \rightarrow \text{—CHOHCH}_2\text{—NH—} \tag{2}$$

Figure 3c,d reproduces the high-resolution C1s spectra for GO and GO-PMDA, respectively. The peaks for C–C (285.0 eV), C–O (287.0 eV), C=O (287.9 eV), and COO (289.0 eV) are clearly observed in C1s scan of GO (Figure 3c). As can be seen, the COO peak at 289.0 eV for GO-PMDA almost disappears upon the amide formation with the amine group of PMDA. The peak intensity of the C–O (287.0 eV) and C=O (287.9 eV) in GO-PMDA also significantly decreases (Figure 3d). This is because the nucleophilic substitution between GO and amine groups can cause deoxygenation and reduction of graphene oxide [27].

AFM measurement is performed to investigate the morphology and thickness of the interface layer grafted on the GO sheet surface. Figure 4a represents the tapping mode AFM images of GO, showing the average height of ~1 nm for a single-layer GO sheet. After grafting PMDA, the height of a single-layer GO-PMDA becomes ~4 nm (Figure 4b), which is much higher than that of GO. The thicker sheet is possibly due to the PMDA chain grafted on GO sheet surface which indicates that the GO-PMDA is successfully obtained in our work.

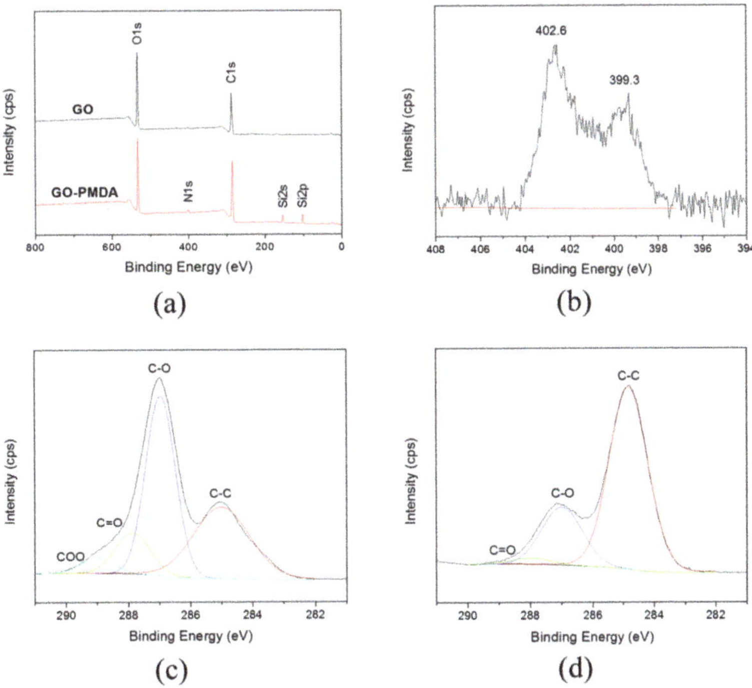

Figure 3. XPS survey spectra of the GO and GO-PMDA (**a**) and high-resolution XPS spectra of N1s for GO-PMDA (**b**), C1s for GO (**c**), C1s for GO-PMDA (**d**).

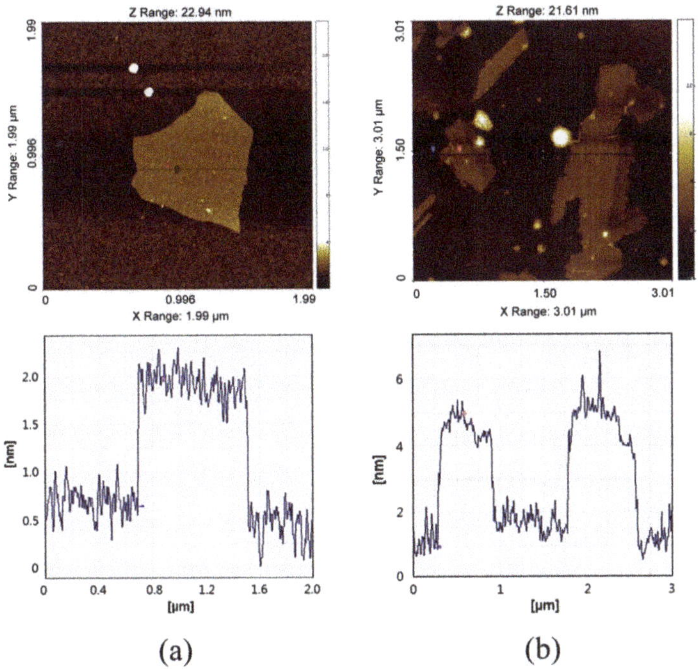

Figure 4. AFM images and corresponding height profiles of GO (**a**), GO-PMDA (**b**).

TGA curves of GO, PMDA, and GO-PMDA are shown in Figure 5. The initial 7.6 wt% weight loss seen for GO up to 100 °C is associated with the thermal desorption of water molecules physically adsorbed onto the hydrophilic GO surface. The main weight loss of GO is found around 190–200 °C because of the decomposition of oxygen-containing functional groups to CO, CO_2, and H_2O [27]. The weight loss between 200 °C and 600 °C is about 8.5 wt%, associated with the removal of more thermally stable oxygen functionalities and thermal decomposition of GO. Compared with the onset degradation temperatures ($T_{5wt\%}$, 78.5 °C) and the char yield (16.5 wt% at 600 °C) of GO, the TGA curve of GO-PMDA exhibits a much better thermal stability with both higher of $T_{5wt\%}$ (141.9 °C) and char yield (61.5 wt% at 600 °C). These results can be attributed to the grafting of PMDA, which is a good char layer stabilizer. Therefore, the char-forming performance of graphene can be greatly improved, so as to better play the role of heat insulation and isolation of combustible substances.

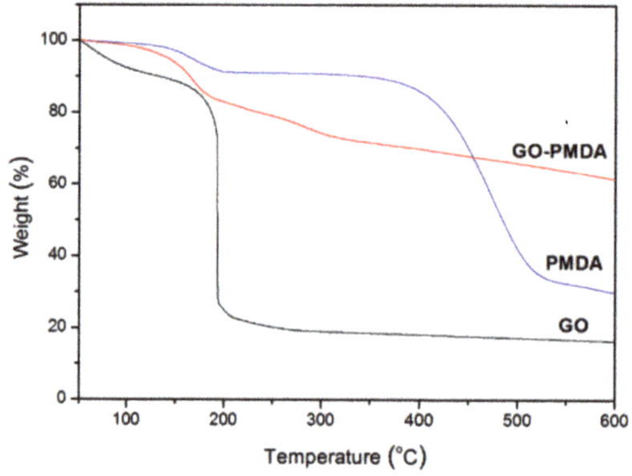

Figure 5. TGA curves of GO, PMDA, and GO-PMDA.

3.2. Dispersion

DMA test is always used to evaluate the interfacial interaction between additives and the polymer matrix. Figure 6 shows the temperature dependence of the storage modulus and tan delta of EP and its nanocomposites. The storage modulus of pure EP in Figure 6a at 80 °C is 928.2 MPa. When adding 2 wt% of GO, the storage modulus of EP composites is decreased by 57.7% and reaches 392.9 MPa below the T_g. The high hydrophilic of GO and its large aspect ratio have a significant effect on the decrease of storage modulus. This is ascribed to that GO affects the cross-linked structure between the EP molecular chains and the curing agent, and a plasticizer role played by GO increases the flexibility of chain segments of EP matrix, which means the reduced cross-linking density of EP will lead to decreased mechanical properties [31]. However, the storage modulus of EP/GO-PMDA shows much higher increase compared with EP/GO. The same tendency is found in T_g values. The addition of 2 wt% GO decreases T_g of the EP matrix from 92.7 °C to 84.8 °C, compared with which, the addition of GO-PMDA increases T_g of EP/GO from 84.8 °C to 89.5 °C. This indicates that the grafting of PMDA improves the dispersion and solubility of GO sheets in the EP matrix.

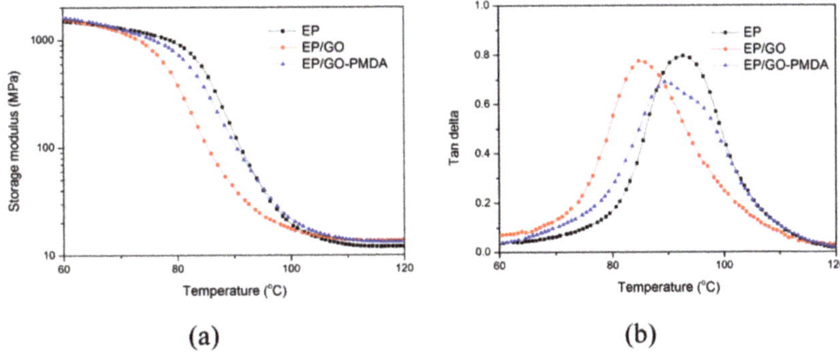

Figure 6. Storage modulus (**a**) and Tan delta (**b**) curves for EP, EP/GO, and EP/GO-PMDA.

3.3. Thermal Stability

The thermal stability of EP, EP/GO, and EP/GO-PMDA under nitrogen atmosphere is measured by TGA (Figure 7). Related data are listed in Table 1. The $T_{5wt\%}$ of EP is 346 °C, compared with which, the $T_{5wt\%}$ of EP/GO is lower than pure EP, since GO is thermally unstable and its major weight loss occurs below 200 °C due to the decomposition of the oxygen-contained functional moieties. The addition of GO-PMDA exhibits a same trend in the $T_{5wt\%}$ compared with GO. Even though, the $T_{5wt\%}$ of EP/GO-PMDA is increased by 15.4 °C compared with that of EP/GO. The T_{max} of the EP/GO and EP/GO-PMDA is quite similar to EP, however, DTG peak rates of the EP/GO and EP/GO-PMDA, which indicate thermal degradation rates, are visibly decreased. Furthermore, the residual char obtained from EP/GO-PMDA is remarkably higher than EP and EP/GO, increased by 1.6 wt% with only 2 wt% addition. The rich char yield formed during decomposition is due to the condensed phase flame retardant mechanism of silicon element in GO-PMDA, which can block the fuel and oxygen between composites and the environment as well as hinder the heat transfer.

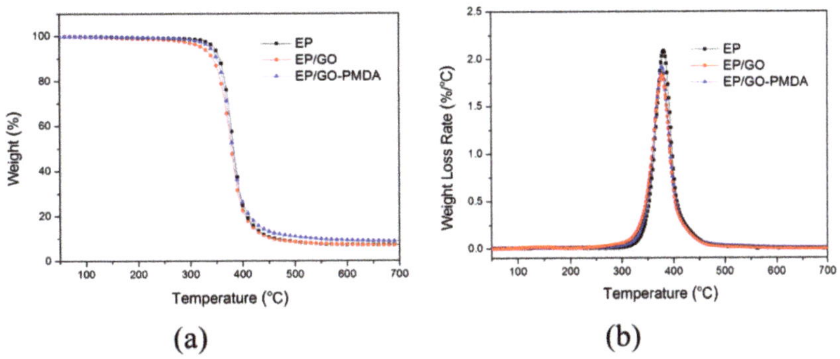

Figure 7. TGA (**a**) and DTG (**b**) curves of EP, EP/GO, and EP/GO-PMDA.

Table 1. TGA data of EP, EP/GO, and EP/GO-PMDA.

Sample	Temperature (°C)			Peak Rate (wt%/°C)	Residues (wt%)
	$T_{5wt\%}$	$T_{50wt\%}$	T_{max}		
EP	346.0	383.5	381.4	2.09	7.02
EP/GO	322.5	378.6	378.2	1.84	6.95
EP/GO-PMDA	337.9	381.3	378.2	1.92	8.76

3.4. Flame Retardancy

The cone calorimeter is one of the most effective methods to evaluate the flammability of various composites in real-world fire conditions. The heat release rate (HRR) and total heat release (THR) obtained from cone calorimeter have been found to be important parameters to evaluate fire safety. Figure 8 shows the HRR and THR versus time curves of EP, EP/GO, and EP/GO-PMDA. In comparison to pure EP, the peak heat release rate (pHRR) and THR of nanocomposites with the incorporation of 2 wt% GO was reduced by 21.6% and 8.8% respectively. The superior flame retardancy of EP/GO over EP could be attributed to the barrier effect of GO, which retards the permeation of heat and the escape of volatile degradation products. Moreover, the pHRR of EP/GO-PMDA in Figure 8a exhibits further reduction (30.5%, compared with EP), even though THR in Figure 8b displays little change (10.0%, compared with EP). The best flame-retardant properties of EP/GO-PMDA could be attributed to two aspects: first, the reduction of GO by PMDA occurs to convert GO into a more stable form, reduced-GO; second, PMDA can create a stable silica layer on the char surface of EP, which reinforces the barrier effect of graphene, simultaneous reduction and surface functionalization of graphene oxide with PMDA for reducing fire hazards in epoxy composites.

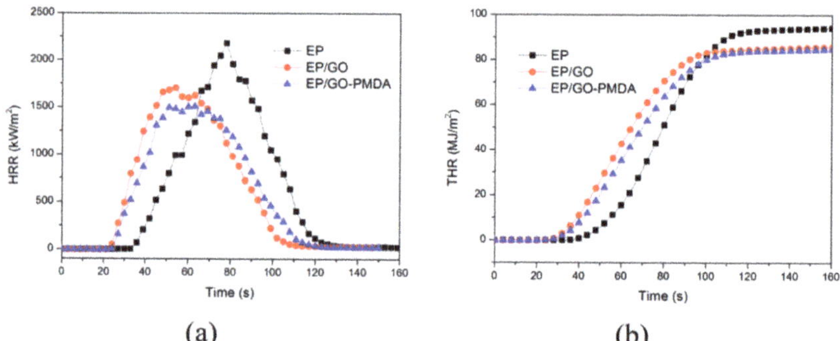

Figure 8. HRR (**a**) and THR (**b**) curves of EP, EP/GO, and EP/GO-PMDA.

4. Conclusions

In this paper, PMDA has been covalently grafted onto the surfaces of GO to prepare GO-PMDA. The results from FTIR, XPS, AFM, and TGA measurements showed that GO-PMDA was successfully grafted onto the surface of GO. The functionalization of GO with PMDA made the hydrophilic GO hydrophobic. DMA test exhibited that the grafting of PMDA improved the dispersion of GO sheets in EP matrix. Furthermore, GO-PMDA could significantly improve thermal stability and flame retardancy of EP compared with the GO. The peak heat release rate (pHRR) and total heat release (THR) of EP/GO-PMDA were reduced respectively. This greatly enhanced flame retardancy of EP by GO-PMDA which was mainly attributed to the synergistic effect of polysilicone and graphene. The PMDA grafting method used in this study can simultaneously improve the dispersion and flame retardancy of graphene in epoxy materials. This will help to further promote the application process of graphene and develop high-performance polymer materials.

Funding: This research was funded by the Ningbo Natural Science Foundation (2019A610032). This work was also supported by the Open Fund of Shanghai Key Laboratory of Multiphase Materials Chemical Engineering.

Data Availability Statement: The data used to support the findings of this study are available from the corresponding author upon request.

Acknowledgments: We gratefully acknowledge the financial support of the above funds and thank the researchers of all the reports cited in our paper.

Conflicts of Interest: The author declare no conflict of interest.

References

1. Gu, H.; Ma, C.; Gu, J.; Guo, J.; Yan, X.; Huang, J.; Zhang, Q.; Guo, Z. An overview of multifunctional epoxy nanocomposites. *J. Mater. Chem. C* **2016**, *4*, 5890–5906. [CrossRef]
2. Huang, H.; Zhang, K.; Jiang, J.; Li, J.; Liu, Y. Highly dispersed melamine cyanurate flame-retardant epoxy resin composites. *Polym. Int.* **2017**, *66*, 85–91. [CrossRef]
3. Zhao, X.; Babu, H.; Llorca, J.; Wang, D. Impact of halogen-free flame retardant with varied phosphorus chemical surrounding on the properties of diglycidyl ether of bisphenol-A type epoxy resin: Synthesis, fire behaviour, flame-retardant mechanism and mechanical properties. *RSC Adv.* **2016**, *6*, 59226–59236. [CrossRef]
4. Zhou, T.; Chen, W.; Duan, W.; Liu, Y.; Wang, Q. In situ synthesized and dispersed melamine polyphosphate flame retardant epoxy resin composites. *J. Appl. Polym. Sci.* **2019**, *136*, 47194. [CrossRef]
5. Yu, P.; Manalo, A.; Ferdous, W.; Abousnina, R.; Schubel, P. Investigation on the physical, mechanical and microstructural properties of epoxy polymer matrix with crumb rubber and short fibres for composite railway sleepers. *Constr. Build. Mater.* **2021**, *295*, 123700. [CrossRef]
6. Yan, W.; Yu, J.; Zhang, M.; Qin, S.; Wang, T.; Huang, W.; Long, L. Flame-retardant effect of a phenethyl-bridged DOPO derivative and layered double hydroxides for epoxy resin. *RSC Adv.* **2017**, *7*, 46236–46245. [CrossRef]
7. Jian, R.; Wang, P.; Duan, W.; Wang, J.; Zheng, X.; Weng, J. Synthesis of a novel P/N/S-containing flame retardant and its application in epoxy resin: Thermal property, flame retardance, and pyrolysis behavior. *Ind. Eng. Chem. Res.* **2016**, *55*, 11520–11527. [CrossRef]
8. Khotbehsara, M.; Manalo, A.; Aravinthan, T.; Turner, J.; Ferdous, W.; Hota, G. Effects of ultraviolet solar radiation on the properties of particulate-filled epoxy based polymer coating. *Polym. Degrad. Stab.* **2020**, *181*, 109352. [CrossRef]
9. Hu, K.; Kulkarni, D.; Choi, I.; Tsukruk, V. Graphene-polymer nanocomposites for structural and functional applications. *Prog. Polym. Sci.* **2014**, *39*, 1934–1972. [CrossRef]
10. Hersam, M. The reemergence of chemistry for post-graphene two-dimensional nanomaterials. *ACS Nano* **2015**, *9*, 4661–4663. [CrossRef]
11. Guo, Y.; Bao, C.; Song, L.; Yuan, B.; Hu, Y. In situ polymerization of graphene, graphite oxide, and functionalized graphite oxide into epoxy resin and comparison study of on-the-flame behavior. *Ind. Eng. Chem. Res.* **2011**, *50*, 7772–7783. [CrossRef]
12. Potts, J.; Dreyer, D.; Bielawski, C.; Ruoff, R. Graphene-based polymer nanocomposites. *Polymer* **2011**, *52*, 5–25.
13. Kim, H.; Abdala, A.; Macosko, C. Graphene/polymer nanocomposites. *Macromolecules* **2010**, *43*, 6515–6530. [CrossRef]
14. Sun, Y.; Li, C.; Xu, Y.; Bai, H.; Yao, Z.; Shi, G. Chemically converted graphene as substrate for immobilizing and enhancing the activity of a polymeric catalyst. *Chem. Commun.* **2010**, *46*, 4740–4742. [CrossRef] [PubMed]
15. Gui, H.; Xu, P.; Hu, Y.; Wang, J.; Yang, X.; Bahader, A.; Ding, Y. Synergistic effect of graphene and an ionic liquid containing phosphonium on the thermal stability and flame retardancy of polylactide. *RSC Adv.* **2015**, *5*, 27814–27822. [CrossRef]
16. Ramanathan, T.; Abdala, A.; Stankovich, S.; Dikin, D.; Herrera-Alonso, M.; Piner, R.; Adamson, D.; Schniepp, H.; Chen, X.; Ruoff, R. Functionalized graphene sheets for polymer nanocomposites. *Nat. Nanotechnol.* **2008**, *3*, 327–331. [CrossRef] [PubMed]
17. Zhu, J. Graphene production: New solutions to a new problem. *Nat. Nanotechnol.* **2008**, *3*, 528–529. [CrossRef] [PubMed]
18. Li, Y.; Kuan, C.; Chen, C.; Kuan, H.; Yip, M.; Chiu, S.; Chiang, C. Preparation, thermal stability and electrical properties of PMMA functionalized graphene oxide nanosheets composites. *Mater. Chem. Phys.* **2012**, *134*, 677–695. [CrossRef]
19. Wang, X.; Xing, W.; Zhang, P.; Song, L.; Yang, H.; Hu, Y. Covalent functionalization of graphene with organosilane and its use as a reinforcement in epoxy composites. *Compos. Sci. Technol.* **2012**, *72*, 737–743. [CrossRef]
20. Qian, X.; Yu, B.; Bao, C.; Song, L.; Wang, B.; Xing, W.; Hu, Y.; Yuen, R. Silicon nanoparticle decorated graphene composites: Preparation and their reinforcement on the fire safety and mechanical properties of polyurea. *J. Mater. Chem. A* **2013**, *1*, 9827–9836. [CrossRef]
21. Wang, X.; Song, L.; Yang, H.; Xing, W.; Kandola, B.; Hu, Y. Simultaneous reduction and surface functionalization of graphene oxide with POSS for reducing fire hazards in epoxy composites. *J. Mater. Chem.* **2012**, *22*, 22037–22043. [CrossRef]
22. Hu, W.; Zhan, J.; Wang, X. Effect of Functionalized graphene oxide with hyper-branched flame retardant on flammability and thermal stability of cross-linked polyethylene. *Ind. Eng. Chem. Res.* **2014**, *53*, 3073–3083. [CrossRef]
23. Attia, N.; Abd El-Aal, N.; Hassan, M. Facile synthesis of graphene sheets decorated nanoparticles and flammability of their polymer nanocomposites. *Polym. Degrad. Stab.* **2016**, *126*, 65–74. [CrossRef]
24. Liao, S.; Liu, P.; Hsiao, M.; Teng, C.; Wang, C.; Ger, M.; Chiang, C. One-step reduction and functionalization of graphene oxide with phosphorus-based compound to produce flame-retardant epoxy nanocomposite. *Ind. Eng. Chem. Res.* **2012**, *51*, 4573–4581. [CrossRef]
25. Yu, B.; Shi, Y.; Yuan, B.; Qiu, S.; Xing, W.; Hu, W.; Song, L.; Lo, S.; Hu, Y. Enhanced thermal and flame retardant properties of flame-retardant-wrapped graphene/epoxy resin nanocomposites. *J. Mater. Chem. A* **2015**, *3*, 8034–8044. [CrossRef]
26. Georgakilas, V.; Otyepka, M.; Bourlinos, A.; Chandra, V.; Kim, N.; Kemp, K.; Hobza, P.; Zboril, R.; Kim, K. Functionalization of graphene: Covalent and non-covalent approaches, derivatives and applications. *Chem. Rev.* **2012**, *112*, 6156–6214. [PubMed]
27. Xue, Y.; Liu, Y.; Lu, F.; Qu, J.; Chen, H.; Dai, L. Functionalization of graphene oxide with polyhedral oligomeric silsesquioxane (POSS) for multifunctional applications. *J. Phys. Chem. Lett.* **2012**, *3*, 1607–1612. [CrossRef] [PubMed]

28. Wang, J. Flame retardancy and dispersion of functionalized carbon nanotubes in thiol-ene nanocomposites. *Polymers* **2021**, *13*, 3308. [CrossRef] [PubMed]
29. Bao, X.; Wu, F.; Wang, J. Thermal degradation behavior of epoxy resin containing modified carbon nanotubes. *Polymers* **2021**, *13*, 3332. [CrossRef] [PubMed]
30. Marcano, D.; Kosynkin, D.; Berlin, J.; Sinitskii, A.; Sun, Z.; Slesarev, A.; Alemany, L.; Lu, W.; Tour, J. Improved synthesis of graphene oxide. *ACS Nano* **2010**, *4*, 4806–4814. [CrossRef] [PubMed]
31. Wang, Z.; Wei, P.; Qian, Y.; Liu, J. The synthesis of a novel graphene-based inorganic–organic hybrid flame retardant and its application in epoxy resin. *Compos. Part B Eng.* **2014**, *60*, 341–349. [CrossRef]

MDPI

St. Alban-Anlage 66

4052 Basel

Switzerland

www.mdpi.com

Polymers Editorial Office

E-mail: polymers@mdpi.com

www.mdpi.com/journal/polymers